Allianz

Allianz Versicherungs-AG
Berlin/München

Handbook of Loss Prevention

Springer-Verlag
Berlin Heidelberg New York 1978

Title of the Original German Edition:
Handbuch der Schadenverhütung

Translated from the German by
Peter Cahn-Speyer Dipl.-Ing., Kew-Surrey/Great Britain

ISBN-13: 978-3-642-66422-9 e-ISBN-13: 978-3-642-66420-5
DOI: 10.1007/978-3-642-66420-5

Softcover reprint of the hardcover 1st edition 1978

Typesetting, printing and bookbinding: Universitätsdruckerei H. Stürtz AG, Würzburg.
2362/3321-543210

Preface

The German original of this Handbook of Loss Prevention was compiled during the course of many years' work by the engineers of the Department for Engineering Insurances, the scientists of the Allianz Centre for Technology and representatives of industry. It is based on the loss experience and practical loss research studies of the Allianz over a period of more than five decades.

The Handbook of Loss Prevention is a supplement to the technical literature from the field of engineering in the form of a collective work comprising examples of damage to machinery and technical plant and pertinent pointers on loss prevention. It hus ranks among the recognised handbooks for engineers in the filds of planning, design, manufacture and operation.

The great interest and wide acclaim according the German edition of this handbook by industry in 1972 confirm the traditional aims of the Allianz in placing loss prevention in technical plants in the foreground of their service to clients.

The English edition of the handbook under review here enables this valuable engineering know-how to be made available at international level, with the object of preventing losses by the exchange of ideas and experience. The literature references have been taken over from the German edition in unchanged form, in order to bring to the attention of English-speaking experts a bibliography, which is little known outside Germany.

Munich, April 1978 The Editor

Authors

Authors of friendly companies

Bartel, A.A. (Dow Corning GmbH, München)
Brücher, A. (Kraftwerksunion AG, Mülheim/Ruhr)
Ehrlenspiel, K. (Bayer. Berg-, Hütten- u. Salzwerke AG, Sonthofen)
Göttling, H. (Technical University München, München)
Hebel, G. (Brown, Boveri & Cie AG, Mannheim)
Kranich, L. (Rheinkraftwerk Albbruck-Dogern AG, Albbruck)
Renz, H. (TELA Versicherungs-AG für Techn. Anlagen, München)
Winkler, K. (Rheinkraftwerk Albbruck-Dogern AG, Albbruck)

Authors of the Allianz Versicherungs-AG

Abinger, R.
Blaschke, D.
Bohnstedt, J.
Bollmann, G.
Breidert, H.
Breitfeld, H.
Delp, G.
Effertz, P.-H.
Essler, R.
Fenzl, G.
Franze, H.-G.
Gugat, A.
Haase, W.
Heimsch, R.
Heinz, A.
Horenburg, O.
Huppmann, H.
Jäger, K.
Kirsch, H.
Kleinmann, P.
Kober, A.
Krauß, O.
Krebber, B.
Kugler, H.
Leopold, J.
Maatz, H.
Neumann-Spallart, H.
Reiche, R.H.
Ritter, W.-D.
Schindler, D.
Schmidt, E.
Schulze, M.
Splittgerber, E.
Wissel, Chr.
Witt, D.
Wittrisch, G.
Wuttke, H.

Authors and co-workers of this book want to express their gratitude to the managing director Dr.-Ing. H. Braun for his purposeful promotion and generous support in planning and realizing this work.

Contents

1. Aims and Possibilities of Loss Prevention

The increasing demand for raw materials and energy, as well as capital and consumer goods, has necessitated the development of ever larger and more efficient machines and technical plant. Increasing capital expenditure and costs of operation and maintenance impose the highest demands on the economical running of production plant. Apart from the maintenance of optimum efficiency during its economical utilization, the availability of a machine plays a role of primary importance in economic considerations. Every unforeseen case of damage which results in an unscheduled outage of the plant, and hence of production, can nullify all earlier calculations.

The layout and design of every industrial plant or machine are a compromise between the technically feasible and the economically justifiable. With the present state of technology, it would undoubtedly be possible to build machines possessing absolute operational safety and unlimited life. However, the cost of building such a machine would be prohibitive. Moreover, its economic life would be much shorter than its functional life because new designs, with higher efficiency and productivity, lower costs for operating personnel, smaller demands on space, etc., would be developed in the meantime or because the requirement for better quality could no longer be fulfilled. Therefore, the trend today—as has already largely been realized in the aircraft industry—is increasingly towards machines which exhibit the highest possible efficiency through maximum utilization of materials for a calculated life span. The realization of such designs requires an exact knowledge of stresses and material characteristics and their behaviour in relation to temperature and time, as well as uniformity of materials. Every engineer knows, however, that these conditions are not always fulfilled, despite sophisticated methods of calculation and improved materials-testing techniques.

With a continuous increase in specific stressing, limits are often exceeded without noticing and, in such cases, so far proven methods of calculation, with their assumptions and constants, are no longer valid. In actual service, for example, calculated stresses can be superimposed by chemical influences or can be exceeded as a result of intentional or unintentional additional stresses. Despite all efforts, it is therefore unavoidable that components fail before their calculated life is reached.

The damage-inducing causes, through which technical installations are endangered, can in principle be subdivided into three groups, i.e.,

- technical faults present in the installation, e.g., planning, design, manufacturing, materials or assembly faults;
- faults appearing during operation, e.g., through ageing, wear, corrosion, erosion or extraneous influences;
- human fallibility, e.g., operating or maintenance faults.

Accordingly, loss-prevention measures have been designed, their purpose being:

- to utilize the experience gained from cases of damage in planning, calculation, design and manufacture, so as to prevent damage of a similar kind;
- to discover places of weakness in machine components early enough to enable them to be rectified in good time before failure occurs;
- to reduce additional stresses to a minimum by servicing and maintenance of machines, as well as by supervision and adherence to operating instructions.

Basically, all efforts, from initial planning to final scrapping of a machine, are directed towards achieving faultless operation, freedom from damage, good efficiency and optimum life. A prerequisite for *purposefully* applied loss-prevention measures is in the first place to recognize true centres of damage initiation and their causes for certain groups of machines by systematic evaluation of as many cases of damage as possible which have already occurred.

Primarily, the users and manufacturers of technical equipment are devoted to these efforts. However, they are subject to certain limitations because, to the user, every instance of comparatively severe damage appears more or less as an individual case, while the manufacturer can utilize in his design, calculation and manufacture only those experiences he has gained from faulty operation and cases of damage relating to his own products.

On the other hand, a technical insurer, such as Allianz, has of necessity at his disposal an extensive, heterogeneous collection of cases of damage from almost all branches of technology and distributed over all manufacturers. By systematic evaluation, the restriction to random or individual cases is effectively overcome and, even for machines which are encountered relatively rarely, the number of cases of damage is still sufficient to permit, here again, recognition of the typical damage areas for the appropriate application of loss-prevention measures.

1.1. Loss Prevention by Allianz

Activity of the Allianz Insurance Company in the field of technical loss prevention by their own insurance engineers originated about 55 years ago with the inspection of steam-turbine installations. At that time, the liability of steam turbines to damage was so high that the question had to be asked whether it was at all possible to insure them against machine damage. After the good experience gained with the inspections of steam boilers, as prescribed by law, continued insurance of steam turbines was made dependent on inspections to be carried out at regular intervals. The success of this measure, introduced often in the face of initial resistance from users, resulted in the acknowledgment that the overhaul of important units now constitutes the most important factor in loss prevention. At Allianz, since that time, loss prevention, operational advice and loss research in nearly all fields of technology form an indivisible part of the Engineering insurances.

At Allianz, the engineering office and the Allianz Centre for Technology, meaningfully complementing each other and in close co-operation with manufacturers and users, devote themselves to the widespread duties of loss prevention.

Figure 1. One of the approximately 30,000 cases of damage per year.

1.1.1. The Allianz Engineering Office

The main activities of the engineering office are as follows:

- Processing of cases of damage which have occurred to insured machines and technical plants. This work includes:
  Visual inspection of damage to determine its extent (Figure 1).
  Collection and evaluation of all data and circumstances relevant to the course the damage has taken, with the object of clarifying the cause of damage as precisely as possible.
  Selection and extraction of representative test-pieces, linings and samples of operational and auxiliary materials, in case chemical or physical investigations become necessary.
  Consultation and determination of the nature and extent of the necessary repairs by the user and manufacturer.
  Recommendations, if the need arises, for temporary repairs of limited duration and safe from the point of view of operation, so as to reduce loss of production.
  Preparation of a report and, after the repair has been completed, settlement of the costs involved.
- Loss prevention by appropriate application of findings and experience gained from the evaluation of cases of damage. Measures for this purpose include:
  Carrying out of inspections and regular overhauls of steam, gas and water turbines, diesel engines, compressors, generators, electric motors, transformers and presses (Figures 2 and 3).

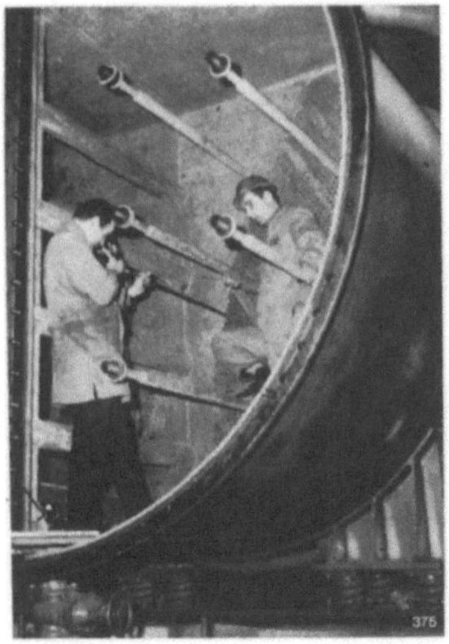

Figure 2. Inspection of condenser tubes, using an endoscope, by an Allianz test team.

Figure 3. Inspection of turbo-generators.

Consultations regarding operation, with proposals for improving attendance, maintenance and servicing of installations, as well as for supplementing measuring, alarm and protective equipment.

Proposals for improving conditions of operation, manufacture and design of installations and components.

- Risk assessment, particularly with regard to new designs and initial designs, with the purpose of estimating dangers as objectively as possible.

Comprising Contractors' All Risks insurance, Erection All Risks, Guarantee, Machinery Breakdown, Electronic Equipment and Loss of Profits following Machinery Breakdown insurance, the Engineering insurances cover not only all the stages in the life of a technical installation, but also virtually the entire gamut of technology. Accordingly, the Allianz Engineering office has a staff of nearly 100 experienced engineers from the most varied fields [1]. Most of these are at the six administrative and operational centres of Allianz in Germany, while some are with the management for France, Italy, Spain and the Netherlands and all are engaged in the tasks mentioned above. At the head office, where all information on cases of damage is received, the engineering office is divided into the following main branches:

driven machines,
civil engineering,
steam generators and pressure vessels,
electrical engineering,
nuclear engineering,
piston machines,
steel structures,
fluid-flow machines,
process engineering,
low-voltage and data-processing installations (by a friendly insurance company).

Here, apart from liaison with the commercial branches and the Allianz Centre for Technology, the activities of key importance are the co-ordination and preparation of knowledge derived from damage evaluation for feedback and dissemination, the acquisition of basic knowledge in special fields, and keeping up to date with developments in economics, technology and science.

1.1.2. The Allianz Centre for Technology

The Allianz Centre for Technology (Allianz-Zentrum für Technik GmbH—AZT) has been in existence for over 40 years as an institute specially devoted to research into and the prevention of

Figure 4. Property insurance for industry, with the risks covered.

Type of insurance	Danger									
	Material damage to the insured object									
	Fire, lightning stroke, explosion	Negligence, mishandling faults	Short-circuit, over-voltage	Accidents during operation, fracture	Faults in materials, design and manufacture	Test run	Assembly or building accidents	Transport accidents	Storm, frost	Acts of God, earthquakes, floods
Fire	×									
Machines		×	×	×	×				×	
Assembly	×	×	×		×	×	×		×	×
Machine guarantee					×					
Low-voltage current	×	×	×	×	×				×	×
Building construction	×	×					×		×	×
Transportation	×	×						×		×
	Financial loss through outage of the insured object (as a result of the cases cited above)									
FLOP [a]	×									
MLOP [b]		×	×	×	×				×	

[a] FLOP = Fire Loss of Profits insurance.
[b] MLOP = Machinery Loss of Profits insurance.
Financial losses caused by failure of the public power supply can be covered by special insurance. The extent of cover of the various types of insurance is determined by insurance conditions in each case and by individual agreements, e.g., regarding self-participation, inclusions and exclusions from liability.

Figure 5. The Allianz Centre for Technology (Allianz-Zentrum für Technik GmbH — AZT) in Ismaning, near Munich. View from Münchner Strasse.

material damage. The work of the institute has the following individual objectives:

- clarification and scientific evaluation of the causes of damage and its effects;
- preparation of proposals aimed at improving materials, design and operating conditions, as a contribution to the prevention of damage;
- application of non-destructive test methods for the prevention and confinement of damage.

The Allianz Centre for Technology is called in if there are cases of damage and faulty operation whose causes cannot be established on site and detailed investigations become necessary. For suspected damage, for the limitation of the extent of damage, for inspections, and for the extended customer service of Allianz, insurance engineers have at their disposal mobile teams for carrying out measurements and analyses of vibrations, as well as for almost all types of non-destructive testing.

Investigations into cases of damage often lead to a dead end, with the result that an unambiguous clarification of the causes of damage is no longer possible, because certain interrelationships remain obscured. Thus, out of daily work emerge important *damage-research tasks,* the solution of which is a prerequisite for purposeful loss-prevention measures.

As in the engineering office, the personnel and technical equipment of the Institute are in accordance with their widespread duties, as engendered by the great variety of types of damage. The personnel are characterized by experts in the following fields:

automotive engineering,
chemistry,
data processing,
design and construction,
electrical equipment,
fluid-flow machines,
gear manufacture,
lifting and mechanical handling equipment,
machine elements,
materials and heat treatment,
measuring techniques,
non-destructive testing,
nuclear reactors,
piston machines,
vibration problems,
welding methods.

Figure 6. A hydropulse installation. The baseplate is supported in such a way as to damp out vibrations, so that shock cannot be transmitted to the surroundings. On the hydropulse cylinder in the foreground is an automobile rear axle being subjected to repetitive loading to check welding seams.

The Allianz Centre for Technology as a whole is divided into three institutes:

Materials Technology

with departments covering:

inorganic chemistry,
organic chemistry,
corrosion,
physical metallurgy,
materials testing.

Preventive Testing Technology

with departments covering:

dynamics of machines,
electrical engineering,
physical testing techniques
calculation,
non-destructive testing.

Both the above institutes are served by two central services: Technology and Administration covering.

transport,
photography,
workshop,
internal service.

Automotive Technology

with departments covering:

demonstration of damage,
training,
examination of vehicles,
research into the causes of accidents.

With regard to the above sections on Materials Technology and Preventive Testing Technology, the duties and important technical equipment, which serve the laboratory as well as being partly mobile, are summarized in the following:

Materials Technology

Inorganic Chemistry

Analysis of metals, deposits, constructional materials, products of corrosion, waters.
The work is mainly carried out by X-ray fluorescent spectrometry. However, other methods of analysis are available, particularly in the field of trace elements. In addition to an electron-beam macro-probe, methods of analysis include spectrophotometry, flame photometry and atomic-absorption photometry, as well as conductometric, potentiometric and coulometric techniques.

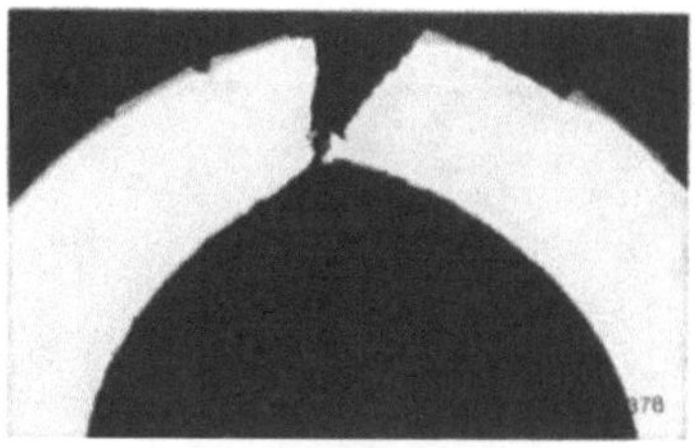

Figure 7. Ruptured superheater tube with scale on the rupture flanks and scale wedges on the exterior of the tube near the rupture.

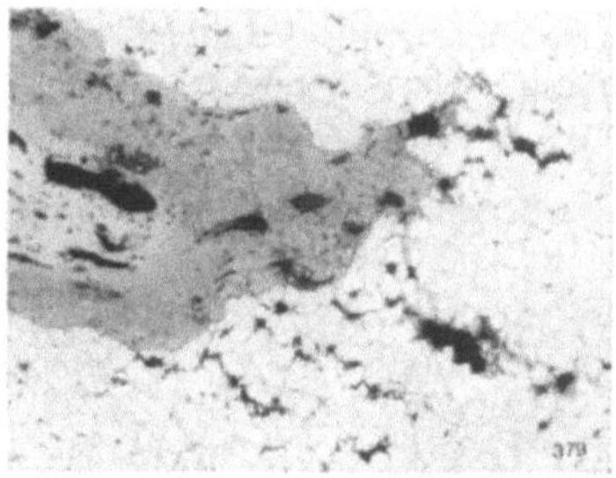

Figure 8. Enlarged scale wedge.

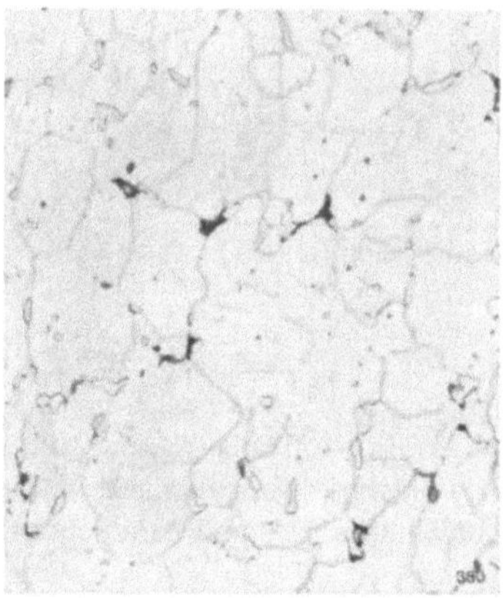

Figure 9. The structure consists of ferrite and carbides which, in the fracture region, are particularly large. In this region, rupture of the structure at the grain boundaries is an indication that the creep rupture strength has been exceeded, as is also borne out by coagulation of the former intermediate structure.

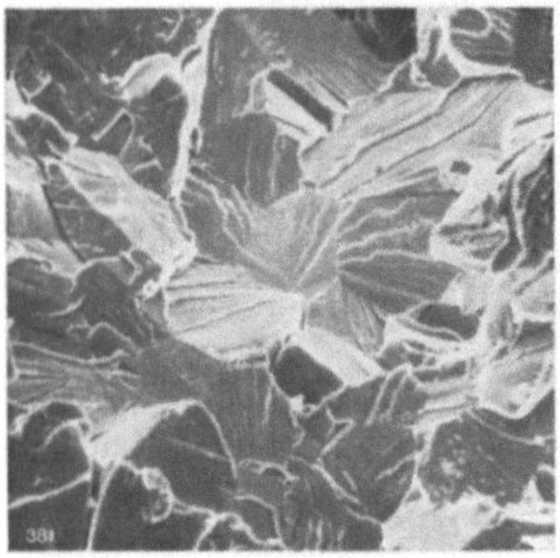

Figure 10. Transcrystalline brittle fractures on the studding of a combustion-chamber tube.

Figure 11. Electron microgram of a longitudinal section of a condenser tube for identifying corrosively acting constituents.

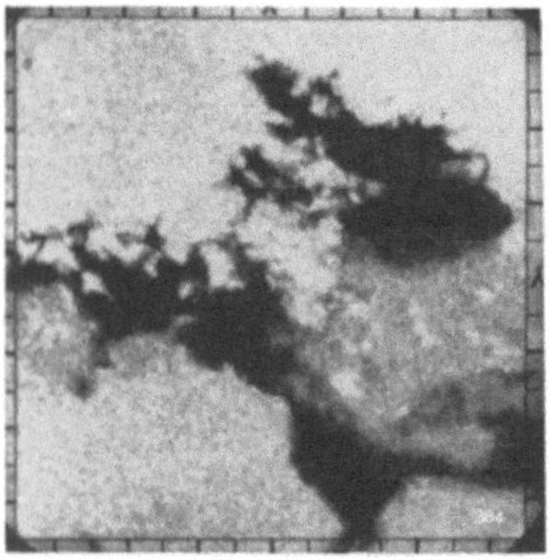

Figure 12. ($Cu_{K\alpha}$)-radiograph (matrix).

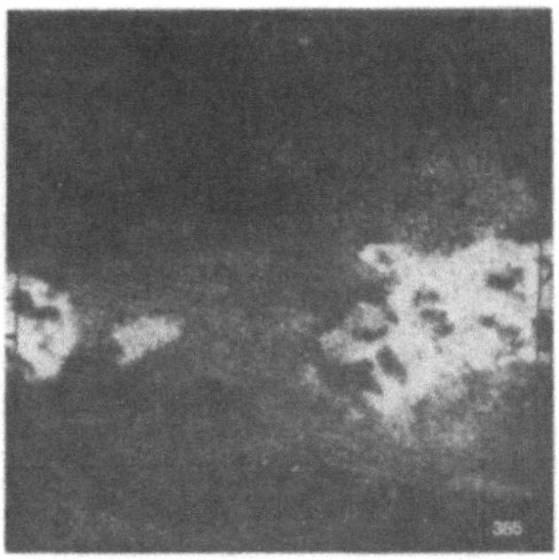

Figure 13. ($Cl_{K\alpha}$)-radiograph (corrosively acting Cl).

Organic Chemistry

Separation and identification of organic substances, particularly plastics, lubricants and fuels, painting media, solvents, fabrics and products of combustion.

For this work, the following are available:

Gas chromatograph with preparative equipment and a step pyrolysis unit, an infrared spectrometer, a steam-pressure osmometer, a rotating-cylinder viscometer, equipment for column chromatography and thin-layer chromatography, and apparatus for the determination of air-separation capability.

For carrying out analyses on site, particularly after a fire in the presence of PVC, there is available a specially designed van (Figure 14), equipped with an atomic-absorption spectrometer, a filter photometer, and an instrument for measuring tan δ, as well as with the appropriate reagents and auxiliary laboratory equipment.

Corrosion

For the examination and duplication of corrosion and scaling processes, there are available:

Autoclaves, furnace combinations with programme-controlled temperature under variable gas atmospheres, together with potentiostatic or potentiokinetic corrosion equipment.

Physical Metallurgy

Preparation of polished metallic sections for the determination of structure and the composition of metallic and non-metallic materials by optical

microscopy (Figures 7–9). Examinations of sections and surfaces, particularly those of fractures, with a scanning electron microscope (Figure 10). Research in the field of microfractography. Examination of the structural and chemical consistency of solids in the microscopic and submicroscopic regions. Examinations of physical properties and states. X-ray fine-structure analysis for the determination of crystal structure, as well as of the type of bond of crystalline materials with the aid of film and goniometric techniques.
Determination of submicroscopic segregations and deformations with a high-resolution transmission type of electron microscope with additional beam-bending equipment. Chemical analysis of microscopically small regions with the aid of electron-beam micro-analysis (Figures 11–13).

Materials Testing

Determination of the mechanical properties of laboratory samples and constructional components, particularly the strength characteristics under static and dynamic stresses.
Available are:
Hardness-testing equipment based on various test methods.
Universal testing machines for different load ranges with force or elongation control. Also, equipment for testing under high and low temperatures.
Pendulum impact-testing machine for determination of bending and tensile impact stresses, with recording of force-time-elongation diagrams.
Shock-testing machines with half-sine, saw-tooth and rectangular shock waves.
Creep-testing equipment.
Facilities for the examination of fracture-mechanics specimens.
Test machines for the examination of alternating strength under high and low temperatures, with or without exposure to corrosive media in rotating bending fatigue tests, and alternating tension and compression tests.
Hydropulse equipment (up to 580 kN) for determination of the strength of structural components and strength in operation (Figures 6 and 15).
For special tests within the framework of research into damage and the reproduction of loss events,

Figure 14. View of part of the interior of the laboratory vehicle, showing the drying oven, crucible furnace, filter photometer, pH- and conductivity-measuring apparatus, and ion-exchange installation.

Figure 15. Investigation of the endurance limit of a plastic instrument-frame.

numerous other types of equipment are available or are being obtained for specific cases.

Preventive Testing Technology

Dynamics of Machines

Loss analysis.
Theoretical and experimental treatment of vibration problems.
On-site measurements of natural-frequency, bearing and shaft vibrations (Figure 16); measurements of displacement.
Special tests on structural components and fluid-flow machines.

Figure 16. On-site determination of the natural frequency of turbine blades with the aid of vibration-measuring equipment.

Electrical Engineering

Electrical measurement of non-electrical parameters, particularly the measurement of static and dynamic forces and displacements (Figure 17).
Measurement of vibrations by means of displacement, velocity, and acceleration sensors.
Measurement and analysis of noise and sound.
Measurement of electrical parameters in high- and low-voltage techniques.
For measuring tasks, analogue and digital measuring procedures are available with the following equipment:
A digital computer, frequency analysers, high-resolution oscillographs, high-quality amplifiers. The computer can also be used on site with the aid of a special vehicle.
Using radiation-proof underwater television cameras, inspection can be made of radioactive components, particularly reactor pressure vessels and their installation (Figures 19 and 20).
For special measurements, the necessary equipment and layout are planned and executed in our own workshop.

Physical Testing Technique

Determination of residual stresses and stresses under load in constructional components with the aid of X-ray and mechanical measuring techniques.
Experimental stress analysis on the machine site by means of a 100-strain-gauge measuring installation (Figure 17).
Measurement of deformations and vibrations by optical holography.
Development of physical testing methods for special problems.

Calculation

Mathematical stress analysis and design assessment.
Loss statistics.
Programmes for on-line evaluation of measurement and analysis systems.

Non-Destructive Testing

Examination of the technical condition of installations, machines and components, especially in respect of hidden faults in the cross-section of materials of statically and dynamically stressed components.
For these examinations the following mobile equipment is available:

X-ray apparatus for coarse-structure examination.

Figure 17. View of the 100-strain-gauge measuring installation and the interior of the special van with computer and peripherals. The installation for rapid interrogation of the 100 measuring points comprises blocks of change-over switches U, the digital read-out D, the control equipment S with measuring amplifier and analogue indicator, and the printer P.

Figure 18. Ultrasonic testing of a crankshaft of a piston compressor. Such investigations are carried out as loss-prevention measures by the measurement and test team of AZT.

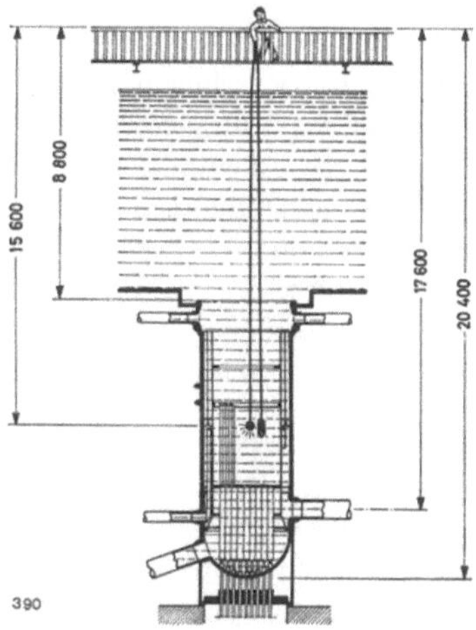

Figure 19. The manipulation distances for the pressure vessel of a nuclear reactor are given by its position. The illustration shows a pressure vessel of a boiling-water reactor. (Conditions for a pressurized-water reactor are similar.)

Ultrasonic test equipment (Figure 18).
Dye-penetrant procedures.
Magnetic and current-circulating apparatus for magnetic-particle testing.
Crack-depth measuring instruments.
Eddy-current test instruments for detecting surface cracks; also, in some cases, special probes.
Rigid endoscopes.
Flexible glass-fibre endoscopes.
Mobile metallographic equipment.
Hardness-testing instruments.
Ultrasonic leakage detector.

With this equipment, the Institute, though primarily at the disposal of those insured by Allianz, can also cater for the non-insured for investigations into the causes of damage and sponsored research. The accumulation of experience of the Institute, amassed over a period of 45 years, forms a broad basis for the investigation of all cases. The results of the investigation into each individual case are presented in the form of a test or research report which is made use of by the Engineering Office of Allianz for consulting the insured in questions of loss prevention, taking special account of the interests of individual plants. Details of the work of the Institute are given in reports of their activities [2, 3].

Automotive Technology

The tasks of this relatively recent sphere of activity of the Institute are as follows:

Demonstration of Damage

Investigations into and development of new methods of repair for vehicles damaged in accidents, together with time studies for reducing damage (Figure 21); studies of repairs by re-spraying.

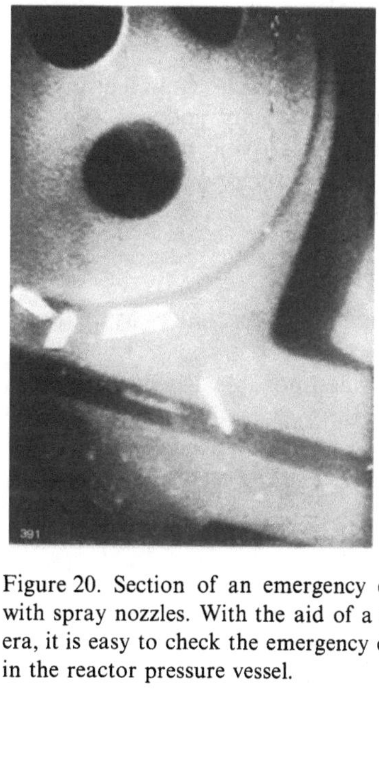

Figure 20. Section of an emergency core-spray ring with spray nozzles. With the aid of a television camera, it is easy to check the emergency core-spray lines in the reactor pressure vessel.

Figure 21. In the damage-demonstration shop of the automotive-technology section of the Institute.

Training

Centre for the basic and advanced training of vehicle and damage experts of Allianz.

Examination of Vehicles

Operation of measuring and test equipment for examining vehicles in connection with their traffic and operational safety.

Research into the Causes of Accidents

Special research into the causes of damage to components of crashed vehicles, as well as investigations into driving stability and adhesion between tyres and road. Driving behaviour of car drivers under certain influences. Impact tests on own crash track.

1.1.3. Exchange of Experience by Allianz

To achieve the widest possible dissemination, Allianz publishes the results of damage evaluations, carried out on a broad basis, in its own technical literature. At the heart of such publications is the journal „Der Maschinenschaden“, which contains reports from insurance engineers and information from the Allianz Centre for Technology on new findings in the field of damage research and loss prevention, as well as articles from friendly institutes, organizations and authors on the subject of maintenance of the serviceability of machines and technical plant [4, 5].

Published in the series of "Allianz Reports on Operational Technology and Loss Prevention" are lectures and discussions of the Allianz technical conferences, as well as important information on loss prevention requiring a more widespread publication [5]. Pamphlets on operational technology and the prevention of damage, technical books, and reprints and brochures complement the editorial programme [6–8].

Loss prevention is also the subject of publications of Allianz engineers in other specialized journals. Furthermore, last but not least, it is the subject of the "Allianz Technology and Insurance Forum", which takes place at intervals of four years and offers interested parties from Germany and abroad the welcome opportunity of extensive exchange of experience in the field of loss prevention.

1.2. Basis and Methods of Loss Prevention

A prerequisite for purposeful loss prevention is a knowledge of the weaknesses which lead to damage to machines and technical equipment. At the Allianz Insurance Company this knowledge is derived from technical damage statistics, which give information on the *cause*, the *location* concerned (component), and *costs* with different *types of machines*. The data have been compiled from about 22,000 cases of damage to machines per year.

Naturally, consideration is given only to those cases of damage which qualify for indemnification under the machinery insurance and are notified. Not taken into account are foreseeable events, i.e., damage through wear or other direct results of permanent operating influences, such as erosion and corrosion. On the other hand, indirect results are included, e.g., the fracture of a crankshaft through additional stresses having their origin in bearing play caused by wear. Moreover, all smaller cases of damage, the repair costs of which are below the agreed deductible are not included in the statistics. Also excluded are those cases of damage whose cause could not be established with certainty. For these reasons, comparison with other statistics is possible only with reservations.

The results of the technical damage statistics of the Allianz Insurance Company and, for electronic data-processing (EDP) equipment (Section 3), the experience of the TELA Insurance Company form the basis of this handbook. Consequently, discussion of causes of damage and measures to be taken to prevent loss is mainly concerned here with those causes which have repeatedly led to damage. It is therefore quite possible that important basic considerations or preventive measures relating to planning, design, manufacture and assembly, supervision, servicing and maintenance have only briefly been touched upon in the various chapters, or even have not been mentioned at all. This in no way reflects adversely on their importance, but merely indicates that they occur more rarely as causes of damage and hence that Allianz has insufficient experience to be able to report on suitable countermeasures.

An attempt to discuss loss prevention for all machines and equipment within the broad spectrum of engineering technology would be beyond the scope of this book. For this reason, a selection has been made in the first place from those fields in which Allianz has the greatest experience. In the following list is given a breakdown of the numbers of machines currently insured in the groups of machines selected.

- Electrical machines: roughly 92,000,
  This group comprises a.c. generators, d.c. machines, electric motors, transformers.
- Steam generators: roughly 6,500.
  This group comprises water-tube boilers, fire-tube boilers, low-pressure boilers, high-speed steam boilers and others.
- Fluid-flow machines: roughly 13,500.
  This group comprises steam turbines, gas turbines, water turbines, turbo-compressors and blowers, centrifugal pumps.
- Piston machines: roughly 7,700.
  This group comprises steam engines, diesel engines, Otto-cycle engines, compressors, pumps.
- Mechanical handling and lifting equipment, earth-moving gear: roughly 6,600.
  This group comprises conveyor-belt installations, rotary-tower cranes, truck cranes, travelling cranes, excavators, caterpillar bulldozers, loading bridges.
- Industrial furnaces of different types:
  No information on the plant insured, since evaluation of damage was limited to that which occurred during assembly, comissioning, and the warranty period; in other words, insurance cover was for a relatively short time.
- Printing machines of all types: roughly 12,450.

With all kinds of machines, as with industrial furnaces, it is necessary to include in the collection of experiences those gained on installations which are insured for only a short period, from assembly to commissioning to the end of the warranty.

As an example, Figure 22 reviews the percentage distribution of three causes of damage to five groups of machines, as derived from damage statistics. For each group, the left-hand bar always refers to the percentage number of cases of damage, and the right-hand bar to the percent-

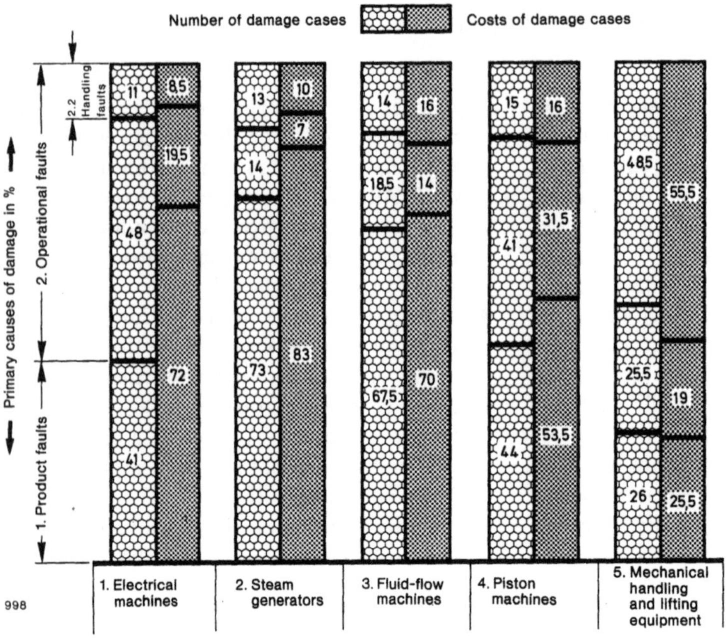

Figure 22. Relative proportions of three causes of damage to five types of machines.

age costs of repair apportioned to the given causes of damage. It must be mentioned in connection with this evaluation that the results give average values and that the proportion of the causes of damage in each group of machines varies greatly with different manufacturers, the power of the installation and its age, and users [9].

Whilst the special chapters dealing with the various groups contain detailed evaluations, as the need arises, the causes of damage upon which Figure 22 is based can be divided into the following main groups:

Product Faults

All causes of damage which are the results of manufacture. These causes include faults in planning and design, such as deficient layout and calculation, incorrect choice of materials, and unsuitable geometry; also, faults in processing, such as incorrect heat treatment, machining errors, assembly faults and, last not least, faulty materials.

Operational Faults

All causes of damage which originate during operation of the installation, such as loosening of components, failure or non-response of protective devices, servicing faults, damage arising from wear, corrosion, erosion, ageing, etc., as well as causes that can be traced back to external influences, e.g., natural forces, foreign bodies, and overvoltages from the grid.

Handling Faults

Handling or attendance faults can be regarded as a subsection of operational faults. In about 95% of the damage cases, no part was played by external influences.

Total Evaluation

Altogether, 16,005 cases of damage, which occurred in the years 1969 to 1974, were evaluated. These cases were distributed among the various groups of machines as follows:

9,135 cases of damage to electrical machines,
1,045 cases of damage to steam generators,
1,215 cases of damage to fluid-flow machines,
1,000 cases of damage to piston machines,
3,610 cases of damage to mechanical handling and lifting equipment.

The proportions of the groups of causes of damage included vary widely with the types of machines considered; accordingly, the key measures necessary for successful loss prevention also vary.

Comments on Product Faults

In terms of numbers and costs, the proportion of damage by product faults is most pronounced with fluid-flow machines and steam boilers. It is also most pronounced in terms of cost with electrical machines and installations, but is relatively small in terms of numbers, indicating high repair costs per individual case of damage.

With these three types of equipment, the trend of technical development towards ever larger and more powerful equipment is particularly noticeable. It happens with increasing frequency that newly developed components, which have not been sufficiently tested in practical operation, are simultaneously installed in large numbers of machines.

Methods of preventing damage, with product faults as the primary cause, involve utilization of the knowledge from loss events in calculation, design and manufacture by way of:

- close co-operation with manufacturers in special cases;
- early publication of experiences in specialized journals;
- open exchange of experience at conferences and seminars.

In this way, an important contribution can be made towards ensuring that, with different manufacturers, damage arising from identical causes will be prevented from being repeated at certain intervals.

Further support in the prevention of damage is offered by:

- purposeful inspection and replacement, at the appropriate time, of components which have led to damage to similar machines as a result of basic deficiencies.

Comments on Operational Faults

The proportion of damage from operational causes, in terms of numbers and costs, is greatest with piston machines, followed by electrical machines and installations. With the latter the high number of cases of damage with relatively low costs stands out, indicating that there are many cases of small damage (see Figure 22).

Methods of loss prevention in connection with operational faults are as follows:

- Optimum supervision of the installation, when not only sudden but also gradual changes in operational data (pressure and temperature, efficiencies, power input, etc.) must be noted and their causes established without delay.
- Constant control and inspection of measuring, control and protective devices.
- Constant control of the condition of operational and auxiliary materials (lubricating, cooling, refrigerating media, etc.).
- Regular servicing, maintenance and overhaul.
- Overhauls and inspections at regular intervals, so that components which have reached the end of their lives through wear, corrosion, erosion or temperature influences can be replaced before failure occurs.

On the basis of systematic evaluations of damage and inspections, optimum intervals between inspections have been established for most machines, or reliable supervisory and control systems have been introduced for determination of the most favourable points in time for carrying out inspections.

Comments on Handling Faults

The largest proportion of damage from faults arising from mishandling is found in the group of machines concerned with mechanical handling and lifting. However, in the other groups of machines dealt with here, this cause of damage is also noteworthy.

For loss prevention, the following procedures are recommended:

- Selection of suitable trained personnel for operating, maintaining, and servicing machines.
- Continual training and instruction of operating personnel, especially with regard to the actions to be taken in the event of machinery breakdown, by means of courses, lectures, literature, and pamphlets.
- Installation of easily operated fittings and control elements, warning notices and instruction plates.
- Provision of easily understood, detailed operating instructions.

Key Measures for Prevention of Damage

Summarizing, it is already possible to deduce from the relatively moderately subdivided statistical information where the key points of the procedures for loss prevention will lie for the groups of machines considered.

- *With electrical machines,* the key points are utilization of damage experience for calculation, design and manufacture; revisions, if possible, at regular intervals; and the training of personnel.
- With *fluid-flow machines* and *steam generators,* the utilization of damage experience for the manufacture of machines, and also inspections and overhauls are in the forefront. In the evaluation (Figure 22) it is essential to consider that the distribution of the damage causes is the result of inspections carried out on turbines by Allianz and on steam generators by TÜV (German Technical Control Board).
- With *piston machines,* the key to loss prevention lies in the improvement of servicing and maintenance, as well as in inspections.
- With *mechanical handling and lifting equipment,* damage arising from handling faults predominates. In this case, therefore, loss prevention lies primarily in training personnel by demonstrating examples of damage, as well as by pamphlets and instruction plates. In addition, nothing must be omitted in the attempt to reduce the proportion of handling faults by improvement of the man/machine system.

1.3. General Recommendations for Loss Prevention

Many basic rules for loss prevention are equally applicable to almost all machines and technical equipment. To avoid excessive repetition in Chapters 2 to 8, generally valid points of view and possiblities for loss prevention are briefly summarized in the following.

1.3.1. Loss Prevention in the Manufacturer's Works

The life of technical installations is to a large extent predetermined by the manufacturer during planning and layout. The manufacturer will make an effort to recognize all influences which are of importance for economically optimum production and utilization. At the same time, however, it is necessary to establish priorities whose coverages are different, inasmuch as for quantity production they are subject to official and internal standards, whereas single-piece production is often custom-designed under the influence of the purchaser. Whilst, with mass-produced items, their future application and the resulting limiting conditions must be planned for the statistically highest possible number, an optimum agreement between manufacturer and user is necessary in the case of one-off or special manufacture.

1.3.1.1. Loss Prevention during Planning

Following the desire for optimum economy is the trend towards larger and larger capacities per unit and higher materials utilization. This involves a danger, because engineers are forced, from known and proven basic data, to extrapolate into unknown regions. Numerous cases of damage to newly developed or initial designs show that permissible limits have often been exceeded. The costs in money and time for careful experiments in the laboratory, on the test bed or on pilot plant are nearly always lower than repair costs, penalties, and loss of prestige through damage – in some instances, a series of damage cases during test runs or the warranty period.

From the inception of planning, the local conditions at the erection site must be taken into account. These conditions primarily concern natural hazards, e.g., earthquakes, landslides, storms, floods, fire and lightning, and also by climatic influences, such as extremes of temperature, humidity, dust (sand storms), frost and drifting ice.

Even with anticipatory planning is it possible to protect installations against such dangers only to a limited extent. The criterion for the extent of safety measures to be adopted in each case should be the optimum compromise between the costs involved and commercial efficiency.

Frequently, however, easily avoidable causes, overlooked during planning, have led to serious damage, such as the destruction of cranes because of insufficient protection from storms; sinking of erection equipment or of vehicles into the ground because the supporting capacity of the subsoil was insufficiently well known; flooding because not enough account was taken of local conditions regarding drainage or because of inexpertly laid-out drainage [10].

For the avoidance of *frost damage* it is essential, apart from providing the requisite technical measures, to give clear handling, maintenance and preservation instructions to personnel in good time before the start of frosty wheather [11].

Fire Protection

Right from the start of planning an installation particular attention must be paid to the provision of suitable measures for fire prevention. It is not enough to pay early attention to preventing a fire from spreading to neighbouring buildings and installations by suitable structural means (fire-resistant walls, doors, etc.). It is also necessary to bear in mind that substantial damage can be caused by the effects of smoke, dust and gases [12]. Highly sensitive measuring and control equipment, a wide variety of electronic modules and data-processing equipment are particularly susceptible to damage by gases containing hydrochloric acid generated by the combustion of polyvinyl chloride (PVC cable insulation, floor coverings, claddings, etc.). These gases form a corrosive solution of hydrochloric acid in contact with quench water or humid air [13–16].

Considerate planning can contribute to a reduction in the possible extent of damage if, for example, automatic ionization-type smoke-alarm systems are installed and if, in the event of a fire, the spread of noxious fumes can be prevented by sealing off heating, ventilation, service and cable ducts vertically and horizontally through bulkheads. Moreover, it is important to lay control and power cables in separate ducts and to provide fire-blocking sections, so that, in the case of a fire, proper running of the installation is possible via the measuring and control system [17–19].

To be able to extinguish a fire quickly and unimpeded, it is important to provide fire-extinguishing equipment which, together with associated connections, must be readily accessible in the event of an actual fire. Suitable ventilation equipment (e.g., extractors, air inlet and outlet ducts, vents and flaps) must be used to ensure adequate discharge of the combustion gases to avoid to a large extent endangering fire-fighting personnel by toxic constitutents and to preserve a clear view for effective fire-fighting.

Fire protection is often neglected at building sites, particularly with regard to assembly items themselves, such as steel structures, and, above all, buildings under construction [20, 21] (see Section 1.3.1.3.6, page 26). In the case of high-rise buildings, consideration must be given to injury to personnel, as well as to material damage. The most important fire-prevention measure is that fire-protection equipment (hydrants, sprinklers, etc.) should always conform with the construction progress and should be ready for operation, i.e., they should be under pressure, throughout erection of the building.

Secondary damage arising from the uncontrolled spread of quench water (e.g., in data-processing installations) can be substantially reduced by simple constructional and design measures, such as sealing of cable ducts and openings to rooms adjacent or underneath, plinths 100 mm high at floor break-throughs for instrument cubicles, etc.

If any industrial installations in the vicinity are endangered by fire and explosion, sensitive and important parts for operation of the equipment should also be protected against potential external hazards.

Space Requirements

Every technical installation must be overhauled or inspected at certain intervals and, even with the greatest care, damage can never be entirely excluded. For this reason, it is essential, during the planning stage, to ensure the economical optimum of opportunities for carrying out inspections, overhauls, and repairs. For this purpose, facilities should include, for example, provision of good access and adequate space, by constructional measures, for depositing heavy components. Also available should be additional lifting gear which, in some circumstances, may at first be utilized only very little, but with whose aid losses as a result of long outages may well be reduced substantially at a later date.

1.3.1.2. Loss Prevention during Construction and Execution

Manufacturing is to a large extent determined by the constructional form of components and the materials. It is therefore the responsibility of the designer to limit possible damage that can originate as a result of design configuration, e.g., avoidable blowholes, welding stresses (cracks, distortions), incorrect heat treatment, non-permissibly high stress peaks arising from a design that does not take proper account of the flow of forces, notches, sharp transitions and accumulations of material. These basically simple and known principles are all too often overlooked, so that over and over again a cause is created for a large variety of damage. The same applies to elongations due to temperature changes, the effects of which are often underestimated and lead to additional stresses, distortions, fractures, and fouling by bridging clearances. Expansion joints, slide ways and similar items can undergo changes in operation and can later impede elongation (dust, slag or seizure).

With the increasing size, complexity and integration of technical installations, design must strive more and more to ensure that components, groups of components, and complete machines are so constructed that inspections, overhauls and repairs can be carried out with a minimum of disassembly work and that adequate facilities are available for inspection without dismantling.

Materials Selection

The criteria for materials selection are the stresses in components as determined by calculation and geometry. Experience shows that, as the use of high-quality materials increases, materials utilization is being driven very far towards the limit of permissible values. In this way, normally negligible fault locations, inclusions or inhomogeneities become the starting points of cracks or fractures.

For components subjected to fatigue stresses, use must often be made of materials whose long-term behaviour is not known but is extrapolated from the results of short-time tests. Consequently, service life varies very considerably. Frequently, insufficient account is taken of the compatibility of materials with each other, with operating and auxiliary materials, as well as with environmental conditions. In particular, the influence of water, vapour and gases can lead to strong corrosive attack. For the exact evaluation of external influences it is often necessary to collect characteristic data over long periods of time so as to establish representative statistical bases (see 5.2.2, page 177).

Subsequent corrosion can to a large extent be prevented by appropriate constructional measures whereby provision is made for the possibility of carrying out repetitive cleaning, derusting, and protection against corrosion with the maximum of simplicity and the minimum loss of time.

Stoppage corrosion periods can be prevented if the design offers the possibility of appropriate protective measures [22].

Planning for Maintenance

Similar considerations apply to ensuring proper maintenance. Thus, the disposition of lubricating points should be designed in such a way that they cannot be neglected or ignored altogether by maintenance personnel. The same precaution applies to devices for the removal of soiling, foreign bodies and dust. If plastics are used, account should be taken of accumulations of dust produced by electrostatic charges.

Servicing and Supervisory Equipment

Damage caused by human failure, such as operating errors, lack of skill and negligence, can often be obviated by psychologically and ergonomically based design improvements to control elements [23]. It is recommended that those parts of an installation which are sources of danger or which are themselves particularly at risk should be provided with key-switches, locking devices, photoelectric barrier systems or similar equipment or should be marked with distinguishing colours. By well thought-out grouping of important instruments and measuring, control and alarm devices, it is possible to facilitate inspection, while faulty starting up and shutting down can be prevented by interlocking.

1.3.1.3. Transport and Storage, Assembly, Commissioning and Test Runs

From the moment components leave the manufacturer's works up to the end of the test run, the possibilities of loss prevention that are applicable to all types of machines are so varied that they will be treated here in greater detail.

Expenditure for measures which contribute to smooth accomplishment of transport, assembly, and test-run operations is compensated by commissioning at the scheduled time. All possibilities of negative influences cannot be eliminated, but experience has shown that damage can be reduced by:

- selection of packaging commensurate with anticipated loadings,
- a transport chain relatively free from risk,
- care regarding storage,
- safe transport with lifting gear,
- attention to stressing by environmental conditions,
- safeguarding of erection sites against theft,
- safeguarding of erection sites against fire and explosion,
- easily understood, unambiguous instructions for assembly and test run, and ensuring that these instructions are being followed.

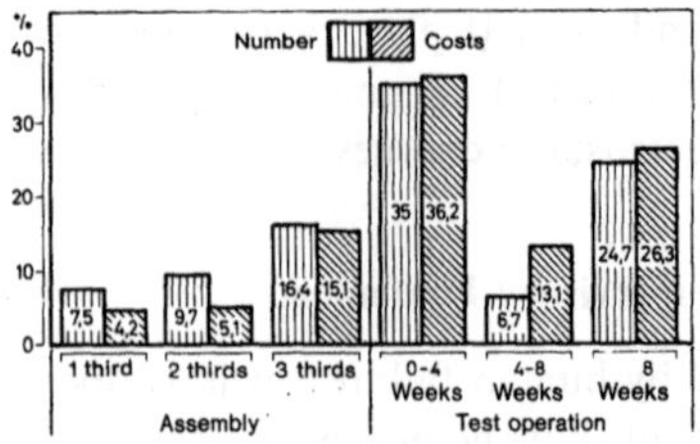

769

Figure 23. Distribution of cases of damage amounting to more than DM 50,000 over various phases of construction [32].

Figure 24. These condensers were in a crate, properly packed and labelled. Ignorant of the meaning of the symbols used, transport workers turned the crate over. The condensers were so badly damaged that they could not be repaired.

1.3.1.3.1. Transport

Packing firms, insurers and also most carriers give information on appropriate *packaging*. External and internal packing materials, closure of crates and corrosion protection must be chosen in accordance with anticipated transport stress on the basis of:

- the transport medium;
- the transport chain (DIN 30780) (duration of transport and storage, transshipment, intermediate storage, storage conditions);
- climatic influence during transport and at the destination (DIN 50010).

The examination of packings in accordance with DIN 55438, DIN E 30799 and DIN 50010 can be agreed upon with packing suppliers.

For transport, equipment and components must be packed, wedged and lashed up in such a way that they are safeguarded against dropping down, shock and vibration, as well as against environmental influences. Components liable to sag must be supported, antifriction bearings must be relieved, and movable parts must be secured during transport.

Particular care must be taken when shipping overseas.

If goods are carried on the deck of ships or are unloaded in ports into barges, there is an increased risk of corrosion or wetting by seawater, especially with such materials as cement, refractory bricks and unpackaged steels. In such cases, special packings or protective coatings are therefore necessary.

Also, the remains of chemicals (e.g., artificial fertilizer) left behind from earlier shipments in the holds of ships can, in combination with condensed water, cause appreciable corrosion.

Labelling of packed components and of unpacked parts must be carried out according to DIN 55402. Indistinct labels or inscriptions that have become illegible through weathering are likely to be disregarded. Good writing alone is useless if personnel are not given instructions regarding the symbols employed and are not told to follow these instructions implicitly.

It is also recommended that packages with a gross weight of less than 1000 kg should be provided with weight data and that all packing units should contain a list of their contents and an indication of the equipment to which they belong. This procedure facilitates the supply of replacements in the event of damage. The goods supplied should be checked for completeness and condition immediately after unloading.

If damage is visible, and also if it is suspected, the transport insurer or a surveyor and the carrier

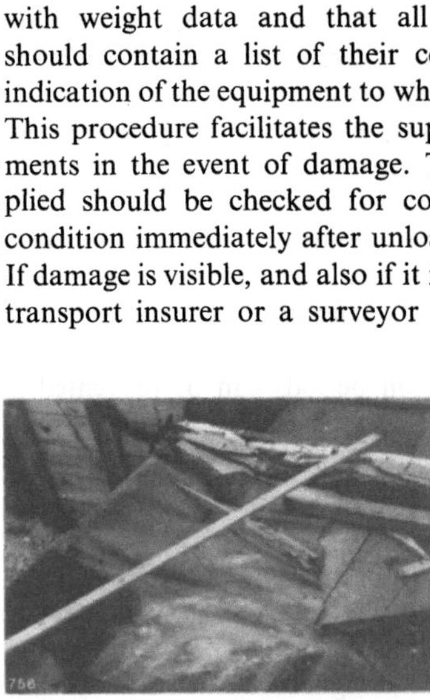

Figure 25. When damage is visible and also when it is suspected, the transport insurer or the damage assessor must be informed immediately.

Figure 26. A hydraulic excavator on a low-bed trailer was caught by its jib under a railway bridge having a clearance height of 4 m. Although the trailer was travelling at a speed of only 20 km/hr, considerable damage was caused to the excavator, the bridge and the trailer.

should immediately be called together and invited to a joint survey to assess the damage and secure later recourse action.

Short *transport chains*, i.e., direct transport, reduce the hazards, whereas transshipments and intermediate storage have the effect of increasing them.

Before despatching goods, it should be ascertained whether conditions are suitable for assembly at the erection site. Delivery to the port of loading should not take place before the date arranged for receipt there and stowage, so as to avoid unnecessary storage, usually in the open. If transpotation involves transshipment, it is necessary, in the case of heavy pieces, to establish, prior to despatch, whether at the points of intermediate storage transshipping facilities are adequate and whether storage facilities are suitable for the type of packing used. With heavy pieces, the transshipment and intermediate storage areas should first be inspected (quay installations, cranes, barges, etc. should be capable of accepting about 20% to 30% more than the weight of the heaviest pieces). Responsible safeguarding of the transport of heavy pieces also requires, before the start of transportation, ascertaining the load-bearing capacities of roads and bridges, as well as examination of the heights and widths of underpasses, bridges and canals, the provision of appropriate warning signs and, above all, the associated repeated instruction of personnel.

1.3.1.3.2. Erection and Assembly Site

The principal should be obliged to make available the final roadway system (without top surface). Should this not be possible, *temporary roads* must be constructed economically with building materials available at the site. In such a case, it is necessary to take into account:

- load-carrying capacity according to the maximum expected loading and also during periods of bad weather,
- adequacy of the road network,
- inclusion in the road plan of locations for lifting gear.

Inadequate roadways and locations for lifting gear very often cause damage to equipment and assembly components, drag out the construction period and hence reduce the economic efficiency.

If the provision of *rail installations* is intended, they should be ready in good time, so that goods can be delivered by rail during the period of construction. The installation, including the connecting track to the nearest railway station, must then

Figure 27. A road for the building site was made from prefabricated concrete slabs, without levelling the subsoil. Gradients and transverse tilting caused a 125-ton mobile crane to topple over.

Figure 28. Covering components in the open with tarpaulins or plastic sheets is effective only if they are tied down in such a way that they also offer protection in bad weather.

Figure 29. With insufficient air circulation, water condensation can form under coverings. Waterproof tops for crates are preferable to tarpaulins or plastic sheets.

be fully operational at the beginning of construction work. Provisional unloading tracks and loading ramps are potential sources of damage.

Earth and concrete work, steel assemblies, pipeline construction and cable laying will overlap during the period of construction and assembly. This fact must be considered in the selection and distribution of *storage areas* in the same way as a sufficient reserve for unforeseeable high demands and perhaps the possibility of later expansion; otherwise, changes in storage will be necessary, with a consequent increase in risk. Storage areas must be strengthened and drained or situated on a slight incline. Heavy constructional components sink into sandy or stiff loam soil, so that they always require a supporting base. Transport goods may topple over in loose sand.

Whether storage should be in the open or under cover will depend on the nature and delicacy of the stored goods and on the expected climatic influences. The protection of crates in the open by tarpaulins is effective only if the tarpaulins are securely tied down. Preferable to canvas or plastic coverings is a well-waterproofed top (sheet metal or roofing felt) for the crates.

Delicate machine components and, above all, electronic equipment should be stored in closed rooms. The storage area must offer protection against rain, frost, water, heat, wind, dust and sand, as well as against insects and rodents.

To prevent damage through condensation, bright metal parts require regreasing if they are stored for longer periods. Protective paint coats peel off and must be renewed. Vacuum packing does not offer protection indefinitely; it is therefore essential, with valuable items, to carry out periodical spot checks.

For the erection of *electrical power plant* on building and assembly sites, VDE 0100 applies, as also do the following comments.

The incorporation of reliable instrumentation and selectively operating protective equipment is absolutely essential for all installations and also for the supply of current to the building site itself. The necessary expenditure must be regarded in relation to possible injury to personnel or damage to materials.

To keep voltage drops in supply cables within limits, the transformer stations should be as central as possible.

The risk of harmful voltage drops is particularly pronounced when using long rubber-sheathed or similar types of cables laid in the open and having a cross-section that does not match the actual needs of the connected consumers.

It must also be remembered that it is possible for several large consumers connected to the grid

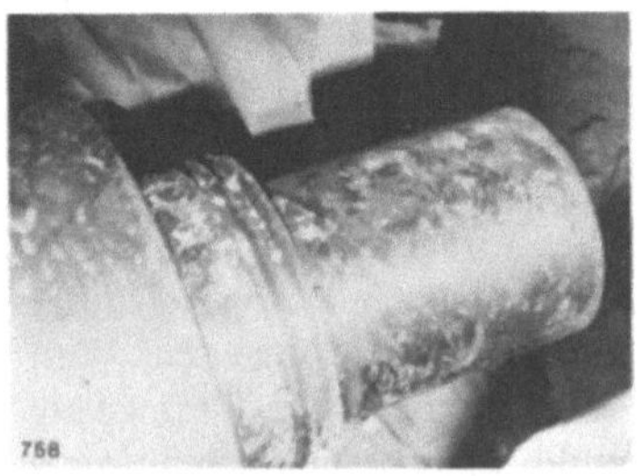

Figure 30. This journal has been corroded by condensation of water. Bright components must be regreased if they are stored for long periods.

Figure 31. This swivelling erection crane fell over when a wire rope was attached to it.

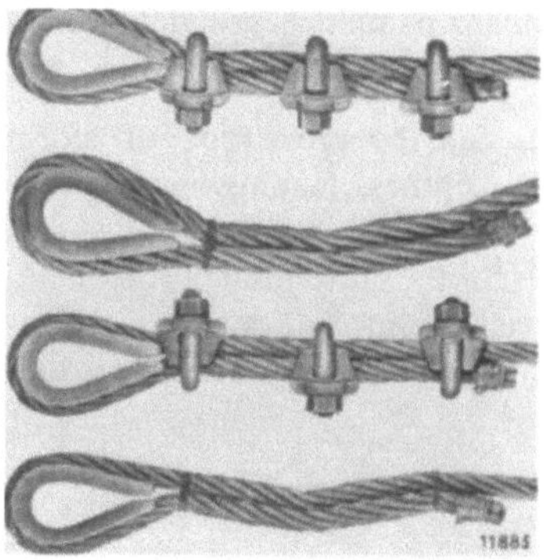

Figure 32. Incorrect assembly of cable clamps results in deformed cables. Cable breakages can originate from places where there are kinks. The upper picture shows all three clamps, and the lower picture the one in the centre, incorrectly fitted.

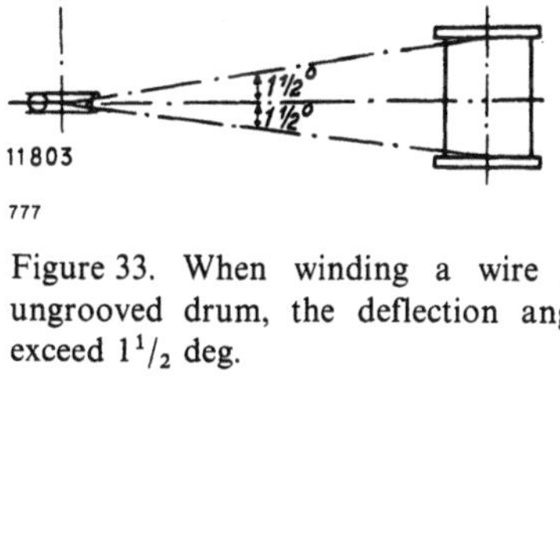

Figure 33. When winding a wire rope onto an ungrooved drum, the deflection angle should not exceed $1^1/_2$ deg.

to draw current at the same time. In such a case, the voltage must not drop to a level which could cause harm to personnel and materials, e.g., during lifting operations.

If the water table is high, care should be taken to ensure that the cable end joints in the cable ducts are as high up as possible. For the cable ducts, there must be provided a drainage pump installation, with whose operation personnel must be familiar.

A decisive factor in the choice of masts for overhead lines is the anticipated climatic stressing.

Lighting of the roads on the construction and assembly sites must be in accordance with the prescribed regulations for proper illumination. Insufficient lighting is tiring and reduces attentiveness.

1.3.1.3.3. Transport and Lifting Gear

Every temporary expedient, even if it appears to be time-saving, increases danger to personnel and materials.

Timber joints must be secured by building clamps, steel-girder supports must be bolted together and secured to a baseplate.

When transporting goods over an incline, they must be kept secure by counterweights.

Particular attention must be paid to *ropes and chains.*

Many cases of damage, caused by falling down of assembly equipment, have been due to loosening or incorrect application of rope clamps. When transporting and lifting heavy components, spliced ropes must be used rather than clamps.

For the handling and application of cranes, see Section 6.2 of this handbook. Lifting gear at assembly sites is, as has been shown by cases of damage, often too weak for the particular application. Overdimensioning by 30% has proved to be suitable as a means of protecting against imponderables and deviations from the planned procedure, such as canting, slipping and shock.

Oblique tension causes overloading of booms, ropes slip out of the grooves in cable drums, cable strands are squeezed and separated, and load hooks are bent out of shape.

The frequency of cases of damage arising when lifting heavy loads by means of several cranes must be particularly stressed. Difficult and critical

Figure 34. While lifting the generator by means of two cranes and two auxiliary hoists, a link in the chain of one hoist broke. The generator crashed down.

Figure 35. This suspension device was used to lift a steel scaffolding. The diameters of the holes for receiving the locking bolts proved to be too small and, with the aid of a flame-cutter, the holes were opened out to an oval shape. Later, during lifting, the lugs fractured in the region of these holes as a result of the large reduction in cross-section and also because of hardening of the material during flame-cutting.

hoisting operations must be under the command of a responsible supervisor, and care must be taken with the selection of lifting gear and personnel. Good means of communication must be provided, e.g., by a radio link.

Forged *load hooks* must be checked once a year, and multi-plate hooks once every three years, for freedom from cracks, deformation (measured from a fixed point) and wear in the mouth of the hook. Welding and heat treatment of hooks and chains are strictly forbidden. Particular care is necessary to avoid damage to lifting tackle when welding is carried out on a suspended load or in the vicinity of removed hooks and chains.

Cross-beams must be dimensioned and produced by experienced manufactures. They must not be changed on site and, should a change become necessary, it must be carried out by the manufacturer. The temptation to use eyebolts for lifting and transport operations should be firmly resisted. For example, the eyes fitted on the top of gearbox covers and designed to carry only the weight of the covers will rupture if the complete gearbox is suspended on them.

1.3.1.3.4. Climatic Conditions

Climatic stressing can cause heavy damage which often occurs because of insufficient prior information.

The following Allianz pamphlets give detailed information on preventive safety measures:

No. 1: Measures for the prevention of storm damage to cranes, loading installations and steel structures.

No. 2: Measures for the prevention of frost damage to machine installations.

No. 3: Reconditioning of soaked machine installations.

Figure 36. With the aid of a floating crane, it was intended to lift a large and heavy pipe, weighing 45 tons and 40 m in length, into position on a pipe-supporting bridge under construction. The cross-beam, specially manufactured for this purpose, fractured in the welding seam of the suspension lug because, when setting down the load, it was stressed in a direction perpendicular to the longitudinal axis.

Figure 37. Cooling jacket of a compressor split open by the effects of frost.

Figure 38. The tubular jib of this excavator was badly damaged by pulling up over the head. Unexpectedly, the cable winch started to rotate because the clutch plates had frozen together during stoppage.

Figure 39. Extensive damage to installed machines was caused by flooding and rubble carried into the machine hall.

Extensive information on extreme and average climatic conditions prevailing during the past ten years in the regions of interest are given for the Federal Republic of Germany by regional meteorological offices, for the European continent by the central meteorological office in Offenbach (Main), and for overseas by the maritime me-

Figure 40. Because of inadequate safety measures, this tank was badly deformed by a sudden squall during assembly.

teorological office in Hamburg. From these reports, the weather influences to be expected in a particular season can be deduced.

To meet time schedules for critical phases of construction, it is also necessary to relay information continuously to the building site from a meteorological office and, in some cases, from a river authority as well. A short-term forecast can almost always make a safe prognosis for a particular region; medium-term forecasts, on the other hand, can only be accepted with reservations.

Extreme values to be reckoned with on a statistical basis, as supplied by meteorologists, should be taken into account not only when planning the construction work itself, but also for the equipment at the building site. Accordingly, general safety measures must be provided for both the structure to be erected and its plant.

The amounts of rainfall and the danger from floods must be taken into account when planning roads and building sites and when dimensioning high-water drainage and protective walls. In addition, consideration must be given to the ready availability of personnel and equipment for snow clearance when calculating the load-carrying capacity required to sustain snow loads, as well as to an abundant supply of water during periods of drought.

Data regarding expected wind forces must also be noted. Moreover, all intermediate phases of construction must be proved to be sound from the point of view of statics, and the safety of incomplete buildings must comply with the requirements of statics. With structures sensitive to wind, abnormal wind conditions must be indicated by an acoustic alarm.

Climatic conditions can affect human beings; under difficult assembly conditions, consideration should therefore be given to the fact that the carefulness and reliability of personnel are reduced during periods of abnormal conditions as a result of fatigue, diminished ability to concentrate and nervousness.

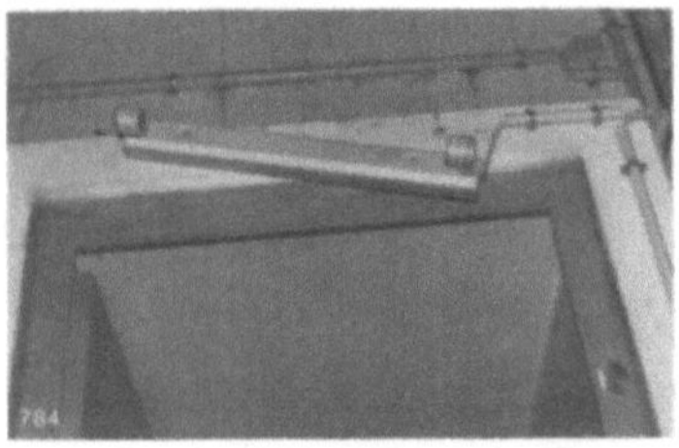

Figure 41. During the night, thieves forcibly dismantled fluorescent lamps which had been fitted in the daytime. In addition, leads were torn away, and plugs, sockets and switches were wantonly destroyed.

1.3.1.3.5. Safeguarding of Building Sites against Theft

Supervision of the erection site is facilitated by fencing it in, leaving only one open but guarded place of access, if possible.
Easily portable valuable items, such as measuring instruments, calculators and cases containing valuables should be taken to the living quarters after work or concealed among other valuable objects which are difficult to move.
During the period of excavation and skeleton construction, the risk of theft is small, but becomes greatest after assembly of the general electrical and sanitary installations.
Effective means of loss prevention are as follows:

- fencing in;
- planning installation work on a time basis with the object of establishing closed security sections which are secured by locking up;
- installing transportable photoelectric alarm barriers;
- transferring responsibility for looking after the tools in a group to a single person (inventory, weekly check);
- checking the work site, when work is finished, for tools and equipment left behind;
- using qualified watchmen, sufficient in number in relation to the size of the assembly site, particularly during non-working periods;
- arranging irregular police patrols in the vicinity of the erection site, especially at night;
- reporting each established case of theft to the local police.

Figure 42. A tar heater, left unattended during a break period, fell over on the roof. As a result, this machine hall of approximately 7,000 m^2 caught fire and was badly damaged.

1.3.1.3.6. Safety Measures against Fire [30, 31]

Information as to which technical fire-protective, constructional and organizational measures can contribute to the prevention of damage by fire is given by the results of the following investigations:

- At which locations on the building site is a fire most likely to occur?
- With what means and equipment can such a fire be extinguished in its initial stages?
- How far can a fire spread?

Particularly endangered are areas where

- flammable liquids, paints, tar, paper, cardboard, foamed plastics, wood and similar substances are stored or processed;
- welding or flame-cutting work is carried out.

The later a fire is fought, the fewer are the chances of preventing it from spreading. During the initial stages, effective fire-fighting is possible with *manual fire extinguishers*. These extinguishers should therefore be available for rapid application in operations involving a fire hazard (e.g., welding, cutting, soldering, and tarring) and in installations where there exists a risk of fire, such as drying ovens and tar heaters and paint and fuel stores.

Within the area of normally exposed external assembly and storage regions, a distribution of at least three portable dry-chemical (12-kg capacity) extinguishers per 1,000 m^2 is advisable. Regarding the application and number of manual fire extinguishers to be held in readiness, see Technical Information, published by the Munich Reinsurance Company "Installation of Portable Fire Extinguishers" (Fire-Fighting Facilities B1–B3).

The effectiveness of manual fire extinguishers is limited by their content of extinguishing medium. If a fire has progressed beyond its initial stages, larger extinghuishing equipment, such as motorized sprays and fire engines, will be required.

When fighting fires on electrical installations and machines and equipment sensitive to extinguishing media, particular care must be taken to protect personnel. When fires break out, it is necessary to contend not only with heat and vision-impeding smoke, but also with the evolution of toxic gases. Fire-fighting without an effective breathing apparatus is therefore often impossible. *Self-contained respiratory equipment*, such as compressed-air breathing apparatus, may be used only by specially trained and suitable personnel.

Quench-water can be held in readiness in ponds or tanks or can be supplied from fire mains. These pipes should be provided at the height of the building level, and a distance of 80 to 100 m is recommended between hydrants (with two outlets and shut-off valves, and allsynthetic hoses, in accordance with DIN 14375 and 3222).

The quench-water supply depends on the anticipated demand, which is determined by a fire-prevention expert.

Fire-fighting with quench-water delivered through a multi-purpose nozzle (adjustable for spraying and full jet) makes it possible to increase the range of operation. The spray-type jet is preferably used with materials such as burning liquids.

Figure 43. This vehicle was used for carrying building rubbish to a pit, where it was burned. The vehicle came too close to the fire and was destroyed by burning.

In areas where quench-water cannot be applied, but where smouldering materials play a part in the start of a fire, transportable dry-chemical extinguishing equipment is suitable, preferably with a special powder medium for smouldering fires.

Account should be taken in the *distribution of storage areas* of the danger from fire and explosion. Depending on the degree of risk from explosion, flammability and combustibility, particularly dangerous goods should be stored separately. Even with goods of low combustibility, arrangements should be made to provide aisles in sufficient number and widths between complexes of stacked goods to act as firebreaks and permit fire-fighting vehicles to be driven in.

During the entire period of construction, care should be taken, for the prevention of fire from spreading as a result of draughts or chimney effects, to ensure that firebreaks are maintained on each unit under construction.

From the start of construction and from one building stage to the next, *fire-prevention measures* should be discussed with the local fire brigade, because the effectiveness of local brigades in combating a fire is often overestimated. Fire-fighting teams recruited from the works personnel offer the advantage that they can be in action quickly and that they possess a more exact knowledge of the

Figure 44. During welding on a bridge-building site, damage from fire occurred. Large amounts of scaffold timber were burned up.

site. However, they cannot carry out the required task effectively unless they are properly trained. Care must be taken to ensure that all personnel at the building site (even if there are changes in personnel) are kept informed with respect to:

- giving an alarm at the building site;
- conduct in the event of fire and disaster;
- safeguarding of living quarters and storehouses against fire;
- safety measures for operations exposed to the fire hazard;
- marking of fire-extinguisher locations;
- marking of underfloor hydrants (they must not be made inaccessible by stored building materials).

Many fires at construction sites have been caused by *welding operations* (see also ATI-International DE6-111).

Flying sparks can start a fire, even from far away. Often, the after-effects of heat generated during welding is the starting point of a fire. Prescribed protective measures, such as the use of cover-plates and cooling down of a working area, are frequently neglected deliberately by personnel, particularly in the case of minor repair jobs. Furthermore, this neglectful conduct is often not objected to by supervisory staff if the protective measures concerned would appear to have a delaying effect on the work to be carried out.

Cleaning rags, sawdust, remains of packing materials, paint residues, wood, etc. often catch fire only after prolonged smouldering and after the welding job has been completed. The welding team must therefore adhere to the following five protective measures:

1. Avoidance of any flammable material in the danger zone.
2. Covering of flammable items which cannot be removed from the danger zone.
3. Sealing of all openings which communicate with neighbouring rooms.
4. Provision of fire-watching personnel.
5. Repeated inspection after the completion of work.

After the completion of welding and cutting operations, a special watching service must be on duty for 3 to 4 hours, especially after the end of work. If there is any suspicion that a fire has broken out, the fire alarm must always be put in action.

Cylinders of pressurized gases, such as oxygen and acetylene, must be protected in all circumstances from heat effects (e.g., from direct sunlight, radiant heat from heating installations) and from mechanical damage. Also, at the place where they are used, they must be secured against toppling over and being bumped by vehicles employed at the building site.

Welding hoses must not be porous; nor must they be temporarily repaired. The burners must be provided with flash-back suppressors.

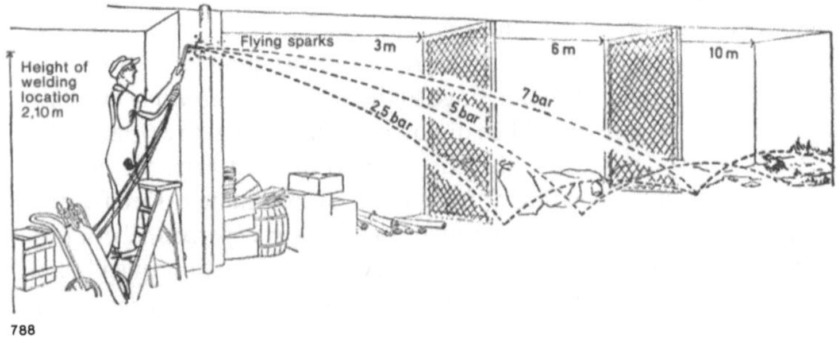

Figure 45. A special fire hazard is caused by open flames and flying sparks.

Figure 46. A steel-reinforced concrete beam, weighing 85 tons and 40 m in length, was suspended on the hook of a mobile crane by a special suspension device. Even before starting the actual lifting operation, the beam tilted to one side and in its slanting position became caught in the suspension, sprang back and tore in web and flanges in several places.

Figure 47. Similar cases of damage also occur when setting down steel-reinforced concrete beams if no measures have been taken to prevent lateral tilting.

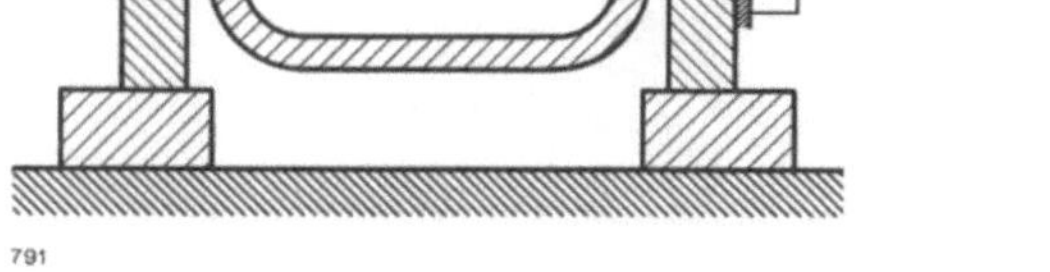

Figure 48. During welding work, earthing cables must be led as close as possible to the welding location. Care should be taken to ensure that in no circumstances can reverse currents flow away via bearings, joints, track wheels and wire ropes.

Gas-cylinder carriers and single cylinders must be stored outside the working area at the end of a working day.

More than 10% of all fire damage at industrial assembly sites occurs during test operation. In the majority of cases, the cause is burning oil. The reason can often be ascribed to the assembly of couplings and flanges in pressure lines in a manner that does not conform with regulations, or to laying the lines in such a way that they are subjected to vibrations, with the result that they break. The escaping oil then catches fire on hot machine parts.

1.3.1.3.7. Easily Understandable, Unambiguous Instructions and Assurance that they will be Followed during the Assembly Process and during Test Operation [32 to 35]

Assembly faults having serious consequences and operating faults are caused by lack of skill, negligence, language difficulties, and also by insufficient technical knowledge. These faults can be reduced during the assembly process itself and during test operation by unambiguous instructions and repeated training sessions for the personnel concerned.

Insufficient co-operation and shortage of time always have the effect of increasing the danger.

Assembly

The causes of damage occurring with lifting processes during assembly are not always due to faults in the lifting equipment, lifting tackle, or the foundation. Such damage can also be caused by the constructional components to be lifted, e.g., if the actual weight of the component to be lifted does not correspond to its estimated weight, or if a component is overstressed during lifting and lowering as a result of the method of suspension used or incorrect handling.

Rotors must be stored in such a way that their journals are supported on rollers, wood or lead foil. When applying ropes, the bearing points must be wrapped in protective coverings. Steel-reinforced concrete beams must be secured against lateral tilting by additional supports.

Good access to welding locations contributes to loss prevention. Incorrect electrode guidance, e.g., during overhead welding; damaged or damp electrode sheathing; inadequate seam prepara-

Figure 49. Temporary hall pillars, wedged-in only in the longitudinal direction of the hall got loose from their anchorage even under a low wind loading and fell over.

tion; welding during excessively low ambient temperatures—all cause damage.
Pre-ignition of electrodes is permissible only on an ignition plate. Penetration notches and craters at the seam edges must be removed by grinding, with the plane of the grinding wheel at right-angles to the welding seams.
Earthing cables must be led as close as possible to the welding location and clamped, so that in no circumstances can current flow back via bearings, shafts, joints, wire ropes, etc.
Tack welds must not be made on highly stressed, load-carrying structures, since they may later become the origins of crack formation. For attaching pipelines and cables at such places, clamping or adhesive bonding is recommended.
Building components, initially *attached provisionally* during the course of assembly work, must be secured in such a way that, even under unfavourable climatic conditions, further construction work in the vicinity should not be endangered.
Incorrect wiring in electrical installations often leads to severe damage.
When assembling distribution equipment, incorrect wiring can also endanger current and voltage transformers. With current transformers, the secondary winding must always be loaded or short-circuited (no fuses provided).
With open secondary circuits, it is possible, apart from very high overheating of the iron core, that voltages of a magnitude fatal to humans will arise. To avoid high voltages on the secondary side, secondary windings must therefore always be earthed. With voltage transformers, however, the secondary winding may be left open or, if it is closed, only via a high resistance (voltmeter, voltage winding with wattmeters, etc.). The secondary circuit of voltage transformers must always be provided with fuses as a protection against overloads, short-circuits and faulty wiring.
"Flying leads" constitute a hazard to personnel and equipment.
Pipelines and casings must be checked before final closure. An extremely wide variety of foreign bodies can easily enter them during assembly and can cause severe damage when operation is started. If such parts of an installation have been checked, they should be closed up immediately after they have been found to be in perfect condition and should be marked as having been inspected.

Figure 50. Packing paper from the relief valve of an oil tank with a label indicating that the paper must be removed before assembly. This injunction was forgotten, and a few days after being put into operation, the tank fractured as a result of a excessive pressure rise.

Figure 51. A steel rope was left in this gearbox. When taken into operation, severe tooth damage occurred.

Figure 52. When filling the tank, water overflowed through the open vent fitting and impeded painting work. A fitter, called in by the painters, flanged on a cover. When the tank was drained, it imploded.

Test Operation

Before starting up a machine installation for the first time, transport stops, assembly aids, etc. must be removed. A check should be made to ascertain whether all the necessary auxiliary and secondary installations are ready for operation. Lubrication points and oilers should be checked for adequate supply. This check must be carried out by expert fitters. Equipment which has been checked for its functional proficiency should be clearly marked. The time taken for such a check, even for small items of equipment, should be taken into account when planning the time schedule.

The insulation of electric motors which have been in storage for comparatively long periods should be checked before they are first switched on. The terminal boards should be given a second check for correct connections before the motor is taken into service. Also, to ensure the correct direction of rotation, the phase sequence should be noted.

To check whether they rotate freely, rotors must be turned by hand. If a machine exhibits abnormal running behaviour after first being connected, the cause must be established immediately. Do not wait to see if there is an improvement before taking action!

When heating up pipelines for the first time, all fixed points, sliding locations and suspensions must be carefully checked to see whether they function properly.

A check of excess-pressure and under-pressure safety elements must precede tests to establish whether closed containers are leakproof. Valves that have been checked should be protected from unauthorized actuation by warning plates. Only by constant supervision of pressure during the entire test period is it possible to take immediate loss-prevention action in an emergency.

To be able to recognize irregularities quickly during the commissioning phase, the use of *additional measuring and testing equipment* for supervising the starting-up procedure should be considered. This equipment, too, should be installed in accordance with regulations.

The danger of incorrect operation of a new installation during the initial commissioning phase is considerable. No plant should ever be put into operation without fully functional safety equipment.

A substantial contribution to the reduction of damage during delivery, storage, assembly and test operation can be made by careful planning, calculation and design, the application of good co-ordination between design, workshop and assembly (as well as on the assembly site itself), purposeful and repeated training of assembly personnel, and the application of modern inspection techniques.

1.3.2. Loss Prevention by the User

The stressing of a component or machine, as determined during planning, calculation and design, is often subjected to additional influences which have not been taken into account (overloads, corrosion, erosion, etc.), with the result that their anticipated life can be greatly reduced to an unknown degree. Numerous preventive measures make it possible for the user to protect the installation from such additional stresses, to maintain operating efficiency during its useful life, and to recognize and replace parts liable to damage in good time. In general, it can be said that the expenditure for loss prevention should be weighed against its effectiveness, and an optimum value should be established.

1.3.2.1. Loss Prevention before Acceptance (Test Operation)

The first duties of the user in connection with loss prevention arise before acceptance of the installation and involve ensuring that all necessary conditions for operation of the plant are satisfied, such as available space, energy supply, illumination, ventilation, and noise suppression.

Before acceptance of a machine or plant, the service and maintenance personnel, including reserves, must be so instructed and trained that they will be capable of running and maintaining the plant on their own. It has always been found expedient to employ maintenance personnel whose services will be required at a later date (e.g., fitters, electricians and control technicians) early enough to make them take part in the assembly of the installation. The resulting knowledge can later be invaluable in the event of malfunctioning.

During the commissioning phase and test operation, the user's own service personnel should be made to co-operate, the responsibility remaining with the supplier. From test operation and unavoidable teething troubles, corrective adjustments, etc., a better knowledge of the operational behaviour of the installation can be gained. Operating instructions must be available, must be carefully checked and, if necessary, changed or complemented. Also, they must be as simple as possible and understandable to personnel in general. Arrangements must be made for clear marking of valves and measuring points.

The customer must insist—even if he has to put up with some delay in commissioning—that, before initial start-up, all the necessary measurement, supervision, protection and control equipment has been properly installed, ready for operation, and, as far as is technically possible, set or adjusted.

Maintenance and inspection schemes should form part of the delivery specification. Apart from the contractual agreement of the supply specification and acceptance conditions, the question of spare parts must also be clarified at any early stage, and orders for spare parts must be placed. In addition, there must be at least two complete lists of spare parts, one for the plant and one for the machine.

1.3.2.2. Loss Prevention during Operation

Within a plant, it is possible to apply loss-prevention measures in relation to operation, supervision, maintenance and repair. These various measures change during the useful life of a machine and depend, on the one hand, on technical developments and the economic situation and, on the other hand, on the way the installation is used. If machines are out of operation for long periods, special inhibiting measures must be taken (see special chapter).

Measures for the maintenance and assurance of the operational safety of machines and installations, as prescribed by law, are not discussed here.

Correct Operation

Personnel responsible for operating machines should be informed at regular intervals of the content and application of operating instructions, and should also be made cognisant of the significance and effects of mishandling or faulty operation.

The fully automatic and semi-automatic operating devices increasingly being used has the inherent risk that, even in the event of minor irregularities, the entire installation will shut down and, when starting up again manually, serious errors may be made. The reason is either that facilities for manual operation are not available or, if they are, personnel are not trained to cope with irregularities. If, technically, manual operation is possible in automatic installations, personnel must be trained regularly in the manual operation of the installation by simulation of irregularities.

Optimum Supervision

The purpose of supervision is to ensure adherence to technical specifications required by the operation. Supervision can be effected by personnel and/or by automatic measuring and control devices and must ensure that the maximum permissible stressing through internal or external influences is not exceeded and that the quality and quantity of products are guaranteed.

The parameters to be supervised (e.g., vibration, rotational speed, pressure and temperature, stresses and flow) are very varied. Because of economic considerations and for ease of inspection, it is possible only to supervise permanently the most important operational parameters. It is essential to find the correct choice and allocation of stresses in relation to the measured value, to know the maximum permissible value and, if the permissible stress is exceeded, to bring into effect suitable control measures to prevent damage or stop it from spreading.

When supervision is carried out by personnel only, gradual changes in technical operational parameters are particularly dangerous because, as distinct from sudden changes, they often remain unrecognized. This problem can be overcome by applying markings to measuring instruments, making lists of limiting values, and by visual or audible warnings. Alarm annunciators should call attention to danger conditions to the largest possible number of personnel and must be available in sufficient quantitiy, even if supervision is effected by recording instruments. In each particular case, consideration should be given to whether protective equipment, e.g., switch-off gear, should be employed in place of an alarm.

All operational records are valueless unless they are evaluated constantly by adequately trained personnel who are conscious of their responsibility and are in a position to initiate suitable countermeasures immediately.

In a largely automatically operated and supervised plant, the small number of personnel remaining must not be allowed to feel that they are superfluous and lose interest in their job. Such personnel should be entrusted with systematic tours of inspection and with looking out for abnormal occurrences which cannot be detected by measurement. Thus, unusual noises, odours, leaky steam or product pipelines, and leakages of flammable liquids (e.g., oil) must be reported immediately and rectified.

All supervisory equipment must be checked on a rota basis for proper functioning and adjustment. The most important supervisory and safety installations should be designed in such a way that they can be checked for proper operation and, if necessary, calibrated not only during periods of shut-down of the plant, but also (as, for example, the emergency trip gear of large steam turbines) during operation.

Measuring instruments or protective devices which have been found on several occasions to be responsible for incorrect readings, false alarms or faulty tripping constitute a special hazard. Thus, because they are regarded by personnel as being unreliable, they may well be ignored when they operate correctly in the event of a real emergency. Such unreliable devices must be replaced immediately and calibrated or properly adjusted in relation to the prescribed response range.

Optimum Maintenance

The object of maintenance is to keep equipment in good order and provide it with auxiliary materials. An obligatory maintenance schedule and trained personnel must be available for all maintenance work. The time intervals are determined on the basis of experience and prescribed by the manufacturer or user. These intervals must always be adhered to as strictly as possible; in no circumstances, however, must they be exceeded by more than 10%, since auxiliary materials usually require supplementing or renewing and parts subject to wear need to be replaced before they fail.

In the interest of preventive maintenance, all abnormal occurrences during the preceding operational period must be carefully evaluated, and the preventive measures (repairs, modifications, improvements) indicated by the evaluation must be adopted in the course of maintenance work [36–38].

All maintenance work, as well as the types, quantities and operational lives of auxiliary materials and replacement parts, must be recorded in machine cards or log-books (commercially available). After completion of work, appropriate checks and functional tests must be carried out.

If several maintenance teams are in action, it becomes necessary to draw up precise lines of demarcation between their fields of duty and to agree on functional tests with each other. Numerous cases of damage have been caused through shift-work, e.g., by tools or auxiliary equipment inadvertently left behind in the installation, or because the following shift has not been informed or only insufficiently informed by the relieved shift of the work carried out previously.

Inspections

The purpose of inspections is to check parts of a machine or installation without major dismantling or even in the operating condition without any dismantling at all, following unusual occurrences (suspected damage) or in accordance with planned schedules. They therefore serve to establish the technical condition of certain components.

As a rule, inspection is effected by visual observations with or without technical aids. The use of rigid or flexible endoscopes, television techniques and other test and measuring equipment has made it possible in recent years to widen substantially the scope and meaningfulness of inspections. By the provision of suitable openings, already catered for in the design, to permit the entry of instruments, the inspection of many machines can be greatly facilitated [39].

Careful inspection before scheduled overhauls makes it possible to spot defects and hence to order important replacement parts in good time. In this way, an extension of the time for overhaul, resulting from delayed delivery of replacement parts, can be avoided.

In all cases, inspections should be carried out only by specially trained personnel of the user's plant, the manufacturer or the insurer. The latter should, in particular, be called in if inspections are to be carried out following unusual, unexplained occurrences or if there is a suspicion of damage.

Overhauls

The scheduled overhaul of a machine installation in a disassembled condition, apart from routine maintenance work, constitutes the oldest and best-known method of loss prevention. Measuring and supervisory equipment, despite improvements and greater sophistication, can do no more than show up changes and damage that have already taken place during operation and perhaps prevent extension of damage by appropriate protective circuits or rapid manual action. In the course of overhaul, however, all important components can be subjected to careful visual and/or instrumental checks, and shortcomings and damage can be detected in their initial stages and rectified.

The overhaul of a machine or plant entails a scheduled shut-down at a time when the effects of non-operation are as small as possible; it should prevent unforeseen outages at an unfavourable time with great losses in production. In other words, a limited non-availability as a result of overhaul is accepted in order to improve overall availability.

The costs and loss of availability engendered by overhauls depend essentially on the *frequency of overhauls*, i.e., the period between two *overhauls*—the *overhaul* cycle—and on the *time required for overhaul* [40].

The *overhaul cycle* should be as long as possible without adversely affecting its loss-prevention function. Determination of the cycle can be based only on empirical values obtained from statistical evaluations by suppliers, users, technical supervision institutions and technical insurers. These values will vary with different types of machine and, moreover, depend in each case on the mode of operation (partial or full load, starting or switching-on frequency, etc.) and the special operating conditions involved [41, 42]. With steam-turbine plant, equipping the machine with measuring and supervisory equipment can also play a decisive role in determining the overhaul cycle [43].

With newly installed plant or machines, particularly with prototypes, a theoretical determination of the interval between overhauls is fraught with uncertainties, because relevant experience is not available. It has always proved reasonable from the technical and economical points of view to carry out the *first* overhaul shortly before expiry of the warranty period given by the supplier. The condition of the machine established during this overhaul, together with experience gained with similar structural components, then makes it possible to decide on a realistic plan for future overhaul cycles.

Shortening of the time required for overhaul reduces the costs of overhaul to the same extent as lengthening the overhaul cycle. Since, however, the latter procedure increases the probability of unforeseen damage, particular attention should be devoted to shortening the time required for overhaul. This can be brought about by careful preparation, if necessary by means of network analysis, so as to explore fully the possibilities of carrying out, during overhauls, repair jobs on many components largely at the same time. Another prerequisite is the ready availability of the necessary tools, personnel and replacement parts needed [44]. To be able to order replacements in good time, experience with earlier overhauls or/and inspections (if necessary, with endoscopes) should be drawn upon, and also the results of overhauls of other similar machines. If, on suitable components of a large installation, *partial* or *intermediate overhauls* are carried out during planned or unscheduled stoppages, a decisive contribution can be made to shortening the time required for overhaul [45] (see Section 5).

The success of an overhaul depends to a large extent on whether all faulty components that can lead to damage are recognized and eliminated. For this purpose, it is necessary to know the possible weak points of the machine. It is therefore always recommended that consultations should be made with experts of the manufacturer and the technical insurer, whose special task it is to find weak points, damage and its causes.

Proper Repairs

Every case of damage that has occurred indicates an existing weakness in the installation or constructional component affected. When carrying out repairs, every effort must therefore be directed towards eliminating weak points, so as to avoid the repetition of damage. In this regard, the most important prerequisite is an exact knowledge of the cause of damage, making it necessary to analyse carefully the phases of damage and their effects. Only on the basis of the results of this analysis will it be possible to decide, from the technical aspect, in which way repairs must be effected, i.e.,

- by replacing the damaged parts of the structure or plant by similar parts or
- by mending the damaged components and, in certain circumstances, introducing improvements at the same time.

The use of *spare parts* is always to be preferred in cases where recourse can be made to a suitable spare-parts store — be it at the user's or manufacturer's premises — and where economical advantage can be gained by savings in cost and time. At the same time, it is absolutely essential to ensure that the cause of damage — with the exception of accidental faults in the material — does not lie in the very component to be replaced (faults in design and calculation, incorrect material, etc.).

When repairing an installation, the relevant protective devices and interlocking circuits must be checked and, if the need arises, changed or complemented. By using more suitable auxiliary materials (lubricants, coolants, etc.) a repetition of damage may also be prevented.

If the cause of damage is a fault on the part of the operator, better training of personnel is often not enough; operating and maintenance instructions should be examined to see whether they cannot be made clearer and more to the point. In many cases it will also be necessary to check the disposition of the operating controls to ascertain whether they could not be made ergonomically more suited to the man/machine system [23].

Preference should be given to the *repair* of a damaged component in cases

- where the requisition of a replacement part entails a substantially larger expenditure in time and/or money than that required for repair;
- where spare parts (e.g., for older machines) are no longer available;
- where improvements in design and materials can be made in the course of repair — a possibility which would not exist when fitting spare parts.

Before each repair, the area adjacent to the location of damage must first be examined for cracks, material faults or detrimental structural changes. If nothing objectionable is found, the stresses in

the component, which, after all, have led to failure, must be investigated with regard to their magnitude and type (static, dynamic, uni-axial or multi-axial, etc.).
To make full use of the manifold possibilities, knowledge of and proficiency in the various methods of repair, with their specific advantages and disadvantages, is necessary. It is always recommended that, before and during repair work, an expert of the manufacturer, a reputable repair firm or the technical insurer should be consulted [46–48].

Exchange of Experience Regarding Operational Behaviour and Causes of Damage

In an endeavour to prevent damage to an installation, the part played by exchange of experience regarding cases of damage and operational behaviour should not be discounted. Collaboration between users, manufacturers, specialist associations, insurers and other organizations is particularly important, bearing in mind the trend towards specialization. At the same time, care should be taken to select, from the mass of possible sources of information, only those which are most relevant to individual plants.
Apart from the exchange of experience within an organization, great importance attaches to the internal passing on of information useful for the prevention of damage. In this regard, it should be stressed that relevant specialized journals, communications, and information should not remain with management but should be passed on to those who can utilize them in practical terms. Moreover, use should be made of the numerous possibilities of further training within an organization and of sending suitable persons to classes, seminars and courses.
By recording and collecting cases of abnormal occurrences, with all data and manifestations, it is possible to prepare statistics of trouble and damage, from which comprehensive measures for loss prevention can be derived [36, 49]. The wider is the basis of such statistics, the more successful will be the results of evaluations. It is therefore recommended to make use of the possibilities offered by specialist associations and insurance companies for the collection of trouble and damage data on a broad basis.
Last but not least, users themselves should contribute to the intensification of exchange of experience by reporting their own operational and damage incidents in the specialized technical press as quickly and as comprehensively as possible.

Bibliography to 1.1–1.3

[1] Abinger, R.: Der Versicherungs-Ingenieur. Der Maschinenschaden 43 (1970), No. 6, pp. 205/211.
[2] Allianz-Zentrum für Technik. Information 1973/74. Technische Information der Allianz-Gesellschaften.
[3] Tätigkeitsbericht 1969/71 des Allianz-Zentrum für Technik GmbH. Allianz-Berichte für Betriebstechnik und Schadenverhütung No. 18, January 1972.
[4] Der Maschinenschaden. Ausgewähltes Schrifttum 1949–1960. München and Berlin: Allianz Versicherungs-AG 1961.
[5] Der Maschinenschaden. Inhaltsverzeichnis 1961–1970. München and Berlin: Allianz Versicherungs AG 1971.
[6] Zerstörungsfreie Prüfung in der Schadenverhütung. Compiled and revised by K.G. Schmitt-Thomas. München and Berlin: Allianz Versicherungs-AG 1968.
[7] Wege zur Schadenverhütung im Maschinenbetrieb. Revised by E.J. Pohl, R. Bark. München and Berlin: Allianz Versicherungs-AG 1964.
[8] Das Gesicht des Bruches metallischer Werkstoffe, Vol. III (Vol. I/II out of print). Revised by E.J. Pohl. München and Berlin: Allianz Versicherungs AG 1960.
[9] Splittgerber, E.: Die Schadenentwicklung in Dampfkraftwerken – Übersicht über die letzten Jahre. Allianz-Berichte für Betriebstechnik und Schadenverhütung No. 14, April 1970.
[10] Gombart, H., Huppmann, H.: Schäden an Dampfturbinenanlagen durch Planungsmängel und unzureichende Betriebsführung. Allianz-Berichte für Betriebstechnik und Schadenverhütung, No. 14, April 1970.
[11] Maßnahmen zur Verhütung von Frostschäden an Maschinenanlagen. Allianz-Merkblatt No. 2, edition 1969.
[12] Schöllhammer, F.: Der Brand in dem Elektronikraum des Kraftwerkes Pleinting der Ilse-Bayernwerk-Energieanlagen GmbH. Mitteilungen der VGB 49 (April 1969), No. 2, pp. 106/112.
[13] Einwirkung chlorwasserstoffhaltiger Brandgase auf Gebäude und Einrichtungen. Allianz-Berichte für Betriebstechnik und Schadenverhütung. No. 19, November 1973.

[14] Feuerschadenverhütungsdienst. Allianz-Report No. 3. München: Allianz Versicherungs-AG. April 1973.

[15] Effertz, P.H., Grupp, H., Jach, W.: Einwirkung chlorwasserstoffhaltiger Brandgase auf Werkstoffe und Bauteile. Der Maschinenschaden 43 (1970), No. 3, pp. 89/99.

[16] Effertz, P.H., Grupp, H.: Einwirkung chlorwasserstoffhaltiger Brandgase auf metallische Werkstoffe sowie auf Bauteile der Feinmechanik und Elektronik (2nd part): Der Maschinenschaden 45 (1972), No. 1, pp. 7/19.

[17] Lutz, H.R.: Der Turbinenölbrand im Kernkraftwerk Mühleberg. Maschinenschaden 45 (1972), No. 3, pp. 96/102.

[18] Hagn, L., Huppmann, H.: Untersuchungen zur Klärung der Schadenursache des Ölbrandes im Maschinenhaus des Kernkraftwerkes Mühleberg. Empfehlungen zur Schadenverhütung. Der Maschinenschaden 45 (1972), No. 3, pp. 103/110.

[19] Christian, H., Grupp, H.: Maßnahmen zur Sanierung der beim Ölbrand im Kernkraftwerk Mühleberg geschädigten Teile. Der Maschinenschaden 45 (1972), No. 3, pp. 111/120.

[20] Becker, W.: Bauaufsichtliche Bestimmungen – Internationale Harmonisierung. Mitteilungen Institut für Bautechnik 5 (1974), No. 6, pp. 161 ff.

[21] Klement, E., Knublauch, E.: Brandschäden auf Baustellen. Zeitschrift Bauwirtschaft 1972, No. 21, pp. 793/798.

[22] Maßnahmen zur Verhütung von Stillstandskorrosionen an Dampferzeugern und Warmwasserheizungen. Suppl. of „Der Maschinenschaden" 36 (1963), No. 3/4.

[23] Rodenacker, W.G.: Bedienungsfehler im System Mensch und Maschine. Allianz Versicherungs-AG, Reprint (1972).

[24] Empfehlung zur Schadenverhütung für den Exporthandel. Deutscher Transportversicherungsverband.

[25] DIN 30780 Transportkette, Begriff.

[26] DIN 50010 Klimabeanspruchung, allgemein.

[27] DIN 55402 Blatt 1 und 2, Markierung für den Versand von Gütern.

[28] Braun, R.: Schäden an elektrischen Schaltanlagen. Der Maschinenschaden (1968).

[29] Maßnahmen zur Verhütung von Sturmschäden an Kränen, Verladeanlagen und Stahlkonstruktionen. Allianz-Merkblatt No. 1 (1969).

[30] Merkblatt für Ausrüstung der Betriebe mit Handfeuerlöschern. Verband der Sachversicherer, Form 133, 1971.

[31] Schwürzer, H.: Brände durch menschliches Fehlverhalten beim Schweißen und Brennschneiden. Der Maschinenschaden 48 (1975), No. 1.

[32] Splittgerber, E.: Schadenerfahrungen bei Großmontagen. Der Maschinenschaden 45 (1972), No. 4, pp. 125/132.

[33] Steinbrück, G.: Bemerkenswerte Montageschäden und ihre Verhütung. 6. Works Manager Conference of Allianz 1964. Reprints of papers, Vol. I.

[34] Schäden durch Implosion. Der Maschinenschaden 34 (1961), No. 11/12, pp. 187/189.

[35] Engelke, H.E.: Alltägliche Montageschäden. Der Maschinenschaden 47 (1974), No. 2, pp. 44/58.

[36] Gappenberger, K.: Schadenkontrolle – optimaler Nutzen für Arbeitssicherheit und Instandhaltung. Z. „Sicherheitsingenieur" No. 5/6/7, 1974. Reported in: Der Maschinenschaden 47 (1974), No. 5, pp. 186/187.

[37] Marx, H.J.: „Neue Wege und Ziele der Instandhaltung". Z. „Rationalisierung" 26 (1975), No. 7/8, pp. 177/180.

[38] Hoffmann, F.: Mehr Sicherheit bei geplanter Instandhaltung. Der Maschinenschaden 46 (1973), No. 2, pp. 47/52.

[39] Splittgerber, E.: Probleme bei Schadenuntersuchungen und -auswertungen. Mitt. der VGB 51 (1971), No. 6, pp. 471/479.

[40] Bark, R.: Die wirtschaftliche Bedeutung der Schadenverhütung. Der Maschinenschaden 41 (1968), No. 4, pp. 125/134.

[41] Krauß, O.: Die praktische Bedeutung der Revision großer elektrischer Maschinen. Der Maschinenschaden 45 (1972), No. 6, pp. 218/230.

[42] Leopold, J.: Die Gasturbine im stationären Einsatz. Der Maschinenschaden 46 (1973), No. 4, pp. 109/113.

[43] Empfehlungen für die Betriebsüberwachung von Dampfturbinen, insbesondere zum Bestimmen des Revisionszeitpunktes. Frankfurt/Main: Verlags- und Wirtschaftsgesellschaft der E-Werke 1966.

[44] Empfehlungen zum Verkürzen der Revisionszeit von Dampfturbinenanlagen. Frankfurt/Main: Verlags- und Wirtschaftsgesellschaft der E-Werke 1970.

[45] Huppmann, H., Jäger, K.: Ist die große Revision von Dampfturbinen noch zeitgemäß? Der Maschinenschaden 45 (1972), No. 5, pp. 169/178.

[46] Löbert, P.: Schäden an geschweißten Bauteilen aus Stahl. Der Maschinenschaden 35 (1962), No. 7/8, pp. 119/124.

[47] Essler, R.: Erfolge und Mißerfolge bei Reparaturen. Der Maschinenschaden 38 (1965), pp. 11/23.

[48] Pahl, E.: Moderne Reparaturverfahren. Der Maschinenschaden 46 (1973), No. 2, pp. 35/46.

[49] Mewes, K.-F., Schäfer, P.: Auswertung durch Schwachstellenzählung – eine Methode der Schadenverhütung. Der Maschinenschaden 45 (1972), No. 1, pp. 28/32.

2. Electrical Machines, Power Cables and Supervisory Installations

2.1. Turbo-Generators

Turbo-generators must be built and operated in accordance with VDE 0530, Part 3/1.66. Table 1 surveys all reported cases of damage which occurred in the years 1969 and 1970 to turbo-generators (50 Hz, 3,000 rpm) with an output equal to or more than 20 MVA after they were first taken into service.
In this section on Electrical Machines, Power Cables and Supervisory Installations, the definition of the causes of damage takes into account the peculiarities of electrical machines, as follows:

Product Faults

This category comprises all damage that can be traced back to faults in the product and includes not only damage that has originated during production, such as from faulty design, manufacture and materials, but also all imperfections which can be traced to the state of technological development, e.g., the quality of the winding insulation and winding-head support at the time of manufacture.

Operational Faults

Operational faults cover those cases of damage which have been brought about by inadequate supervision not directly connected with supervision of the turbo-generator itself. Exceptions are cases of damage through short-circuits which have occurred during the warranty period, i.e., it is assumed that, during this period, generators are proof against short-circuits in accordance with VDE 0530, Part 3/1.66.

External Influences

This category embraces all extraneous harmful influences which are in no way the result of inadequate supervision.

Winding Faults through Ageing

Under this heading are all cases of damage in which ageing of the insulation is not necessarily the sole cause but certainly the fundamental cause. It has been decided, quite deliberately, not

Table 1. Distribution of damage cases in accordance with their causes and locations.

Cause of damage	Distribution of the number of damage cases [%]	Location of damage	Distribution of the number of damage cases [%]
Product faults	47	Stator winding	27
Operational faults	16	Stator laminations	2
External influences	14	Casings	8
Winding faults through ageing	23	Rotor winding	29
		Other rotor parts	34

Figure 1. Apart from marks left on the end-bell seat by leakage currents, the initiation of a crack (arrowed) can be seen on one tooth of this rotor.

to restrict cases in this category to those taking place only after attainment of the average life of the insulation; otherwise, any faults which occurred earlier – apart from those caused by inadequate supervision – would, of necessity, have to be included in the category of product faults.

Discussion of Damage Locations

More than half of all cases of damage concern the rotor, and in this connection it has been shown that older machines are more prone to damage. A large proportion of cases of damage to the rotor winding is caused by displacement of the winding in the winding head as a result of ageing, the method of cooling, and also the mode of operation (peak load or base load). A major proportion of damage to rotors is concerned with damage to end-bells and their seat on the rotor body, this damage primarily manifesting itself in the form of chipping, cracks or current leakage. As regards the stator winding, it is principally the winding heads that are endangered by excessive current-induced stresses which cannot be taken up by the supports, possibly because they have aged too much (resulting in slackening or shrinking) or because the stresses have become too high through bad synchronizing or similar causes.

2.1.1. Planning of Turbo-Generator Installations

The dynamic behaviour of the foundation is of great importance for the operation of the turbo-generator. By calculation and measurement it is essential to ensure that the natural frequency of vibration of the foundation does not fall within the region of the operating speed.

The dimensions of the machine room should provide sufficient space not only for the assembly of the turbo-generator, but also for taking down individual machine parts during overhaul.

The electrical data of the generator must be based on expected mains and operating conditions.

Sufficient protective devices are indispensable. Since, in this respect, the experience of manufacturers, users and insurers can hardly differ, and since views regarding the appropriate extent of the required protection can involve only minor differences, it should suffice to give here a brief reference to relevant protective measures. These measures are concerned with differential, overcurrent duration, impedance, stator earth faults, stator winding faults, asymmetrical loads, reverse power, voltage increase, underexcitation and rotor earth faults, as well as loss of excitation.

Figure 2. Slide rail of a rotor melted through by current leakage.

It frequently happens that considerable damage is caused to turbo-generators by overvoltages coming from the grid. The installation of overvoltage-protection devices at the transformer, as well as, in certain circumstances, additional capacitors on its low-tension side, is therefore essential.

2.1.2. Commissioning and Operation of Turbo-Generators

Before the first revolution of the generator shaft it is necessary to recheck carefully whether any foreign bodies have been left inside the machine after assembly operations. Severe damage can result from tools, fitting aids, and other foreign bodies inadvertently left behind.
The first mechanical test run reveals whether the complete set has been properly installed and whether the measured values of the parameters which are important for an assessment of the machine are satisfactory, such as quietness of operation in accordance with DIN 2056, and proper indication of temperature, oil flow, etc.
When this mechanical test run, with measurement of the set parameters, has been completed, but not less than 100 hours after testing, a level check of the bearings should be carried out.
The effectiveness of the insulation of the individual windings of the generator must be checked with a megohmmeter before the first excitation. When checking the excitation winding, the test voltage must under no circumstances be higher than the operating voltage. This requirement applies particularly in the case where inadequate insulation values are to be expected, e.g., with damp windings.
As an approximation, it can be assumed that the insulation resistance, converted to 75 °C, is adequate if a value of 1 MΩ/kV is achieved. To estimate the effect of temperature on insulation resistance R_i, a simple rule-of-thumb conversion is to halve the resistance for each 10 °C rise in temperature. For example, if R_i is 20 MΩ at 45 °C, it can be taken as 10 MΩ at 55 °C, 5 MΩ at 65 °C, and so on.
The duration of measurement should not be less than 1 min. This limiting value should, in general, be achieved easily, even after prolonged periods

Figure 3. This chisel was left in the stator bore during assembly of a generator, resulting in severe damage to the lamination stack when the generator was taken into service.

out of service. Difficulties can arise only with slow-running water-turbine generators, owing to their larger number of slots.
Since the condition of the insulation is of decisive importance for the operational safety of the machine, it should periodically be measured and recorded, at least at times when the machine is out of operation. If the machine is non-operational for a longer period, measurements should be made immediately before it is started up again.
If a particular value of insulation resistance is unsatisfactory, the machine must be dried out, as discussed below. If the insulation values of the generator are satisfactory (should there be any doubt, these values should be compared with those supplied by the manufacturer), it can be excited. If, during non-operational periods, changes of any kind are made in the protective or synchronizing equipment, basic tests must again be carried out before the machine is put into operation. (For such work, specialists from the manufacturer should be called in.)

Drying (Generators and Motors)

Insulations based on synthetic resins can absorb moisture only if they have been damaged by

Figure 4. Crinkled shellac-micafolium insulation of a 60-MVA generator. In the position shown, electrical breakdown as a result of moisture absorption occurred when the generator was restarted after a prolonged stand still.

cracking, peeling, etc. Drying of machines with synthetic-resin insulation is therefore restricted to the surface and can, for example, be achieved by a current of air produced by turning the rotor without excitation.

On the other hand, windings with insulations based on shellac and other varnishes or on bituminous materials are very sensitive to moisture. Almost all windings produced in 1965 and earlier fall into this category. For drying windings of this type, four methods can be applied:

- Drying under short-circuit.
- Drying with alternating current.
- Drying with direct current.
- Drying with warm air.

With all four methods of drying, the temperature of the windings should not exceed 75 °C and moisture should be removed by air circulation.

Drying under short-circuit is most frequently applied to generators, partly because no expensive preparations are required and partly because drive power is available. To carry out this method, the machine should be run at nearly nominal speed and should be excited in such a way that the maximum winding temperature of 75 °C is reached gradually—about 8 hours in the case of large machines.

Drying with alternating current is most commonly used for asynchronous motors. This method prevents the rotor from turning and, in the case of slip-ring motors, short-circuit it. A voltage of from 10 to 20% of the nominal voltage is generally required for the stator winding. With slip-ring motors, care must be taken to ensure that the wrapping temperature should in no circumstances exceed 100 °C. Hand-sized openings in the cooling circuit are adequate for air circulation. With totally enclosed motors, particular attention must be paid to the heating procedure to avoid the risk of overheating.

Drying with direct current is used only if the two above-mentioned methods cannot be applied. This method of drying needs sources of current that can continuously supply from 25 to 50% of the nominal current of the machine to be dried out.

Moreover, with large powers, switching from one winding to another is very costly. With this method of drying, there is the additional risk of shaft magnetization, and it is therefore essential to ensure that the circuit of the individual windings and their connecting leads does not have loops that surround the shaft and subject it to the action of the current required for drying. In general, this will not occur if the connections are as follows:

- Positive pole to U.
- X to Y.
- Y to W.
- Z to negative pole.

The validity of the above circuit connections can be checked on the basis of the disposition of the circuit connections of the winding on the winding head of the machine.

With turbo-generators, the rotor should be turned through 180° every two or three hours to prevent asymmetrical heating. If rotation is continuous, care should be taken that no field is generated in the bore, as it could lead to overloading the drive.

If, in the case of smaller machines, it is not possible to disconnect the neutral point or the interlinkage, drying current can be fed to one phase and returned through the other two phases. To achieve even heating of all parts, connections must, in this instance, be changed every hour.

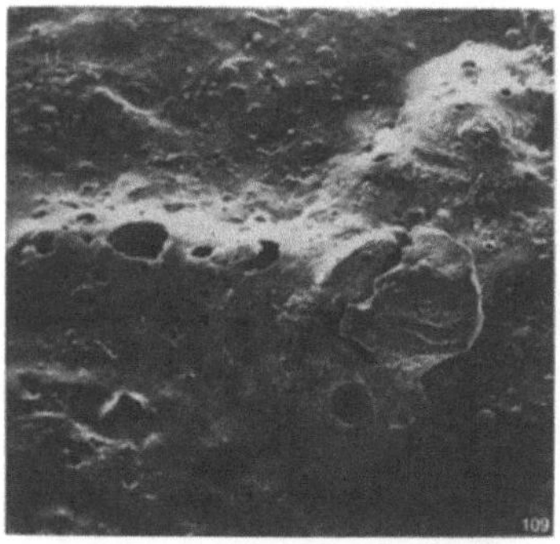

Figure 5. Examination of current leakage in plain bearings. To be able to distinguish with certainty damage by current leakage from mechanically induced defects in plain bearings, an electric current was applied to the bearings. The above scanning electron micrograph shows the place where current leakage occurred. Gas generation, with crater formation in the bearing metal, resulted from local overheating.

Irrespective of the circuit selected, it is appropriate to estimate the heating power on the basis of ohmic resistances.

Drying with warm air requires a relatively high expenditure in terms of the number of heating devices and time involved, depending on the size of the machine, because, in this case, not only the winding to be dried out but the entire mass of machine must be heated. For this reason, the method of drying with warm air is mainly applied to small motors.

Insulation of Bearings

In the construction of electrical machines, it is not possible to eliminate entirely small asymmetries in their active parts. Thus, a potential is created in the shaft, leading to undesirable current flow through the bearings if their insulation is defective. The satisfactory condition of the insulation must therefore be checked thoroughly after the first excitation of the machine. If an earthing brush is used to conduct away static charges originating in the turbine, it must not be positioned on the exciter side of the generator.

Maintenance

Maintenance personnel should check the generator every hour. On their rounds, maintenance men must follow the directions given in the operating instructions. In general, all measuring stations for temperature, pressure and vibration must be visited and the readings recorded. With larger machines, these values are more and more frequently being registered on chart recorders and, in this case, it is important to note the trend of the curves plotted and to check for any irregularities.

As before, the condition of the brushes, their running properties, and, perhaps, their sparking, as well as the condition of the commutator and slip-rings and of the leakage air filter, must be checked on site. Also, any abnormal noise generation should be noted. This applies particularly to the less extensive measuring and supervisory equipment of the smaller and older machines; on the other hand, access to the exciter equipment and to the slip-rings is easier. Moreover, any vibrations of individual machines should be noted, e.g., by placing a hand on the part, since there are only a few vital parts which are permanently supervised by vibration transducers.

Figure 6. The sort of thing that can be caused by a short-circuit. These torn-away fan blades were found in the interior of a machine after a short-circuit.

2.1.3. Inspection of Turbo-Generators

When an inspection is carried out at the proper time, there is a good chance of revealing defects in their early stages of development and hence to prevent loss, as distinct from the mode of operation of most protective devices.

The purpose of every inspection is to recognize and eliminate faults which could lead to damage. A distinction can be made between the first inspection, scheduled inspections, and unscheduled inspections.

The first inspection reveals objectionable changes in the machine that can be traced back to shortcomings in design, manufacture, transportation, assembly or special conditions of operation. This inspection should be carried out as soon as possible, one year after commissioning is recommended.

Scheduled inspections should be carried out at intervals not exceeding four years. With older machines, such an interval is sufficient since adequate experience exists regarding the weak points that need watching. With newer machines, this experience must still be accumulated by inspections. The situation has become more difficult because of the large increase in output which has taken place in recent years in the field of turbo-generator construction.

Unscheduled inspections become necessary after short-circuits, after operation with asymmetrical loading or asynchronous running-up with high slip. Particularly critical is asychronous switching. In the most unfavourable case (phase opposition), forces could be four times higher than those with terminal short-circuit, and the rotor would also be heavily overloaded. In all such cases, an immediate inspection is indispensable.

Figure 7. During an inspection it was discovered, before major damage could occur, that a fixing screw was missing from the radial fan.

General Directives for Inspection

Abnormalities indicative of faults are a reduction in smoothness of operation, a change in noise level, and variations in the appearance of certain machine parts. Variations of the latter type include deposits of dust or of abraded particles, fretting corrosion, discoloration through heat, incrustations, and damp or oily patches, as well as completely missing constructional components.

Figure 8. Strongly pressed-in synthetic-resin insulation of a stator bar of a 150-MVA machine, caused by loose wedges at the root of the slots.

The cause and origin of the above abnormalities, if they arise, must be investigated. In the course of inspection, the machine must be cleaned, particularly if it is air-cooled, since accumulations of dirt impede heat transfer to the cooling medium. Cleaning and inspection must encompass the entire cooling circuit.

Particular attention must be paid to the loosening of components and to foreign bodies in the machine. In this connection, it should be mentioned that, as a matter of principle, personnel should be prohibited from entering a machine until they have first emptied their pockets. All tools to be used should be counted and listed before inspection takes place, and any additional tools that may be required should be added to the list. After completion of the inspection, a check must be made to determine whether all the tools are complete and that the listed number has been removed.

Figure 9. Axially displaced slot wedges bridge the end-bell gap and lead to undesirable current leakage in the case of asymmetrical loading.

Inspection of Stator Windings

Stator windings with insulation based on varnish or bituminous-type insulation are the elements of a generator most likely to be damaged. Furthermore, their repair costs are high. Thermal stressing of the stator winding is often a contributory factor, as it involves substantial relative movements between the conductors and insulation, so that the ability of the winding to fulfil its purpose is endangered not only by thermal but also by mechanical stresses. With insulation based on synthetic resins, introduced about six years ago and now more and more used, an improvement in the operational safety of these machine components can be expected.

Particular attention should be paid during inspection to firm seating of the winding. Loose windings give rise to local abrasion which reduces their dielectric strength. Loosening can be recognized by the presence of abrasive deposits or, in the case of loose slot wedges, by a simple sonic technique involving light tapping with a hammer. The sound produced with loose wedges being different from that with tight wedges. Loose wedges must be renewed or shimmed.

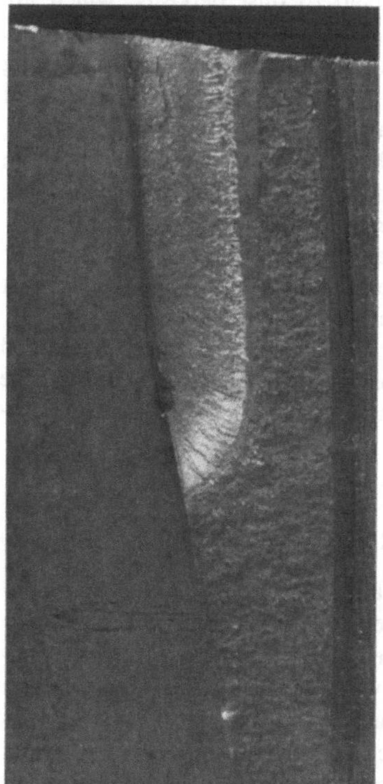

Figure 10. Fracture surface of a cracked fan blade. The advanced fatigue fracture is clearly visible. The remaining fracture was caused in the course of examination.

Windings with varnish- or bituminous-type insulation have a tendency to swell during operation. They therefore not only press firmly against the flanks of the slots but create voids in the insulation which lead to glow discharge. Measurements of tan δ merely give an indication of the extent of these voids and, with respect to the life expectancy of the winding, their meaningfulness is very limited.

Firm seating in the winding head is as important as firm seating of the winding in the slot. The winding-head reinforcement must therefore be examined carefully and, if loosening appears at this position, it is necessary to apply additional wrappings or shims. During this work, care must be taken not to damage other parts of the insulation.

Cracks, bubbles or discoloration of insulating lacquer indicate overheating, due either to insufficient cooling or to overloading.

Figure 11. Short-circuit, as a result of carbon deposits between the two through-bolts of a slip-ring supply.

Lamination Stacks and Casings

When inspecting lamination stacks, special attention must be paid to dust deposits, which are indicative of abrasions in the bar insulation. These abrasions are caused by a loss of tightening pressure and are located mainly at the end of the stack. Apart from damaging the insulation, loosening of the lamination stack can result in the breakage of teeth which, as foreign bodies, can inflict severe damage.

All screwed casing joints must be checked for tightness. Sheet metal for air or gas ducts must be firmly attached and free from cracks.

Water and oil coolers must be emptied, cleaned and, if required, sealed. The water side of the oil cooler must be checked for corrosion and erosion.

Inspection of Rotors

Only the special tools supplied by the manufacturer should be used for removal of the rotor.

After the rotor has been removed from the machine, the air vents in the end-bell (particularly in the case of air-cooled machines) must be thoroughly cleaned to obviate the risk of thermal imbalance. Slight loosening of the wedges is unimportant, as they are pressed in during operation by centrifugal force.

Wedges which have risen at the rotor ends are indicative of large asymmetrical loads. In such a case, renewal of the wedges is necessary. Unfortunately, there is no practical means (other than by using an endoscope) of checking the condition of the winding heads of the rotor winding and the position of the winding-head gaps without withdrawing the end-bells. This procedure which would be very desirable, as demonstrated by damage statistics, at present only can be effected at the manufacturer.

The most highly stressed constructional components of a generator are the two end-bells. Their accurate seating on the rotor body must be checked, i.e., by measuring the gap between the end-bell and rotor and comparing it with previous records. In general, a gap up to a maximum of 3 mm is permissible. In addition, the end-bell gap should be examined carefully under suitable illumination for loose parts of the insulation, shims, and hot spots. If the gap is bridged by axially displaced wedges, undesirable leakage of current will occur at these locations when the load is unbalanced. The wedges must therefore be knocked back into position. In the case of substantial damage concerning one of the abovementioned points, the manufacturer and insurer should be contacted.

The tightness of the rotor-balancing screws and the safety of their locking devices must be checked individually. Fan blades can be inspected visually and tested for freedom from cracks by tapping them and listening to the sounds produced.

Carbon deposits in the region of the slip-rings can lead to flash-over. Immediate cleaning is essential.

Loosening of the slip-rings can also be detected sonically. Slip-rings which are out of round (dial-gauge check) must be rectified with grinding equipment, taking care not to go below the minimum permissible diameter. Reground slip-rings must be run in, with the temperature of the rings and brushes controlled until a film has been formed. If the temperature rises during running-in to a higher extent than that stipulated as permissible by the manufacturer, the number of brushes and, hence, the generator output must be reduced until normal temperature conditions are re-established.

Bibliography to 2.1

[1] Bonfert, K.: Betriebsverhalten der Synchronmaschine. Berlin, Heidelberg, New York: Springer-Verlag 1962.

[2] Mayer, H.: Die Isolierung großer elektrischer Maschinen. Berlin, Heidelberg, New York: Springer-Verlag 1962.

[3] Krauß, O.: Die Revision von Turbogeneratoren. Der Maschinenschaden 43 (1970), No. 6, pp. 213/225.

[4] Wiedemann, E.: Internationale Entwicklung von Großturbogeneratoren. Elektrizitätswirtschaft 67 (1968), pp. 249/255.

[5] Krick, N.: Entwicklungsstand der Turbogeneratoren. BBC-Mitteilungen 56 (1969), pp. 368/379.

[6] Braun, R.: Entstehung, Beurteilung und Vermeidung von Staub in Turbogeneratoren. Der Maschinenschaden 42 (1969), pp. 91/93.

[7] Neugebauer, H.: Selektivschutz. Berlin, Göttingen, Heidelberg: Springer-Verlag 1958.

[8] Haase, W.: Wichtige Schutzeinrichtungen an Elektromotoren und Generatoren. Paper No. 25, 7. Works Manager Conference of Allianz 1968.

[9] Wanke, K.: Turbogeneratoren für Kernkraftwerke. Technische Mitteilungen AEG-Telefunken 59 (1969), pp. 1/13.

[10] Lambrecht, D.: Strömungs- und Erwärmungsprobleme bei Turbogeneratoren mit wassergekühlter Läuferwicklung. Konstruktion 22 (1970), pp. 265/275.

[11] Empfehlungen für die Betriebsüberwachung von Turbogeneratoren, insbesondere zum Bestimmen des Revisionszeitpunktes. Frankfurt/Main: Vereinigung Deutscher Elektrizitätswerke 1968.

2.2. Three-Phase Motors

The following survey (Table 1) is concerned with cases of damage suffered by three-phase motors during 1970 and the first six months of 1971.

Winding Faults through Ageing

All cases of damage under this heading are to windings which have been in operation for more than 15 years. This time limit has been selected on the basis of tests proving that insulation ageing is the decisive factor in damage to the windings. Moreover, an additional meaning is given to the column "Location of Damage", which already contains the total number of damage cases to windings.

Discussion of Damage Locations

Table 1 shows that the stator winding is the component most liable to suffer damage. Here, a particularly strong influence is exerted by the stresses produced during switching-on, as well as during mains switching. Damage to the rotor winding is preponderantly the result of inadequate attention during service, e.g., too many consecutive starts and excessively long start-up periods. These cases of damage, which are brought about by external influences, could, at least, partly be reduced by suitable protective devices. The large number of cases of damage to windings through ageing indicates the need to replace windings weakened through old age by new windings in good time.
Bearing damage listed in Table 1 is characterized by the fact that, in most cases, it causes heavy additional damage to lamination stacks and windings.

2.2.1. Planning of High-Voltage Three-Phase Motor Installations

Before planning the installation of a large three-phase squirrel-cage motor as a prime mover, the relevant electricity authority should be consulted, and it should be established whether the existing mains conditions permit direct switch-on.
During the planning stage, the torque characteristic of the driven machine should already be matched with the characteristic of the motor. Particular attention should be paid to the starting procedure. Motor torque at all speeds must be higher than the characteristic of the driven machine, details of which are available from the manufacturer. When matching the motor with the driven machine, the nominal power in continuous operation must not be exceeded. A further point to be observed when planning the drive is that the stated torque values are subject to deviations which have their origin in manufacture. For example, it is possible that saddle moments generated at certain speeds by higher harmonics will prevent the achievement of nominal speed.
During starting, the rotor and stator are subject to severe thermal stressing. The starting process should, for this reason, take as little time as possible (from 2 to about 10 sec) according to the requirements of the driven machine. Longer

Table 1. Distribution of damage cases in accordance with their causes and locations.

Cause of damage	Distribution of the number of damage cases [%]	Location of damage	Distribution of the number of damage cases [%]
Product faults	47	Stator winding	75
Operational faults	9	Stator laminations	3
External influences	18	Rotor winding	12
Winding faults through ageing	26	Bearings	8
		Other parts	2

Figure 1. Because of soilage of this winding head, heat dissipation was impeded, resulting in excessive heating of the winding and unserviceability of the insulation through premature ageing.

starting periods require special rotors. A further criterion during planning is the number of starts required in a given unit of time. Normally, a cold motor can be switched on and off three times in succession, and a warm motor only twice. Higher starting frequencies must be expressly sanctioned by the supplier.

If a motor is to sustain mains switching, it must be provided with stronger winding-head reinforcements, compared to those of a normal motor. Mains switching is regularly carried out in power stations and large industrial undertakings.

Motor-Protection Devices

Each motor must at least be provided with its own protective switch, relay, or other cut-out device, to ensure that single-phase operation and unintentional overloads will switch off the motor before non-permissible and dangerous temperature rises can affect its windings. Fuses alone, without a motor-protection device in series, are not sufficient.

After actuation of a motor-protection device with thermal overcurrent and short-circuit quick release, the cause of actuation must in all circumstances be established. Under no condition should the sensitivity of the device be reduced; otherwise, its protective efficiency would be lowered.

(For low-voltage motors, it is recommended that full motor protection should be installed, as it constitutes the best control over the stator winding.)

Waterproof switchgear is advantageous as it can be used, not only in humid rooms, but can also be operated safely with wet or dirty hands.

For thermal release, a lockable resetting device is recommended, to ensure that it can be operated only by authorized personnel, e.g., a works electrician or foreman.

An undervoltage release must always be provided for slip-ring motors to prevent destruction of the windings when current is restored after interruption of the mains supply. An undervoltage release is also recommended for squirrel-cage motors to prevent them from starting on their own after an interrupted mains supply.

If non-magnetic quick-release switching devices are used, e.g., contactors with bimetal relays,

Figure 2. This winding head of a high-voltage motor became overheated through overloading. The impregnating medium was expelled, and the winding head, as well as the fixtures, became embrittled.

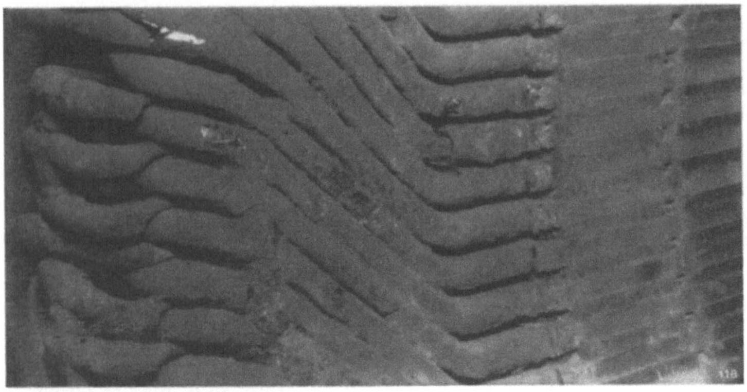

Figure 3. Traces of flash-over can be recognized in this heavily soiled winding head of a three-phase motor. Because of this soilage, adequate heat dissipation was no longer possible, causing the winding to age prematurely and to be unable to withstand the movements of the winding head during switching-on.

fuses must always be connected in series on the incoming side. Fuses are also necessary in cases where the short-circuit current at the switching position can become greater than the current rating of the protective switching device.

Setting of the protective device should be in accordance with the data on the motor name plate. For instance, if the plate indicates a voltage U of 220/380 V and a current I of 2.9/1.7 A, the value of 2.9 A applies when connecting to a 220-V three-phase supply. In other words, the lower voltage corresponds to a higher value of current, and vice versa.

2.2.2. Installation and Commissioning of Three-Phase Motors

During installation, care should be taken that the shafts of the motor and driven machine are aligned in accordance with installation instructions. The use of a dial gauge is strongly recommended.

When connecting up, the phase sequence should be noted (to obviate the risk of counter-rotation). Terminal connections must be tight, and the machine must be earthed in accordance with regulations.

Before coupling up a motor provided with rolling bearings, the shaft-clamping fixture must be removed and a test made by hand to ascertain whether the rotor turns freely. All bearings must be lubricated as prescribed.

The oil supply to plain bearings, particularly in the case of forced lubrication, must be tested before starting up to guarantee proper functioning and freedom from leaks.

When installing oil pipelines and other metallic auxiliary lines, no bridging of the bearing insulation must take place.

Before switching on the motor for the first time, the insulation resistance of the winding must be checked with a megohmmeter. To eliminate the risk of damaging insulation, the voltage of the measuring instrument should not be too high, the recommended value being equal to or less than two-thirds of the voltage at the terminals.

As a rule of thumb, the minimum value of insulation resistance R_i [Ω] of the stator windings of high-voltage and low-voltage motors can be derived in terms of the nominal voltage U and the

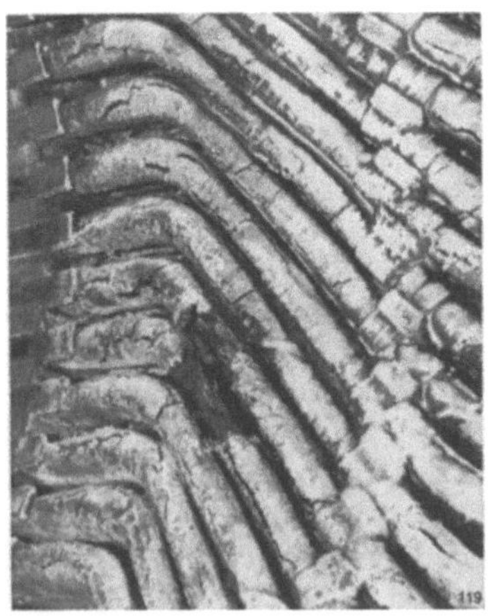

Figure 4. This winding head was unable to withstand the stresses caused by frequent switching. The winding suffered damage, after the fixing components became loose and the wrappings were torn.

rotor diameter D [m] from the following relationships:

High-voltage motors: $R_i \geqq 1{,}000\ U/D$
Low-voltage motors: $R_i \geqq 1{,}000\ U$

If these minimum values are not attained, the winding must be dried out (see also Section 2.1.2). If the measured value of insulation resistance is near but still above the lower limit, it is appropriate to run the motor for about two hours without load or to drive it from an external source. Motors having type P 33 protection must be installed in such a way that the drain hole for condensed water faces downwards.

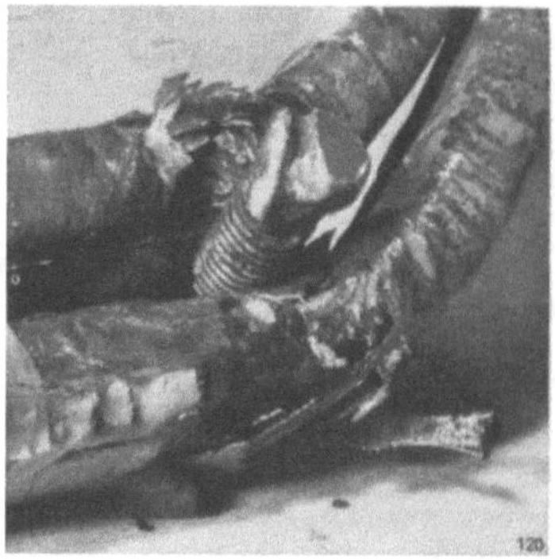

Figure 5. This screw fell into the winding and caused a short-circuit to earth. The fact that the screw got loose would appear to indicate a faulty locking device. During inspection, particular attention should be paid to the tightness of screws, pins, etc. The locking arrangements of screwed connections should also be checked with extra care.

2.2.3. Operation and Maintenance of Three-Phase Motors

When operating the motor it is absolutely essential to adhere to the starting frequencies permitted by the manufacturer (for guidelines, see Section 2.2.1).

The motor must not be overloaded in continuous operation; otherwise, the windings will overheat. An excessively high winding temperature results in its premature ageing and destruction.

The limiting excess temperatures are determined by the class of insulation used in the winding, as indicated in Table 2.

The maximum permissible temperature (limiting temperature) is equal to the sum of the temperature of the cooling medium (40 °C), the limiting excess temperature (heating limit), and the safety factor for the hottest location.

The safety factor [°C] for the hottest location depends on the class of insulation, as follows:

A	E	B	F	H
5	5	10	15	15

If current changes without changes in load, or if there are short-term fluctuations in input current, the cause may be due to the following damaging effects:

- Rotor-bar fractures.
- Winding short-circuits.

A longer starting period under otherwise equal operating conditions indicates fracture of a rotor bar. Starting times must be measured at intervals of about two months and recorded for the purpose of comparison. Deviations from the normal starting period must not exceed 10%.

Special instruments are required for the supervision of motor-protection devices. Manufacturers recommend that, depending on the method of

Table 2. Limiting excess temperatures of electrical machines (extract from VDE 0530, Part 1/1.66, para. 24).

	Permissible limit of excess temperature [°C] with insulation of Class				
	A	E	B	F	H
All windings, except field windings	60	75	80	100	125
Field windings	65	80	90	100	125
Iron cores with embedded windings	60	75	80	100	125
Commutators and slip-rings	60	70	80	80	80
Plain and rolling bearings	50	50	50	50	50
Rolling bearings with special greases	60	60	60	60	60

Table 3. Classification of insulating materials in accordance with their heat resistance (after VDE 0530, Part 1/1.66, para. 13).

Class	Insulating material	Maximum permissible continuous temperature [°C]
A	Cotton, silk, cellulose, paper, wood, oil-based wire lacquer; impregnated or as a filler	105
E	Wire lacquers based on polyvinyl acetate, polyurethane or epoxy resins	120
B	Fibreglass, asbestos, and mica products with Class B impregnating media (shellac, bituminous materials, resin lacquers, etc.)	130
F	Fibreglass, asbestos, and mica products with Class F impregnating media (epoxy resins)	155
H	Fibreglass, asbestos, mica, and silicone rubber with Class H impregnating media (e.g., silicone resins)	180

Figure 6. In this high-voltage motor, a short-circuit to earth occurred because the slot wedges became loose and the winding in the slot started vibrating. Firm seating of slot wedges is important, and particular attention should be paid to this requirement during inspection.

protection, a check should be made at intervals of 3 to 24 months.

Contact-pieces of switchgear showing signs of welding and wear should be renewed in good time; replacement pieces should be held in store, for this purpose. With oil-filled contactors, the quantity and condition of the oil should be checked regularly. The fixed and moving contacts of starters should be examined regularly for pitting and, if necessary, should be smoothed.

Supervision of the Bearings of Three-Phase Motors

The very small air gap of asynchronous motors is the reason why damage to a bearing is usually not necessarily restricted to the bearing itself, but can also, as a result of lowering the rotor, cause severe damage through fouling, leading to the destruction of windings and lamination stacks. The bearings must therefore receive special attention.

Insufficiently lubricated bearings, or bearings operated with spent lubricants, are extremely hazardous, and their life is shortened very quickly. For this reason, it is important to meticulously observe the maker's instructions regarding the replacement of grease in rolling bearings and the changing of oil in plain bearings. It is recommended that the prescribed intervals for grease replacement and oil changing should be specified in writing.

The nominal life of a rolling bearing for large electric motors is more than 20,000 operating hours. Since the nominal life is defined as the period in which 90% of all bearings of a particular type can survive without damage under identical

Figure 7. The rotor fouled the stator in this high-voltage motor. Distortion of the frame was established as the cause. Such distortional changes must also be looked for during inspection.

conditions, it will be found that, in practice, a longer operating life than that stated above can be expected. Removal of a bearing before it has reached the 20,000-hour period is not recommended; however, should it become necessary for one reason or another, a new bearing should be installed, i.e., repeated disassembly and re-assembly of a bearing should be avoided.

By observing the appearance of grease exuded from a bearing, an indication can be obtained of the condition of the bearing. Normally, fresh grease is light and transparent, but it gets dark when contaminated. For instance, wear particles from steel impart to the grease a grey to black colour, while wear particles from brass give it a green colour.

Bearing wear is caused by the penetration of particles of dirt from the outside through faulty seals, or by lack of grease. Also, if unsuitable qualities of grease are used, signs of wear can appear.

With plain bearings, a life of more than 100,000 hours of operation can be achieved.

Lowering or settling of the rotor of a motor equipped with plain bearings takes place very slowly during operation and can be determined by measuring the air gap between the stator and rotor. The changes in air gap arising from settling of bearings must be very small, their magnitude depending on the dimensions of the motor. A comparison with the results of the initial measurement of air gap must show a deviation of not more than 0.2 mm, this value applying to journals up to 100 mm in diameter.

Leakage currents in bearings can be responsible for considerable damage. It is therefore necessary to check bearings for any traces of these currents. Faulty oil seals must also be checked, as they permit the ingress of oil from the bearings into the winding.

Oil rings may be damaged by improper treatment and by operational influences. For instance, they may become out-of-round, in which case they will no longer rotate and will grind themselves into the shaft. In operation, a sign of an out-of-round oil ring is expulsion of oil from the bearing.

Figure 8. In this high-voltage motor (a section of the stator lamination stack is shown), the rotor fouled the stator, resulting in severe damage. It was established that the cause of damage was an error in erection, in that the base plate was able to become loose.

Figure 9. Shrinkage of the insulating materials caused axially displaced intermediate layers of this stator winding.

An important consideration is that the oil supply must be monitored in such a way that control instruments measure real values. For example, the operation of an oil pump should not be checked merely by ascertaining whether the motor is running, but by flow indicators.

Examination of the oil makes it possible to reveal signs of wear. Such an examination should be concerned with impurities in and ageing of the oil, and should include indications as to the presence of water in the oil.

Since screws on bearings can also become loose and shims can "give", all screws must be checked for tightness. If irregular running is observed, it is necessary to open up the bearing and to scrutinize the load pattern.

Even with perfect bearings having oil-ring lubrication, the oil should be changed regularly in accordance with the motor manufacturer's instructions, using oil of a suitable quality. Details of the correct oil to be used should be specified in writing.

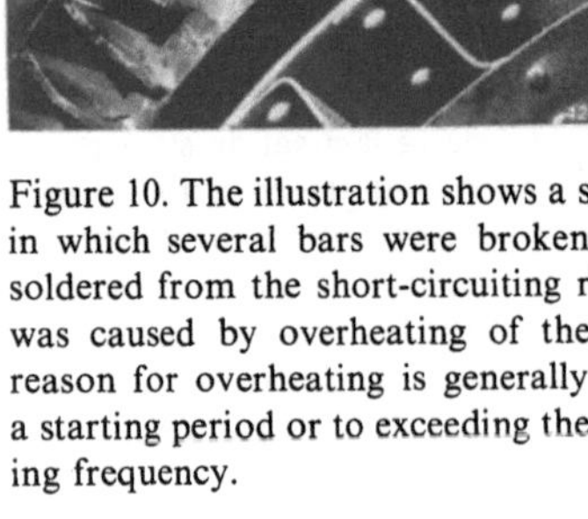

Figure 10. The illustration shows a squirrel-cage rotor in which several bars were broken and became desoldered from the short-circuiting ring. This damage was caused by overheating of the rotor bars. The reason for overheating is generally due to too long a starting period or to exceeding the permissible starting frequency.

2.2.4. Inspection of Three-Phase Motors

First Inspection

With a new motor, the first inspection should, if possible, take place within the first year of operation, to establish whether there are any faults in manufacture and installation and whether the motor has been overstressed in operation.

Scheduled Inspections

With motors of more than 750 kW, an inspection is required after 500 switching cycles or 8,000 hours in operation and, at least, every two years. With motors of smaller power, inspection at the same intervals is strongly recommended. Particular attention must be paid during inspection to the removal of dirt and to the tightening of loose parts.

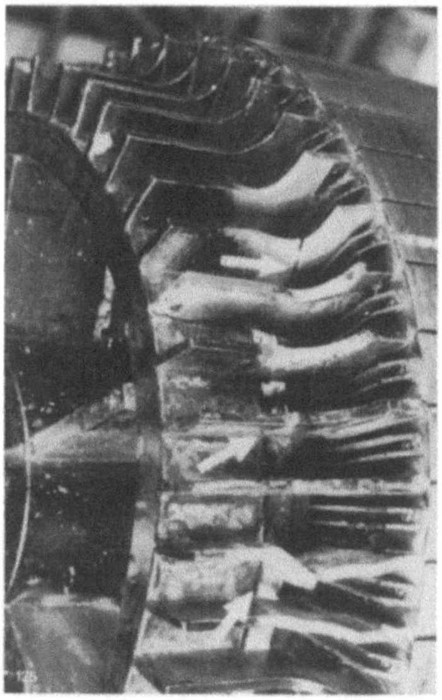

Figure 11. This squirrel-cage rotor contains a number of broken bars. The cause of breakage was vibration of the bars, which can be generated if the bars are not firmly embedded in the slots. During inspection, the rotor cage must in all circumstances be examined for fractures, and the faultless seating of the squirrel cage must be ensured.

Unscheduled Inspections

An unscheduled inspection must be carried out when operating conditions are arduous and after the occurrence of irregularities, such as sudden braking or blockage of the machine set.

Inspection of Stators

The most heavily stressed parts of a motor are the windings. In particular, the winding heads of the stator winding are heavily stressed by surge currents during starting, since every start-up of the motor represents a short-circuit. Stressing of the winding heads is therefore considerably higher with three-phase motors in comparison to generators. In the course of time, the winding-head fixings become slack, the coil ends become loose, and the individual coil components start vibrating. The winding heads must be securely fixed again and put into a safe condition for operation. In this connection, attention must be paid not only to soilage, but also to oil smears. Cracks are a sign that the winding has been overworked.

Before applying a new coating of lacquer to the winding heads, the old layer must be removed, as thick layers of lacquer reduce heat dissipation.

During inspection, seating of the slot wedges must be checked for firmness and, if necessary, rectified. Slackening of the slot wedges in the stator can generally be recognized by the wedges standing proud in relation to the slots and by the formation of dust-like abrasive particles from wedges and insulating materials. With sufficient experience, loose slot wedges can also be detected by the simple sonic technique mentioned in connection with the inspection of generator stator windings in Section 2.1.3.

All screwed connections of the stator and bearings must be checked for tightness. A check must also be made to determine the tightness of the motor connections in the terminal box.

Existing protective devices, e.g., slot thermometers, must be proved to be operational.

The winding heads must be checked during inspection for distortion, and any loosened parts must be secured.

Inspection of Rotors

When inspecting a rotor, it must be established whether the rotor has retained its original shape or whether it has been displaced axially.

Figure 12. Heavy soilage of the squirrel cage of a high-voltage motor. Because of clogged cooling slots, adequate cooling of the rotor can no longer be guaranteed.

Figure 13. Short-circuit of the winding at the slot exit of a 4-MW slip-ring motor.

Even if the rotor is removed from the machine, individual bar fractures are not easy to ascertain by simple means. However, some indication of the condition of a bar can be obtained by the following procedures: An optical check will show up locations of overheating. The presence of iron burns and the appearance of small raised projections between the iron and bar are indicative of a bar fracture. The sound of bars when tapped with a hammer or flat piece of iron can also be used for assessment, while tapping the short-circuiting ring all over with a rubber or wooden hammer makes it possible to find fault locations by sound. Finally, it is necessary to tap the end-bells all over and assess their condition by sound. An optical check of the inner surface of a bar is not possible with all designs but, in certain circumstances, it can be observed by means of a mirror.

When examining the bars, care should be taken to ensure that they are firmly embedded in the slots to prevent the bars from vibrating.

Other Points

Water coolers should be seen to be leak-proof and should be cleaned. Any changes in the foundation or deviations from the proper alignment of the machine set should be noted.

Bibliography to 2.2

[1] Andresen, E.Ch.: Der Entwicklungsstand von Drehstrom-Käfigläufermotoren für große Leistungen, AEG-Mitteilungen 54 (1964), No. 1/2, pp. 8/23.

[2] Bödefeld, Th., Sequenz, H.: Elektrische Maschinen. Wien: Springer-Verlag 1942.

[3] Gahleitner, A.: Eine neue Reihe von Drehstrom-Asynchronmotoren großer Leistung. Siemens-Z. 40 (1966), Suppl. Motoren für industrielle Antriebe, pp. 6/13.

[4] Huppmann, H.: Schäden an Gleit- und Wälzlagern. VDI-Berichte No. 141, Gleit- und Wälzlager, pp. 97/105.

[5] Krauß, O.: Die Revision von Elektromotoren großer Leistung. Der Maschinenschaden 40 (1967), No. 5, pp. 153/62.

[6] Krauß, O.: Anlaufprobleme bei Drehstrommotoren mit Kurzschlußläufer. Der Maschinenschaden 40 (1967), No. 2, pp. 41/51.

[7] Krauß, O.: Beanspruchungen und Schäden an Drehstrom-Kurzschlußläufermotoren durch Schaltvorgänge. Der Maschinenschaden 42 (1969), pp. 133/141.

[8] Stiebler, M.: Die Berechnung von Übergangsvorgängen bei Induktionsmaschinen mit Stromverdrängungsläufern. Archiv für Elektrotechnik 51 (1966), No. 1, pp. 23/37.

[9] Wiedemann, E., Kellenberger, W.: Konstruktion elektrischer Maschinen. Berlin, Heidelberg, New York: Springer-Verlag 1967.

[10] Woydt, G.: Beanspruchung und Versteifung der Wickelköpfe von Drehstrommotoren. Siemens-Z. 40 (1966), Suppl. Motoren für industrielle Antriebe, pp. 28/33.

[11] Zaar, G.: Elektrische Maschinen. ETZ-A 89 (1968), pp. 476/482.

[12] Allianz-Merkblatt Nr. 3. Maßnahmen zur Erhaltung der Betriebssicherheit von Elektromotoren. 3rd edition 1969.

2.3. Salient-Pole Machines

Salient-pole machines differ essentially from the three-phase machines previously considered in their rotor configuration. The information given below is therefore concerned mainly with the rotor field system, inasmuch as all other constructional components of salient-pole machines must be maintained and inspected in the same way as three-phase machines.
Table 1 is concerned with all cases of damage which occurred to salient-pole generators after commissioning in the years 1970 and 1971.

Discussion of the Causes and Locations of Damage

Many of the high percentage of damage cases to stators were caused by winding faults through ageing and were greatly influenced by the high average age of the machines. In contrast, the percentage of cases of rotor damage was markedly lower, when compared with the percentage for turbo-generators (see Section 2.1). This decrease can largely be explained by the greater possibility of checking the rotor windings of salient-pole machines during inspection; as such, the comparatively low number of cases of rotor damage can be regarded as one of the benefits conferred by inspection.

2.3.1. Operation of Salient-Pole Machines

With salient-pole machines, soilage is one of the factors most likely to cause damage. Apart from soilage in the form of liquid and dust, foreign bodies of external and internal origin can also be present and constitute a hazard to moving and stationary components.

Soilage

Since most salient-pole machines are air-cooled, there is also a risk that dirt and other foreign particles will enter the interior of the machine via the cooling system. Deposits of dirt in the cooling path or the cooling slots can cause excessive temperatures which prematurely ages the winding. In every case where soilage is discovered, its cause should be established and removed, e.g., by changes in the cooling system. The same considerations apply to checking for leaks in the air paths of the cooling circuit and in oil pipelines.
Any soilage found must be carefully removed. During cleaning, precautions should be taken to avoid damage, particularly to the insulation. Strongly adherent soilage can be removed only by using a suitable cleaning medium, the choice of which should be subject to the agreement of the manufacturer. In this regard, it is important to ensure that the cleaning medium used will not damage the insulation.

Leakage Currents

To avoid leakage currents, which are particularly destructive in bearings, insulating materials are incorporated. Should cratering or signs of arcing appear on bearings, screws or bolts, their cause must be eliminated.

Table 1. Distribution of cases of damage to salient-pole generators.

Cause of damage	Distribution of the number of damage cases [%]	Location of damage	Distribution of the number of damage cases [%]
Product faults	33	Stator	63
Operational faults	7	Rotor	37
External influences	19		
Winding faults through ageing	41		

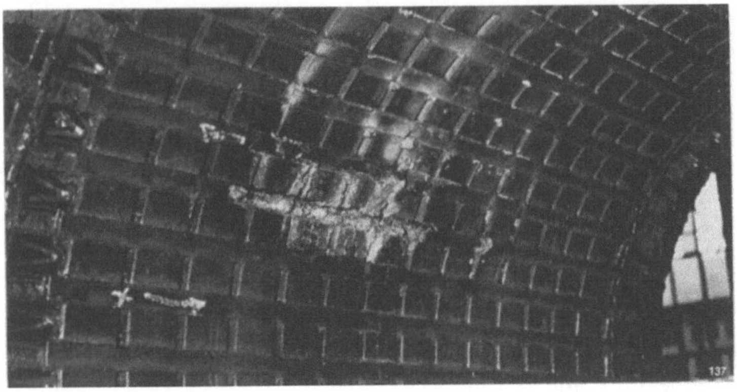

Figure 1. Damage caused through breakage of part of a tooth. The electrical breakdown at the damage location resulted from a short-circuit to earth outside the machine.

Operating Noises and Quietness of Running

A change in operating noise is usually a sign of faults and damage. With salient-pole machines, clearly audible rattling sounds, especially during run-down, can often be traced to loose insulating components or to slackening of the whole pole winding. Changes in magnetic noise indicate comminution of the joint insulation in the stator.

If an increase in excitation current results in deterioration of running quietness, it is a sign of a short-circuit in the pole winding, and tests should be carried out with different active power values. Further information on winding short-circuits can be obtained by plotting the short-circuit characteristic and comparing it with the original characteristic of the manufacturer.

To locate short-circuits in windings, the use of the a.c. test method is recommended. In this method, the pole winding is fed with current at as high a frequency as possible, and the voltage drops are measured. If there is a marked difference in voltage in comparison with the other poles, a winding short-circuit exists in the measured pole. To obtain a more precise location of the fault, this comparative test method is applied to individual windings.

For such measurements, voltages should be selected which, yield unambiguous values, but do not endanger the insulation, e.g., 220 V.

2.3.2. Inspection of Salient-Pole Machines

First Inspection

The first inspection should throw light on the influence of local operating conditions on the existing design and its shortcomings. The most favourable time for inspection is six months or at the latest one year after commissioning. During this first inspection, voltage tests should not be carried out. Rebalancing of the pole-wheel should be undertaken only if replacement of essential components has become necessary.

Scheduled Inspections

Depending on local operating conditions, scheduled inspections are necessary at intervals of two to four years. As in the case of asynchronous motors, synchronous motors should be inspected after 500 switching cycles or after a maximum of 8,000 hours of operation, but at least after two years.

Unscheduled Inspections

An unscheduled inspection must be undertaken after all operating conditions deviating from normal, such as short-circuits near the terminals, asynchronous running and asynchronous connections, or after non-permissible overspeeding. In general, it can be said that all rapid changes in rotational speed, such as sudden braking by an extremely heavy load, make an unscheduled inspection necessary.

Inspection of Stator Windings

The stator winding should be so firmly embedded in the slots of the lamination stack and the winding heads that the stresses induced by the current cannot cause displacements and vibrations. Insulating materials, however, are subject to shrinkage and wear, which manifest themselves as slackening. During inspection, it is therefore necessary to check the firm seating of all those parts which hold the winding.
As regards the stator, the winding is held in place by well-fitting slot wedges. Loose slot wedges can be detected not only by displacements and slackness of the wedges, but also by the presence of powdered wedge material. In the winding heads, distance pieces, ties and interlacing must be checked. Here slackness would also permit undesirable movements. The first signs of slackening manifest themselves by cracks in the insulating lacquer.
For testing the insulation resistance, measurement with a megohmmeter is sufficient. A high-voltage test is not necessary.

Inspection of Stator Lamination Stacks

In most cases, stator lamination stacks are provided with cooling slots disposed lengthwise between individual stacks. Compression along the total height of the laminations should everywhere be sufficient to prevent the possibility of individual lamination stacks, e.g., in the cooling slots, from being pressed apart. With loose sections in individual lamination stacks or in end-stacks, there is a risk that, as a result of vibration, parts of teeth will break off. If these lamination pieces are allowed to enter the interior of the machine as foreign bodies, severe damage may be caused.
Compression checks must include the pressure fingers at the end lamination stacks. If slackening is observed, it must be rectified [1, 2]. The compression system, screwed connections, tie-rods, etc. must also be checked for tightness.
Changes in the magnetic noise of large synchronous generators, which are provided with joint insulation, give an indication, in the form of a characteristic humming noise, of grinding and wear of the insulation between stator components. As far as it is possible to see, fretting

Figure 2. Slackening in a lamination stack as a result of ageing. In individual parts of the lamination stack can be seen gaps (arrowed) of about 1.5 mm in width.

Figure 3. Fluttering of this piece of a tooth (arrowed) in a cooling duct could, if it broke away, cause severe damage to windings.

Figure 4. Slackening in a lamination stack was remedied by driving additional cross-pieces into the cooling slots. Subsequently, the old and new pieces were electrically welded together, using the back-up plate visible in the illustration.

Figure 5. The joint insulation of a horizontal hydraulic-turbine-driven generator was almost completely smashed to pieces during operation, particularly at the centre of the lamination stack. It was still possible to remove the insulation in large chunks from the lamination stack. However, individual laminations had worked themselves deeply into the insulating material.

corrosion appears at the joints if the insulation is destroyed. If none of the insulation remains, there is a danger of severe damage through iron burns.

Inspection of Casings

The casing, which carries the stator lamination stacks, including the winding, is fixed to structural parts of the building. Even support of the casing by the foundation is important. In the event of extremely high loading, e.g., short-circuits near the terminals, the fixings are very heavily stressed and displacements can occur. In such cases, bolts, studs and screws exhibit signs of shearing. A careful check (an unscheduled inspection) then becomes necessary.

Inspection of Coolers

In air-cooled synchronous machines with air recirculation, leakages in the air-circulation path lead to soilage, mainly in the stator windings and in the cooling system. After a long period in service, this soilage causes increasing blockage of the cross-sectional flow area in the cooler and hence a reduction in the flow of cooling air. An increase in the temperature of the winding and its harmful effects on insulating materials must then be expected.

On the water side, substances dissolved in the cooling water or carried along with it may attack the interior surfaces of the tubes and the tube plates. If this occurs, the tubes and tube plates must be replaced before water can penetrate the cooling circuit. In high-voltage machines, insulations which are damp or soaked through have so low a dielectric strength that failure will occur within a very short time.

For modern machines, particularly those of high power, direct water cooling of the conductors has been adopted. With such designs, pipelines, including couplings and ancillary equipment, as well as the water-treatment plant, should be included in the inspection.

Synchronous machines with hydrogen cooling, used, for example, with large phase-changers, require especially careful examination of the hydrogen circuit.

Figure 6. Coil frame of a pole whose sheet-mica insulation stands proud by several centimetres.

Inspection of Rotors

When a scheduled inspection is being carried out, the rotor must be removed from the stator. It is only possible to eliminate removal of the rotor with multi-pole machines which have short coil heights and permit visual inspection of the pole windings and pole attachments after removal of all shrouding. When inspecting the bore of the stator, a number of poles are removed.

Figure 7. The lower coil frame has been completely torn apart after 20,000 hours in service. For operational reasons, recompression at the proper time had been omitted.

Poles and Damper Windings

Screws, keys and longitudinal bolts (including their locking arrangements) used for attaching pole shoes and poles must be checked for tightness. Slackness manifests itself by the ejection of material particles from fretting corrosion and, in smaller machines (e.g., 1 MVA) by clicking noises during run-down. Changes in or overloading of the damper winding of pole-wheels are indicated by discoloration of the layer of insulating lacquer or of the conductor material. The connections of the ring sections between the poles, as well as between the bars and rings, should be checked and, if necessary, rectified.

Figure 8. Scorching of these slip-rings was caused by badly worn carbon brushes. This damage occurred despite the fact that several persons were entrusted with inspection of the brushes at weekly intervals.

Inspection of Pole Windings

All insulating materials used in the pole winding are subject to ageing, depending on time in service and on operational conditions. This ageing manifests itself as a reduction in the radial height of the winding. In the first place, it is the insulating material of the coil frames that is affected by this shrinkage, which results in tangential displacement of the coil frames. Further displacements of individual windings or a number of windings follow, particularly at the coil ends. If pole-wheels are designed for post-compression in a radial direction, it is possible to eliminate shrinkage simply by tightening up the compression screws.

Pole supports are provided in the space between neighbouring poles and serve to hold the partly long coils in a tangential direction. Shrinkage of the insulation between the supports and pole coil is compensated by tightening the pole-support screws. By tightening the pole supports it is not possible to eliminate shrinkage of the radial height of the winding. If the tangential compression of the pole supports is no longer sufficient to hold the coils firmly, radial shrinkage must be compensated by adding coil frames. To insert these additional frames, the poles, or the pole shoes, must be removed.

The connections between neighbouring coils and the conductors between slip-rings and coil entry must be free from cracks, since heavy damage can otherwise occur as a result of arcing. These pole connecting leads are, in many generators, rigid. Slack poles give rise to high stresses, which causes cracking. When overhauling the older types of pole-wheel, it is recommended that they should be rebuilt with more flexible pole connections. Perfect contact must be assured with all screwed connections in the exciter circuit, including the connection between slip-rings and supply leads.

Inspection of Slip-Rings

It is very important that the running surface of slip-rings, which are of split design in larger machines, be in faultless condition. Pronounced scoring and burn and etch marks must be removed by grinding over. If etch marks form under the carbon brushes while the machine is not running, the brushes—provided that the selection of a different grade of carbon brush does not effect a remedy—should be lifted clear of the rings after the machine comes to a stop and repositioned when it is started up again.

Bibliography to 2.3

[1] Kellenberger, W.: Auskippen von Statorblechen infolge Wärmespannung bei großen Wasserkraftgeneratoren. BBC-Mitteilungen 53 (1966), No. 9, pp. 538/545.

[2] Lühl, H.: Beseitigung von Druckfinger- und Zahnblechlockerungen an Wasserkraftgeneratoren. Der Maschinenschaden 25 (1952), No. 1/2, pp. 7/11.

[3] Krauß, O.: Die Revision von Synchronmaschinen mit Schenkelpolläufern. Der Maschinenschaden 44 (1971), No. 4, pp. 117/125.

[4] Humburg, K.: Die synchrone Maschine. Sammlung Göschen, Vol. 1146. Berlin: Verlag de Gruyter 1962.

[5] Kraft, E.: Die Vielfältigkeit von Wasserkraftgeneratoren – ihre Anpassung an die Gegebenheiten der Wasserturbine und des Netzes. Elektrizitätswirtschaft 61 (1962), No. 10, pp. 341/347.

[6] Dabbous, A.A.: Grafische Verfahren zur Ermittlung der Ortskurven und Kennlinien einer Synchronmaschine, ETZ-A 87 (1966), No. 6, pp. 200/204.

[7] Liebe, W.: Kühlung von Großmaschinen ETZ-A 87 (1966), No. 13, pp. 434/442.

2.4. Direct-Current Machines

Table 1 surveys all cases of damage that occurred in the years 1966 to 1968 to d.c. machines having a power rating of more than 300 kW.

The high percentage of product faults in Table 1 is explained by the fact that they include damage caused during erection and damage during warranty, which account for a large proportion of the total.

Discussion of Damage Locations

From Table 1 it can be seen that the locations of damage are predominantly on the rotor, and especially on the armature winding.

The main problem with armature windings is dissipation of heat losses. Since, according to Montsinger's rule, the life of the insulation is halved if the permissible temperature is exceeded under continuous operating conditions by 10 °C, overheating must be avoided.

With larger d.c. machines, however, the operator must concern himself primarily (apart from routine maintenance involving cleaning and changing of brushes) with problems of commutation, since damage to the armature winding, as well as to the commutator, often brings about a deterioration of commutation characteristics. The following directives for keeping d.c. machines in working order are therefore centred on commutation. Further directives for commissioning, maintenance, operation and inspection of d.c. machines can, with a few exceptions, be taken from Sections 2.1 and 2.2.

Figure 1. Burning away of iron as a result of a double short-circuit to earth.

2.4.1. Supervision and Maintenance of Direct-Current Machines, with Particular Reference to Commutation

With d.c. machines, especially those of higher powers, the main concern of the operator must be the assurance of satisfactory commutation. If commutation is found to be unsatisfactory, it is often necessary to discuss its cause and remedy between the machine manufacturer, the brush manufacturer and the operator.

Supervision

With modern drives, at minimum armature voltage, armature current, short-circuit to earth, rotational speed, excitation and ventilation should

Table 1. Distribution of cases of damage in accordance with their causes and locations.

Cause of damage	Distribution of the number of damage cases [%]	Location of damage	Distribution of the number of damage cases [%]
Product faults	74	Stator winding	8
Operational faults	8	Other stator parts	3
External influences	8	Rotor winding	59
Winding faults through ageing	10	Commutator	22
		Other rotor parts	8

be supervised. Release of the overvoltage relay or corresponding electronic supervisory installation often takes place when the nominal voltage has been exceeded by 10% because, on the one hand, higher overvoltages endanger the commutator by flash-over (exceeding the permissible segment potential) while, on the other hand, a 10% reserve is required for speed regulation.

As already mentioned in the discussion of damage locations, special attention must be paid to overheating of armature windings. Overheating is best determined by direct measurement of armature temperature (using resistance thermometers or thermocouples via signal generators or slip-rings, with a control unit in series) or by means of bimetallic measuring instruments fed with current proportional to the armature current. According to VDE 0530, the highest permissible continuous temperature when using Class B insulation is 130 °C, while with Class F insulation it is 155 °C (see also Section 2.2.3, Table 3).

The permissible instantaneous values of current and the effective currents must be supervised in accordance with the instructions on the nameplate. If no instructions are given, VDE 0530, Part 3/1.66, is valid, i.e., a maximum of 1.5 times the nominal current is permissible for 2 min at nominal voltage. Overcurrents, when they occur, should bring about disconnection of the drive from the mains, e.g., through quick-release magnetic relays, to prevent damage by short-circuits. The effective current is a measure of heat generation in the armature winding. This current, or better still the temperature of the armature winding itself, should always be determined if there is any suspicion of overloading. Conventional bimetallic measuring instruments generally require a nominal current of 5 A. Since, however, their internal resistance is so high that the supply via a shunt is often impossible, an additional impedance transformer is required; alternatively, the current/time relationship can be determined oscillographically and the value of the effective current I_{eff} calculated in accordance with the formula

$$I_{eff} = \sqrt{\frac{1}{T} \int_{t=0}^{T} i^2(t)\,dt}.$$

The temperature of the commutator, which usually should not exceed 100 °C under nominal

Fig. 2. Burning away of iron through a winding and earth short-circuit caused by overheating.

operating conditions, also depends on the current loading of the drive, provided, of course, that ventilation is as it should be. The surface temperature of the commutator can be determined either with colour-changing crayons, paints, etc. or, after the machine is at rest, with temperature sensors, in which case the measuring point must be thermally insulated from the environment.

The supervision of ventilation with externally ventilated machines is carried out with the aid of wind-vane switches and appropriate signal or release equipment. Measuring equipment for earth short-circuits and overspeeding will, when actuated, cut off the drive from the mains.

For the prevention of damage to d.c. machines, the greatest importance attaches to supervision of commutation by operating personnel. This supervision should be effected during operational and stationary periods and mainly consists of a visual check on the necessary spark-free commutation. If there is an increase in sparking, indicating deterioration of commutation, the supervisory personnel must inform the engineer in charge, who should satisfy himself at regular intervals that commutation is satisfactory. In addition to watching commutation, personnel must look out for excessive brush wear.

Maintenance of Brushes

Supervisory personnel must watch for excessive brush wear and must written records. Brush wear can be said to be normal if, it is about 1.5 mm per 1,000 hours of operation, in the case of good commutation with smaller machines,

Figure 3. Torn-away commutator and compensating connections as a result of overspeeding.

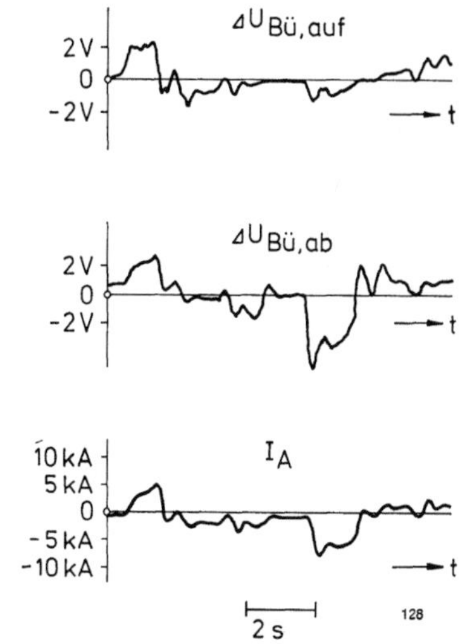

Figure 4. Brush voltage $\Delta U_{Bü, auf}$ (leading edge) and $\Delta U_{Bü, ab}$ (trailing edge), armature current I_A as a function of time t.

whereas values higher than 5 mm per 1,000 hours are often encountered with larger drives. In contrast to the brushes, the commutator wears much more slowly, i.e., about 1 mm per year of operation. The period between replacements of brush sets depends on the amount of wear and the length of the brushes.

Excessive brush wear can be established only by comparison with other machines of the same type and mode of operation, or by comparison with their previous operating life. Too high brush wear can be ascribed to mechanical and electrical causes, and an examination should first be concerned with establishing mechanical causes and then electrical causes.

The Most Frequent Mechanical Causes of Excessive Brush Wear are:

Commutator out-of-round or eccentrically running.
External vibrations.
Mica protruding from the intermediate micanite layers of the commutator.

The Most Frequent Electrical Causes of Excessive Brush Wear are:

Insufficient brush pressure: Accepted values of brush pressure vary from 150 to 350 g/cm^2; the lower values refer to machines for stationary applications and the larger to mobile machines (traction motors). The manufacturer's instructions should always be observed. Brush pressure can be checked with spring balances.

When exchanging brush-holders, identical springs should be used, and the distance between the brush-holder and commutator (1.5 to 2.0 mm) should be maintained.

Overloading through excessive current: This may be brought about by the circulation of currents from inadequate commutation or by external overcurrent.

Uneven current distribution across the brushes (to be checked, for instance, with a clamp-type ammeter): Some causes of uneven current distribution uneven brush pressure, uneven brush spacing, uneven air gap between the armature and main poles, the use of different grades of brushes, and jamming of the brushes in their holders as a result of dust and/or heat.

If more than 20% of all the brushes of a commutator or spindle have to be replaced (in which case care should be taken to ensure that all brushes are of the same quality), they must be ground in. This operation is carried out by interposing an abrasive cloth between the brushes (with the pressure fingers applied) and the commutator, first with a coarser and then with a finer abrasive medium. It is also possible to grind in the brushes by roughing the commuta-

Figure 5. Burnt edges of commutator bars caused by unsatisfactory commutation. The blackened edges suggest under-commutation.

tor with grindstones. After grinding and polishing, and also after each scheduled brush change, it is necessary to wipe away all traces of grinding media and abraded brush dust from the commutator area.

It is also the duty of operating personnel to inspect visually the surface of the commutator and to report any deviations in normal appearance to the engineer in charge. If there are any such deviations which indicate some deficiency or damage not yet recognized, an inspection should always follow.

Inspection of Commutators

The first object of inspection is to look for any possible changes in the surface of the commutator. It is often very difficult to determine their causes and the following recommendations can therefore only serve as a general guide.

Appearance and Possible Causes of Abnormal Commutator Surfaces

Blackened areas irregularly distributed over the commutator: The causes are often local mechanical flattening of the commutator engendered, for example, by worn-out bearings, by out-of-round running of machines, by distortion of the commutator through unsatisfactory mechanical connection to the driven machine, or by improper grinding of the commutator.

Blackened leading or trailing edges: These often happen as a consequence of overloading for an excessively long period, because commutation is at an optimum only for the nominal load, i.e., for a particular load point.

Burnt bars: Connections that are loose or are of high ohmic resistance between the armature winding and commutator in the preceding darkened bar (in the running direction) or faults in the compensating winding cause burnt bars. Prolonged operation with such faults also darkens other regions of the commutator, separated by one more pole pitches.

Darker or lighter markings on individual groups of bars: If the number of marked bars per group corresponds to the number of conductors per slot, the cause is over- or under-commutation.

Formation of grooves and burrs: The formation of grooves or score marks on commutators and/or burrs on bars is often caused by protruding mica, by dust-laden air, or by prolonged operation under low load. Scoring is also possible through incorrect positioning of the brushes on the commutator. Guidelines for the correct staggering of the brushes in the axial and peripheral directions are included in the following section on the inspection of d.c. machines.

During an inspection of the appearance of the running surfaces of the brushes should always be assessed. If it is found that inadequate commutation or a changed commutator surface exists, remedies in accordance with the suspected causes should be devised. Should these remedies fail to lead to improved commutation during the subsequent operating period, a thorough inspection is advised.

Figure 6. Cracks (arrowed) can be seen in the armature conductors just above the soldered joints. During operation, the insulation peeled off from adjacent segments and risers as a result of shrinkage. This shrinkage and the ensuing shrinkage cracks, reduces the strength of the segments, so that vibrations during rotation led to incipient cracks or complete rupture of the conductors.

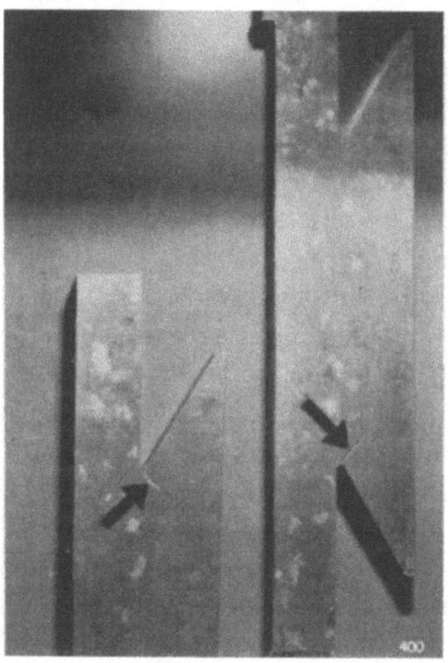

Figure 7. Cracks (arrowed) in commutator bars caused by excessive stressing during assembly.

2.4.2. Inspection of Direct-Current Machines

First Inspection

Provided that the running period is normal and that no unusual events, such as short-circuits, take place, the first inspection of the machine should be made after 2,000 or 3,000 hours of operation. The first inspection should occur after one year at the very latest, even if these operation periods have not been reached.

Scheduled Inspections

The factors determining scheduled inspection intervals for d.c. machines are operating time and operating conditions. Generally, the period between inspections of d.c. machines is the same as that for three-phase machines, i.e., about 8,000 hours in operation.
For rolling-mill motors and rolling-mill auxiliary drives, a period of one year is often demanded between inspections. The same applies to marine motors.
The biennual inspection of exciter machines must be dispensed with in those cases where the main machine cannot be taken out of service. An increase in the length of time between inspections can here be justified, inasmuch as exciter machines always run at a constant speed and are often subjected to only small load variations.

Unscheduled Inspections

After short-circuits or sparking around the commutator, an immediate, partial inspection must take place, irrespective of the mode of operation.

General Directives for the Inspection of D.C. Machines

The following directives for inspection deal only with those constructional components which are typical and important for d.c. machines; the focal point lies with commutation. For such components as bearings, cooling system, etc., as well

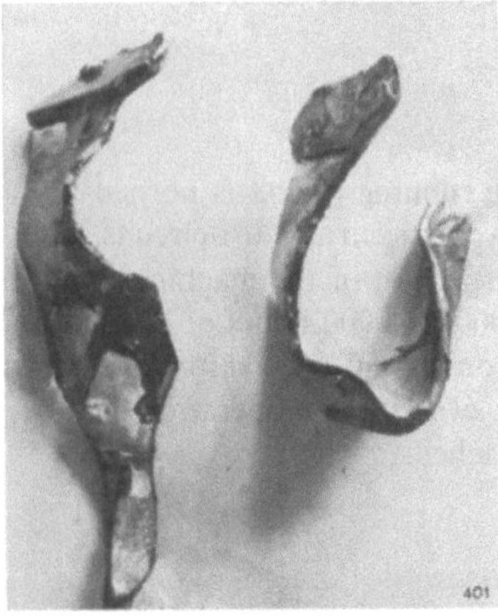

Figure 8. Deformed commutator connections after damage through overspeeding.

Figure 9. Rupture of banding as a result of damage to the armature winding.

as general problems arising from soilage, slackening of components, and other causes, refer to Sections 2.1, 2.2, and 2.3.

During an inspection, those components which transfer torque, such as keys, shrink-fit connections and welds, should undergo examination for perfect seating and operational safety. With shrinkage connections, it is useful to make initial reference marks of position, so that, during an inspection, any displacement of the shrinkage fit is recognizable. Also, the attachment of the lamination stack must be checked.

Cracked lacquer on insulations often indicates mechanically slack windings. Apart from repairing any such cracks, it is necessary to establish their cause.

Soldered connections on commutator risers and compensating windings may become loose and hence increase their ohmic resistance. It is therefore important to examine such connections during an inspection of d.c. machines, using mirrors, endoscopes, or measurements of resistance.

Dust and dirt, mixed with oil and carbon dust (generated by brush wear), will lead to overheating and the creation of leakage paths, with the consequent risk of electrical flash-over, which can endanger the armature winding directly or indirectly by loosening the taping. The taping, including soldered connections and locking arrangements, must be inspected regularly.

In addition, a careful check must be made of the rotor banding, which may consist of a band made of steel or of synthetic fibre/synthetic resin. This check must also include soldered connections and locking arrangements. Flaking off of banding lacquer or cracks in the lacquer are indicative of displacement of the banding. Such displacement or slackening of the banding can occur if the insulating material underneath it has aged or dried out. When checking soldered connections, any tin spatter, should be noted, as this indicates a defective soldered connection. If the compensating circuitry is accessible, its connections should also be examined with, for example, the aid of mirrors. If it is inaccessible, it can be checked by resistance measurements.

If large machines require an examination to detect foreign bodies, it may be possible to make a visual inspection though the air slots of the armature winding, using a lamp or an endoscope. At the same time, the crosspieces in the air gaps can be checked for firm seating. It is sometimes possible to carry out these inspections with mirrors.

During an inspection, it is also necessary to check the thightness of all screwed connections, fastening arrangements, and washers made from insulating materials. In addition, the commutating pole screws must be inspected for slackness, and a

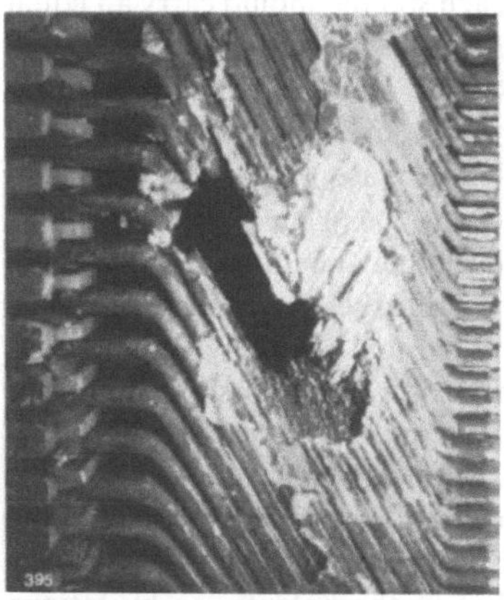

Figure 10. Burnt-out area in a winding head, caused by damage to the insulation and by dirt bridges.

check should be made to ascertain that there is no displacement of shims.

Since insulation components, spacers and interlacing can shrink and distance-pieces can break, they must all be checked during an inspection, as should screwed connections of the winding elements and soldered joints. Since commutating pole coils can expand when subjected to surge currents and cause short-circuits or earth connections in the main pole winding by fouling, they must undergo particularly careful examination. In addition, proper seating of the commutator winding must be ensured.

Commutation

Unsatisfactory commutation can be due to mechanical and electrical causes, and the initial investigation should be concerned with the possible presence of mechanical faults. Then, if no irregularities can be found during this investigation, it must be extended to electrical causes. In this regard, it should be mentioned that appropriate electrical tests are more costly.

Mechanical Causes of Unsatisfactory Commutation

These causes are incorrect brush pressure, badly bedded brushes, out-of-round commutator, protruding pieces of mica, distortion of the commutator through heat, axial misalignment of brush spindles, deficient soldered joints on the compensating or armature winding, or incorrect brush position.

Electrical Causes of Unsatisfactory Commutation

In principle, the electrical causes of unsatisfactory commutation are over- or under-commutation, which can be determined, for example, by finding oscillographically the voltage between a brush spindle and the commutator, e.g., with the aid of inclined brushes insulated against the brush-holder, in the region of current transfer of the spindle as a function of the distance between the leading and trailing edges under certain operating conditions (if possible, under nominal load). Over-commutation can be recognized from the brush-voltage curve by a reduction of the voltage at the trailing edge. With under-commutation, the curve of brush potential rises towards the trailing edge. When recording these curves, note the sign of the voltages measured because, with extremely bad commutation, opposite signs are possible within the same curve.

Slight under-commutation at full load is desirable. Too high a degree of over-commutation can be eliminated by increasing the air gap of the commutating poles, by reducing overlap or by shunting of the commutating pole winding. In the same way, under-commutation can be reduced by increasing the air gap of the commutating poles or by decreasing overlap. Since, however, with increased overlap, there is a risk that a commutating coil will come within the region of the main field, the advice of the manufacturer should be sought.

Stagger

If during the brush-holders are removed an inspection, difficulties often arise with regard to stagger when re-assembling, because the previously used stagger arrangement is no longer known. For correct stagger of the brush-holders and brushes in the axial and peripheral directions, the following basic principles apply:

Axial Stagger

The attempt should be made to ensure that each peripheral commutator path is swept by an equal

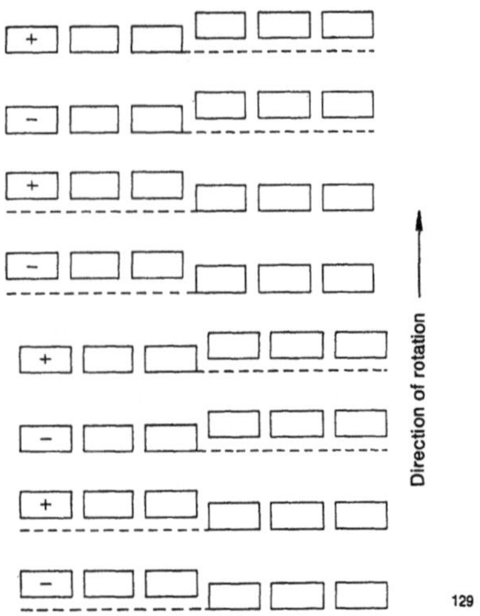

Figure 11. Correct symmetrical stagger in the axial and peripheral directions for an eight-pole machine having six brushes per spindle.

number of negative and positive brushes. It is recommended that each two neighbouring brush spindles be equally spaced.

Peripheral Stagger

Recourse is often made to stagger in a peripheral direction to improve commutation. As a matter of principle, efforts should made to ensure that on each peripheral path there is an equal number of brushes with forward and with backward stagger. Moreover, each spindle should carry an equal number of brushes with forward and backward stagger. If however, there are contradictory demands for optimum axial and peripheral stagger, as is often the case, those involving axial stagger should always be given preference.

During inspection, it may be found that the cable connections of the brushes have corroded. The reason for this corrosion is usually an acid-containing atmosphere; if it is not possible to remove the corrosive medium, tin the cables or sheath them in plastics. Burnt-through cable connections are caused by uneven or insufficient pressure of the brush-holders, by loose terminal screws, by sticking of the brushes in the holders, or by fatigue fracture as a result of out-of-round commutators.

Bibliography to 2.4

[1] Nürnberg, W.: Die Prüfung elektrischer Maschinen. 5th edition. Berlin, Heidelberg, New York: Springer-Verlag 1965.

[2] Meyer, H.: Die Isolierung großer elektrischer Maschinen. Berlin, Heidelberg, New York: Springer-Verlag 1962.

[3] Kohlebürsten und elektrische Maschinen. Morganite Carbon Ltd., London 1961.

[4] Krauß, O.: Die Revision von großen Gleichstrommaschinen. Der Maschinenschaden 41 (1968), No. 5, pp. 174/176.

[5] Krauß, O.: Die Revision großer Gleichstrommaschinen. Reprint, published by Allianz Versicherungs-AG.

2.5. Three-Phase Transformers

Table 1 surveys all cases of damage that occurred during the years 1969 and 1970 to transformers with powers equal to or more than 16 MVA and voltages equal to or more than 110 kV.

Discussion of the Causes of Damage

Product Faults

In Table 1, product faults as a cause of damage have been broken down into three categories: pure product faults, faults in adjusting devices and winding faults. The purpose of this subdivision is to clarify how great is the share of *pure product faults* as a cause of damage. It would be unreasonable, for example, to attribute to a transformer that has been in service for 40 years a product fault which has been caused by a short-circuit external to the transformer.

Under the heading product faults as the cause of damage are all those cases which, irrespective of the time when they occurred, could be proved to have resulted from deficiencies in production or which took place within the warranty period, but could not be ascribed to any errors on the part of operators or to any external influences. Therefore, as indicated in Table 1, 74% of the total number of damage cases were due to product faults and, of these, 68% were *pure* product faults, that about 50% of all cases of damage in the years 1969 and 1970 must be assigned to pure product faults. *Faults in adjusting devices* comprise all cases of damage on tap-changers and stepping switches which occurred after expiration of the warranty period and which have not been determined as pure product faults. In these cases, the problem is, almost without exception, a matter of deficient contracts resulting from carbonization of oil. *Winding faults* under the heading of product faults include a winding faults discovered after expiration of the warranty period. The damage picture usually shows ageing or disintegration of the insulation.

Operational Faults

Operational faults (14%) are mainly due to incorrect switching or reconditioning of oil.

External Influences

In this group, overvoltages are predominantly the cause of damage, on the premise that these overvoltages are certainly above the insulation level of the transformer. It is possible to establish that the number of damage cases from external influences depends largely on the extent of the protective devices provided, as well as on the weather. The number of damage cases can differ by as much as 100%.

2.5.1. Planning and Commissioning of Transformers

The supply contract between the user and manufacturer of a transformer must be adapted in each instance to the prevailing operational

Table 1. Distribution of cases of damage to transformers in accordance with their causes and breakdown of the term *product faults*.

Cause of damage	Distribution of the number of damage cases [%]	Breakdown of product faults	Distribution of the number of damage cases [%]
Product faults	74	Pure product faults	68
Operational faults	14	Faults in adjusting devices	19
External influences	12	Winding faults (including faults through ageing of the insulation)	13

conditions. In Germany, the rules laid down by VDE and DIN are valid for the construction of transformers. In particular, there are regulations VDE 0532/8.69 for transformers and reactors; regulations VDE 0370/10.66x for transformers and transformer and circuit-breaker oils; regulations VDE 0371 (in preparation) for Askarels; regulations VDE 0531/12.69 for stepping switches for transformers and reactors; and DIN leaflets DIN 42500 *et seq.* for transformers and ancillary equipment. Moreover, the association of German electric power-stations – VDEW e.V. – has issued technical supply contracts as guidelines for manufacturers and users of three-phase oil transformers with a nominal power of over 2 MVA.

A transformer which has been designed and built in accordance with these guidelines can generally be relied upon for satisfactory operation, provided that the limiting values of operating voltage, overload capacity, loading of the neutral point, and permissible temperatures are not exceeded. These values should be given in the operating instructions, in addition to technical data and details of protective equipment. Examples of switching sequences and actions to be taken in the event of transformer troubles should also be included in the operating instructions.

When erecting a transformer, care should be taken to ensure that seating on all points of support or on the wheels is even, i.e., that the points of support are on the same level, by checking with a spirit level. To make certain that gas bubbles will migrate rapidly into the Buchholz relay, in the event of weak gas evolution due to a fault in the transformer, the latter should be erected with a 1% inclination. This inclination must be arranged in such a way that the connection to the expansion vessel is located at the highest point of the cover. The transformer should be connected in accordance with the circuit diagram.

If the voltage is adjustable, care should be taken to select the correct tapping. With parallel operation, it is essential to be absolutely satisfied, by taking measurements, that there is no difference in potential between the parts to be connected. Before commissioning, transformers must be earthed according to regulations. The rod gap setting on bushings must be checked before switching on and must conform with the values given in VDE 0532/8.69. If a transformer that has been in storage for a long period prior to commission and has therefore absorbed large amounts of moisture, the condition of the oil must be examined to ascertain whether its properties possess the minimum values necessary for safe operation. This examination is also useful with new transformers because the measured characteristics of the oil serve as comparative values for later measurements (see also the section on *Examination of Oil* on page 76 *et seq.*). After checking the protective and supervisory equipment, the transformer can then be switched on.

Protective Equipment

The extent and type of protective and supervisory equipment depend, for reasons of economy, on the power of the transformer (see also [1] in the

Figure 1. Part of a 31.5-MVA tap-changing transformer, in which atmospheric overvoltages led to flash-over between the entry shielding ring of the high-tension winding and the iron core. The cause of flash-over was that the protective gap of the surge arrester was not properly adjusted and the earth-wire protection was incomplete.

Figure 2. In a 1-MVA tap-changing transformer, moisture in the insulating oil (dielectric strength = 42 kV/cm) and the resulting deterioration of the insulation caused flash-over between the conductors of different phases along a wooden block which served as a distance-piece. The iron components of the "active" part showed signs of rusting.

Bibliography). The usual protective devices are the Buchholz relay, tap-changer and differential protective units, overcurrent and overvoltage relays, and air-drying equipment.

A large proportion of damage to transformers is due to overvoltages. When such cases are examined, it is found repeatedly that the desired protective action of the equipment installed is nullified by improper application.

A transformer can be protected to a large extent from atmospheric overvoltages, e.g., from lightning, by suitable surge arresters installed immediately at the mains terminals. For this, however, a proviso is maintenance of the usual protective distances between the arrester and the transformer terminal [2]. The transformer station and the incoming and outgoing overhead lines must be provided with adequate earth conductors. In stead of using such conductors above the transformer station, the latter can also be protected by means of lightning rods.

A transformer also requires a surge arrester if the high-tension side is connected to an overhead line via a cable. In such a case, the overhead line adjoining the cable must be protected from direct lightning strokes or reverse discharges by earth wires and by low earth resistances of the supporting masts (VDE 0141/2.64xx, paras. 32, 33) over a distance corresponding to twice the cable length. The surge arrester is positioned near the cable head on the side of the overhead line. Booster and auto-transformers require overvoltage protection for the series winding; the arresters must be located between the terminals of a winding strand, as well as between each terminal and earth.

The neutral point of a transformer must also be protected by surge arresters to limit possible internal overvoltages generated by a three-pole impulse. These arresters also provide protection from internal overvoltages resulting from switching operations.

Another important protective device which, in practice, has not always been given the necessary attention is the air drier, which basically comprises a transparent unit containing a suitable desiccant (silica gel or calcium chloride) and an oil seal. The drier is installed between the air space in the expansion vessel and the outside air in such a way that, if the volume of the oil increases or decreases with changes in temperature, air will either be forced out or sucked in through the drier. Depending on load fluctuations or nominal power, oil transformers can have drying units with one or more desiccant containers. Moreover, an air drier must be provided for each enclosed compartment in the expansion vessel.

Apart from these air driers, which are generally used today on transformers as a protection against the ingress of moisture, other methods, based on the use of a nitrogen cushion or a rubber diaphragm, have been developed for excluding air. However, these methods have now fallen into disuse because disturbances in automatic procedures and leakages that developed have reduced their success and have not justified their expense.

Figure 3. Oil on the cover of a 5.6-MVA furnace transformer was set on fire by flying sparks during a melting process. This oil had penetrated through leaking high-tension bushings, and it was only because of flash-over between the bushings that the fault was brought to light.

In addition on the ingress of moisture from the outside, it is possible for a transformer to be endangered, after a long period of operation, by internal moisture released during ageing of the cellulose in the insulating paper [3]. This moisture usually settles in the expansion vessel and can be eliminated by an air-circulation system incorporating a drying unit. In certain circumstances, such a system makes it possible to dispense with oil-drying equipment throughout the life of the transformer [4].

2.5.2. Supervision of Transformers

A routine visual inspection, according to a maintenance schedule, is necessary. Leakages manifest themselves through oil patches. Leaky flanges can be repaired by tightening the bolts or by renewing the gaskets. Burst welding seams can be repaired by rewelding.Particular attention must be paid to leakproof bushings, since an exudation of oil forms a film which absorbs dust particles from the air. As a result, the insulating strength is reduced to such an extent that flash-over can occur. Particularly dangerous are oil films on the cover because they can catch fire from flash-over or flying sparks.

Discoloration or flaking of paint films over large areas indicates overtemperatures through stray currents, which reduce efficiency. A possible remedy is appropriate shielding [5, 6].

Corrosion on transformer tanks must be removed. If it is extensive, repainting must be carried out.

The temperature of the oil must be supervised. The maximum value (according to operating instructions) should be marked on the thermometer. Frequent causes of excessively high tem-

Figure 4. In this 5-MVA transformer, stray fields induced eddy currents in the walls of the tank. Hot spots led to flaking of the paint.

Figure 5. Low oil level in a 5-MVA tap-changing transformer caused puncturing of and flash-overs between the high-tension bushings.

perature are:

- Overloading through increased power consumption by the user.
- Changes in iron losses, restricted or faulty cooling.

It is therefore necessary to include the cooling system in the supervisory and maintenance schedule. In particular, oil circulation and the flow of water and air should be checked. Also, in the case of a transformer operating inside a building, adequate ventilation should be ensured.

For the control of oil content, oil-level gauges are installed. When there is insufficient oil, its cause should be established and removed, and the transformer topped up with dry oil. Expansion vessels with several oil chambers are provided with several gauges to indicate the differences in oil levels. Tap-changing transformers also require checking oil levels of the motor drives.

Figure 6. The conductors to the bushings in the interior of this 100-kVA transformer were burned through as a result of arcing because the oil level only covered the windings. On the upper yoke, corrosion products had already formed.

The expansion vessel is provided with air-drying equipment, in which silica gel or calcium chloride removes the moisture from air passing through the air breather of the transformer. The need to renew the charge of desiccant is indicated by a change of colour from blue (dry) to red (wet). The blue (active) layer must be at least 7 mm thick. If the desiccant is not changed in time, moisture will not be trapped and will pass into the insulating oil of the transformer.

Moisture in the oil can be recognized by a red coloration of the desiccant layer on the oil side. Oil containing moisture must be examined and reconditioned in accordance with the values obtained for its properties.

Gas Formation

If the presence of gas is announced by the Buchholz relay, it is necessary to establish whether or not the gas has been generated by a fault in the transformer. For this purpose, the gas in the Buchholz relay can be analysed. (If this is not possible, a check with a gas-testing instrument will suffice.)

If the analysis shows that only air is present, the transformer can be reconnected. If breakdown of the oil is revealed (particularly in connection with a thunderstorm), it can be assumed with relative certainty that an oil-discharge gap has been punctured by an overvoltage. In most cases, after such an event, these gaps possess, a dielectric strength that is adequate for the operating voltage, so that, here again, the transformer can be reconnected. However, when doing so, the necessary care must, of course, be taken to ensure that all protective devices are set back to the minimum values and times corresponding to the reconnection test. With large transformers, it is recommended that the voltage be increased gradually until it reaches the operating voltage. This recommendation regarding Buchholz gas generation is naturally valid only if, apart from the Buchholz relay, no other protective device has responded.

Figure 7. In a 10-MVA tap-changing transformer, leakages developed at the welding seams of these welded-in radiator tubes after about 50,000 hours in operation.

2.5.3. Inspection of Transformers

As distinct from rotating electrical machines, transformers are not subject to the influence of mechanical motions on their "active" parts, such as windings, iron core, etc. Disassembly of the active parts is not necessary when inspecting a transformer because the condition of the insulation of an oil transformer can essentially be determined by examining the insulating oil. The inspection schedule for transformers contains inspection times from three months to five years, depending on the component.

Every three months, but at least once per year, the protective and supervisory equipment must be tested for proper functioning. It is recommended that this functional test be carried out during non-operational periods. At intervals of one to two years – or more frequently with dirty cooling air or water – cleaning of the cooling installation becomes necessary. An indication of the proper time for cleaning can be obtained by measuring the oil temperature and comparing it with values measured earlier. With dirty coolers, oil temperature increases for the same cooling-air or cooling-water inlet temperature.

Examination of Oil

Regular testing of oil indicates the course of ageing of the oil and windings of the transformer, as well as incipient faults. These tests are carried out according to [10] and the maintenance instructions of the transformer manufacturer.

A test for oil purity and a test for breakdown voltage are performed at yearly intervals. Every two years these tests are supplemented by determination of neutralization value, saponification value, electric loss factor at 90 °C, 50 Hz, and inhibitor content.

The neutralization and saponification values measure the content of acid ageing products in the insulating oil. The loss factor of the oil increases as a result of the formation of oxidation products in the oil ageing. Corresponding to the proportion of oil by volume, the loss factor of oil-impregnated paper insulation, in aged oil, is determined by the loss factor of the oil. If dissipation of heat losses from the solid dielectric is impeded through inappropriate design, e.g., unfavourable distribution of oil-impregnated insulation in the leakage channel of a transformer,

Figure 8. Part of a 31.5-MVA tap-changing transformer, in which a high loss factor of the insulating oil (1934×10^{-3} at 90 °C) led to thermal instability and a breakdown of the main insulation in the region of the upper shielding ring.

a breakdown due to thermal instability, will occur in the region of the highest field strength at the periphery of the windings when the temperature exceeds a critical value. If the loss factor tan δ exceeds a value of $1{,}000 \times 10^{-3}$ at 90 °C, the transformer manufacturer should be consulted with regard to the permissible limiting value.

The dielectric strength of transformer oil is reduced by moisture, fibres, dust, soot, and sludge.

An insulating oil whose properties have values lying outside the following limits is no longer operationally safe:

Neutralization value:	>0.6 mg KOH/g of oil
Saponification value:	>1.5 mg KOH/g of oil
Electric strength:	<30 kV for $R \leqq 110$
	<40 kV for $R > 110 \leqq 220$
	<45 kV for $R > 220$
Loss factor tan δ:	$>$ limiting value according to manufacturer

If one of the values for an oil is found to be outside the above limits, the oil needs reconditioning. This work must be carried out carefully by highly skilled personnel because, as indicated in Table 1, 14% of the cases of damage to transformers in the years 1969 and 1970 were due to operational faults, most of which involved faulty oil reconditioning.

Reconditioning of Oil

Oils whose electric strength reaches the limiting values given above or lies below them, but whose chemical properties and loss factor have values below the critical limit, can be re-used after mechanical purification in a filter press and drying. Whilst large transformers of over 1 MVA can be cleaned and dried on site by mobile equipment, it is more economical, in lower power transformers, to drain off the oil in the workshop for cleaning and drying in a stationary installation. With transformers having a high moisture content, it must be assumed that the core

Figure 9. The input stepping switch of a 75-MVA autotransformer, showing the formation of oil carbon (generated by the carbonization of oil) on the selector contacts after about 24,500 switching operations within a period of seven years. The transformer was operated at only three settings.

and insulating materials are also moist. In such cases, drying on site is tedious and difficult.
Transformer oils, if their chemical and electrical values have reached or lie outside the prescribed limits, must be regenerated chemically or replaced by new oil. Provided that the oil is capable of being regenerated, economic concerns will determine whether regeneration with fuller's earth, combined with moisture removal, is carried out on site or whether the oil should be changed. Regeneration or change of oil can be performed on site or in the workshop.
The new oil used for oil changing is normally inhibited to neutralize any residual ageing products which cannot be removed from the active part of the transformer, and hence to delay ageing of the new charge of oil.
If oil renewal or regeneration takes place in the workshop, ageing products can be removed at that time from the active part. The windings and iron core must be checked for operation safety. Slackness in aged windings can be eliminated by recompression, carefully done to avoid the danger of fracture.

Tap-Changing Equipment

About 14% of all outages of transformers during 1969 and 1970 were due to faults in adjusting equipment. The majority of these outages were attributable to difficulties with contacts as a result of carbon generated by the carbonization of oil in the course of operation. Stepping switches and tap changers are prone to the formation of such carbon on the contacts of tap-changing transformers that are only rarely switched. At least once a year, therefore, the tap changers should be moved over all positions and the stepping switches over the entire adjustment range.
Inspection of stepping switches must be carried out after about 40,000 switching operations or, at the latest, after five years. Inspection should take place sooner if, in the course of operation, irregularities are noted on the on-load tap changer, e.g., pronounced overheating of the tap-changer oil, sluggish operation, etc. Old, usually sooty, tap-changer oil should be renewed. Contacts must be checked for burning and must be replaced in the event of deep pitting. Contact springs and energy-storage springs must be inspected for fractures. Resistors and reactors should be in good condition. All insulating components must be carefully cleaned.

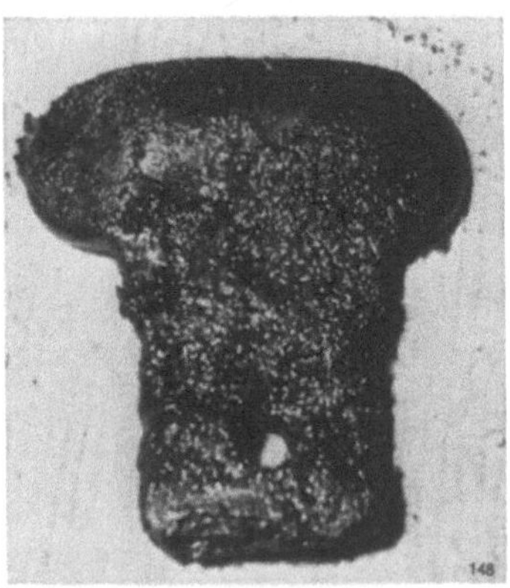

Figure 10. Movable contact of a tap changer of a 32-MVA autotransformer, showing formation of oil carbon after the contact had not been actuated for years. Gas generation led to a Buchholz-relay alarm.

Figure 11. Following the fracture of an energy-storage spring, this stepping switch from a 100-MVA autotransformer stopped between two contact positions and would not lock in the end position. A short-circuit between steps was the result.

Bibliography to 2.5

[1] Technische Liefervereinbarungen für Drehstrom-Transformatoren. Issued by the Vereinigung Deutscher Elektrizitätswerke. Frankfurt/Main: Verlags- und Wirtschaftsgesellschaft der Elektrizitätswerke 1969.
[2] Rabus, W.: Der Überspannungsschutz von Mittel- und Großtransformatoren. Technische Mitteilungen AEG-Telefunken 59 (1969), No. 2, pp. 138/143.

[3] Kugler, H.: Der Einfluß von Wasser auf die Betriebssicherheit von Großumspannern. Der Maschinenschaden 44 (1971), No. 2, pp. 61/65.
[4] Hutzel, O.: Untersuchung und Reinigung von Isolieröl in der Elektrizitätswirtschaft. Elektrizitätswirtschaft 66 (1967), No. 12, pp. 345/347.
[5] Dietrich, W.: Schirmwirkung und Verluste elektrisch gut leitender Abschirmungen von Transformatorkesseln. ETZ 88 (1967), No. 22, pp. 545/550.
[6] Matthes, W.: Beitrag zum Problem der Zusatzverluste durch streuende magnetische Wechselfelder. ETZ 90 (1969), No. 4, pp. 75/80.
[7] Potthof, K., Uhrig, H.: Betriebsverhalten von Großtransformatoren, Fehlerdiagnose durch Gasanalyse. AEG-Mitteilungen 57 (1967), No. 6, pp. 320/322.
[8] Schermann, W.: Gaschromatographische Untersuchungen an Transformator-Gasen. Energie 21 (1969), No. 6, pp. 202/205.
[9] Soldner, K.: Gasprüfgerät zum Buchholzrelais von Transformatoren. Elektrizitätswirtschaft 70 (1971), No. 22, pp. 638/643.
[10] Ölbuch. 4th edition 1963. Published by Vereinigung Deutscher Elektrizitätswerke.
[11] Potthoff, K.: Zustandsprüfung von Transformatoren durch Prüfung des Öls. Elektrizitätswirtschaft 68 (1969), No. 23, pp. 764/767.
[12] Bauer, K., Soldner, K.: Zur Frage der Inhibierung von Isolierölen für Transformatoren. Elektrizitätswirtschaft 67 (1968), No. 25, pp. 735/740.

2.6. High-Power Cables

Consideration is given below to plastic- and paper-insulated lead-covered cables for voltages up to 110 kV. For higher voltages, special cables are used.

Table 1 represents the causal distribution of cable damage cases examined from 1966 to 1971.

By comparison with other machine elements, the unusually high percentage of damage cases from external influences is largely attributable to earth-moving constructional work by third parties. For the protection of cables, the greatest attention should therefore be paid to the possibility of damage by such construction.

Table 1. Percentage distribution of damage cases in accordance with their causes.

Cause of damage	Distribution of the number of damage cases [%]
Product faults	25
Operational faults	10
External influences	65

2.6.1. Planning of Cable Networks

When planning (and also when extending) cable networks, the load distribution is best determined by models of the distribution networks and electronic data-processing (EDP) installations. The outage of certain cable lines must be simulated. Damage through overheating can then be eliminated.

With paper-insulated cables impregnated throughout, the permissible rise of the line (Table 2) or, with a vertical layout, the maximum permissible height (Table 3) must be observed.

When determining a cable route, care should be taken to ensure that no structural work is planned above it. The depth of cable-laying should be chosen in such a way that possible damage by earth levelling, etc. is virtually eliminated. In the case of road crossings, railways and waterways, it is necessary to provide for an extension of cable ducts at a later date.

Lengths of cable built into overhead grids act as condensers in the event of incoming travelling waves, e.g., through atmospheric disturbances. Depending on the cable length and the disposition of the earth-wire protection, one or two non-linear-resistance arresters must here be provided. In practice, lengths of cable are usually protected at both ends by such arresters. However, if only one arrester is used at terminal stations, it must be installed at the station end of the cable. This location is determined by the different characteristic impedances of the cable and overhead line, which cause a reflection at the end of the cable. If it is desired that the incoming wave be initially reduced, an arrester

With cables of normal impregnation (paper-insulated) for 1 to 30 kV	
up to a cable length of 500 m:	10% maximum rise
without length limitations:	4% maximum rise
For plastic-insulated cables, these restrictions are not valid	

Table 2. Permissible rise of the line according to *Siemens Montageschrift MS 320.*

50 m maximum length with belted cable	up to 6 kV
15 m maximum length with belted cable	up to 10–15 kV
30 m maximum length with three-jacket cable	up to 10–15 kV
15 m maximum length with three-jacket cable	up to 20–30 kV
Greater lengths are possible with plastic-insulated cables, particularly those of tension-resistant design.	

Table 3. Permissible lengths of cables arranged vertically, according to *Siemens Montageschrift MS 320.*

must be provided on the second- or third-last mast. In such a case, it must be ensured by an overvoltage earth wire at the end of the overhead line so that no direct lightning stroke can occur at this part of the line.

2.6.2. Storage of Cables

When storing cables, care must be taken that the ends are well sealed. Lead-covered cables can be sealed by soldering over or by dipping the end into a vessel containing a heated liquid cable-sealing compound; after solidification of the compound, the cable end should, if possible, point upwards. With both sealing methods, the cable surface must first be cleaned carefully with petrol or kerosene and mechanically polished, to promote good adhesion and hence prevent the ingress of moisture. In cases of damage, it has been found that, as a result of bad adhesion, moisture can be discovered at a distance of about 10 m from the cable end.

If used cable drums are employed, particular attention must be paid to ensuring that there are no nails protruding from the inside which could damage cable surfaces.

Figure 1. Part of a compound-impregnated cable which was not sealed during storage. To prevent damage from any moisture that might possibly have penetrated, 1 m of cable was cut off before it was put into service. After two weeks in operation, the damage shown in the illustration occurred – moisture had penetrated further than had been expected.

2.6.3. Laying of Cables

Cable tensioning is increasingly effected with the aid of winches. If tensioning is carried out by machine, the corresponding tension diagrams should be prepared. These diagrams are very useful for clarifying cases of damage, even after some years. Moreover, it is essential that cable-laying drums, relative to cable diameter, should not be too far apart. The maximum permissible tensions P [kgf] for the various types of cable indicated in Table 4 should not be exceeded.

If, instead of using a cable grip, a pull is exerted directly on the conductors of a cable by means of a pulling head, the maximum permissible tension P, depending on the conductor material and the cable diameter, can be taken as

3 kgf/mm² for aluminium conductors; 5 kgf/mm² for copper conductors. (For oil and gas pressure cables, different values apply.)

The pulling heads, which are mainly used for paper-insulated cables, must make a waterproof seal on the cable end and are manufactured specially for each cable type. Their field of application is actually very limited, and only with paper-insulated cable and copper conductors do they provide a higher permissible tension than that of a cable grip.

Table 4. Maximum permissible tension P [kgf] of different cable types, according to [3] in Bibliography. D [mm] = external diameter of the cable.

Plastic-insulated control cable	NYY up to 2.5 mm²	$P=0.5\ D^2$	
Plastic-insulated power cable	NYCY, NAYCY, NAYY	$P=0.5\ D^2$	
Plastic-insulated power cable	NYCEY, NAYCEY	$P=0.5\ D^2$	
Plastic-insulated power cable	NYY, NYFGbY, NAYFGbY	$P=1.2\ D^2$	with
Paper-insulated control cable	NKBA(Y), NAKbA(Y)	$P=0.33\ D^2$	cable
Paper-insulated power cable	NKBA(Y), NAKBA(Y)	$P=0.33\ D^2$	grip
Paper-insulated power cable	NEKBA(Y), NAEKBA(Y)	$P=0.33\ D^2$	
Paper-insulated power cable	NKLEY, NAKLEY	$P=0.33\ D^2$	

Example of Calculation of Highest Permissible Tension

Cable type: NAKBA $3 \times 95\ mm^2$; 5.8/10 kV, with outer diameter $D = 50$ mm. In accordance with Table 4, the maximum permissible tension P with a cable grip is $P = 0.33 \times 50^2 = 825$ kgf. With a pulling head, the maximum permissible tension, based on aluminium conductors, will be $P = 3 \times 95 \times 3 = 855$ kgf. (With a pulling head and copper conductors, the maximum permissible tension would be $P = 3 \times 95 \times 5 = 1{,}425$ kgf.)
From values obtained by experience [14], the tensile forces P required from winches can be assumed to have the following relationships to cable weight:

In trenches without great radii of curvature:
P = about 15 to 20% of the cable weight.

In trenches with one or two arcs, each of 90°:
P = about 20 to 40% of the cable weight.

In trenches with 3 arcs of 90° each:
P = about 40 to 60% of the cable weight.

In pipes with arcs totalling 300°:
P = about 100% of the cable weight.

The validity of these relationships is subject to the proviso that easy-running and direction-changing rolls are available.

Cable-Laying in a Bed of Sand

Before embedding cables in sand, the cable jackets must be examined carefully for injuries. Even the smallest damage to jackets, particularly with aluminium-sheathed cables, can lead to corrosion within a short time and consequently to cable failure. Such injuries can best be prevented by sleeves.
The sand used for the bed must be free from stones and foreign bodies. The protective cover over the cable should be without gaps and should ensure good heat dissipation. For example, bricks are preferable to tiles.
The use of a special *cable-line warning band* introduced during the last few years has proved of value. This luminescent yellow plastic band is positioned in the soil at a distance of from 20 to 30 cm above the cable, and serves to give a clear indication of the presence of the cable during subsequent excavating work.

Cable-Laying at Low Temperatures

If the outside temperature is less than +3 °C, cable-laying should not be carried out because of the risk that the insulation will break. An exception here concerns cables with polyethylene insulation, which can be laid at ambient temperatures down to 0 °C.
In particularly urgent cases, the possibility exists of laying cables at still lower temperatures. For this purpose, the whole cable must be taken to a heatable store-room, where its temperature is raised to from about 20 to 25 °C (direct exposure to heat should be avoided), after which it must be installed immediately.

Cable Terminals and Sleeves

The critical locations in cable lines are, as indicated in statistics of power-supply undertakings [16], the terminal chamber and the sleeve. Since there is a large number of qualitatively equivalent designs of terminal chamber, the following comments are limited to those which apply to all designs.
With soldered cable terminals, the permissible temperature for the solder, as well as for the cable sheath, must be carefully maintained to. Excessive temperatures and even too long an exposure to permissible temperatures will reduce the strength of the insulation.
The cable compound must always be matched to the cable, i.e., the cable manufacturer determines the compound to be used. Furthermore, attention should be paid to maintaining the

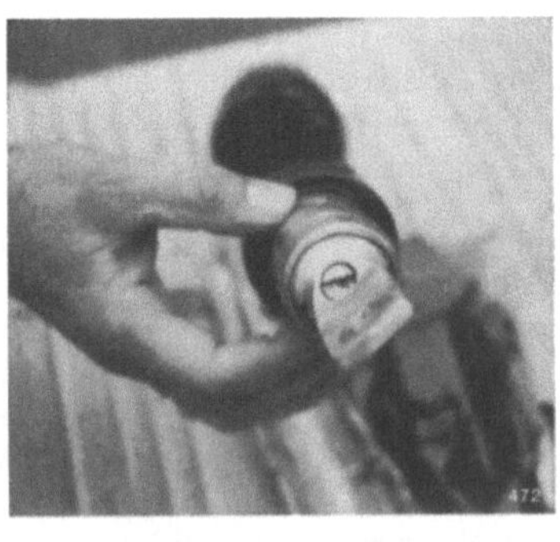

Figure 2. This cable end was not properly soldered all over. Clearly visible is an annular gap through which moisture can penetrate, causing dielectric breakdown.

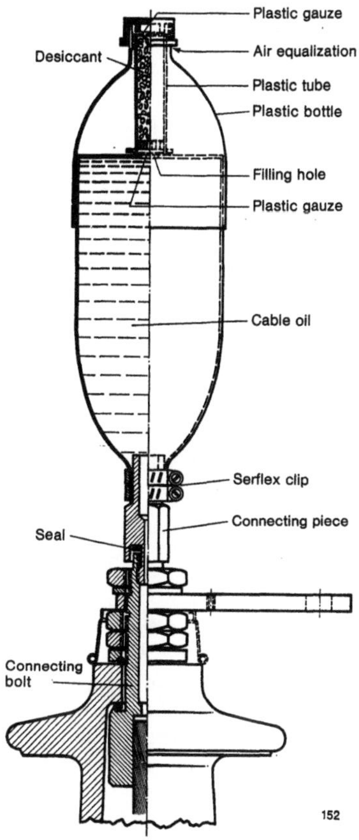

Figure 3. Terminal chamber with a transparent oil-equalizing vessel. (Courtesy of Mittelschwäbische Überlandzentrale AG, Gingen.)

cable compound permanently at the prescribed level.

The comments above also apply to the preparation of cable sleeves. Particularly with aluminium-sheathed cables, there are many cases of damage at the cable joints. The causes of this damage is inadequate connection of individual conductors within the sleeve, irrespective of whether the joint is made by welding, soldering or pressing.

In regard to connection, there are also many cases of damage at the cable terminals. The design of the cable terminal shown in Figure 3 seems to be advantageous, inasmuch as sometimes difficult soldering work is replaced by a screwed connection. However, operational experience over a number of years is not yet available. With terminals of lead-sheathed cables, flexible copper or bronze tapes are frequently used as protection. In practice, these tapes cannot be kept completely free from dirt and moisture, so that electrolytic reactions take place, leading to pitting of the lead sheath and future breakdown. Such damage can, however, be prevented by using glass-fibre tapes.

2.6.4. Commissioning of Cables

Before a voltage is applied to a cable line for the first time, the quick-acting distance protection must be adjusted to approximately the nominal current of the cable, if such protection is provided. A d.c. test is then recommended, during which leakage currents are supervised. In this test, a maximum voltage of 4.5 U_N should not be exceeded, taking into account the flash-over voltage of the cable terminals. The permissible leakage current (order of magnitude: μA) depends on the type of cable and the length of cable tested. It is essential that the leakage current should not increase during the test period (about 15 min per conductor); if it does increase, there may faulty insulation, and further measures to be taken must be discussed with the cable manufacturer.

A large proportion of damage cases which are caused, for instance, when laying a cable and assembling the terminals and sleeves and remain undected, cannot be revealed during the d.c. test. An a.c. test with measurement of tan δ is therefore recommended before commissioning the cable. However, in most cases, such a test is too costly, although the loss factor trend does permit additional conclusions to be drawn regarding the expected life of the cable. Since it is possible that the location of a fault may become necessary at a later date, a record of line lag measurements should be prepared with the aid of a *"reflectograph"* and filed for future use.

2.6.5. Operation and Maintenance of Cables

During operation and maintenance, attention should initially be given to the cable terminals, where the danger exists that thermal fluctuations

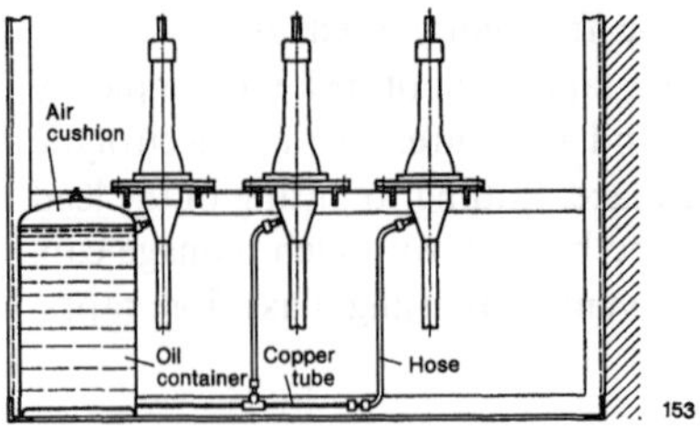

Figure 4. Proposal for a topping-up installation in a switching unit. (Courtesy of Südkabel, Mannheim.)

will cause moist air to be sucked in, resulting in the condensation. This, in turn, leads to flash-over below the terminal in the region of conductor separation—a source of damage that occurs very frequently.

With cable terminals made of porcelain, as often used, it is not possible to check the level of the compound; for this reason, transparent terminals, which permit examination from the outside at all times, are usually provided today. If one is unwilling to replace porcelain terminals already installed with transparent terminals, it is possible to add a topping-up installation. Different designs of such installations are available, two of which are shown in Figures 3 and 4.

The life of insulating materials also decreases with increasing temperature. The heating up of cable as a result of damage or short-circuits, however, has shown that short-time heating of paper-insulated cables above the recommended permissible maximum temperatures of 140 and 160 °C is possible (according to VDE 0103/1.61) without causing damage that would necessitate renewal of the cable. Plastic-insulated cables exhibit a similar behaviour.

A series of tests was carried out on a 10-kV paper-insulated lead cable (belted type, 3×95 mm^2) with a voltage of 8 kV, to determine the influence

Curve	Number of tests	Final short-circuit temperature	Mean value
a	-	Cable as new	
b	1	154	154
c	1	200	200
d	3	194 ... 202	199
e	3	194 ... 202	199

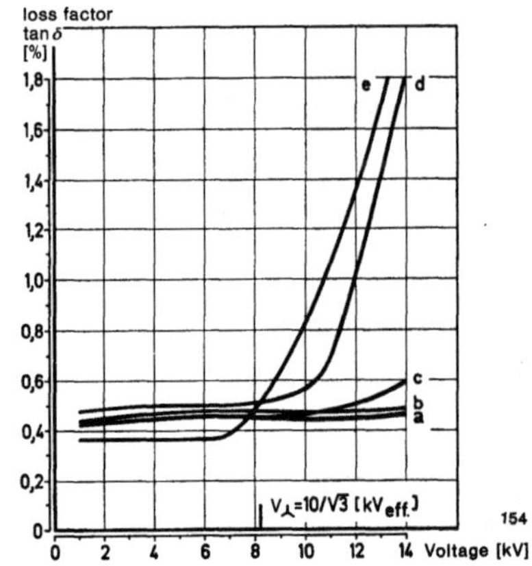

Figure 5. Curves of loss factor tan δ of a paper-insulated lead-sheathed cable in a series of tests conducted to determine the influence of temperature and time under temperature. Curve (e) was recorded three weeks after curve (d).

Figure 6. Flash-over on a 10-kV belted cable.

of temperature and time under temperature on its serviceability. The results of these tests (Figure 5) indicate that, with temperatures in the range between 150 and 163 °C, e.g., 154 °C, as in curve (b), there is virtually no change in the value of the loss factor tan δ. However, if the temperature is raised, for example, to 200 °C, the loss factor increases [curve (c)]. The influence of the number of tests (time under temperature) is shown by curves (d) and (e), in which much greater deviations of tan δ can be observed. Cables that have been subjected to such loading are no longer serviceable.

In summary, it can be said that the greatest potential danger to cables is excavation work performed without proper care. Despite the fact that strict regulations exist for such work, it happens again and again that cables are badly damaged by excavation work. Information regarding the exact location of cable lines is therefore of great importance. For electricity supply undertakings it has proved useful to issue informative instructions for the benefit of constructional firms concerned. In addition, memoranda should be published, as prepared, for instance, by order of the German Post and Telecommunications Ministry [17].

Bibliography to 2.6

[1] AEG-Hilfsbuch. Handbuch der Elektrotechnik. 10th edition. Gütersloh: Verlag C. Bertelsmann 1967.

[2] Hoppner, A.: BBC-Handbuch für Planung, Konstruktion und Montage von Schaltanlagen. 4th edition. Essen: Verlag W. Girardet.

[3] Siemens-Handbuch. Kabel und Leitungen für Starkstrom. Published by Siemens-Schuckertwerke AG.

[4] Taschenbuch für Kabel und Leitungen. Published by Süddeutsche Kabelwerke, Zweigniederlassung der Vereinigten Deutschen Metallwerke AG, Mannheim.

[5] Taschenbuch Kabellegung und Kabelmontage. Published by Süddeutsche Kabelwerke, branch of Vereinigte Deutsche Metallwerke AG, Mannheim.

[6] Einführung in die Starkstromkabel-Technik. Vol. 1: Aufbauelemente. Vol. 2: Typenkunde I und III. Published by Kabel- und Metallwerk, Gutehoffnungshütte AG.

[7] Baatz, H.: Überspannungen in Energieversorgungsnetzen. Berlin, Göttingen, Heidelberg: Springer-Verlag 1956.

[8] Fritz, L.: Über Versuche zur Bestimmung der Kurzschlußbelastbarkeit von 10 kV Papierbleikabeln. Technical paper of Studiengesellschaft für Hochspannungsanlagen (Aug. 1960), group K-45-4.

[9] Gröber, R., Stein, B., Weitzel, H.G.: Versuche zur Ermittlung des Kurzerwärmungsfaktors und der zulässigen Kurzschluß-Endtemperatur bei PVC-Kabeln mit Kupferleitern. Elektrizitätswirtschaft 69 (1970), pp. 578/582.

[10] VDE 0103/1.61: Leitsätze für die Bemessung von Starkstromanlagen auf mechanische und thermische Kurzschlußfestigkeit, § 11, table 5.

[11] VDE 0255/1.68: Bestimmungen für Kabel mit massegetränkter Papierisolierung und Metallmantel für Starkstromanlagen (ausgenommen Gasdruck- und Ölkabel).

[12] VDE 0271/3.69: Bestimmungen für Kabel mit Isolierung und Mantel aus Kunststoff auf der Basis von Polyvinylchlorid für Starkstromanlagen.

[13] VDE 0286/10.56: Vorschriften für probeweise verwendbare Starkstromkabel mit Metallmantel.

[14] VDE 0472/6.65: Leitsätze für die Durchführung von Prüfungen an isolierten Leitungen und Kabeln.

[15] Siemens-Montage-Vorschrift Ms 320.

[16] VDEW Störungs- und Schadensstatistik, 1970.

[17] Anweisung zum Schutze unterirdischer Fernmeldeanlagen der Deutschen Bundespost bei Arbeiten andere (Kabelschutzanweisung). Edition September 1962.

2.7. Equipment for Protection, Measurement, Regulation and Control

The details given in this section for maintaining the serviceability of protective, measuring, regulating and control equipment indicate the crucial points to be observed for the prevention of damage and are based on experience gained from cases of damage that have actually occurred. Thus, some hints are given to the operators of electrical installations, from which they can recognize the features that merit special attention.

2.7.1. Planning and Design

If experience shows that a certain protective device installed was never brought into action, this should not be taken as a reason for excluding it when planning a new installation. On the other hand, protective devices which have proved their value should not necessarily be accepted without critical examination; system improvements, with substantial advantages, might well have been introduced during the past years, e.g., electronic relays.

If protective, regulating or control equipment is housed in air-conditioned rooms, the water-carrying sections of the air-conditioning plant should be located in a separate room situated below the air-conditioned room. When planning, the risk of a water-carrying pipe breakage must always be borne in mind. Unsupervised switch-rooms should be provided with a smoke detector. For wiring and cabling, an adequate number of terminal clamps and terminal strips must be held in reserve. Telephonic connections between switch-rooms and control rooms are the preferred means of communication in the event of damage.

Crammed construction not only makes it more difficult to carry out repairs in the case of failures, but can also greatly increase the extent of damage there is a fire, unless flame-retarding bulkheads are provided.

To prevent faulty release, premature ageing or loosening, all equipment should be mounted in such a way as to be unaffected by vibrations. Care should be taken to protect devices from dust. In aggressive or heavily dust-laden atmospheres, the air in switch-rooms should be pressurized and air cleaners should be provided. Sun blinds should be installed to prevent direct sunlight from causing overheating of equipment. The heat generated, including that in control cables, must be dissipated. With cable ducts, particularly if they run vertically, adequate ventilation must be available. If devices must be cooled for power-loss dissipation, the temperature and amount of the cooling medium must be controlled.

Figure 1. Flooding of an air-conditioned switch-room. The height of the water level is indicated by the line in the illustration.

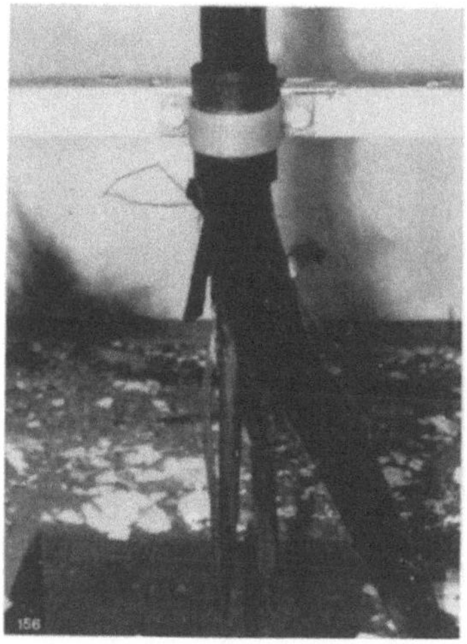

Figure 2. This supply cable for a 400-kW motor was destroyed by a short-circuit current. Because of a fault in the wiring, the short-circuit member of the protective relay was not released.

2.7.2. Commissioning and Operation

After assembly, tests should be scheduled as the servicability of all devices must be checked. Prior to these tests, it is necessary to ascertain that any portable instruments and accessories to be used are fully operational, as faulty instruments can cause damage. If tests are carried out in groups, the particular groups tested must be clearly marked, so that there can be no omissions. Also, care must be taken to reconnect, after testing, any protective devices that have been disconnected.

Multi-conductor control cables should be measured right through, even if they are identified by colours. Prior to these measurements, they should be tested for earth and interconductor short-circuits.

Figure 3. After the application of a heavy asymmetrical load, these slot wedges became soft and were bent outwards under the influence of centrifugal force.

Colour-changing paints at contact locations are reliable temperature indicators.

The contacts of limit switches and interlocking mechanisms must be protected from external influences. When laying out suspended cables, due account must be taken of loop formation, bearing in mind local conditions. Clamping racks for supporting bundles of cables in a parallel run must not have protruding edges.

The load capacity (voltage, current, power) of devices must be clearly and unambiguously marked on the nameplate.

Plastic covers are preferable to metallic covers.

In precision mechanical components, soldered connections are better than screwed connections, which present the danger of wire breakage.

Cable glands should not be located in housing covers. A robust housing rear wall prevents distortion of the chassis during assembly and hence jamming of built-in components. Keys, levers or rocker arms must be as strong as possible.

In switchgear cubicles, shelf space should be provided for components subject to wear (contacts, fuses, quench-chambers, etc.). In this way, time is saved when replacements become necessary, and mix-ups or makeshift arrangements are avoided.

Figure 4. Damage to the winding and lamination stack of a generator as a result of failure of the earth short-circuit protective device.

Before applying a test current, its frequency, voltage and polarity must be determined and matched with the requirements of the device under test.
In the event of blown fuses, the cause must first be investigated and then remedied. Only the prescribed replacement fuses (slow- or quick-acting, correct amperage) must be used.
Switchgear cubicles or individual items of equipment must be stored only in protected rooms, even if short storage periods are involved.
Adverse influences are frost, heat, dust and moisture.
It is particularly important that operating personal be instructed by the manufacturer regarding any innovations in the devices installed in a good time.
The external security of equipment includes protection from damage through external influences, such as moisture, dust, bumps or impacts from transport vehicles, as well as protection from unauthorized operation.
No components must be removed from a completed installation, not even on loan. Faults can occur when these components are being replaced.
During operation, switch-rooms and switchgear cubicles must be kept closed. Emergency switches or control devices must be secured against inadvertent release (lead seals, cover-plates or key-operated switches should be used). In addition, all markings should be clear and unambiguous.
An ocassional check must be made of the independent auxiliary voltage for relcase, actuation and other switching operations. Outage of batteries must be announced at the central control station.
Damage statistics, started when the plant is first put into operation, can indicate weak points and can help with investigations of the location of similar damage.
Cable routes, changes in switch circuits and exchange of individual items of equipment must be recorded in the circuit diagrams.
In co-operation with the manufacturer, a maintenance schedule should be devised and should

Figure 5. This starting resistance of a slip ring motor was permanently switched on, the damage shown was caused by bridging of an auxiliary relay with a piece of wire.

Figure 6. As a result of exposure to excessive radiant heat, the insulation of a part of crane control leads became brittle after a short period in operation.

include the intervals between servicing. This schedule should specify which parts can be serviced only by the manufacturer and which parts can be serviced by the user.
By agreement with the manufacturer, a store of spare parts must be established, and spare parts taken from the store must be replaced in good time.
Regular maintenance of protective devices is of particular importance. In the case of larger plants, the installation is recommended of test equipment that permits serviceability to be supervised during operation. Planned disconnections are good opportunities for checking the chain of protective devices.

Equipment which must be removed for overhaul must be checked for serviceability after being reinstalled, in conjunction with the installation as a whole.
Limit switches, interlocking switches and auxiliary relays are often loaded above their permissible capacity, with the result that they fall short of their expected life. Intermediate inspections are therefore necessary. When checking main contacts for burn-out, auxiliary contacts must also be checked. In addition, the interior of quench-chambers (contactors) must be inspected for metallic coatings.

Bibliography to 2.7

[1] Haase, W.: Wichtige Schutzeinrichtungen an Elektromotoren und Generatoren. Paper No. 25. 7th Works Manager Conference of Allianz, May 1968.
[2] Spieser, R.: Krankheiten elektrischer Maschinen, Transformatoren und Apparate. Berlin, Göttingen, Heidelberg: Springer-Verlag 1960.
[3] Walter, M.: Relaisbuch, Stuttgart: Franckh'sche Verlagshandlung 1968.
[4] AEG-Hilfsbuch, 10th edition Berlin 1967.
[5] Löslein, E.: Pflege und Überwachung von Schutzeinrichtungen. Der Maschinenschaden 26 (1953), No. 5/8, pp. 69/74.
[6] Baatz, H.: Überspannungen in Energieversorgungsnetzen. Berlin, Göttingen, Heidelberg: Springer-Verlag 1959.

3. Electronic Data-Processing (EDP) Equipment

The increasing allocation of management tasks to EDP equipment and the progressive take-over by EDP installations of work in the field of production control have heightened the dependence of an enterprise on the smooth operation of its EDP equipment. The trend to combine, for economic reasons, all activities in the data-processing field in a computer centre, and there, perhaps, carry out all work on a single large and powerful data-processing system, further increases this dependence.

Precautionary measures which, if possible, entirely prevent damage from occurring or limit its extent, as well as provisions for the early detection and averting of its risk, therefore become increasingly important. Greater safety is always more costly, but the profitability of such an expenditure cannot be assessed on the basis of normal economic considerations, as it may be decisive for the survival of an enterprise.

Which of the measures recommended in the following, and how they should be applied in a particular case, can be decided only by consideration of the circumstances peculiar to each case.

Survey of Cases of Damage

The data given in Tables 1 and 2 primarily refer to an examination of 132 cases of small and medium damage which occurred during a period of $2^1/_2$ years. This examination is limited to damage cases amounting to a maximum of DM 100,000 and does not include more costly larger cases taking place during this period. In view of the relatively high costs of these large damage cases, it is felt that their comparatively low number, on a statistical basis, and their associated random nature would tend to bias the results of the examination. However, to give some idea of the importance which attaches to cases of large damage, column (3) in Table 2 includes, in addition to small and medium cases of damage, a reference to more costly cases of large damage, recorded over a longer period but reduced proportionately to the $2^1/_2$-year period of the examination.

Discussion of the Causes of Damage

The subdivision of the causes of damage in Table 1 had to be carried out in accordance with

Table 1. Distribution of damage cases in accordance with their causes and costs (small and medium damage up to DM 100,000).

Cause of damage	Distribution of the number of damage cases [%]	Distribution of the costs of damage cases [%]
Operating errors and other careless actions	45	25
Short-circuits, overvoltages, supply failures	20	25
Water damage of all kinds, breakdown and outage of air-conditioning plant	15	15
Fire, lightning, and explosions of all kinds	15	20
Theft, sabotage, acts of God, and other causes	5	15

Figure 1. One of the many cases of minor damage resulting from operating errors. The illustration shows a section of one of the components of the ruined die of a card punch.

the numerical data available, so that the information in this Table does not permit a comparison with the corresponding information given elsewhere in this handbook.
From the more or less even distribution of the proportions of costs, it can be concluded that, in the region of small and medium damage, no specific causes of damage to EDP equipment predominate. In terms of the distribution of the number of damage cases, the percentage of those caused by operating errors and other careless actions was easily the highest (45%). However, the majority of these cases of damage were small and could mainly be traced to water damage of all kinds, breakdown and outage of air-conditioning plant, and fire. In this regard, the fire damage in Table 1 is chiefly concerned with that resulting from smouldering fires and restricted to individual sections of the plant.

Discussion of the Extent and Cost of Damage

The division adopted in column (2) of Table 2 into small damage (up to DM 10,000) and medium damage (over DM 10,000 and up to DM 100,000) was based on an examination involving a subdivision even smaller than that in column (1). Small damage is mainly concerned with repairs restricted to the replacement of a few components. Medium damage repairs requires the replacement of a larger number of modules or essential elements of equipment, or smaller repairs to a number of pieces of equipment.
The significance of large, more costly damage to EDP installations is indicated in column (3), which includes for comparison cases of large damage recorded, as previously mentioned, over a longer period but reduced proportionately to the $2^1/_2$-year period of the examination.
The low number of damage cases upon which this examination is based does not permit generally valid conclusions to be drawn. More extensive material for a detailed and more exact examination is not at present available. The possibility of larger fluctuations therefore cannot be excluded.

Cost of damage [DM]	Distribution of total damage		
	(1)	(2)	(3)
up to 2,500	10%	small	small
over 2,500 up to 5,000	15%	damage	damage
over 5,000 up to 10,000	10%	35%	10%
over 10,000 up to 25,000	25%	medium	medium
over 25,000 up to 50,000	20%	damage	damage
over 50,000 up to 100,000	20%	65%	20%
over 100,000			large damage 70%

Table 2. Percentage distribution of total damage in relation to cost.

3.1. Prevention of Damage to EDP Equipment by the Manufacturer

The major proportion of damage to EDP equipment can be traced to the effects of external influences. As a result, manufacturers of EDP installations should make efforts to ensure that their equipment is as safe and reliable as possible.

Within their scope of influence, there are possibilities for manufacturers to make contributions to the prevention of damage, particularly in regard to the following points:

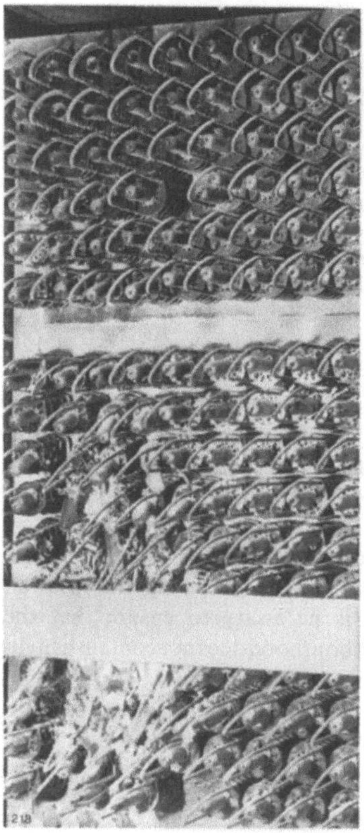

Figure 2. In this first generation EDP installation combustible materials were used. A short-circuit caused fire, which broke out in one section of the installation and spread quickly to other sections.

Development and Planning

- Further improvements in respect of insensitivity to ambient influences and failure of electrical supplies.
- Regular incorporation of protective devices for excessive temperatures, overvoltages and operating errors.
- Maintenance and extension of an organization for advising customers on the planning, erection and installation of computer centres (see [1] in Bibliography).

Design and Production

The danger of fire originating from a device in an EDP installation can be reduced decisively by the selection of suitable materials and proper disposition of component elements. All materials should, at least, be difficult to ignite and should be self-extinguishing, particularly printed-circuit boards [6].

Resistances and other elements which, during operation or in the event of a disturbance, can become very warm or hot must be mounted on printed-circuit boards and correctly spaced so that the latter cannot be damaged or ignited.

Installation and Commissioning

The installation of EDP equipment must not be undertaken before all constructional work in the computer centre has been completed. In a test run lasting several days, the air-conditioning plant must prove its ability to function satisfactorily and, even if time is at a premium, this test run must not be dispensed with.

3.2. Prevention of Damage to EDP Equipment by the User

3.2.1. Planning, Erection and Installation of the Computer Centre

Regulations and Guidelines

The laws, regulations and rules of building authorities and other governmental departments also apply to buildings used for computer installations. Thus far there are no regulations which apply specifically to computer installations [11].
The technical conditions stipulated for the connection and operation of a data-processing installation, as well as proposals for the design and configuration of the computer plant, are contained in the specifications for the installation of EDP systems issued by the manufacturers [9, 10].
Appropriate recommendations are made by Verband der Sachversicherer e.V., Cologne, in its memorandum on "Fire Prevention in the Premises for Electronic Data-Processing Installations" [12]. In addition, the requirements and recommendations of the relevant insurance companies for insuring the EDP installation must be observed [13].
Moreover, it must be established in each individual case whether internal safety regulations exist which are also valid for the erection and operation of the computer plant.

Figure 3. The required environmental conditions could be obtained only after an electric booster heating installation had been built into the double floor. Overheating caused a fire and led to large-scale damage.

Building and Location within the Building

As a matter of principle, EDP equipment should be housed only in fire-resistant buildings (fire-resistant supporting structures, ceilings and roof). It is recommended that a large computer plant be installed in a building specifically selected for the purpose. In one's own building it is much better and easier to provide safety measures. The building should be provided with protection from lightning.
Because of the possible danger of flooding, EDP installations should not be situated in premises below street level, but should be located high enough to remain unaffected by floods, breakages of water mains, or reverse flow from drains. On the other hand, installation on the upper floors of multi-storey buildings is not recommended because of the risk that a fire in the lower floors may cause damage, e.g., by dense smoke rising through staircases, lift shafts, pipes and cable ducts. Also, the recovery of data carriers is made more difficult in the case of a fire.
The best position in a building is on the ground floor or first floor. When selecting the location of premises, it is necessary to ensure that the immediate neighbourhood does not contain a high-risk area. These safety considerations should also extend to neighbouring buildings that do not form part of one's own organization.

Structural Measures for Protection from External Damaging Influences

The rooms housing the EDP installation should be separated from other parts of the building and from other rooms in such a way as to confer fire resistance in accordance with DIN 4102. Doors and closures of other openings should at least be flame-retardant. Particular attention should be paid to the sealing of cable conduits by fire-resistant building materials. Air-conditioning ducts at the entry to and exit from the EDP installation must be provided with closures that are operated automatically from the fire-alarm system.

The fire-resistant enclosure should include the smallest possible number of auxiliary rooms in order to achieve optimum protection.

Rooms which are intended for expansion at a later date should be included in the protected area right from the beginning.

In cases of increased risk, e.g., through the immediate vicinity of workshops or stores, the EDP installation should be located in a separate fire-protected area.

By means of appropriate structural specifications, which must be matched specifically to the particular location under consideration, any water from surrounding rooms must be prevented from penetrating into the room housing the installation, especially into its double or raised floor. Particular sources of danger are water from fire hoses cascading down staircases, water from burst pipes, water overflow from plant in surrounding rooms, water from burst mains, and rainwater during storms. The ceilings of computer rooms should be as watertight as possible, thereby avoiding of ceiling break-throughs. If it is not feasible to exclude the possibility of water penetrating through the ceiling, it is necessary to provide means for rapid drainage of water. Conduits, ventilation ducts, and supply shafts for other purposes of the building should not be led through EDP rooms. If, however, these items have already been installed in existing buildings, suitable precautions should be taken to safeguard the EDP equipment. The same proviso also applies to supply lines for operation of the EDP installation, e.g., the water supply for the water-cooled central processor unit.

Internal Construction and Fittings

The fire risk of the internal construction and fittings of EDP rooms must be reduced as far as possible by the selection of suitable materials. In addition, large computer installations should be subdivided by fire-retardant partitions to the maximum extent practicable from the operational and structural point of view.

Dividing walls, as well as wall and ceiling panelling, should be made of non-flammable materials—a requirement that merits particular attention when selecting sound-absorbing materials.

Figure 4. Partition of an EDP room by a dividing wall had the effect of restricting this fire damage mainly to the equipment in the room.

When selecting materials for the raised floor and the floor covering, use should be made of materials which, as far as possible, do not accumulate electrostatic charges. Also, some firms offer floor coverings which, in the event of fire, emit only small quantities of corrosive gases.

Computer installations are normally equipped with fluorescent lighting. Faulty choke coils and condensers, employed inside the lighting fixtures as ballasts for fluorescent lamps, constitute a fire hazard. For this reason, only ballasts in accordance with VDE 0560 and identifiable by the sign (F) or (FP) should be used. The lighting equipment must be attached only to non-flammable materials. As a precautionary measure, it is recommended that ballasts should be exchanged every five years.

Internal fittings should be made from materials that are at least difficult to ignite. Curtains should not be fitted; comfort in rooms can be achieved by other, e.g., architectural, means.

Air Conditioning

EDP installations have special requirements with regard to maintaining certain ambient conditions, i.e., temperature, relative humidity, and freedom

Figure 5. Water from the air-conditioning plant installed above the computer centre trickled through the ceiling, damaging mainly the core memory located behind the magnetic-tape units.

from dust, and the only way of achieving these conditions is by air conditioning. The particular operational requirements of a computer installation demand the provision of its own air-conditioning plant, tailored to fulfil its special needs [8]. To include a computer installation in the normal office air-conditioning system cannot fail to create difficulties in maintaining the specific environmental requirements of the installation and, in any case, should not be contemplated for safety reasons.

When determining the required capacity of the air-conditioning plant, possible extensions of the EDP installation at a later date must be taken into account, so that, in such an event, environmental conditions can be reliably maintained. The provision, as an afterthought, of additional equipment which is not an integral part of the air-conditioning system as a whole often leads to an increased safety risk.

The air-conditioning plant should be located in the vicinity of the EDP premises—if possible in the floor beneath, which may mean in the cellar. If the plant is on the same level as or above the EDP installation, special precautions should be taken to avert the danger of water escaping from the mains supplying the air-conditioning unit, e.g., by placing the unit in a tank with ample drainage facilities. The inclusion of the air-conditioning unit in the EDP room itself is not acceptable.

Adequate cooling and heating facilities must be safeguarded throughout the year. If not, the resulting deviations from the required ambient conditions may cause operation of the equipment to be suspended and may perhaps damage it as well.

The intake side of the air-conditioning plant should be remote from the discharge side. The choice of air filters will depend on the external ambient air conditions and on the requirements of the EDP installation. If the ambient air contains corrosive or noxious gases or solids in suspension, special filtration measures, e.g., the use of activated-carbon filters, will be necessary in accordance with an air analysis which must first be carried out.

When deciding the type of air-conditioning plant to be used, it is advisable to consult the manufacturer of the EDP equipment, who is, on the basis of his extensive experience, in the best position to judge whether the planned air-conditioning plant will meet all requirements.

The air-conditioning plant should be provided with the requisite control elements to ensure that, in the event of a fault, it will shut itself down automatically. However, there still remains the need for additional independent monitoring equipment, to be described later.

Electricity Supplies

EDP installations are normally connected directly to the mains. Compliance with certain regulations for connection, as well as careful planning and execution of the electrical installation, are essential prerequisites for operational safety. The tolerances prescribed by the manufacturers for voltage, frequency, harmonic content and phase symmetry must be adhered to for all conditions of the mains supplies. Even short deviations of a few milliseconds duration can lead to system faults, disruption of the program run and, in certain circumstances, even to damage of individual electronic modules and data carriers.

A separate supply from the mains for the EDP installation is emphatically recommended. Other consumers of electricity, such as the air-conditioning plant, data-recognition equipment, sorting machines, and room supplies, must not be fed from the supply for the EDP installation. The mains supply should at least come from a low-tension distributor and should, if possible, be directly connected to a transformer which does not feed any other consumers of electricity.

In this way, it is possible to eliminate disturbances in the vicinity of EDP installations arising from switching on and off high current consumers (e.g., lifts), from equipment rich in higher harmonics (e.g., rectifiers) and from superimposed high frequencies from a variety of industrial, medical or communications equipment. Regular noise suppression not only limits high-frequency interference from the EDP installation itself, but also prevents incoming interference from users in the vicinity. The distributor cubicle of the EDP system must be electrically isolated from other distributor cubicles, must be made completely of metal, and must be easily recognizable by personnel. Monitoring of supplies of electric current is recommended, and the necessary equipment and installation can be combined with the distribution system. Wall sockets and lighting circuits should be connected to the normal house supply. The entire electrical installation must conform to VDE regulations.

Regarded over a long period and by conventional means of observation, the mains supply can be said to remain constant within the prescribed limits. In Germany, it is calculated that there are on an average two supply interruptions of longer duration per year in urban grids, and more frequent interruptions in rural grids, these interruptions mainly being caused by thunderstorms and hoar-frost. Short-time reductions to below 85% of the nominal voltage are relatively frequent – up to ten times per month. How extensive interruptions in mains supplies are will depend on the various measures that can be taken to circumvent their effects.

If the number of interruptions is above average, it is recommended that the supply be put under special observation for a longer period. The results of this observation will be helpful when deciding on the measures to be taken. If, on the basis of other operational experience, frequent interruptions in mains supplies are suspected, this observation procedure should be carried out before the installation of the EDP equipment. Manufacturers of EDP equipment have at their disposal the necessary measuring equipment.

Figure 6. Interruption of the electrical supply led to damage of several printed-circuit boards of an EDP installation. One area of damage is illustrated here.

Emergency supply systems can deliver current of sufficiently constant voltage and frequency, even if there are large fluctuations in the mains supply or if it is interrupted altogether. Depending on the duration of the interruptions which must be bridged and the demand for uninterrupted serviceability of the EDP equipment, different types of emergency system can be used. Some examples are given in the following:

"Flywheel" generating sets without a battery can bridge short supply failures of from about 0.4–1 sec. In this case, the EDP equipment is continuously supplied from a converter set comprising a motor, a flywheel (an energy-storage device) and a generator. In the event of longer-lasting mains interruptions, the converter set switches itself off automatically. Experience has shown that, with such installations, more than 90% of all mains interruptions can generally be bridged.

In principle, the only difference between the system just described and a converter set with a battery is that the battery serves as an additional energy-storage device, making it possible to bridge mains failures lasting up to a few minutes. During this period, the EDP installation can usually be switched off in discrete steps, so that the mains failure cannot inflict any damage. Such a set with a thyristor inverter supplies voltage with a high harmonic content. It must therefore be established in every case whether the EDP equipment to be installed is suitable for connection to such a supply.

Various applications of EDP installations demand an uninterrupted electrical supply, even if there is a mains failure of longer duration. An emergency supply system for this purpose can be regarded as a combination of the system described above with a stand-by emergency diesel generating set. If the mains supply fails for any length of time, the diesel generator takes over the supply.

The choice of emergency system to be used in a particular case will depend on economic considerations. The relatively high initial costs of emergency supply systems should be compared with those arising from possible damage and operational disruptions of the EDP equipment. Manufacturers of these emergency supply systems will give their advice on specific technical details of design.

3.2.2. Measures for Preventing Damage to EDP Equipment during Installation

3.2.2.1. Operational Measures

Restriction of Admittance to EDP Installations

Admittance to the EDP room, as well as to data-carrier archives and program information, should be restricted to those who work in these areas or to those possessing a special permit. Outsiders should be denied admittance, except in very special cases and only with the express permission of the manager of the computer centre. Adherence to these precautionary measures can be assisted by the installation of door locks with call or bell devices [5].

During maintenance work carried out by outside firms, members of the computer-centre staff should always be present.

Visits to the EDP installation by outsiders should be restricted as far as possible.

Storage of Flammable Materials

The quantities of flammable materials kept in EDP rooms must be limited to those covering daily needs. Large amounts of paper or data carriers should be stored in rooms specially assigned for the purpose and located outside the fire-protection area of the EDP rooms.

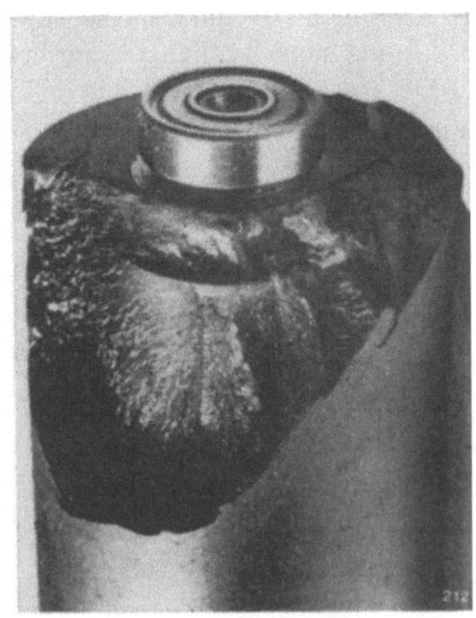

Figure 7. Minor damage resulting from a mishandling fault. A cleaning cloth, caught between rotating rollers, wrecked the colour roller of a high-speed printer.

Figure 8. Smoke from a fire caused considerable damage in an EDP installation. The extent of soiling is clearly visible on this worktable. The light, non-soiled areas in the background are places where stacks of paper had been kept.

Only small quantities of highly flammable liquids, as used for the cleaning of machines, may be kept in EDP rooms and only in small quantities in individual containers of limited volume (less than 200 ml). Other highly flammable liquids must not be stored in EDP rooms and, if their use is essential for special work, they must be removed from the rooms at least at the end of working hours.

Waste Paper

Waste paper always entails a special fire risk. Care should therfore be taken, through regular emptying (at least daily) of waste-paper baskets, to ensure that large quantities of waste paper does not accumulate in EDP rooms. The storage of waste paper collected should be located as remotely as possible from the EDP room.

Waste-paper containers should be made of non-flammable materials and should be provided with self-closing lids. Waste-paper baskets made of plastics are unsuitable.

Any covered spaces in which waste could accumulate, including cable ducts in the floor, must be examined regularly.

Prohibition of Smoking

Smoking in EDP rooms must be prohibited. This ban not only reduces the risk of fire caused by emptying ashtrays containing still-smouldering cigarette ends into waste-paper containers, but also prevents false alarms from the fire-alarm system and the generation of noxious tobacco fumes. To obviate the danger of smoking in secret, the operating personnel should always be given the opportunity during certain break periods to smoke without restriction in an adjacent special room.

Other Fire Risks

Naked flames, e.g., from Advent candles, must be prohibited in EDP rooms and their adjacent rooms. Similarly, the use of portable heating equipment, such as radiant heaters under writing desks and immersion heaters for making coffee, should be banned as a matter of principle.

Experience has shown that welding, which may be necessary for carrying out repairs, constitutes a serious fire hazard. When welding, it is therefore essential to take special precautions [11], the meticulous observance of which is justified by the extremely costly damage that may otherwise result and, indeed, has resulted, as confirmed by companies, insurers and fire brigades. All materials that are combustible must be removed from the welding site, and parts of installations or machines which may be endangered by sparks or radiant heat must first be shielded with fire-resistant coverings.

Wherever welding is carried out, a fully trained fire-fighter must be on duty and must be provided with a sufficient number of fire extinguishers. Immediately after completion of the welding work, this fire-officer must make a thorough search to ascertain whether any material has been ignited, e.g., by welding sparks, and must remain on duty until the welded joint has cooled sufficiently to be touched with the hand. Gas cylinders and welding transformers must be placed outside the EDP room and, if the welding operation is interrupted, it is always necessary to remove welding torches, hoses or cables.

Internal regulations (such as obtaining a permit to weld from the appropriate engineer responsible for safety) must be adhered to strictly.

Cleaning and Maintenance of Rooms

Cleaning staff must be carefully selected and must be made aware of the importance and value of the installation. Instructions should include references to safety equipment and particularly delicate machinery. These instructions should be repeated at regular intervals.

Only suitable non-corrosive materials should be employed and it is recommended that their harmlessness be confirmed by their manufacturers, as well as by the manufacturer of the EDP equipment concerned. In this way it is possible to avoid the apparently inexplicable appearance of corrosion as a side effect of the cleaning media used. This corrosion has on several occasions necessitated extensive repairs involving the replacement of complete items of equipment.

To avoid damage to particularly shock-sensitive equipment (e.g., drum-type storage devices) heavy electrical machines should not be used for cleaning purposes.

Instruction of Personnel Regarding Emergency Procedures

Operating personnel should be told what they should do in the event of an emergency and should be conversant with the operation of the protective and safety equipment provided for such contingencies. Instructions in this respect should be given at regular intervals and should include exercises in the use of the available fire extinguishers. The manufacturer of the extinguishers or the local fire brigade should be consulted as to how these exercises should be carried out.

Easily legible notice boards should indicate briefly the necessary measures to be taken in the case of a fire.

It is well worth setting up an emergency plan in advance, giving directions regarding which of the centre's own departments and which of the service departments of the firms that have supplied technical products to the computer centre should be informed in a given case.

Figure 9. Result of the lack of a fire-alarm system. During a routine patrol, a fire in the computer centre was discovered early enough to restrict catastrophic damage to only one part of the EDP installation. If a fire-alarm system had been installed, such a large amount of damage would not have occurred.

The local fire brigade should be familiar with the whereabouts and nature of the installations in the computer centre, so that, in the event of a fire in the centre or its neighbourhood, their methods of fire-fighting can be designed in such a way that, if at all possible, damage due to the application of extinguishing media to parts of the installation not directly affected by the fire will be avoided. An inspection of the computer centre by the fire brigade offers an opportunity for the mutual exchange of practical experience.

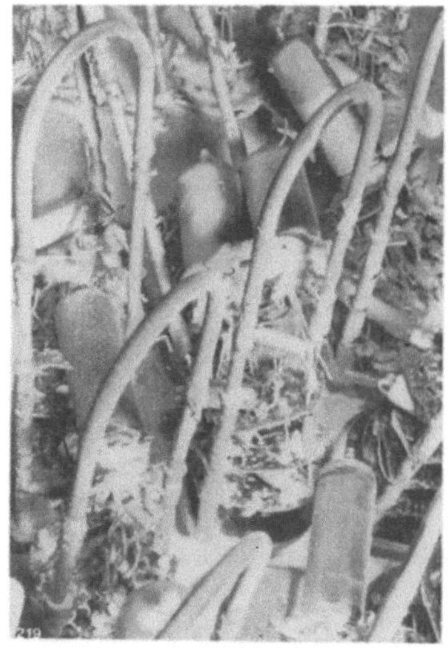

Figure 10. The use of a powder fire-extinguishing medium not only leads to residues of powder on parts directly affected by fire damage (as illustrated), but also on parts not affected by the fire.

3.2.2.2. Automatic Supervisory and Protective Equipment

Fire-Alarm System

The time that elapses between the outbreak of a fire and its discovery is critical for effective fire-fighting. In its initial stages, a fire can be tackled satisfactorily by means of fire extinguishers. If a fire is not detected and counteracted quickly, damage to delicate machine parts can be catastrophic, even after a short time.

An early warning of fire can be guaranteed by smoke detectors [2]. The most frequently used detectors are an ionization type which gives a warning as soon as combustion gases are present, even if they are not actually visible in the form of smoke. The adjustment of detectors should be as sensitive as possible. Unpleasant false alarms can be avoided by the use of a two-loop system, in which an alarm is given only if at least two detectors in different loops respond.

When deciding on the location of these detectors, the complicated air-flow conditions resulting from the air-conditioning system must be taken into account. For this reason, the installation of a fire-alarm system for an EDP installation should be entrusted to firms that are experienced in this field.

The detectors should be positioned not only on the ceiling of the EDP room, but also in the double floor (cable-fire detectors) and in the inlet and outlet ducts of the air-conditioning plant (wind-tunnel or test-chamber detectors). Also, the auxiliary rooms of a computer centre, particularly those containing the air-conditioning plant and the electrical supply installation, should be provided with fire alarms to enable a fire originating in these rooms to be detected and extinguished quickly and to ensure that no smoke will penetrate into the EDP room.

The alarm system should also incorporate push-buttons for manual operation.

Both audible and visual warning indications of a fire should be provided for EDP rooms. If the central alarm station is not located in the computer room, it has been found expedient to fit an indicator panel at the entrance to the EDP room, from which it should immediately become clear which of the detectors of the alarm system has been activated. At the same time, an alarm indication must be transmitted to the responsible fire station, e.g., the plant station, or to the commissionaire at the main gate. Whether a direct connection to the public fire-fighting service is possible will depend on local conditions. The fire-alarm system can, moreover, be used to switch off equipment automatically, as will be discussed later.

In Germany, special regulations exist regarding the construction and operation of fire-alarm installations, issued by the Association of Insurance Companies, which also offers an expert advisory service.

Figure 11. An explosion and the consequent outbreak of fire led to the total destruction of a process control computer. The illustration shows one of the printed-circuit boards from the computer.

Fire Extinguishers

For taking immediate action against an outbreak of fire, an adequate number of fire extinguishers must be provided, not only in the rooms housing the EDP installation, but also in neighbouring rooms. When selecting extinguishing media, it is necessary to make sure that they are suitable for combating the possible types of fire which may break out in these rooms and that, at the same time, their use will not cause unpredictable damage to machines which have not actually caught fire.

According to prevailing opinion, only carbon dioxide extinguishers should be employed in EDP rooms. With these extinguishers, if properly used, an outbreak of fire on machines can be fought effectively, without causing damage to other equipment. The resulting cooling effect is of no practical importance.

For combating smouldering fires, such as those which can occur with paper or wood, CO_2 extinguishers are less effective. Suitable extinguishing media for smouldering fires would be powder extinguishers or water. However, with the use of these powder extinguishers, although they are very effective, extensive soiling must be expected, even of equipment which is not directly affected. Removal of the powder residues is difficult and time-consuming, and is sometimes even impossible. The use of powder extinguishers is therefore justifiable only if other extinguishing media have failed.

The application of water as an extinguishing medium is recommended only in cases where it can be carried out expertly. The jet of water must be capable, if required, of being transformed into a spray. With proper application, the damage caused by the water can be restricted to the burning item, whose destruction must be accepted, if necessary, to save other parts of the installation [7].

Recently, halons, now on the market, have been recommended as suitable extinguishing media for EDP installations. Their good extinguishing capabilities are known, but it has not yet been proved that, when using halons, there are no risks to fire-fighting personnel and that they do not cause damage to equipment that has not actually caught fire. Also, their admission as fire extinguishers has not yet been confirmed for fires of class A.

Recognizing the problems involved, insurers of EDP installations stipulate that only CO_2 extinguishers must be provided for EDP rooms. Other types of fire extinguishers must be kept only in adjoining rooms and must be marked as being unsuitable for use on machines.

Automatic Fire-Extinguishing Plant

The provision of automatic fire-extinguishing plant in EDP rooms is an obvious precautionary measure to ensure that, at times when the computer centre is unoccupied, the inception of a fire will be counteracted immediately and hence severe damage will be prevented. However, which particular type of automatic extinguishing plant is the most suitable is often the subject of heated discussion. While the sprinkler type of installation has been recommended or stipulated in the U.S.A. [3], the prevailing European opinion is that this kind of installation should not be used in EDP rooms because of the impending risk of severe damage by water. The recommendation here is to flood the computer centre automatically with carbon dioxide. Recently, an automatic extinguishing plant using halon as the medium has been under discussion, but the same reservations as those for fire extinguishers also apply in this case.

A fire-extinguishing plant of the sprinkler type is based on the use of a number of individual sprinkler units which act selectively and which are effective only in the immediate area of a fire [2]. Each sprinkler unit has a spray-type nozzle provided with a valve that reacts to heat, the valve opening only when the temperature of a sprinkler unit is raised to a predetermined level by the rising heat from a fire. The pressure drop thus created in the pipeline system can also be used to trigger the fire alarm at the same time. The area protected by a single sprinkler covers from 8 to 20 m^2, depending on layout. It is therefore evident that, in the event of accidental release, a large amount of equipment can be destroyed. If the installation of a sprinkler plant cannot be avoided because of other strict regulations, at least the so-called "Preaction" system should be applied [4]. In this system, the sprinkler pipelines are filled automatically with water only if the fire-alarm installation has detected a fire. In this way it is easier to avoid damage by broken pipes or faulty releases.

Under normal conditions of risk, sprinkler installations have well proved their value and have shown up to advantage. However, whether their application in EDP installations does more good than harm can, in the end, be determined only in a particular case of fire and, even then, only if the specific conditions prevailing are taken into account.

The extinguishing efficiency of an automatic stationary CO_2 fire-extinguishing plant depends in the first place on displacement of the oxygen necessary for combustion by the carbon dioxide, which is harmless to equipment. For effective fire-fighting, a substantial quantity of CO_2 is required, its storage is expensive, involving high investment costs. For this reason, a satisfactory compromise in many computer centres is merely to flood the double floor, so as to be able, at least, to combat fires in areas of difficult access for conventional extinguishing methods. With automatic release, special precautions must be taken to ensure that any personnel who happen to be in the affected room are given time to leave and hence cannot come to harm. Apart from the provision of audible and visual alarms which indicate the response of the automatic fire-estinguishing plant and warn personnel to leave immediately, the installation should be wired in such a way that it comes into action only if two fire-alarm detector loops respond.

With automatic extinguishing plant using halon as the medium, the cost of storage is much less than with carbon dioxide, with a consequent effect on the investment costs. The conditions for the installation of such a plant, however, are not as yet determined for the reasons given earlier.

If a stationary fire-extinguishing plant is built into EDP rooms, the particular regulations of the Association of Insurance Companies must be taken into account.

Supervision of Air-Conditioning Parameters

The safe operation of an EDP system depends on the maintenance of certain air-conditioning parameters. Large and long-term deviations outside permissible limits have, in the past, led to the equipment damage to an extent which, apart from other considerations, can result in interrupted operation of the installation for long periods. The values of air-conditioning parameters must therefore be monitored constantly by an automatic supervisory and alarm system. This system must satisfy the following requirements:

The temperature and relative humidity of room air and, in some cases, of machine air, must be monitored independently of the control of the air-conditioning plant. If the EDP rooms are large, supervision at several positions is recommended.

Deviations from permissible high and low limits of temperature and relative humidity must be announced by visual and audible warnings in the computer centre. The acoustic alarm should be capable of being switched off by hand, but the optical indication should persist as long as conditions are outside the permissible limits. To ensure that, on occasions when the computer centre is unmanned, the necessary damage-prevention measures are taken if an alarm is given, the warning indications should be transmitted during non-working hours to a permanently occupied post.

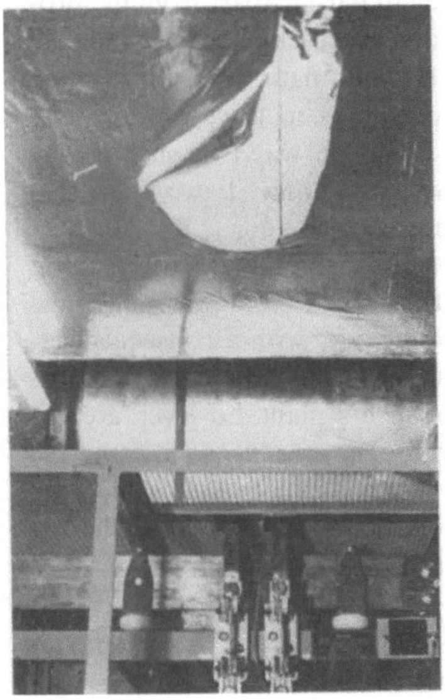

Figure 12. Excessive moisture in the air produced condensate in the air-conditioning duct above the control cubicles of an emergency installation for uninterrupted current supply. Water coming through the gap seen in the illustration caused short-circuits in one of the cubicles, putting the rectifiers out of action and leading to non-permissible heavy discharge of the battery sets.

Figure 13. All batteries had to be replaced. The illustration shows one battery cut open.

The air conditions should be registered on an electric recorder. Important data are then made readily available in the event of faulty operation or for use as evidence to substantiate a warranty claim.

The supervisory and recording equipment should be connected to the mains in such a way that, if the air-conditioning plant is switched off or is otherwise non-operational, these instruments will continue to function as long as the mains supply is maintained. Their connection to the lighting circuit, rather than to the power circuit used for the air-conditioning plant, is recommended.

The supervisory equipment should be linked with the summation signal generated by individual fault indicators from the air-conditioning plant, so that irregularities will be announced immediately and not only after deviations occur in the ambient conditions of the EDP room.

The requirements and recommendations of reputable manufacturers of EDP installations are in general agreement with those given above regarding supervisory equipment. Moreover, it should be realized that insurers of data-processing plant make such equipment (which is obtainable from the manufacturer of the air-conditioning plant) mandatory before granting full insurance protection. It is also recommended that, during the planning stage of the supervisory equipment, the whole installation should be agreed upon with the manufacturer of the EDP plant, particularly if the automatic switching-off procedure, described below, is to be actuated by the supervisory system.

Automatic Switching-off

It is impossible to make quite sure, especially during non-operating periods, that immediate action will be taken on initiation of alarm signals generated by the supervisory equipment. It is therefore expedient in the circumstances to let the supervisory system shut down the EDP installation and the air-conditioning plant automatically. Shut-down of the latter plant will necessitate switching off the EDP installation within a short period of time. On the other hand, an EDP installation should not be shut down abruptly during a program run since the consequences in such a case are likely to be similar to those resulting from an electrical power failure (damage to hardware and loss of data). Time-delayed switching-off, which offers the possibility of a manual override to prevent the switch-off, can provide a solution that takes account of the desirability for a maximum of uninterrupted operation. This solution, despite the fact that it entails a higher investment in instruments and equipment, is preferable to a day/night switch, since manual switch-over can easily be forgotten and automatic switch-over is out of the question because of the generally irregular operating times. A practical solution could be as follows:

The fire-alarm system should switch off the air-conditioning plant, as well as the EDP installation, with a delay of about 2 min. During operation, such an interval should give enough time to decide whether a false alarm has been given or whether a fire has actually broken out; in the former case, switching-off can be prevented manually and, in the latter case, it can be effected earlier by an emergency switch to be described later. Automatic switching-off should prevent smoke-laden air, circulated by the air-conditioning plant, from reaching all EDP equipment. In addition, if a fire breaks out in a machine, it should not be aggravated by an electrically induced fire, and such a fire can be prevented by automatically switching off the supply current.

At the same time, the fire-protection safety flaps of the air-conditioning ducts should be closed automatically.

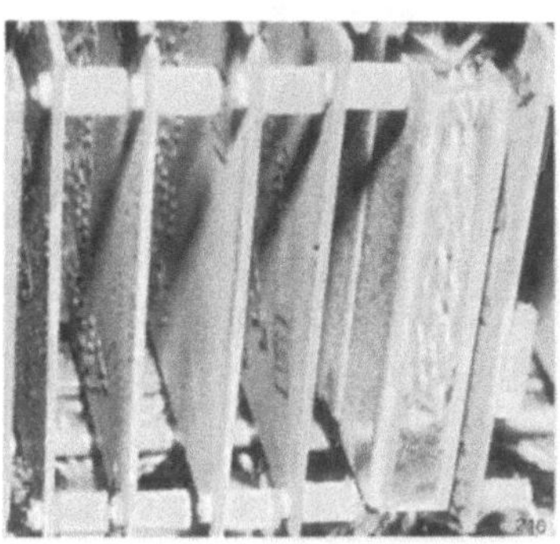

Figure 14. The unnoticed removal of the fine filters from the air-conditioning plant caused, within a short time, the deposition of dust on the EDP installation, making it unserviceable. Only after expensive cleaning and repair work lasting several days could operation be started again. All signs pointed to sabotage.

In any case, the air-conditioning and EDP plants should be switched off immediately when an automatic fire-extinguishing system, if one is installed, becomes operative. The fire-protection flaps also must be closed.

Moreover, the supervisory equipment for air-conditioning parameters should cut off the air-conditioning and EDP plants in a similar manner, with a delay period which may, depending on the type of EDP plant involved, be chosen somewhat longer, so that the program run can be interrupted at a definite point. Automatic switching off of the EDP installation is especially necessary if it is left switched on even when the computer centre is unattended.

Emergency Switchgear

In emergency situations (electrical mishaps, accidents caused by moving mechanical parts, fire, etc.), EDP and air-conditioning plants must be switched off quickly, with complete disconnection of the circuits from the source of current. For this purpose, it is desirable to provide separate emergency switchgear for the EDP and air-conditioning plants. This switchgear should be capable of being operated not only by personnel at their work positions, but also from the emergency exits, and the complete switching system should include all devices in the danger area. The subdivision of a computer centre into several danger areas, e.g., if several systems have been installed, is permissible only if these areas are clearly discernible.

The emergency switching installation can be combined with the master switch in the distribution system to form the protective circuit-breaker in the main distribution cubicle. The operating position of the switch must be clearly visible, and the circuit-breaking elements must be protected against misuse and accidental operation.

Protection of Computer Centres against Unauthorized Entry

The installation of a burglar alarm or other suitable system for protection against unauthorized entry is recommended for computer centres carrying out work of a highly secret nature or where attempts at sabotage cannot be ruled out.

3.2.2.3. Supervisory Duties of Personnel

The supervisory functions carried out by operating personnel during office hours should not be restricted merely to the faultless operation of the EDP installation and adherence to operating instructions. Many breakdowns of other technical auxiliary equipment give warnings well in advance through changes in their operating parameters.

All control instrumentation of the air-conditioning plant, power-supply installation, fire-alarm system, etc. must therefore be in full view of operating personnel. The causes of changes in operating parameters must be established immediately. The careful observation of some apparently unimportant details has sometimes contributed to the avoidance of severe damage at a later date.

Daily visual inspection of the air-conditioning plant is recommended, so as to discover in good time leakages, abnormally dirty filters, and whether filter elements and operating media need replacing. This task can also be assigned to the internal maintenance staff.

Particular attention must be paid to ensuring the permanent operation readiness of existing installations, which should, if possible, never be out of action, not even temporarily. Therefore, if repairs are required, care must be taken to carry them out with the minimum of delay and to render the installation operative again until the next possible period of overhaul.

In addition, it is desirable to keep an eye on the immediate vicinity of the computer centre to see whether there is any work in progress that might be detrimental to the operation of the centre, whether there are easily combustible materials kept in storage, or whether there is any constructional work in progress. Attention should also be given to the suction inlet of the air-conditioning plant.

The computer centre should be patrolled regularly during non-operational periods, and the control personnel should be instructed in regard to those points which require particular attention. If an internal control service does not exist, it can be entrusted to an outside private firm. In such a case, the control personnel should have a knowledge of whom to advise in the event of unusual occurrences. It is recommended that a notice board at the entrance to the computer centre should list the appropriate service stations or persons and their telephone numbers.

Outside normal working hours, alarm signals from automatic supervisory equipment must be transmitted to a suitable permanently occupied post (night gate-keeper, factory guard, fire brigade). It is essential to ensure that all persons who are on duty at these posts understand the meaning of the alarm signals and that they are given detailed instructions regarding the specific actions to be taken. Also, practical experience has shown that it is absolutely necessary to check periodically whether these instructions are being obeyed and whether they are kept up to date.

For installations in which there are no permanently occupied posts, automatic emergency telephone equipment, as also used for burglar-alarm systems, offers the possibility of transmitting alarm signals automatically by telephone to the private residence of a person responsible for the operation of the computer centre.

3.2.2.4. Maintenance

A prerequisite for the safe operation of all installations in a computer centre is regular and consistent maintenance work. The normal method of ensuring regular maintenance is to sign a contract with the manufacturers or suppliers concerned. When making such a contract in respect to air-conditioning plant, it is necessary to make

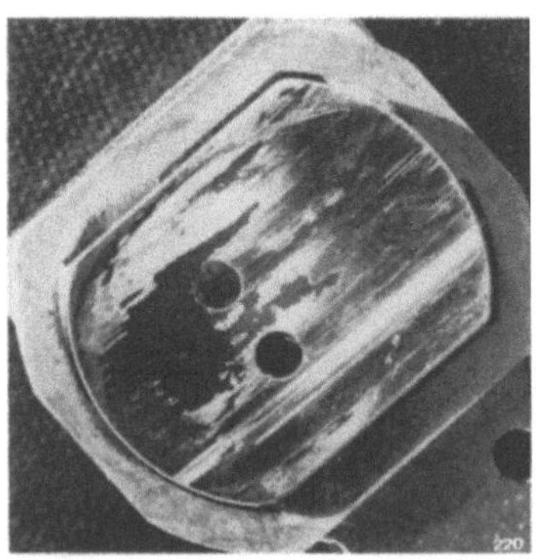

Figure 15. Operating instructions must always be observed. If, for example, the prescribed cleaning of magnetic disc stacks is omitted, it inevitably leads to the damage depicted on the write/read heads. The disc stack itself becomes unusable.

certain that the maintenance firm can also provide an emergency service outside normal working hours. The execution of preventive maintenance work must be supervised.
Maintenance can be entrusted to members of one's own staff only if they have been specially trained and are available in sufficient numbers, and if regularity of maintenance is ensured by proper organization.
For all maintenance work that is not necessarily concerned with the machines themselves, sufficient space should be provided in the vicinity of the machine room. With larger data-processing installations, it is recommended that a special room be provided for this purpose.
Maintenance contracts are particularly necessary for:

- The EDP installation itself;
  (In the case of a hired installation, provision for maintenance will already be included in the leasing agreement. If peripheral equipment (mixed hardware) from various manufacturers is employed, a cooperative relationship should be established between the maintenance staff and the different firms.)
- The air-conditioning plant;
  (Maintenance every three months is customary and should include the measuring and recording instruments for the air-conditioning parameters. Recalibration of the measuring instruments, depending on their type, will be necessary from time to time.)
- The power-supply installation;
- The fire-alarm and automatic fire-extinguishing installation;
- All other technical equipment;
  (Small freight elevators and other equipment for which a maintenance agreement is possible.)

Figure 16. Constructional work gives rise to additional hazards. The cigarette end (arrowed) found in this cable conduit and the broken connector in the foreground are proof that the necessary care had not been taken.

The signing of a maintenance contract involving the execution of regular maintenance work does not absolve the computer centre from responsibility for carrying out its own regular supervision of the operational behaviour of the above installations.

3.2.2.5. Inspections

A thorough inspection should be made at least once a year of technical installations which are not regularly maintained. Regular inspection is also necessary for internal components of equipment, and care should be taken, when planning and building the computer centre, that the equipment and components in question are easily accessible. Regular inspections are particularly desirable for all electrical installations, as well as for lighting (including emergency lighting) and circuits for wall sockets. These installations should be examined for signs of wear and for conformity with VDE specifications, with special emphasis on the leads of electrical equipment which is not permanently connected.
The space between the double floor must be checked frequently for dampness and for water that may have leaked in, as well as for dirt, scrap paper and other waste materials that may possibly have been left behind from previous work. For the purpose of inspection, a number of removable floor sections should be provided, and the tools necessary for lifting them clear should always be available for inspection and also for use in an emergency.
The ceiling lining should be inspected about once per year, particularly for abnormal soiling of the sound-absorbing materials.
The fire extinguishers must be checked for missing parts, for filling, and for adherence to the prescribed test discharge distances.

Checking of water and waste-water installations in neighbouring rooms is often overlooked. A number of cases of water damage have shown that special care should be taken to ensure the unimpeded drainage of waste-water.
Finally, it should be remembered that the loss-prevention precautions decided upon should be reviewed every year, and improvements should be introduced or additional measures taken when they become necessary on the basis of changed environmental conditions and enlargement of EDP installations.

Figure 17. For the water supply of a building site in the vicinity, connection was made to an already unsound hose in the cellar of the computer centre. During the weekend, this hose burst under pressure. The cellar was flooded and the electricity supply plant installed there was 40 cm under water. Apart from the damage caused, operation of the EDP installation was disrupted for several days.

3.2.2.6. Constructional Work in the Vicinity—Rebuilding in the Computer Centre

Experience has indicated that computer centres are enlarged frequently. New equipment and plant are installed or parts of the installation are re-erected in different rooms. With all such reconstruction, special attention and caution are required. Rebuilding or changes in the position of equipment should never be undertaken without detailed consultation with the manufacturer. The work proper should at least be carried out under the supervision of a member of the manufacturer's staff, and transportation of equipment should be entrusted to a company experienced in such work. Manual work of any kind should be carried out only after first instructing the workmen and under the supervision of a member of the computer-centre staff.
Constructional work in the vicinity of the computer-centre building or on the building itself can endanger the operation of the centre through disturbances, dirt and dust generation, or accidents. If in doubt, the centre should consult the manufacturer of the EDP installation regarding the effects of such building work, as he has the necessary experience and has at his disposal the requisite measuring instruments for determining the effects of vibration on equipment or for checking impurities in the air, etc.
Reconstructional work in the computer centre or building work in its vicinity can often represent a substantial increase in potential danger. If there is any doubt, the insurer of the EDP installation must always be informed of any such increased risk—firstly to maintain insurance coverage to its full extent, and secondly to take heed of the insurer's expert knowledge in regard to any special precautions that may be required.

3.2.2.7. Protective Measures for Data Carriers

In most cases involving loss of or damage to the EDP installation, it is possible to maintain emergency operations by making use of an outside installation. However, this expedient is impossible if the data carriers have been damaged, so that particular attention must be paid to their protection. Two basically different methods are available for bringing about effective protection from loss of data, both of which should be used at the same time. These methods involve safe keeping of data carriers and providing the basis for a possible reconstruction of important data files.
All data carriers that are not immediately required for processing should be stored in archives separated from the EDP rooms proper by fire-resistant walls. In addition, the basic air-conditioning parameters (temperature and humidity) for the storage of data carriers must be maintained. An air-conditioning plant, separate from

Figure 18. A data-carrier archive destroyed by fire. The lack of preventive measures resulted in the destruction of magnetic tapes. Steel cabinets do not afford sufficient protection.

that serving the EDP rooms, is recommended. All protective measures enumerated earlier for the EDP installation are also applicable to the room housing the archives.

Fire-resistant cabinets, specially designed for storing data carriers, are available on the market. These cabinets offer adequate protection, for a limited period, from fire or water, provided that their doors are always closed after taking data carriers out or putting them back. The storage of data carriers in such cabinets gives, only in certain circumstances, a degree of protection equivalent to that afforded by a fire-resistant separation of the rooms containing the archives.

These cabinets are recommended for storing duplicate data carriers not in use, for particularly important data carriers inside the archive rooms or for bridging the gap until one's own archive rooms are ready.

The removal of data duplicates carriers containing important basic information and programs, data carriers of the previous generation, and time-defined data carriers (e.g., those containing intermediate results) to a neighbouring building, but located outside the danger area of the computer centre, will ensure that the expenditure, in terms of time and money, for reconstructing all existing data files will be kept within reasonable limits. The reconstruction must be capable of being carried out by machine; the acquisition of all data from the original information or from lists in the case of large amounts of data is hardly possible in view of the time required, since regular work must be carried out concurrently. If one's own installation is working to full capacity, reconstruction of data by machine can in some circumstances be undertaken with the co-operation of outside installations.

Details of how to bring into effect such safety measures depend largely on local conditions and on the type, organization and management of the data banks.

Bibliography to 3

[1] Rechenzentren, Planung von Bau und Einrichtung. Berlin, München: Siemens AG.

[2] Brandschutztagung der Bayerischen Versicherungskammer, München 1968. München: Bayerische Versicherungskammer, Landesbrandversicherungsanstalt, pp. 25/39.

[3] Recommended good practice for the protection of electronic data processing and industrial automation. Factory Insurance Association, Hartford, Chicago, San Francisco.

[4] Müller-Lutz, H.L.: Feuerschutzmaßnahmen für elektronische Datenverarbeitungssysteme. From: Automation der Büroarbeiten, 2nd edition. Karlsruhe: Verlag Versicherungswirtschaft 1969.

[5] Müller, W.: Schadensverhütung und Versicherung für Datenverarbeitungsanlagen. Datenträger und Programmierungsunterlagen. IBM-Nachrichten 19 (1969), No. 195.

[6] Brandsicherheit elektrischer Anlagen. ETZ-Report No. 1, pp. 5/7. Berlin: VDE-Verlag 1970.

[7] Brandschutz. Deutsche Feuerwehrzeitung 24 (1970), No. 3. Stuttgart: Verlag W. Kohlhammer.

[8] Heinz, R., Renz, H.: Der EDVA ein sicheres Klima. Maschinenmarkt/MM-Industriejournal 77 (1971), No. 43.

[9] IBM, Installationsplanung, Datenverarbeitungs-system/360 ab Modell 25. IBM Deutschland, March 1970.

[10] Bedingungen für die Installation und den Betrieb einer Datenverarbeitungsanlage. Siemens AG, München.

[11] Betrieblicher Brandschutz. Vorschläge zur Durchführung von Brandschutzmaßnahmen. Köln: Verband der Sachversicherer 1970.

[12] Brandschutzrichtlinien für elektronische Datenverarbeitungsanalgen (EDVA). Herausgegeben vom Verband der Sachversicherer e.V. Köln.

[13] TELA-Merkblätter: „Schadenreport" und „Schadenverhütung". Field Data Processing. TELA-Versicherungsaktiengesellschaft für Technische Anlagen, München.

[14] Effertz, P.-H., Grupp, H., Jach, W.: Einwirkung chlorwasserstoffhaltiger Brandgase auf Werkstoffe und Bauteile. Der Maschinenschaden 43 (1970), No. 3, pp. 89/99.

[15] Effertz, P.-H., Grupp, H.: Einwirkung chlorwasserstoffhaltiger Brandgase auf metallische Werkstoffe sowie auf Bauteile der Feinmechanik und Elektronik. Der Maschinenschaden 45 (1972), No. 1, pp. 7/19.

4. Steam Generators

About 75% of all steam generators installed in West Germany have an output up to 2.5 tonnes/hour and 20% between 2.5 and 25 t/hr. However, the remaining 5% represent 65% of the total installed capacity. Thus that the recommendations given in the following chapters refer to particular output groups.

In most cases, small steam generators are fired by oil or gas. With insufficiently purified water, the susceptibility to damage of this group rises markedly. This fact is partly explained by the specific heat loading, because maximum heat release with liquid or gaseous fuels takes place along a substantially shorter path, compared with, for instance, stoker firing. Moreover, examinations of damage cases to small steam generators confirm repeatedly that up to 50% of all cases registered can be traced to salt and sludge deposits. For this reason, the importance of faultless water treatment and conscientious operation must be stressed.

Other problems arise, however, with industrial and public power stations, where questions of design and planning, manufacture, the selection of materials and planned maintenance of certain parts of the installation are in the foreground. Table 1 indicates those parts of an installation where measures for the prevention of damage promise success.

Discussion of Locations and Causes of Damage

The evaluation in Table 1 of the 490 cases of damage to steam generators was carried out retrospectively, starting from mid-1971. New designs are included. The term "product faults"

Table. 1. Distribution of cases of damage in accordance with location and cause, based on an analysis of 490 cases of damage to steam generators (external influences not perceptible). The group of "high-speed" steam generators is not taken into account here.

Location of damage	Number of cases of damage [%]	Causes of damage	
		Product faults [%]	Operational faults [%]
Economizers and evaporators	23	62	38
Fire tubes	5	12	88
Boiler drums	3	75	25
Headers	3	93	7
Superheaters and reheaters	17	55	45
Coolers	4	84	16
Fittings	4	72	28
Furnaces	16	45	55
Air preheaters	4	68	32
Tube mountings and brickwork	16	66	34
Other parts	5	60	40

refers to faults in planning, design, manufacture, maintenance and materials. In the group of "operational faults" are included faults in upkeep and servicing and, to a lesser extent, such external influences as wear and corrosion.

Outstanding in this evaluation of damage cases is the high proportion of product faults, i.e., faults on constructional components which were in existence before commissioning. The only exception is the damage location "fire tubes". In the high proportion of operational faults, maintenance faults stand out as the cause of damage.

When tabulating the cases of damage according to product faults and operational faults, the following picture for steam generators (the "high-speed" class is excluded) emerges:

Product faults: 60%; operational faults: 40%.

4.1. Water-Tube Boilers

Water-tube boilers can be divided, in accordance with their design principle, into natural-circulation boilers, forced-circulation boilers, and once-through boilers.
With natural-circulation boilers, it is possible, without any additional equipment, to operate under low partial loads. Short-time load increases can be dealt with because of their storage capacity, taking account of the maximum rate of pressure reduction.
A disadvantage of natural-circulation boilers is their slow starting up from cold. The starting-up period can, however, be shortened considerably by heating with steam from an external source.
The rapidly increasing demand and higher investment costs have led to standardization, e.g., packaged construction. With this design, gas-impermeable welded tube plates have proved successful.
In West Germany, forced-circulation boilers are mainly used for waste-heat utilization in process plant.
Once-through boilers are almost exclusively employed for public power stations in West Germany in unit sizes for present and future needs, but are sometimes used for industrial power stations. Their type of circuit depends on requirements (the planned number of hours under full-load operation, cold starts, starting-up time, guaranteed partial load, etc.).
In the following, laws, resolutions and guidelines (see [8–13] in the Bibliography) are not dealt with, but points are stressed which repeatedly lead to damage, as indicated in Table 1.

4.1.1. Manufacture, Assembly and Commissioning of Water-Tube Boilers

4.1.1.1. Demands on Planning and Design

When ordering a steam generator, the at least following points must be clarified:
- Steam consumption curve, with indications of peak and partial loads.
- Starting-up period.
- Anticipated hours of operation per year.
- Fuels to be used.

With regard to planning and design, the following demands, based on past records of damage, should be made:

Materials

The maximum permissible stressing of materials, at the time of planning, should not be fully utilized, but a safety factor, based on past experience, should be taken into account. With various installations, sections under pressure should be replaced in good time if it has been found that
- long-term parameters were estimated too high when planning;
- the temperatures of materials were higher than those determined by thermodynamic calculations;
- substantial high-temperature corrosion took place.

The higher investment costs are negligible when compared with the possible costs of material damage and outages.

Pickling

If, before putting the sections under pressure into operation, they are cleaned by a pickling treatment, sufficient blow-down points must be provided at the lowest points of individual circuits to ensure that no dissolved residues remain. The processing taking place during cleaning must be considered from the fluid-flow and chemical points of view. Constructive hints for satisfactory pickling treatment can be found in [14].

Insulation

Brickwork insulations must be provided with sufficient supports (approximately every 3 m, depending on type) to ensure satisfactory unloading.

Tube Supports

Tube supports must be protected from radiant heat. Many cases of damage are known where a flow of flue gas behind the tubes led to scaling. The material of the supports and the filler

Figure 1. Penetration of flyash between brickwork and supports led to overheating and fracture of the tube supports.

material must be matched to the tube material so as to attain sufficient resistance to scaling without any tendency to embrittlement.

Fly Ash

Damage often originates from fly ash lodging between tubes and brickwork and, by its wedging action, overloading the tube supports. Fly ash must be prevented from accumulating behind the tubes by suitable design measures, e.g., by partitions. Even with dirty boilers, it is important to ensure that the heating surfaces have sufficient room for expansion during operation (Figure 1).

Low-Temperature Corrosion

The danger exists, particularly with installations fired by fuel oil, that flue gases may diffuse as far as the external sheet-metal enclosure, where they can cause low-temperature corrosion. This danger can be averted by proper sealing of all pipes led through brickwork or by keeping the external sheet-metal enclosure warm.

Pulse-Transmitting Lines

In the case of open-air installations, pulse-transmitting lines, e.g., a line for remitting an indication of water level to the control station, must be heated to avert the risk of freezing. Electrical heating (not above 100 °C) is required. If heating is effected with a steam envelope at a higher temperature, it is possible that, in the event of a rapid pressure reduction (from a burst tube), disastrous erroneous indications can be given, because evaporation of the condensate results, and the density of the water/steam mixture no longer corresponds with the correct interpretation [18].

Facilities for Inspection

In the design stage, the provision of viewing sockets on thick-walled components, e.g., headers and spray coolers, for the insertion of rigid or flexible endoscopes should later prove useful for inspection purposes. Thus, quick and safe indications can be obtained regarding the further serviceability of such thick-walled components (Figure 2).

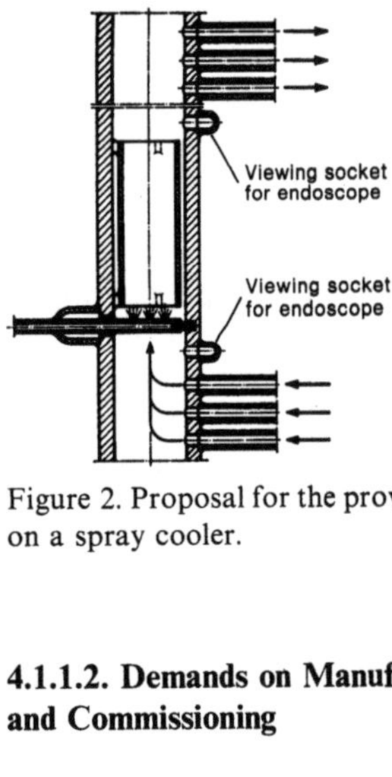

Figure 2. Proposal for the provision of viewing sockets on a spray cooler.

4.1.1.2. Demands on Manufacture, Assembly and Commissioning

Manufacture

If alloyed and unalloyed heat-resistant tubes are used at the same time, a rapid examination by spectrum analysis has proved successful for recognizing the inadvertent mix-up of materials.

Ultrasonic inspection for inclusions, laminations, and reductions in wall thickness has, in a number of installations, prevented damage and disruption of operation. Faults in and mix-ups of materials have the same effect during operation as excessive temperatures (Figure 3).

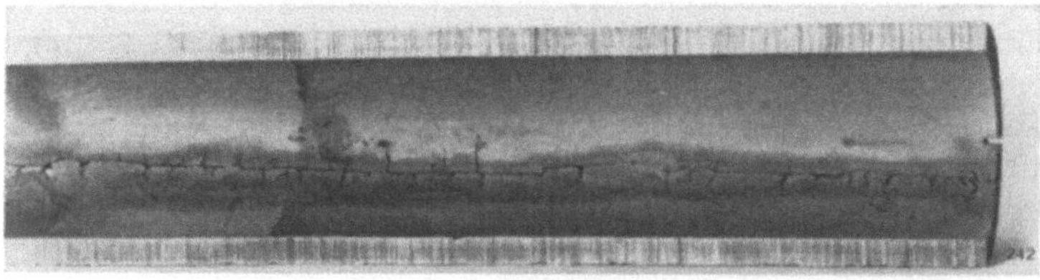

Figure 3. Part of a tube with a longitudinal crack caused by coarse non-metallic inclusions.

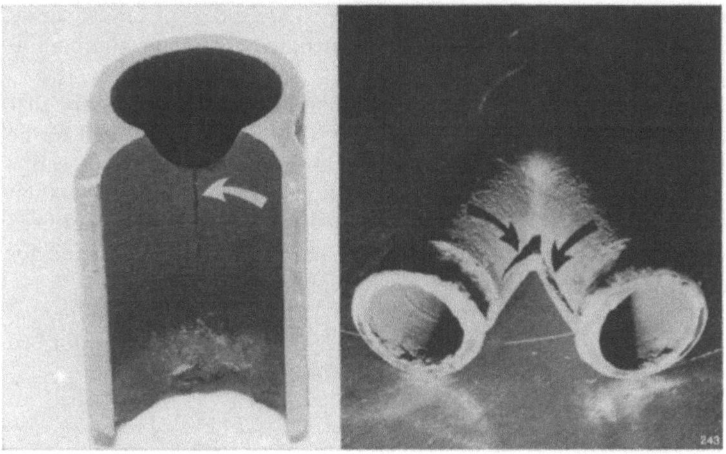

Figure 4. Drop-forged and bored-out pipe branch-pieces from two different installations. Where the two bores meet, a sharp edge is created which acts as a stress raiser. By rounding this edge, stress peaks can be reduced.

For later examination, e.g., of structure, it is expedient to store samples of materials for purposes of comparison.
Difficulties with defective branch-pieces of pipes have been reported from several installations when cracks originated from the very sharp edge formed at the intersection of the two bores (Figure 4). This edge must be ground away, or another method of manufacture must be resorted to for making branch-pieces [9].
With spray coolers, one end of the protective sleeve must be attached by a strong welded or screwed connection, while the other end must be free to expand. Under no circumstances must a protective sleeve, torn away by vibrations, be permitted to be dragged by steam flow into communicating pipelines or to block the free cross-sections of the connecting tubes.

Assembly and Commissioning

An attempt should be made to select future operating personnel at the start of assembly and let them work together on the site. As a result, the necessary familiarity with the plant will be attained at an early stage.

For many years, manufacture and assembly have been carried out in the workshop, as far as this arrangement has been possible in view of transportation. At the erection site, the components supplied should be grouped neatly, and care should be taken that all pipe ends, nipples and headers remain properly sealed.
If any pressure parts are to be bent and welded, it is necessary to recheck drums, headers, etc. for foreign bodies, such as tools, welding electrodes and plugs, before they are closed up for pressure testing. It happens frequently that, after bending tubes of small cross-section on site, sand, paper and other matter remain inside them.
In view of the large number of firms active on an erection site, it is essential that all work be co-ordinated by persons responsible for superintending construction. Only then is it possible to avoid:

- the erroneous attachment of insulations, platforms or scaffolding to pressure-carrying components;
- the use of completed insulations as work platforms and their consequent ruin;

- careless damage to already installed measuring and control equipment during subsequent installation of other components.

All auxiliary machines must be subjected to test runs. Switching, supervisory, safety and control equipment must be checked before commissioning to ascertain possible switching faults and to ensure proper functioning of interlocks.
The risk of fire, makes it necessary, before lighting up for the first time and also during the first hours of operation, to ensure that the extinguishing equipment and the temperature controller of a Ljungström preheater are ready for operation (Figure 5).
Under the pressure of time, a plant is frequently started up with insufficient instrumentation. If damage occurs, the delay in commissioning is often much longer than the time that would have been required for completion of the safety and supervisory equipment.
Before commissioning, all those concerned must be told about the sections of the plant on which no more work is to be carried out. Further measures for commissioning are described in [15].

Figure 5. Severe damage to a Ljungström air preheater. After the first lighting-up tests and preparatory work, the plant was shut down after a few hours in operation. Iron started to burn unnoticed, and the fire could no longer be brought under control with foam extinguishers. Large sections melted completely. The installation of a new rotor required extensive disassembly work in the boiler-house.

4.1.2. Prevention of Damage to Water-Tube Boilers by the Operator

The availability of the fuel at the proper time stipulated in the contract is essential for smooth commissioning and a test run as free as possible from trouble. For banked operation, for instance, it may be decisive to use a coal of lower density than that contracted under guarantee for continuous operation.
To delineate responsibilities, it is necessary that the point in time when risk is transferred to the operator be laid down in writing [10]. Any shortcomings that have come to light during operation before the end of the test run, but have not been rectified, should be recorded and the time for their elimination planned in advance.
Before the test run, i.e., before the time of actual steam delivery, personnel must be fully informed in regard to the functioning of the plant.

In Operation

In the ensuing period, special instructions must be issued for possible operational troubles.

Procedures during Operational Troubles

Strong overfeeding (natural circulation): close feed check-valve, close steam line, extinguish fire, open drum drainage.
Water deficiency (natural circulation): extinguish fire, no further feeding if the water level has sunk below the minimum mark. The danger exists that heated sections have already attained excessive temperatures. Release steam slowly from the boiler through throttled steam valve.
By analogy, the same considerations apply if the *circulating pump of forced-circulation boilers fails* for any length of time. Here, the danger of spontaneous steam generation and uncontrollable pressure rise exists if the circulating pump is again taken into operation.
Once-through boilers require attention to be paid to circulating pumps that are taken into operation at low loads or when starting up. It is essential to make certain that there is no phase of operation during which hot boiler water can penetrate into the winding of the motor and destroy it (Figure 6).

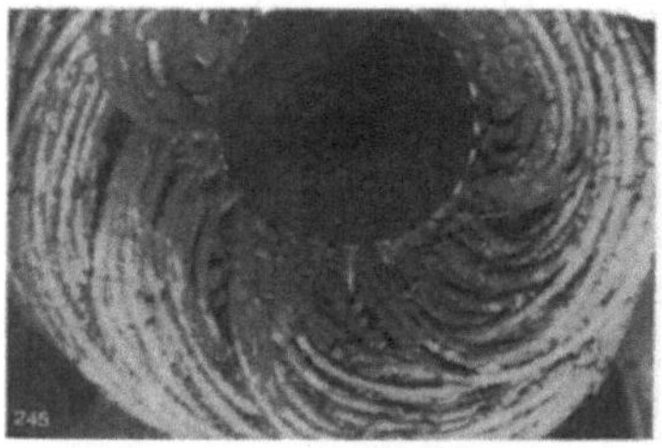

Figure 6. Winding head, on the pump side, of a circulating pump unit without a stuffing box. Due to a leakage in the cooling circuit, hot boiler water penetrated and attacked the PVC insulation, generating hydrochloric acid which destroyed the lamination stack, impeller and bearing.

As a matter of principle, all observations of abnormal operation must be recorded in writing by the operating personnel as soon as possible, so that the causes can be clarified.

Superheaters

The superheater is a constructional component which is frequently damaged, especially with natural-circulation boilers.
The superheater of a cold boiler, if no start-up connection is provided, must be filled with water via the feed line for start-up or, if the design permits, must be shielded from the combustion gases. If a start-up line is provided, the superheater must be supplied with an amount of steam sufficient for cooling. Small and medium-size industrial boiler plants are often provided with a start-up line, but it usually ends up in the open air without a silencer. As a result, the line is used as little as possible to avoid inconveniencing the neighbourhood by objectionable noise. Such misplaced economy (lack of a silencer) and consideration for people living in the vicinity (start-up line remains closed whenever possible) lead to unavoidable damage to superheaters.

Heating Surfaces

It is known that, as a result of progressive fouling of the heating surfaces, the temperature of the flue gases increases towards the end of the boiler. The maximum permissible temperature of each section of the heating surface must be known and measured, so that time for cleaning can be scheduled at the appropriate moment.
However, it is not only on the heat-resistant tubes of the various sections of the heating surface, but also on the sheet-metal of the Ljungström preheaters that excessive flue-gas temperatures have a damaging effect. Thus, heated sheet-metal gets covered with scale, loses its elasticity and, with vibrations, wears through.
If secondary heating surfaces are kept clean by shot-blasting equipment, care should be taken that all shot is recovered. It must not remain in "dead corners" or bake together with ash. Because of the substantially increased weight, hoppers or channels may become overloaded and clogged.

Maintenance

For the expert maintenance of all equipment, it is necessary to prepare a maintenance schedule, which should contain the special operating regulations prescribed by the manufacturer. The schedule should include indications of the intervals (each shift, day, week, etc.) at which certain work should be carried out. All lubricating media should be agreed upon with the manufacturers, using only a few brands, if possible. The observation of operating temperatures is important. The bearings of coal crushers and air preheaters often have a longer life if grease of a higher viscosity is used.
The timely supply of spares for those components which experience has shown to be subject to wear or liable to damage is of decisive importance for trouble-free operation. In this way, risky temporary expedients become superfluous.

Important Checks on Operation

- Daily check and calibration of flue-gas analysing equipment.
- Daily check of the pH-value of boiler water with natural-circulation and forced-circulation boilers. Control of the pH-value of the feed water is not in itself sufficient. An additional colour check based on the membrane-filter technique makes a safe assessment possible and is easy to carry out.

- With various heavily loaded natural-circulation and forced-circulation boilers, severe corrosion on the water side (on-load corrosion) has been reported. This can be overcome by sufficient alkalizing treatment of solids [17].
- Regular checking of the safety devices for preventing water deficiency and/or for indicating water level. Thus, in the case of damage to the superheater of a Benson once-through boiler equipped with water-deficiency safety devices, the cause was clearly a leakage in the economizer which normally should have been recognized immediately. After the damage had taken place, it was established that, although measuring points for these safety devices (for comparison of water flow at the economizer inlet and outlet) had been provided, they had never been incorporated in the measuring and supervisory system.
- Regular checking of the firing interlocks.
- Regular analyses, in the case of coal-fired furnaces, give an early indication of changes in the composition of the fuel. With pulverized-coal firing, each such change can bring about a different flame path. As a result, there may be undesirable ash deposits on heating surfaces. Large lumps of slag that have dropped down present a safety hazard during operation by tearing away from heating surfaces, with consequent damage (Figure 7). The lives of personnel are at risk if excessively large hot lumps of ash fall into the water bath of the slag-removal plant.

Figure 7. Substantial slag formation (20 to 30 tonnes) on the heating surfaces of a superheater, caused by an uncontrolled flame path. Lumps of slag dropped down deformed heating surfaces below and tore many tubes from their mountings.

- Controlled soot-blowing with properly adjusted lances can to a large extent prevent the formation of deposits. Adequate water extraction before blowing (temperature control) is important. When the blowing process has been completed, all lances must be returned to their end position. This control is particularly important with automatic soot-blowing installations.
  If the plant is being checked in the case of damage or during an inspection, the condition of the blowing heads (bent or fractured), the blowing angle and the end position should be controlled.
- Weekly test start-up of the stand-by pump with automatic starting equipment.

4.1.3. Inspection and Revision

The following is not a discussion of the tests prescribed by law. However, damage that occurs repeatedly can often be recognized at an early stage by properly directed checks and can be eliminated with a relatively low expenditure of time and money.

These checks include:

- Control of the reduction in wall thickness of tubes through corrosion and erosion.
- Measurements of bulging. If tube bulges are discovered in time, a lot of damage can be prevented and remedial action taken, once their cause has been established.
- Planned shut-downs for the control of tightness of sections under pressure. It is perfectly feasible to carry out such tests under reduced pressure. During these tests, small and still insignificant leaks (e.g., weld porosity) can be discovered and can usually be remedied without difficulty.
- Examination with an endoscope of parts subjected to pressure, such as collectors and spray coolers (attemperators). Protective sleeves that have become loose or broken injection tubes can cause substantial damage in sections under pressure (Figures 8 and 9).
- Removal of incipient cracks in drums by grinding to eliminate their notch effect.
- Checking of mountings and brickwork, particularly of the ceiling, and suspensions. The ceil-

Figure 8. In this protective sleeve of an attemperator the welded attachment to the outer tube fractured. The sleeve fell down and was forced into one of the outlet openings. Later, the longitudinal seam was torn open. Damage resulted to headers and tubes behind.

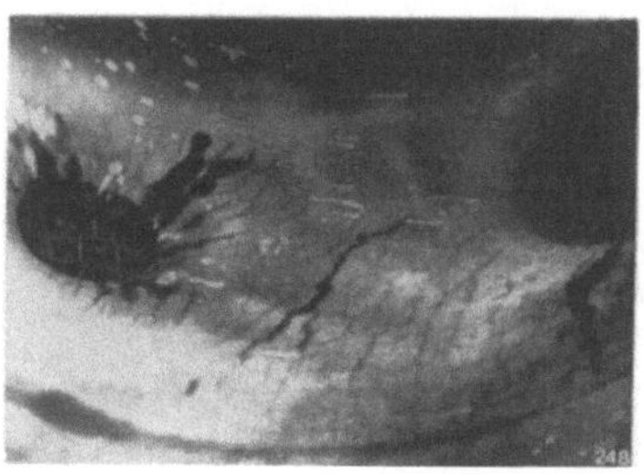

Figure 9. Cracks through thermal shock in the pressure chamber of an attemperator. The cause was tearing away of the protective sleeve, so that injected water impinged directly on the outer wall.

Figure 10. Damage to the intermediate ceiling of a Benson boiler, caused by wear of the ceiling brickwork. The entire mounting arrangements in this region were damaged through the action of flue gases.

ing must be absolutely tight, so that heat accumulations cannot destroy the suspensions (Figure 10). If shortcomings (often design faults) are recognized at an early stage, they can first be repaired at critical locations and then eliminated by incorporating design changes during the next shut-down.

- Checking of the combustion chamber. Studding and refractory material must be in good condition (see Section 4.2. Furnaces).
- With Ljungström air preheaters, the packing of the heating plates becomes loose with time. The plates then flutter during operation and wear through at points of contact unless they are made secure again in good time or defective plates are replaced. Similar damage must be expected if the necessary firm packing of the heating plates is lost through other causes (e.g., failure to fit replacement plates after withdrawing test plates).

4.2. Furnaces

4.2.1. Planning and Design

Virtually stoichiometric combustion is desirable with liquid and gaseous fuels and, to make this possible, a sufficient number of measuring points must be provided. An attempt should be made to measure the fuel and air separately in front of each burner, if possible, but at minimum in front of each plane.

For flue-gas analysis, measuring points for at least carbon dioxide and carbon monoxide should be available. The probes for this purpose should have their sampling locations away from the walls and should be positioned sufficiently deep into the flow cross-section of the flue gases (more than 500 mm) to prevent erroneous readings. Since the viscosity of flue gases is proportional to temperature, the streams of gaseous products of combustion scarcely coalesce. This means that, once gas streams of a reducing or an oxidizing nature are present, they will be maintained as such and can easily produce an incorrect picture of the mode of operation.

Flue-gas analyses should be conducted near the actual location of combustion, and in any event in front of the Ljungström air preheaters; otherwise, results cannot be accepted with any degree of assurance.

The disposition and number of inspection ports must be adequate—even with pressure-firing. Only then is it possible, particularly when lighting up, to adjust combustion properly.

With *pulverized-coal slag-tap furnaces,* the method of application of the lining, the guaranteed fuel and the sintering process, in addition to the quality of the lining, must be agreed upon by contract.

The *pulverized-coal and ash ducting* should be capable of being replaced economically at places of heavy wear.

The *combustion-chamber tubes* are usually studded by machine. So far, the stud material X 10 CrAl 24 (24% Cr) cannot be durably welded to the tube material. Under operating conditions, embrittled zones soon develop between the stud and tube, and these may lead to breaking away of large areas of studding and lining. With the less scale-resistant material X 10 CrAl 7 (7% Cr), this difficulty does not exist. For these reasons, it is better to apply initially studs made of this material and then, when scaling progresses, to attach studs of higher chromium content to the stumps—a procedure that has proved successful on several occasions. There are other methods of avoiding this difficulty (bimetal studs, sleeved studs), but their implementation is unlikely as they are costly.

With *coal-feeding equipment,* there is a risk of corrosion by water vapour. The use of stainless sheet can overcome this risk.

When *mill-drying fuel* with a high water content (lignite), it is always necessary to ensure cooling of the coal crusher rotor, bearing in mind the high temperature of the flue gases sucked back. The cooling-water pressure switch or flow controller should therefore be included in the chain of interlocks of the mill.

The *exits from bunkers* should be designed in such a way that continuous sliding is always assured, to safeguard against the risk of bridging. In addition to a favourable configuration of the hopper, there are various lining processes for improving sliding behaviour. It must be said, however, that thus far there are no lining or covering methods which guarantee a long life with all bunker fuels. With particularly difficult fuels, precautions must be taken to prevent or eliminate bridging, e.g., by needle weirs or pusher installations.

The minimum requirements for the safety of various firing methods are laid down in references [11–13].

4.2.2. Assembly and Commissioning of Furnaces

When assembling and commissioning furnaces, the following points must particularly be borne in mind:

- When assembling, all measuring equipment (e.g., venturis in ducts) must be built in exactly in accordance with the drawing.
- The pressure connections of the measuring points must be flush with the inner wall of the ducts and must be free from burrs.

- The positions of calibrated flowmeters used for the measurement of air or gas quantities must be noted with respect to the direction of flow.
- The notches (markings) on the spindle of stop and control butterfly valves must coincide with the actual position of the flaps.
- The installation of sheet-metal guides in duct bends must be carried out strictly in accordance with the drawing.
- After the rammed lining has been applied to the combustion chamber, entry into the chamber is no longer permissible.

The following measures are recommended:

- Before lighting up in the cold condition, velocity measurements at the burner exit should be carried out from the combustion chamber. In this way, inclined positions can be recognized in time and can be eliminated.
- With pulverized-coal firing, the feeding equipment should be calibrated by delivery tests, as far as the size of the installation permits such a procedure.
- With installations that can be controlled on site, it is desirable to use lead-sealed switches to prevent unauthorized operation.
- The running-up time of the drive motors for the coal mills must be recorded in writing.

4.2.3. Operation of Furnaces

Servicing and Supervision

In every furnace, the *combustion-chamber and flue-gas* paths must always be under careful supervision. All inspection and cleaning openings must remain free from fly ash. Also, the flame path must be such that, even under full load, the flames cannot impinge on the tubes.

Observation of the flue-gas temperature as a function of load indicates of the degree of soiling of the heating surfaces. With coal-fired furnaces, regular fuel analysis permits early recognition of changes in fuel composition. Under proper supervision, it is then possible to check heating surfaces for depositions of slag and, if necessary, to take the requisite action for their removal.

When burning heavy oil, a constant viscosity must be ensured by preheating. To achieve perfect atomization, the atomizing nozzles must receive a great deal of attention. Burners which are not in operation must be protected from scaling by the provision of cooling air.

A *clogged mill* must never be freed by spraying while it is in a hot condition. Such attempts have on various occasions led to the formation of a longitudinal crack in the rotor, resulting in distortion (Figure 11).

With a *milling plant*, the *grader temperature* must always be below the stipulated maximum, which depends essentially on the volatile constituents of the coal. If the grader temperature is higher than this limiting value, any excess of which must be signalled to the control station, there is a danger of a grader fire or an explosion (Figure 12).

In the event of unbalance or replacement of the *beater heads*, these must always be weighed in pairs, and heads of the same weight must be installed diametrically opposite to each other.

Because of their nature, *coal-feeding and grading equipment*, dust and ash conduits, and drop and suction shafts are subject to heavy wear. By re-

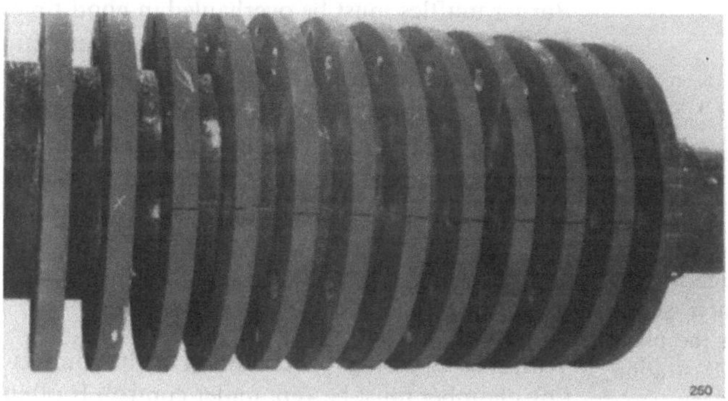

Figure 11. Longitudinal crack in a mill rotor. A tensile stress above that recommended and a low notch toughness (< 1 mkg/cm^2) are indicative of ageing. Spraying of the mill while still warm caused stresses, resulting in the crack.

Figure 12. Damage to grading equipment as a result of an explosion of coal dust in the mill section. The head cover was torn away, as also were the upper covers. The upper part showed deformations and the ceramic cladding was destroyed.

placement at the proper time, unforeseen damage (Figure 13) can be prevented.

A pulverized-coal firing plant must never be run with the *flue-gas filter* switched off. Apart from creating a dust nuisance in the neighbourhood, it has been found that the suction blades of induced-draught fans are eroded within a very short time. Similar difficulties have also been experienced with an insufficient degree of separation of the filter or with dust streams ahead of the induced-draught line. To prevent fly ash from being carried in the stream, it is necessary to ensure that all ash funnels remain unclogged.

The silence in operation of induced-draught fans must be kept under regular supervision. If the blades are hollow, any holes eroded in them will give access to the empty spaces within and allow them to be packed with airborne dust, creating unbalance and consequent severe damage.

Figure 13. The cladding of flue-gas return ducts could not withstand operating stresses. The outer sheet-metal covering was burnt through, so that the plant had to be taken out of operation. The illustration shows the start of damage.

Maintenance, Inspection and Revision

With a firing plant, it is also important to establish maintenance schedules.

Stoker Firing

All moving parts must be overhauled thoroughly at regular intervals (one to two years, depending on their period in use). Grate-cooling bars must be checked for freedom from cracks and for wall thickness. In a reducing atmosphere, there is a risk of erosion.

Pulverized-Coal Firing

Burner muffles must be overhauled in good time. Coal dust and secondary air must be mixed only at the burner outlet, so that a special check for erosion is important in the primary-air section. If mixing takes place too early, there is a danger of a burner fire. If the burner is cooled by water-carrying pipes, the latter must be checked for erosion.

Slag-Tap Firing

With slag-tap firing, the condition of the combustion chamber must be kept under control. It often

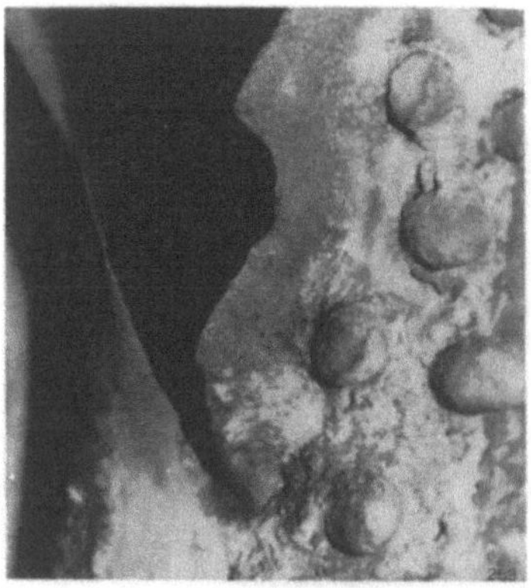

Figure 14. Combustion-chamber tube with the studs worn away. The knob-like roots of the studs indicate impairment of wall thickness.

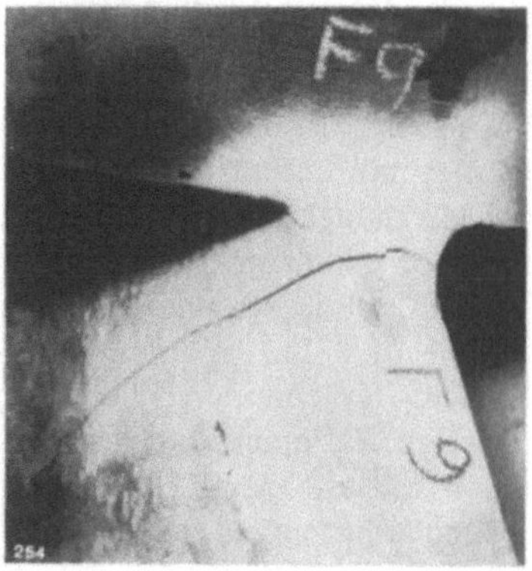

Figure 15. Cracks in the reinforcing ribs of a tube mill. The cracks were caused by welding, i.e., thermal stresses, and resulted in rough running of the mill.

becomes necessary to re-stud periodically and apply a new rammed lining where it has worn away. Before re-studding, the thickness of the tube wall must be checked and, as a rough indication, it can be taken that the wall thickness is still adequate if a stump of the stud has remained. The actual length of the remaining stump then plays a subordinate role. If the stumps are reduced to no more than knobs, it must be assumed that wall thickness has been substantially reduced (Figure 14).

A technique for applying new studs to old stumps has proved successful and has enabled a good grip to be provided for the new rammed lining, particularly if the studs are not in line. A new rammed lining can be applied satisfactorily if the damaged sections are first sand-blasted, and it is not necessary to perform the laborious task of baring the original tube surfaces, if there is satisfactory adherence. After sand-blasting, the new lining can, if a binder is first applied, then be sprayed over the remaining old lining.

Mill Installations

Mill installations are subject to wear and must be checked regularly. The life of impact plates, beater heads, as well as the cladding of pan mills, etc. must be determined for each operating condition.

With tube mills, cracks in the front-wall reinforcing ribs have often been experienced (Figure 15). Regular non-destructive testing, while the mill is stationary, is recommended at least once a year.

4.3. High-Speed Steam Generators

The term "high-speed steam generators" applies to steam boilers of small output (up to about 5 tonnes/hour) which, though occupying a small space, are capable, of supplying steam in the desired quantity and at the required pressure after a start-up time of a few minutes. Once-through boilers are mainly considered here but, because of their design and quick-starting properties, consideration will also be given to some types of natural-circulation boilers.

After high-speed steam generators were put on the market in the German Federal Republic at the end of the 1950's and found quick acceptance, there was, until 1964/65, a steady substantial rise in the frequency of damage cases and in the specific repair costs per case. The majority of damage cases occurred to components subjected to pressure, particularly steam coils, the main reason for this damage trend being underestimation of the necessary servicing and maintenance, as well as of the high demands made on the feed-water. Since, however, users have gained experience and manufacturers have improved design and service to customers. The result has been a small reduction in the frequency of damage.

Figure 1. At the burner muffle and sheet-metal insert of a high-speed steam generator, marked signs of wear repeatedly manifested themselves at short intervals in the course of two years in operation. The vertically disposed burner muffle could not be inspected without disassembly. By improvements in design and material, it was possible to lengthen the life of the burner muffle. In addition, an inspection opening was provided at the top of the boiler, enabling the condition of the muffle to be checked regularly.

4.3.1. Guidelines for the Prevention of Damage to High-Speed Steam Generators

Both the manufacturer and user of high-speed steam-raising plant must realize that maintenance is often not what it should be. For this reason, individual constructional components should be designed for maximum safety. With once-through boilers, pressure-measuring points at the inlet and outlet of heating coils are necessary, so as to be able to have an indication at all times of deposits in the coils that manifest themselves by an increasing pressure loss. Constant measurement of differential pressure, with adjustable maximum contacts and an acoustic alarm, is recommended.

Equally important for the safety of the installation is measurement of flue-gas temperature and, here again, there should be included an acoustic alarm which is actuated when a maximum adjustable value has been exceeded.

Care should be taken to check the boiler at critical locations and ensure that it can be dismantled

Figure 2. In this longitudinally welded tube of a high-speed steam generator, corrosion occurred at places where the longitudinal seam was not welded right through.

Figure 3. Section of a tube coil at the water-inlet side, showing external wear and break-throughs resulting from not reaching the dew-point.

easily. Only in this way is it possible to carry out purposeful inspections (Figure 1) and to execute partial repairs in the event of damage. Even today, it often happens that the entire installation is scrapped because of insufficient accessibility of the location of damage, as the repair costs are too high in comparison with a new installation. In the region of the cold end, i.e., the feed-water inlet, corrosion has been known to occur on the flue-gas side because the dew-point of water has not been reached and, indeed, this corrosion must be expected in view of the fact that operation is often intermittent (Figure 3). This difficulty can be lessened by raising the inlet temperature of the feed-water to between 55 and 60 °C. However, under normal conditions, a higher inlet temperature should not be chosen because, with various types of pumps, the formation of steam bubbles (cavitation) and hence damage to the pump must be expected.

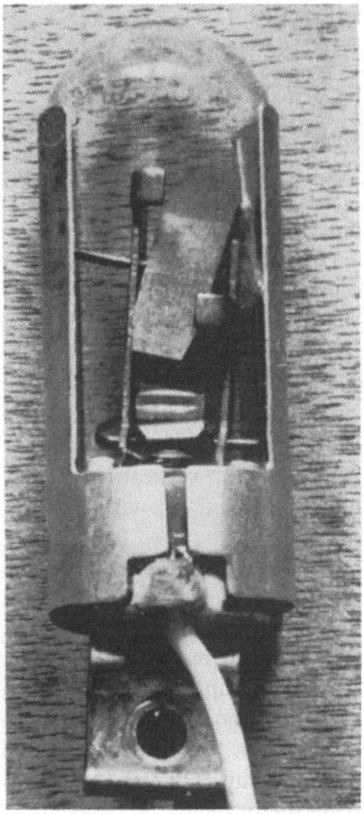

Figure 4. Jamming of a switch rocker of a water-level controller on a natural-circulation boiler contributed substantially to severe damage through lack of water. Through slipping out and canting of a small spindle, the switch rocker did not operate when the water level fell, and the burner was not cut off.

Maintenance

Because high-speed steam generators to a large extent function automatically, regular checking of their operation and regular maintenance are essential. Operating personnel have also become maintenance personnel and must be conversant with the design and operation of the safety and control gear. Even small, sometimes unrecognized, shortcomings frequently lead to extensive damage, often to total loss, from lack of water (Figure 4). An urgent recommendation to every user is therefore to send his personnel to a course of boiler attendants at the Technischer Überwachungs-Verein (TÜV, the Association for Technical Supervision), so that they can be instilled with an understanding of the interplay between various functional groups.

With all high-speed steam-raising plants, the loading of the heating surfaces is high, and great importance therefore attaches to conditioning of the feed-water (Figure 5). The guidelines compiled by the above-mentioned Association [2] must be adhered to strictly. Regeneration must not be postponed until the ion-exchange material is exhausted. Regular control of the condition of the exchange material is recommended.

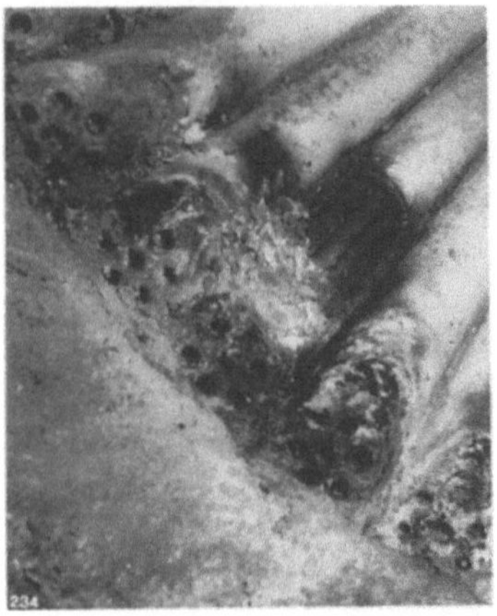

Figure 5. Heavy deposits of boiler scale in the lower part of the water tubes of a high-speed steam generator gave rise to severe thermal overloads and bulging, with crack formation through inadequate cooling. The entire pressure component had to be renewed, when design improvements were also being carried out.

Once-through boilers are particularly sensitive to water of inferior quality, since nearly all entrained salts remain in the tube coil and constrict the tube cross-section. To avoid overheating of the tubes, they should be flushed regularly (daily, if possible).

Natural-circulation boilers are less endangered. However, here again, frequent and sufficient blow-down is absolutely necessary.

To avoid corrosion damage by pitting, de-aeration of feed-water is very important. Since thermal de-aeration is usually too expensive for such small installations, dosing with hydrazine (levoxin) or the addition of sodium sulphite (Na_2SO_3) is required for fixing the oxygen.

To prevent corrosion while the installation is out of operation (Figure 6), "wet inhibition" is recommended. In this procedure, all pressure-sustaining parts are filled with feed-water containing a sufficient excess of hydrazine, and care is taken to ensure that the water is kept in constant circulation. The excess hydrazine must be regularly checked (pH-value > 11) and, if necessary, maintained by further additions of hydrazine.

Frost damage, which occurs frequently to small boiler installations, can be avoided by the simple expedient of emptying completely the boiler and pump when there is a risk of frost. When the installation is out of operation for a longer period, filling with anti-freeze may be appropriate.

4.3.2. Inspection and Revision

From experience it is recommended, particularly for small plants, that a maintenance agreement be concluded with, if possible, the manufacturer. In this agreement, regular checks of control and safety installations, as well as water-conditioning, should be included.

During inspections, fittings must be dismantled and cleaned. When carrying out functional checks, such components as linkages must only be moved slowly to approximate operational conditions as closely as possible. Some safety devices to ensure against water shortage have responded to rapid control movements and have failed in operation with a slowly sinking water level.

If difficulties are encountered with water conditioning and excessive deposits, the boiler must be thoroughly cleaned. Such cleaning on the water side of the tube coils of once-through boilers is always necessary if the pressure loss becomes excessive (20 atm as a guide value).

With this pressure loss, there is always also a rise in the outlet temperature of the flue gas because of the smaller heat extraction of the heating surfaces. For cleaning on the water/steam side, only tested and approved anti-scale and scale-dissolving media must be employed [7]; other-

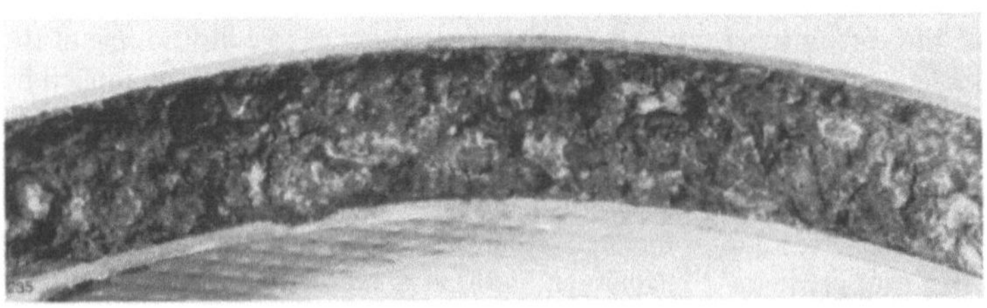

Figure 6. Typical example of corrosion by oxidation in a tube coil of a high-speed steam generator which had not been adequately inhibited whilst out of operation.

Figure 7. This multi-diaphragm feed-water pump of a high-speed steam generator was severely damaged by frost as a result of carelessness on the part of the boiler attendant. The housing was ruptured in several places, and all distributor heads, as well as some of the riser tubes, exhibited crack formation. The pump had to be replaced completely.

wise, extensive damage can occur on the water side of tube coils.

Since, in most installations, experts are not available, regular checking and adjustment of the water-conditioning plant are recommended. The cost of this procedure is substantially lower than that of damage arising from the use of water of inferior quality.

4.4. Large-Water-Space Boilers

Boilers with a large water space are designed as upright boilers, fire-tube boilers, locomotive-type smoke-tube or flue boilers and, most importantly combined fire-tube and flue boilers. Boilers of the latter category, which are produced in large numbers as three-pass types and sometimes also as four-pass types, will be discussed here from the point of view of damage experienced.

The highest attainable operational pressure of these boilers is about 25 atm gauge, and the maximum steam output of the most generally employed single fire-tube design is around 10 tonnes/hr, corresponding to about 6 Gcal/hr with hot-water heaters. Although higher outputs are obtainable, they involve certain problems and, in this connection, it must be appreciated that, for higher outputs, modern compact designs of water-tube boiler are more favourable in regard to capital outlay and space requirements.

Firing

For heating fire-tube and flue boilers, consideration should preferably be given to oil, gas or coal. However, these boilers, with appropriate additional equipment, are also suitable for burning low-grade fuels, such as wood waste or chips. Combined firing, too, can be employed.

With oil firing, there is a higher thermal loading of the fire tube in the vicinity of the burner, as compared with coal firing. Consequently, when converting a furnace from coal to oil, the output of the boiler should in no circumstances be raised. Moreover, a recommended procedure is to limit the maximum oil supply and to check the water-conditioning facilities. In this way, damage by overheating can be prevented.

Tube Plates

Damage by cracking occurs repeatedly to tube plates in the return chamber because protruding ends of the welded-in smoke tubes of the second pass are uncooled. For this reason, these smoke tubes must always be welded in without standing proud. Furthermore, cracks can appear, particularly in welding seams of corner stays that are made too rigid and cannot follow the differential thermal expansion between the shell and base. These shortcomings can be rectified by adopting differently shaped corner stays with a slight inclination and a corresponding extension.

Superheaters

In some designs, the superheater is bypassed during start-up by means of a controlled flap on the flue-gas side. These flaps distort, and experience has shown that they never close tightly, so that the flue gases impinge on the uncooled tubes and cause scaling. It is preferable for the tubes of the superheater to be water-cooled during start-up. Also, it should be mentioned here that the steam space of the boiler should be as large as possible, if for no other reason than to restrict the entrainment of water into the superheater and to prevent its damage by salt deposition.

Control

In view of the rationalization of operations, combined fire-tube and flue boilers are automated to a large extent, and some of them are provided with special equipment to indicate faulty operation. Since 1967, statutory regulations have per-

Figure 1. This fire tube of a three-pass boiler was badly dented as a result of water shortage and exhibited a gaping crack. Because of subsequent evaporation, damage occurred to the boiler and boiler-house.

mitted the indirect supervision of steam generators, irrespective of their size, so that tending of the boiler is no longer the primary consideration, but rather the servicing and maintenance of control, safety and alarm equipment.

In the case of damage by water shortage, which still occurs (Figure 1), the cause is often stated as failure of automatic controls. In fact, such damage can in most cases be traced back to inadequate servicing and maintenance of automatic equipment. Small faults, such as broken wires, faulty operation of relays, and leaking or stuck floats, have repeatedly led to substantial damage. It should also be noted that only equipment whose constructional components have been tested must be employed.

A number of older installations still exist, which are not capable of being indirectly supervised, although they are equipped with automatic water-level control. For such steam generators, particularly if they are oil-fired, the incorporation of an additional water-shortage safety device is urgently recommended. This device cuts off the fuel supply in the event of a failure of the control equipment, and the fuel supply can then only be restored manually. However the boiler supervisor is still by no means relieved of his obligation to exercise careful observation, particularly of the water level.

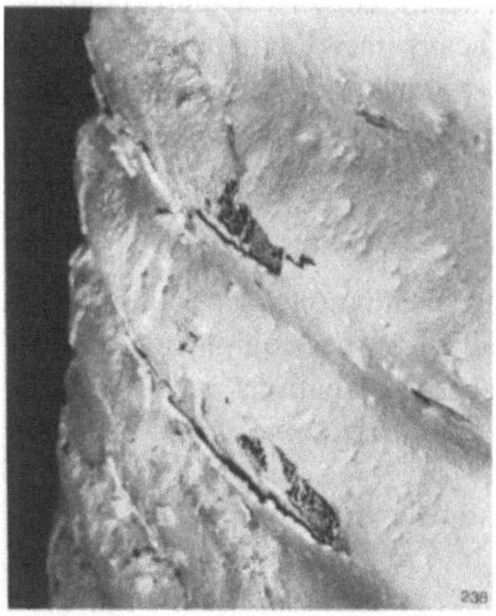

Figure 2. Heavy deposits on the fire tube of a three-pass boiler. Thermal overloading caused shallow indentations.

Figure 3. In addition there were heavy deposits on the water side of the smoke tubes.

Water Conditioning

As a result of inadequate cooling through heat-transfer-resistant deposits, highly loaded heating surfaces are exposed to excessively high heating up and are therefore subject to deformations or the development of cracks (Figures 2 and 3). Particularly endangered are the fire tube and the rear tube plate.

Deposits can also lead to damage in a different way. Functional troubles have repeatedly occurred to safety and control elements by clogging of their connecting lines to the boiler. Further even today the importance of thermal and chemical de-aeration of feed-water has not been sufficiently recognized in installations with boilers of smaller output. The result is rusting of tubes and tube plates.

Insufficient blow-down of the boiler can also be a disadvantage. In certain circumstances, corrosion takes place at the bottom of the boiler shell, mainly underneath the deposits.

Oil in the condensate and feed-water has harmful effects. Depending on their thickness and consistency, oil films on the heating surfaces, particularly those of the fire tube, can impede heat transfer and lead to overheating of the material. These films usually form together with other deposits from the water. To prevent damage, oil must be removed from either the exhaust steam from the machine or the condensate after the machine. In both cases, well-proven designs of oil-removal equipment are available. In the guidelines issued by VdTÜV, a limiting residual oil content of 2 to 3 mg/litre of boiler water is recommended.

In general, the higher the operational pressure, the higher the specific loading of the heating surfaces, and the higher the content of solid materials, hardeners and heavy metal compounds in the feed- and boiler-water are, the lower is the permissible oil content.

Oil Burners

With oil-burning installations, particular attention must be paid to burner equipment. Repeated smaller or larger detonations can usually be traced to faults in burner components (solenoid valves, chokes).
Unsatisfactory adjustment of the burner can result in damage to the fire tube by overheating. Such cases of damage occur if, for example, the burner flame is biased to one side or the burner muffle is locally damaged or if the flame cone is too steep with a muffle that is somewhat too short. To avoid such damage, regular maintenance by the service department of the burner manufacturer has often proved beneficial, in which case the burner muffle, too, must be checked and replaced at certain intervals, as the need arises.

Inhibition

Every boiler is subjected to the risk of corrosion during long lay-off periods. To avoid this risk, several different inhibition procedures are recommended, e.g., wet inhibition, keeping the boiler under pressure with extraneous steam, and dry inhibition—a method preferred for boilers with a large water space.

Bibliography to 4.1–4.4

[1] Richtlinien für die Speise- und Kesselwasserbehandlung bei Dampferzeugern bis 64 atü Betriebsdruck. Edition 1968. Techn. Überwachung 9 (1968), pp. 349/354.

[2] Richtlinien für die Speise- und Kesselwasserbehandlung bei Schnelldampferzeugern. Edition 1966. Techn. Überwachung 8 (1967), pp. 64/65.

[3] Kraus, W.: Schnelldampferzeuger, Einfluß der TRD, der Betriebsbewährung und der Wirtschaftlichkeit auf die Konstruktion. Techn. Mitt. June 1971.

[4] Heßler, R.: Schnelldampferzeuger. Energie No. 9 (1967).

[5] Börsig, F., Buchholz, E.: Schäden an Dampfgeneratoren durch mangelhafte Wasseraufbereitung. Der Maschinenschaden 34 (1961), No. 5/6, pp. 65/73.

[6] Steyrer, H.: Stand und Betriebserfahrungen mit der Automatisierung des Betriebes von kleinen und mittleren Dampferzeugern. 7. Works Manager Conference of Allianz, 1968.

[7] Zugelassene Kesselsteingegen- und -lösemittel. BWK 10 (May 1958) and continuously supplemented in Techn. Überwachung.

[8] VGB-Richtlinien für den Bau und die Bestellung von Heißdampfrohrleitungen und Speisewasserdruckleitungen. Published by Vereinigung der Großkesselbesitzer. 6th edition 1965.

[9] VGB-Richtlinien für die Herstellung und Bauüberwachung von Hochleistungsdampfkesseln. 7th edition. Published by Vereinigung der Großkesselbetreiber. Essen 1968.

[10] VGB-Richtlinien für die Bestellung von Hochleistungsdampfkesseln. 5th edition. Published by Vereinigung der Großkesselbesitzer. Essen 1965.

[11] SR-Öl: Sicherheitstechnische Richtlinien für Ölfeuerungen an Dampfkesseln. Berlin, Köln, Frankfurt (M): Beuth-Vertrieb 1969.

[12] SR-Gas: Sicherheitstechnische Richtlinien für Gasfeuerungen an Dampfkesseln. Berlin, Köln, Frankfurt (M): Beuth-Vertrieb 1967.

[13] SR-Kohlenstaub: Sicherheitstechnische Richtlinien für Kohlenstaubfeuerungen an Hochdruckdampfkesseln. Berlin, Köln, Frankfurt (M): Beuth-Vertrieb 1969.

[14] Pieper, B.: Beizgerechte Kesselkonstruktionen. VGB-Conference „Dampfkessel und Dampfkesselbetrieb" 1968, pp. 137/145.

[15] VGB-Merkblatt No. 11: Vorbereitende Maßnahmen für die Inbetriebnahme neuer Dampferzeuger. Published by Vereinigung der Großkesselbesitzer. 2nd edition. Essen 1964.

[16] VGB-Merkblatt No. 19: Nahstöchiometrischer Betrieb von Ölfeuerungen, Published by Vereinigung der Großkesselbetreiber. Essen 1969.

[17] Köhle, H.: Zur Frage der Festkörperalkalisierung bei Naturumlaufkesseln. Allianz-Berichte für Betriebstechnik und Schadenverhütung, No. 16 (1971), pp. 38/40.

[18] Franze, H.G., Rabe, H.: Großschaden an einem Steilrohrkessel durch Wassermangel. Der Maschinenschaden 44 (1971), No. 3, pp. 92/96.

4.5. Steam-Carrying Pipelines

In this section some points are discussed which can lead to difficulties with steam-carrying pipelines and their components (fittings, expansion joints, suspensions, etc.) in power stations or industrial plants.

Cases of damage recorded by Allianz over a number of years have been evaluated with regard to their causes. It was found that there were about

70% product faults and
30% operational faults.

Although this relation is subject to fluctuations because of the comparatively small number of damage cases involved (about 30 per year), it still shows the starting point for prevention of damage.

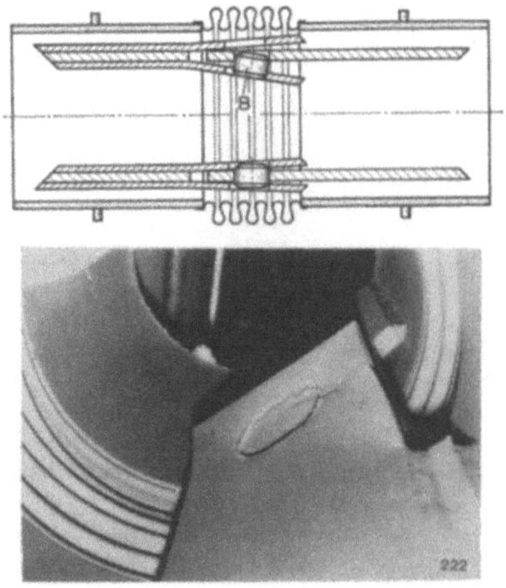

Figure 1. An internally tensioned expansion joint welded into a pipeline. In this joint, the articulation bolts (B) were made too short. In several expansion joints of this type, the tack welds have fractured and the bolts have slipped out of the tensioning straps.

4.5.1. Planning and Design of Pipelines

To establish operational safety, the aim should be the simplest possible layout restricted to only the essential fittings and branch-pieces. For an acceptable layout with adequate gradual fall and the shortest possible line length, precise knowledge of the load capacity of anchorage points and of the joints positions is essential.

For absorbing thermal expansions arising during operation, compensation loops or expansion joints of various designs are provided, and these must withstand the highest operational temperatures, as well as frequent load changes (temperature and pressure) for long periods. Particularly with bellows-type expansion joints, the choice of material and design must match these requirements. From an evaluation of 44 cases of damage to expansion joints [3], it emerges that, without adequate basic knowledge regarding calculation, materials and design, the availability and safety of an entire power-station installation are open to question if such an important pipeline component fails. Reference [3] concludes with the statement that, in particular, methods of calculation need radical improvement.

With other, internally tensioned articulated expansion joints, weak tack welds between the holding bolts and internal tensioner have fractured. As the bolts then slip out, the expansion joint becomes overloaded and leaks, perhaps damaging connected pipelines and their components (Figure 1).

Exact design data concerning pretensioning, suspensions, supports and slide locations, as well as gradual fall, are essential. Perfect drainage and de-aeration must always be ensured. Accumulations of material (at sharp changes in cross-section, flanges and fittings) can, under non-constant conditions, such as starting up and running down, as well as with load changes, lead to substantial temperature gradients and hence to large additional stresses which may cause crack formation.

Heating pipes, based on a bypass with a small internal cross-section compared with that of the main fitting, are appropriate. In this way, it is possible to heat up pipelines gently and protect main fittings from severe erosive attack by throttling.

The installation of safety valves must be carried out in such a way that faultless blow-off is guaranteed at all times. For this purpose, the control lines of pilot-operated valves must be as short

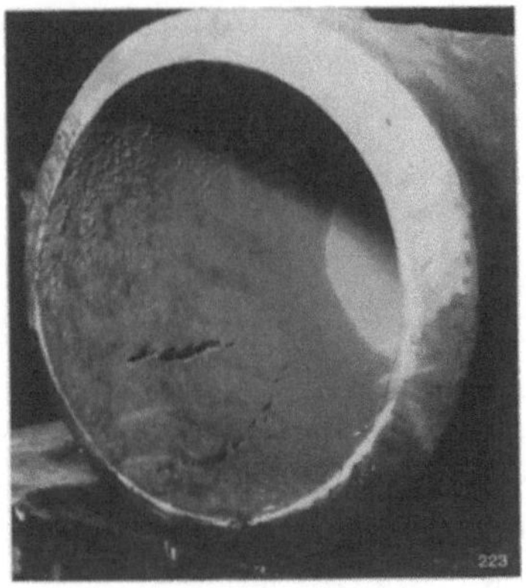

Figure 2. In this pipe bend from a high-pressure line (material 13 CrMo 44), the crack resulted from exceeding the long-time rupture strength after 105,000 hours in operation.

as possible and must have sufficient slope (> 15°)

Many installations have now reached or have even exceeded 100,000 hours in operation (Figure 2). The long-time rupture-strength values have had to be lowered twice in recent years. Since only strength values determined in actual service can provide a safe basis of calculation for the planning of future installations, it is appropriate to incorporate in the layout stage calibrated measuring sections in the pipelines. Measurements of diametral expansion should then be made at intervals of 3,000 hours in operation. In this way it is possible to amass a greater knowledge of the long-term rupture strength of materials.

For later (non-destructive) tests, it is very useful to have, apart from test certificates of pipeline materials, sample pieces for purposes of comparison.

4.5.2. Manufacture, Assembly and Commissioning

In accordance with present technology, it is possible to select rolled, drawn or welded pipes, the method of manufacture being determined by the envisaged application in an installation, dimensions and economic considerations. Insofar as current testing techniques are concerned, ultrasonic testing is recommended in all cases. In this way, differences in wall thickness and faults can be located before delivery. In cases of damage, it has been established frequently that the cause of damage has been a poor degree of purity of the material (slag inclusions, etc.). For example, Figure 3 shows a section from a hot steam line which had slag inclusions and had to be replaced.

When bending pipelines, care should be taken to ensure that the ovality engendered by the bending operation does not exceed permissible limits (< 5%); otherwise, it may be cause damage.

Before delivery, pipe sections must be checked for unobstructed cross-section and their ends must be covered.

When making *preparations for welding*, care should be taken that:

- internal diameters are equal;
- alignment is accurate and there are no radial displacements;
- welding edges are so prepared and their distance fixed that a perfect root seam is created;
- alignment and fixing of the components to be welded are effected in such a way as to ensure that tensile stresses cannot arise during welding.

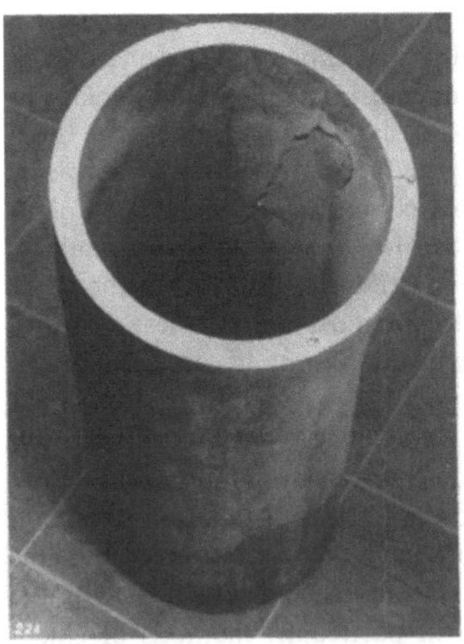

Figure 3. Part of a high-pressure line (material 15 Mo 3), showing splitting open and flaking of the internal wall. The cause of damage was impurity of the material.

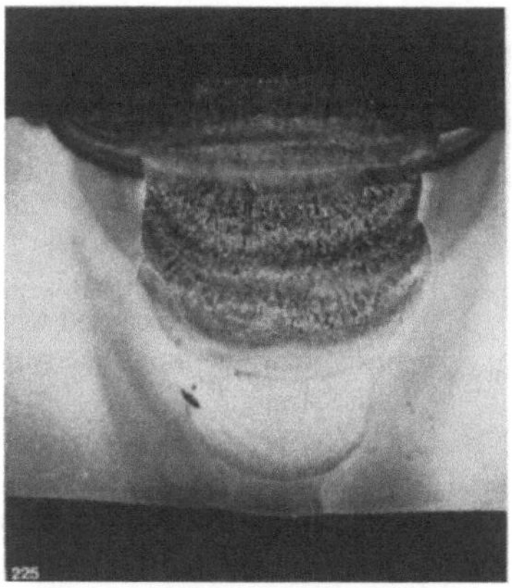

Figure 4. A high-pressure line made from 14 MoV 63. Strength tests of the base material of both pipe sections yielded perfectly satisfactory values. Even small tensile forces in the weld material caused separation without deformation, the structure of the material being coarse-grained.

Quite decisive for operational safety is controlled heat flow during welding itself and annealing in accordance with prescribed procedure after a seam has been completed. Figure 4 shows an etched section through a multi-pass circumferential weld, in the lower layer of which can be seen a fusion fault in the weld-metal/parent-metal interface. In the lower layers, the grain structure is fine, whereas grain size increases markedly towards the top layer, which was no longer influenced by the heat of subsequent welding passes. Coarse radial crystallization, with clear segregations at the crystal boundaries, is characteristic of the structure in this region. It also explains the lower notch impact and strength values of the weld material. During heating up of pipelines, all fixed anchorage points, slide locations, suspensions, etc. must be checked for satisfactory performance. These critical positions must be established accurately before commissioning and, above all, the initial values must be marked in the cold condition.

When purging pipelines, considerable reaction forces are generated. These forces are withstood by temporary but robust supports, the dimensioning of which must not be entrusted to assembly personnel. Instead, the firm responsible for the pipelines should be obliged, in the supply contract, to calculate and prescribe these auxiliary structures.

To check the steam purity attained by these purging procedures, it is possible to attach a copper plate, on which the presence of any foreign bodies will be manifested by dent marks, behind the exit cross-section.

Prevention of Damage by the User

After the pipeline has been completed, allowances on its measurements, as installed, must be recorded in conjunction with the manufacturer. At the same time, the positions of all welding seams must be established, so that, if damage is suspected, it will be possible to find the right location for stripping the insulation from welding seams made during assembly or from those made at the works of the manufacturer.

If at all possible, the personnel of the user should be present during assembly. In each instance, complete familiarity with the layout of the pipeline before starting operation is essential. Clear colour markings [1, 2] greatly facilitate matters and hence reduce the danger of faulty servicing.

When commissioning, slide locations, anchorage points and suspensions must be checked, in conjunction with the manufacturer, for satisfactory operation.

Thorough de-aeration, draining and gentle preheating when starting up are indispensable for the attainment of long operational periods. The maximum temperature gradients prescribed by the manufacturer must not be exceeded. In all circumstances, the entrainment of water or the

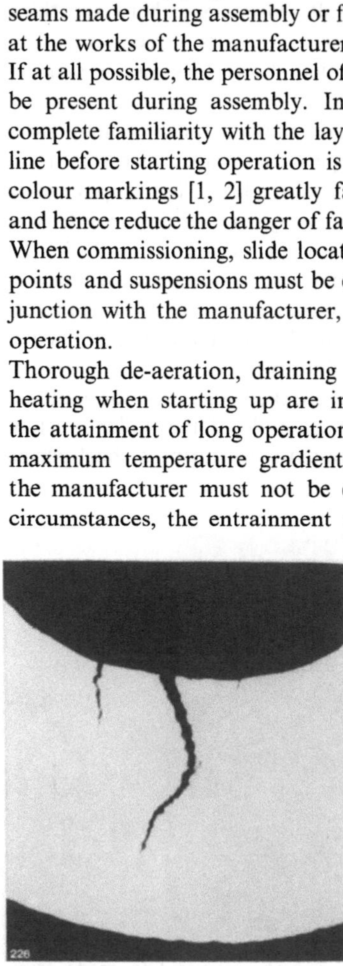

Figure 5. Sample from an intermediate-pressure line (material St 35.8). The cracks are filled with corrosion products.

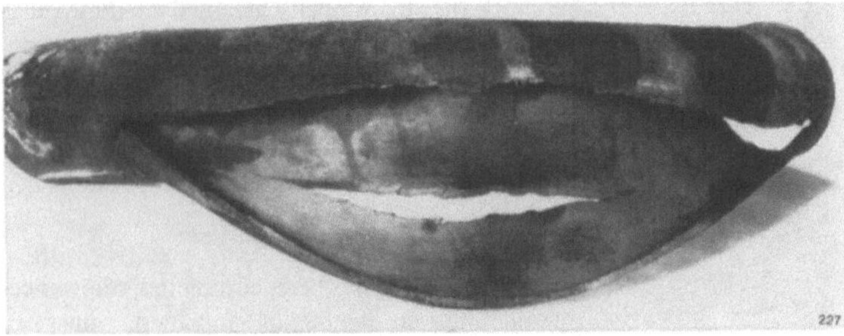

Figure 6. Part of a bypass line (material 15 Mo 3), showing damage caused by exceeding the long-time rupture strength.

drainage of condensate into vertically disposed pipelines must be prevented; if not, the formation of cracks by thermal shock or temperature fluctuations can result. Moreover, there are thick deposits of boiler-water salts present, cases of damage, such as that shown in Figure 5, can occur.

When taking an older plant out of commission, it repeatedly happens that branch-pieces with upward-pointing pipes are left. These pipes create a weak point in the system, in that condensate running down can cause cracks, usually in the region of widening, due to temperature fluctuations or thermal shock. In such cases, it is preferable to remove the branch-piece and replace it by a suitable straight piece.

Fittings must be checked regularly for faultless functioning. For example, if the boiler must be subjected to a pressure test,during a short period out of service, care should be taken that all stop valves ahead of the pipeline close tightly, so that a sudden temperature drop of the still hot pipeline is prevented.

Inspection and Revision

As mentioned earlier, many installations have now been in operation for more than 100,000 hours. A recent publication [4] points out the dangers which exist as a result of the falling-off of strength values, as well as of the ensuing embrittlement processes (see also Figures 6 and 7).

Following an earlier publication [6] summarizing experience with steel 14 MoV 63, difficulties have recently arisen again with pipelines made from this material. Exceeding the long-time rupture strength of weldments and additional stresses in the

Figure 7. Part of a high-pressure line made from 13 CrMo 44, in which the crack was produced by exceeding the long-time rupture strength after about 80,000 hours in operation.

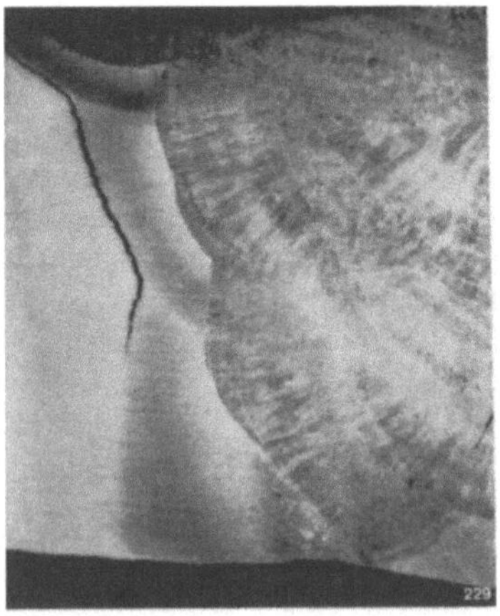

Figure 8. Part of a hot reheater pipeline made from 14 MoV 63. The crack follows the heat-affected zone.

pipeline system which are not amenable to exact calculation are primary causes of damage. Figure 8 shows a cross-section through an intercrystalline crack in the base material, the crack following the heataffected zone. Additional cracks appear at the grain boundaries in the matrix, particularly on the inner wall of the pipe. This aspect of the damage must be associated with exceeding the long-time rupture strength.

To protect human lives and also to maintain real values and increase the availability of the installation, proper checks should be carried out at reasonable intervals, especially if it is to be expected, on the basis of the past period of operation, that the limit of long-time rupture strength has been reached. This consideration always applies to installations which have already attained the 100,000-hours limit. Of currently available non-destructive testing techniques, the following can be applied:

Dye-penetrant testing.
Magnetic-particle testing.
Ultrasonic testing.
"Walk-around" assessment on site of component structure.
Internal inspection by rigid or flexible endoscopes.

With good planning (stripping of insulation at the proper time, erection of scaffolding if necessary), these methods can be applied without undue loss of time. By comparing the results it is often possible to discover damage of a similar nature early enough to prevent unforeseen outages.

Bibliography to 4.5

[1] Rimbach, K.: Kennzeichnung von Rohrleitungen nach dem Durchflußstoff. Der Maschinenschaden 42 (1969), No. 2, pp. 50/51.

[2] DIN 2403. Kennzeichnung von Rohrleitungen nach dem Durchflußstoff.

[3] Schoch, W., Kußmaul, K., Baumüller, F., Hänsel, G.: Erfahrungen mit einschichtigen, hochfesten ferritischen und austenitischen Bälgen von Rohrgelenkkompensatoren in 20-ata-Heißdampfleitungen. Der Maschinenschaden 43 (1970), No. 2, pp. 61/64.

[4] Burchardt, H.: Rohrbogen riß vorzeitig. Planen, Prüfen, Investieren (1971), p. 23/25.

[5] VGB-Richtlinien für den Bau und die Bestellung von Heißdampfrohrleitungen. Published by Vereinigung der Großkesselbetreiber.

[6] Erfahrungen mit dem Stahl 14 MoV 63. Report of VGB-Ausschuß für Materialfragen 1966.

5. Fluid-Flow Machines

5.1. Steam Turbines

Steam turbines, with or without gearing, are designed as generator drives at a fixed operating speed, and as pump and compressor drives for a permissible range of speeds.
Rationalization, automation and increased power capacity of units make it possible to discuss industrial and large turbines together in regard to damage analysis and recommendations for measures to be taken to prevent damage. Industrial turbo-generator sets approach the standard of large public electricity-supply sets not only with respect to their design, but also with respect to supervisory and protective equipment. If necessary, special features of the two designs of turbine will be pointed out. In references [1–8] in the Bibliography, the various standards for the two turbine types are grouped together.
Today, steam turbines have ratings ranging from a few kilowatts when used as an emergency drive for auxiliary oil pumps up to 1,300 MW as saturated-steam turbines in nuclear power stations, such a turbine being the most powerful prime mover. In the following a damage analysis of all the 1,700 steam turbines, with ratings up to 600 MW, insured by Allianz, will be reviewed.

Table 1. Distribution of cases of damage to steam turbines in accordance with their causes and locations.

Cause of damage	Distribution of the number of damage cases [%]	Location of damage	Distribution of the number of damage cases [%]
Product faults	*64.1*	Rotor blades	29.0
Faults in planning, design and calculation	16.5	*Bearings*	*16.7*
		Radial bearings (12.5)	
Assembly faults	16.0	Axial bearings (4.2)	
Technological shortcomings	10.6	Shaft seals, balance pistons	15.6
Manufacturing faults	8.7	Rotors with discs	10.3
Material faults	8.0	Casings with baseplates, screws	9.8
Repair faults	4.3	Strainers, fittings	4.0
		Control	4.0
Operational faults	*15.3*	Guide blades and diaphragms	3.4
Supervisory faults	10.6	Gears, transmissions	2.4
Maintenance faults	4.7	Pipelines	0.8
		Other parts	4.0
External influences	*20.6*		
Foreign bodies	7.2		
From the electrical grid	4.1		
Others	9.3		

Discussion of Damage Investigations

The large proportion of product faults indicated in Table 1 reflects the outstanding development of turbine construction during the past decade. The adoption of very high rotational speeds in the design of industrial turbines and of large powers, with the concurrent relative reduction in dimensions, makes heavy demands on mathematical skills, particularly in connection with the layout of the rotor blading from the point of view of vibrations.

Although a great deal of technological effort is put into the design and manufacture of rotor blades in view of the large number used in a steam turbine, they are still the most critical turbine components with regard to reliability and availability, as indicated by the comparatively high percentage in Table 1 of cases of damage to rotor blades.

Figure 1 illustrates the distribution of damage to rotor blades plotted against time in operation. Figure 2 shows the probability of failure of steam-turbine components in relation to time of operation, based on an analysis of the life of 84 large steam turbines of 100 to 370 MW from the insurance records of Allianz [9].

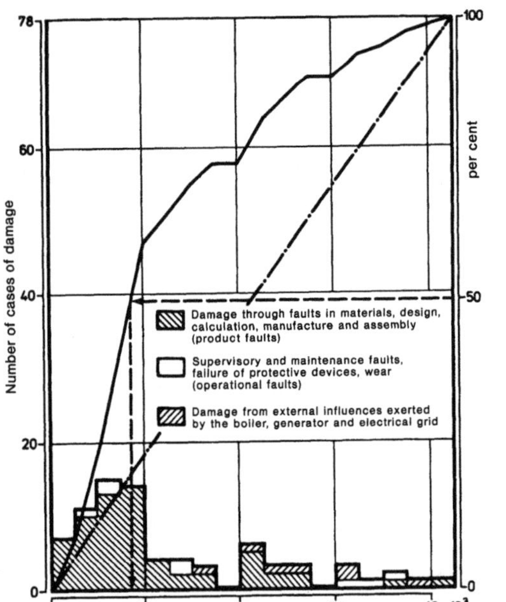

Figure 1. Frequency of damage to rotor-blade rows. About 50% of all damage to these rows occurred within 8,750 hours in operation.

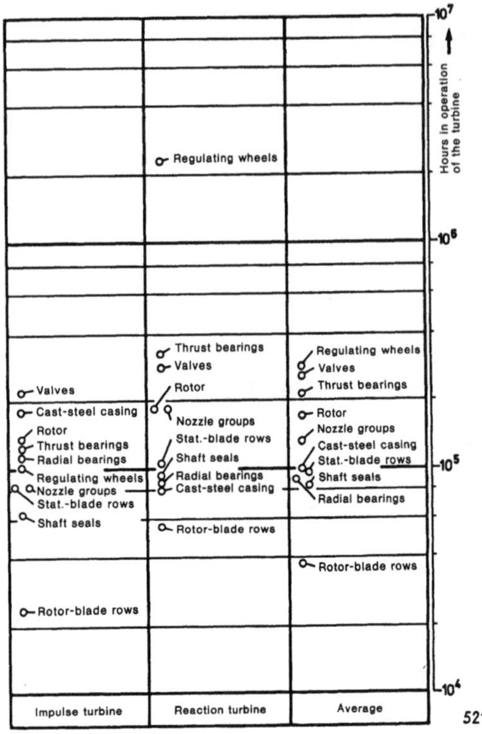

Figure 2. Probability of failure of components examined in relation to the number of hours in operation of impulse and reaction types of steam turbines.

Because of the outstanding importance of damage to rotor blades in connection with the reliability of steam turbines—irrespective of their power, speed and type—all cases of damage to rotor blades which occurred over a period of three years to steam turbines insured in respect of machines, assembly, and warranty were examined in detail [10].

The results of this examination are shown in Figures 3–6. As indicated in Figure 4, the cost of repairing damage to rotor blades amounts to from 40 to 50% of all costs of damage to steam turbines, the reason being the extensive disassembly and reassembly work required.

The second highest percentage of cases of damage in Table 1 occurred to bearings, which, because of the heavy damage that can ensue as a result of fouling by bridging of the radial and axial clearances, can involve a large expenditure in terms of repair work and stoppages [11].

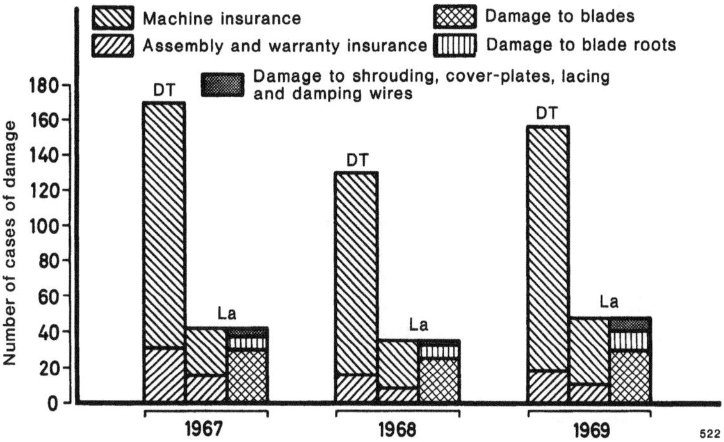

Figure 3. Number of cases of damage to steam turbines and rotor blades. DT = total steam-turbine damage, including rotor-blade damage; La = rotor-blade damage.

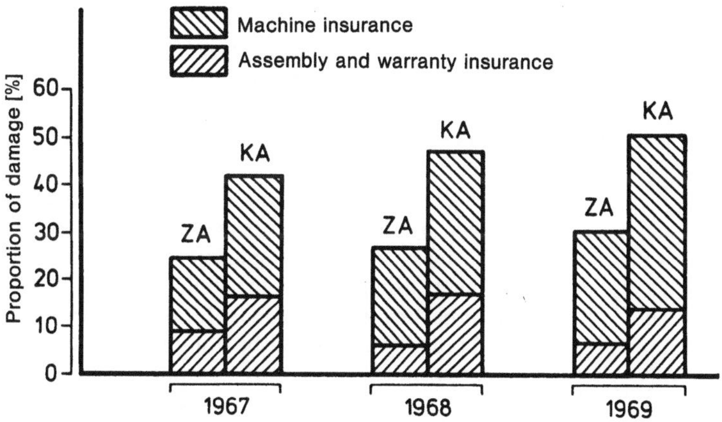

Figure 4. Proportion of rotor-blade damage, in terms of numbers and cost, in relation to all cases of damage to steam turbines. ZA = proportion by number; KA = proportion by cost.

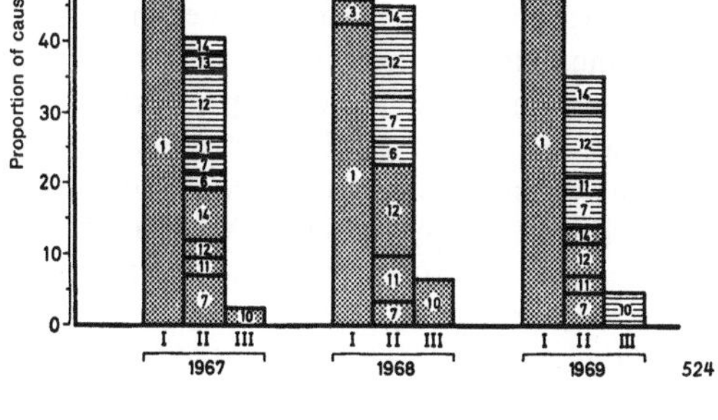

Figure 5. Numerical proportion of individual causes of rotor-blade damage.
Product faults, main group I: 1 Faults in planning, design and manufacture; 2 Faults in materials; 3 Repair faults; 4 Assembly faults.
Operational faults, main group II: 6 Supervisory faults; Maintenance faults; 11 Wear damage suffered during operation; 12 Corrosion fatigue cracking and stress corrosion cracking; 13 Corrosion while out of operation; 14 Failure of control and supervisory installations.
External influences, main group III: 10 Foreign bodies.

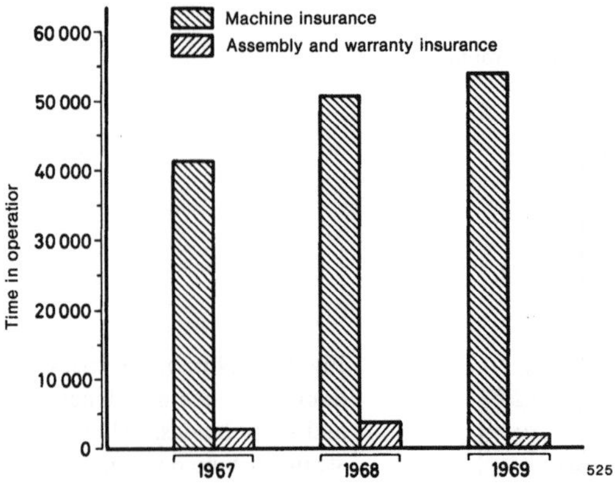

Figure 6. Average number of hours in operation before the inception or recognition of damage to rotor blades.

Table 2. Survey of conditions accompanying the inception of damage to steam turbines.

Attendant conditions	At [%] of all cases of damage
Rough running	16.2
Mechanical overload	15.8
Expansion due to temperature changes	15.5
Jamming	10.4
Wear, corrosion, ageing, erosion, cavitation	8.9
Loosening	8.9
Thermal shock	5.5
Inadequate lubrication	5.2
Water hammer	4.9
Thermal overload	4.0
Failure of control and protective equipment	1.5
Deposits, corrosion while not in operation	1.2
Excessive speed	1.0
Hot ventilation, reverse running	1.0

Table 2 surveys a number of conditions attendant on the inception or recognition of damage to steam turbines.

Supervision of running condition is also the best guarantee of discovering

- damage to rotor blades,
- damage by fouling and
- rotor distortion,

i.e., all damage influencing the balance of rotating parts, before major damage can occur [7.1, 7.2].

Table 3. Principal forms of damage to steam turbines.

Nature of damage	At [%] of all cases of damage
Rubbing	23.0
Fatigue and creep failure	18.5
Damage to bearings	14.6
Thermal stress cracking	11.7
Sudden failure	9.3
Incipient cracks	8.0
Mechanical surface damage	5.4
Corrosion and erosion	3.3
Shaft bending	2.4
Wear	2.3
Abrasion	1.5

For the supervision and protection of steam turbines, the proper procedure is to rely on vibration measurements of bearing pedestals, as well as shafts.

Table 3 indicates the chief types of damage to steam turbines.

Radial and axial bridging of clearances can be caused by thermal distortion of casings and rotors, by changes in alignment or displacement of inner casings, by rough running through deterioration of inertial balance, by damage to axial and radial bearings and by non-permissible elongations under non-steady operating conditions. This type of damage results in the highest losses and the longest stoppages (Figures 7 and 8).

Rotor-blade damage can be caused by dynamic overloading (Figure 9), i.e., by incorrect layout from the vibrational point of view or by additional forces that may occur as impact when a seating becomes loose. While it is possible to trace the cause of blade damage unambiguously in the regions of superheated steam by examining the fracture surface under a scanning electron microscope [10], it is difficult today still to establish, in the region of wet steam, whether the design of the blade or the chloride content of the steam is responsible for the damage incurred. The experience of Allianz with damage, as well as operational experience with proved blade designs in chloride-containing wet steam, recommend the reinforcement of rotor-blade rows susceptable to damage and a better layout from the point of view of vibrations.

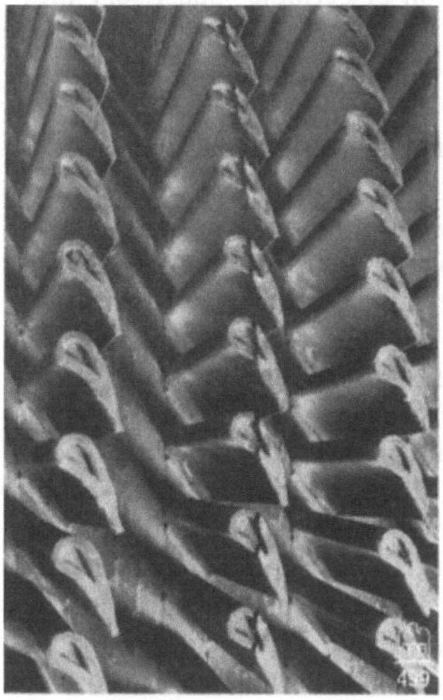

Figure 7. Radial rubbing of rotor blading caused by excessive fluctuations in steam temperature.

Figure 8. Axial rubbing of the blading of a regulating wheel.

Damage to bearings is a third type of damage that must be considered (see Section 8.3 on bearings in this handbook). This damage is caused by faults in the bonding of bearing metals, by incorrect installation of the bearings, by self-induced vibrations of the shaft in the bearings or clearance excitation of the rotor, and by an inadequate supply of lubricant during non-steady operating conditions, particularly when changing over from one pump to another.

Thermal stress cracking nowadays occurs primarily in cast-steel components in the region of high-

Figure 9. Fatigue fracture at the root of a blade in a regulating wheel as a result of dynamic overloading.

est temperatures and where there are changes in wall thickness. Thermal stress cracking of rotors is a rare occurrence. Components affected include the housings of fittings, particularly emergency stop and control valve casings, steam admission chests, and the wheel-chamber section of high-pressure casings.

Based on this analysis of damage cases, it should be possible to give some directives for the prevention of damage and attendant losses.

5.1.1. Directives for the Prevention of Damage during the Planning Stage

It is recommended that planning should be carried out in accordance with the relevant standards (see [1–5], particularly [5.6] in the Bibliography), taking account of the installation as a whole, i.e., steam generation, condensation, the driven machine, and such especially important items as bearings, oil supply, and gearing (see Sections 8.3, 8.5, 8.6 in this handbook).

5.1.1.1. Determination of the Limits of the Delivery Specification

For the clear determination of responsibility, it is necessary to define unambiguously the limits of the supply contract, laying down the permissible changes in pressure and temperature of live and reheat steam.

A list of drawings for planning the construction, assembly, and later repairs must be agreed upon by contract.

Dimensioned drawings facilitate the manufacture of replacement parts and reconditioning work during inspections and repairs, also occasionally in outside workshops. This is of special importance in the case of export to foreign countries.

Assembly drawings and working instructions for the first erection and for later disassembly and reassembly must contain the most important weights and dimensions, with indications of the lifting tackle required and its points of application, so that outside or operating personnel can carry out disassembly work safely in an emergency. Forces, moments and prestress values must be stated on all pipe connections.

Foundation drawings must contain all recesses and must give an indication of weights. The grouting material for embedding the baseplate and foundation must be considered. Only approved grouting materials, which neither shrink nor swell, even under humid conditions, must be selected.

An occupation plan of the machine-house floor indicates the load-bearing capacity for depositing disassembled heavy components, and includes the plan views of heavy components that must be set down during repair work.

Schematic drawings with a maintenance schedule for control, supervisory, protective, shaft-sealing and oil-supply systems are of assistance in the training of operating personnel and should make possible optimum maintenance, fault-finding and fault-elimination during operation.

The machine handbook, apart from a description of the steam-turbine installation, contains all important individual instructions for operation and maintenance, determined during the commissioning period. It must be completed by the supplier of the steam turbine before take-over by the operator.

The inspection book contains all records of measurements after assembly, inspections, revisions, and the evaluation of periodic supervisory measurements.

5.1.1.2. Determination of Measuring Points

The installation, in accordance with standards, of measuring points for pressure, temperature and flow is necessary to prevent faulty measurements at a later date and disagreements between the manufacturer and operator. At flow-measuring points with orifices or nozzles, it is recommended that an inclined endoscope socket with a bore of about 15 mm be fitted behind the throttle position. In this way, it will be possible to check, not only before and after the acceptance tests, but also when unexplained changes in flow quantitiy occur during operation, the mechanical condition of the installation without the need for dismantling.

5.1.1.3. Conceptual Layout of the Machine

The conceptual layout of the machine depends on the demands of future operation. In general, steam-turbine installations should be laid out in such a way that they are suitable for operation at part and peak loads since, in the course of a long operating life, a change-over from base load to part load or peak operation is inevitable. This state of affairs cannot be influenced by the user but is rather brought about by developments in energy demands. Special peak-load units are also known which, in the present energy situation, practically operate at base load, i.e., they attain a large number of running hours per year, thereby underlining the difficulties of a correct prognosis years head of the mode of operation of steam turbines. Steam turbine installations can be designed to be suitable also for peak loads with little extra expenditure; as such, they give greater assurance of damage prevention to casings and fittings than would be the case with installations which do not provide for these operational requirements and then have to run under non-permissible conditions.

While the steam generator is often incapable of fulfilling the demands of the steam turbine in the case of hot and warm starting, difficulties arise in the turbine itself in the rare instances of cold starting. Today, the following possibilities exist for cold, warm and hot starting, based on "block" and "bus-bar" installations.

Block Installations for Cold Starting

The requirement here is for delivery, with the control valves fully open, of live steam and reheat steam to the high-pressure and intermediate-pressure cylinders at a temperature from about 200 °C at the lowest steam pressures, either via a throttled emergency stop valve or a separate start-up valve. This requirement can be met, even at a later time, by suitable layout of the cross-over stations of the high-pressure (HP), intermediate-pressure (IP), and low-pressure (LP) cylinders, or by start-up stations. The turbine is heated by large quantities of steam at the lowest pressure from the beginning of its delivery from the boiler to all cylinders up to the attainment of nominal speed, and it is operated in parallel with the regulating cross-over stations. With this cold-starting method, a modified vacuum protection system is recommended for supervising the temperatures in the LP section since, with a lower vacuum and a temperature at the condenser inlet up to 120 °C, the most favourable conditions exist for the IP section. With an increase in the amounts of cross-over steam, determined by boiler-firing and tube-wall temperatures, the vacuum is gradually increased to the specified value. There is no objection to supplying the turbine with saturated steam when operating with the barring gear and subsequently up to 1,000 rpm.

The complete drainage system, particularly in the pipelines, must be liberally dimensioned for this cold-starting method, and the drains must be led separately to the condenser. The live- and reheat-steam pipes are heated together with the turbine by the flow of steam through them.

Block Installations for Warm and Hot Starting

The temperature of the live steam, independently of that of the reheat steam, must be matched to the metal temperatures of the HP and IP cylinders by separate regulation in such a way as to ensure that the live and reheat steam entering the fittings and casings is from 30 to 50 °C hotter than the supervised walls. The steam turbine, with its high concentration of costly components, notably in regard to the HP and IP casings with their stop and control valves and HP and IP rotors, deserves to be treated with as much care as the banks of tubes of the reheater of the steam generator.

The different, usually faster, cooling of live-steam and reheat-steam pipelines cannot in most cases be adjusted successfully by improved insulation. It is recommended that, according to the disposition of the cross-over stations in the installation as a whole, these long pipelines should, by suitable dimensioning of the drainage system and by simple heat bypasses, be capable of being heated, independently of the turbine, during warm and hot starts by steam of the appropriate condition direct from the boiler. By quickly increasing the speed in from 2 to 5 minutes up to the synchronizing speed of the generator and by rapid load application, heating at the HP outlet to above normal operating temperature can be avoided.

Bus-Bar Installations for Cold and Warm Starting

A start-up station with water injection is recommended to permit warming up the live-steam line, the inlet fittings and the HP section of the turbine, with the control valves fully open and using throttled stop valves for heating or a starting valve, which by-passes the stop valves, at speeds up to the nominal speed. A heating procedure at a moderate vacuum, as described above, is recommended.

Back-pressure, pass-out back-pressure, and pass-out condensing turbines can be warmed up, with a saving in time, from the back-pressure or pass-out line, in which case rotor distortion must be guarded against if no barring gear is installed. A more favourable procedure is simultaneous heating with live steam and full admission, i.e., with a throttled emergency stop valve or start-up valve.

If these requirements are fulfilled in a steam-turbine installation and if, by suitable supervisory equipment for wall temperatures, indications are given to operating personnel of any approach to dangerous limiting values, every turbine set can be employed for peak-load operation, provided that the following criteria are satisfied or at least are largely taken care of in the design of the set.

- The emergency stop valves must be capable of operating as throttle valves for short periods of time. If these valves are of the open/closed type, a start-up valve, included in the trip-gear system, must bypass them up to nominal speed and, better still, until part load is applied.
- The control valves must be capable of being fully open, so as to make it possible to operate the turbine with full admission and throttle control during the heating-up phase.
- The usual pressure-dependent vacuum protection must be implemented by temperature-dependent vacuum protection, so as to exclude the possibility of hot ventilation in the LP section.
- All drainage lines from different pressure regions of a steam-turbine installation, including pipelines, must be led separately to the condenser.
- The axial clearances at shaft seals and blade rows should be large enough to avoid the waiting times that might otherwise become necessary because of difficulties arising from differences in expansion under unfavourable speed and part-load conditions.
- Separate chests for emergency stop and control valves, situated at the side or below floor level and connected to the turbine casing by flexible pipelines, have smaller wall thicknesses and are of simpler construction. They therefore reduce thermal stresses resulting from changes in temperature during start-up, run-down and load changes in walls and between walls of different thickness and different average temperature.
- Because of the stepping of pressure and temperatures, the use of inserted casing components, such as for live- and reheat-steam flow, nozzle chests, inner casings and fixed-blade carriers, leads to smaller wall thicknesses and reduced temperature differences in walls and between different parts of the casings.
- Full admission and the consequent operation under an even pressure gradient with known modifications [15–17] reduce thermal stresses and hence prolong the safe operating lives of turbine casings.
- The required load, usually the nominal load, should, according to [13], be reached with live-steam and reheat temperatures from 450 to 480 °C, since temperature changes in the upper temperature region must take place particularly slowly, so as to avoid non-permissible thermal stresses and hence a reduction in the lives of casings and rotors.

If these starting aids are present in a steam-turbine installation, no difficulties will also be experienced when carefully running down the turbine, if operation with the barring gear should be shortened for repair or maintenance work. It is possible, by slowly reducing live steam under load, followed by the through-flow of cooling steam during run-down and operation with the barring gear, to take steam turbines of even the largest output out of operation for dismantling in about half a day or one day [18, 19]. This cooling operation must be supervised carefully because of possible difficulties arising from relative expansions. All drainage points must be kept open to allow water entrained in the cooling steam to flow away immediately.

After giving these recommendations for making steam turbines suitable for peak-load operation, which also apply to existing installations, suggestions will now be given in connection with the design of important steam-turbine components.

5.1.2. Guidelines for the Prevention of Damage during Design and Manufacture

5.1.2.1. The Steam Strainer

The Strainer Element

Today steam strainers usually form an integral part of the emergency stop valve. Damage through fracture by vibration can occur if the strainer design is too weak (Figure 10). The flow velocity of the steam, together with the pressure differential across the strainer element, i.e., the energy transferred from the flowing steam to the strainer element, must not be increased suddenly. Strongly damped strainer designs and the well-proven perforated type of element are preferable. Foreign bodies can clog the openings of strainer elements or close them up by hammer-

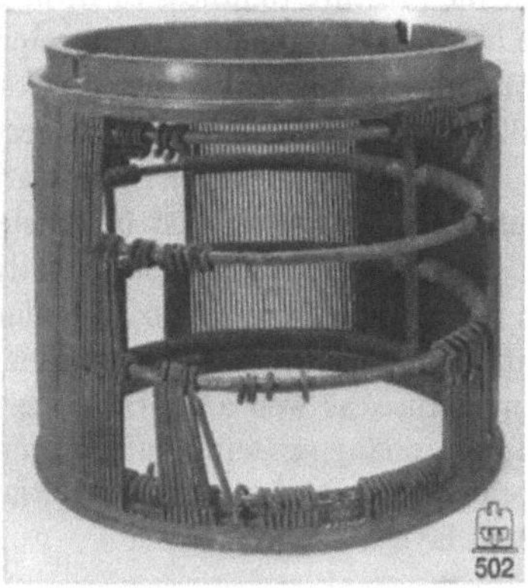

Figure 10. Damage to a steam strainer as a result of embrittlement and vibration of the strainer bars.

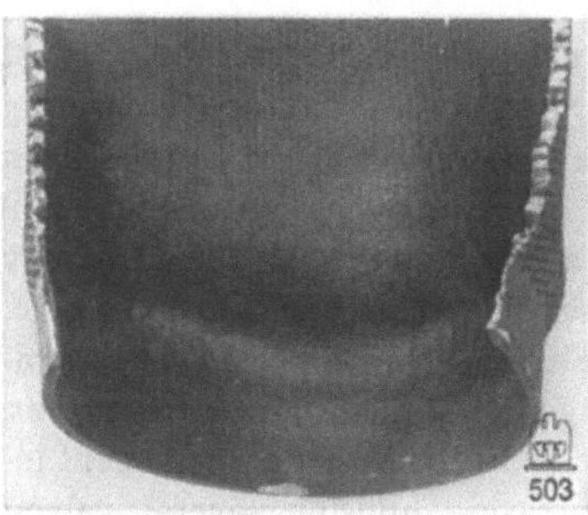

Figure 11a: Internal view.

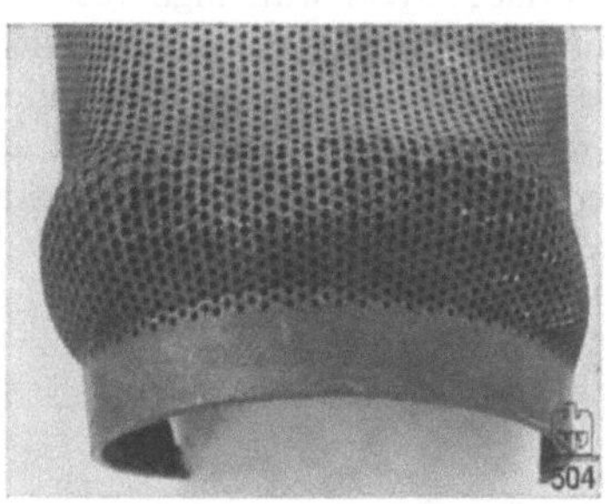

Figure 11b: External view.

Figure 11a and b. Bulging of the element of a steam strainer as a result of hammering.

ing (Figure 11). The pressure difference at positions ahead of and behind the strainer often reveals such damage too late or not at all, because foreign bodies can at the same time also cause bursting of large areas of the strainer element. For this reason, visual inspection, without the need for dismantling, should be made possible by the provision of an endoscope socket [12].

The Strainer Casing

The design of the strainer casing should be such that foreign bodies, after impacting the strainer, can fall back into a dead space, at least during shut-downs, to prevent continuous hammering of the element. With vertically positioned strainers and upward flow, this can easily be arranged by a length of dead pipe. In this way it is also possible to ensure that, when the strainer and emergency stop valve are opened up, parts will be prevented from falling into the live-steam line and, on starting up, from being flung against the strainer.

5.1.2.2. Emergency Stop Valves, Control Valves for Live, Reheat, Cross-over and Pass-out Steam

The following indicates a number of faults and damage that must be prevented by proper design and materials selection [20]:

Damage by jamming and seizure of valve spindles and spindle guides (Figure 12) reduces the safety of shut-down operations.

Since *nitrided valve spindles* can be made only from low-alloy materials, such as 10 CrMo 9 10 or 30 CrMo V 9, their use has declined in recent years, because of their poor surface durability, particularly when interacting with valve packings.

Hard-chromium-plated spindles, made, for instance, from X 20 CrMo 12 1, have also been unsuccessful because of unavoidable micro-cracks in the hard chromium layers.

Today, *stellited spindles* are recommended for the largest dimensions or spindles made completely from stellite for smaller dimensions. The base material, X 20 CrMo V 12 1 or 10 CrMo 9 10, is first covered with a nickel interlayer, onto which is then plasma-sprayed a stellite layer on, for example, a cobalt base. Spindles with a drawn-on interchangeable stellite sleeve have also been employed. The guidance of such spindles with a stellite surface in stellite bushes or stellite rings has been quite successful in achieving maintenance-free spindle sealing.

Figure 12. Traces of seizure in the cone guide of a governor valve.

Jamming of spindles occurs by plastic deformation of valve spindles through one-sided direction of flow, through inclined support of the back seal in the fully open position, or through inclined support of the valve seat in the closed position. By creating thermal symmetry in the configuration of the casing, particularly in the region of the valve seat and the spindle seal, casing distortions are reduced. Maintenance-free ring seals afford a better fit of the valve seat, which is subject to opening and especially large closing forces, than does spindle packing. By concave shaping of the seating surfaces in the casing and convex grinding at the valve cone, bending moments can be kept away from the valve spindle or their intensity can be reduced. This latter procedure is most likely to be successful, since deformations of the casing are inevitable in the long run.

Axial or radial instability of control-valve spindles results in hammering of the spindles and complete seizure of the valve and in breakage of the spindle, with consequent severe damage to the valve seats and sometimes even to blading. Such damage can best be prevented in design by overdimensioned rigid valve spindles, to which the valve cones, with only small balancing cross-sections, have been firmly attached. Loose valve cones, combined with balanced valves, have in the past created more difficulties than compact, rigid, permanently balanced valves.

Cracks and break-outs of the wear-resistant layer on valve seats in casings and on valve cones adversely influence the pressure-tightness of fittings and can cause damage to blading by foreign bodies. Predominant are thermal stress cracks, which can usually be discovered in time by inspection before large break-outs can occur. The seating surfaces must either be on replaceable bushings or, if in the casing, must be thermoelastically separated from the casing as a whole by annular grooves, so that, during heating up and welding, i.e., during manufacture, as well as during operation under rapidly changing temperatures, no cracks can develop between the casing material and wear-resistant layer.

Cracks on the Inside of External Walls and in Internal Separating Walls of the Casings of Fittings

Cracks in external walls endanger the safety of power stations. Cracks in internal separating walls can reduce the strength of the fitting casing as a whole and create a bypass from the steam spaces ahead of and behind the valve seat, thereby endangering safe shut-off of the steam turbine. Individual valve arrangements, possibly a combination of emergency stop and control valves, should be preferred to the earlier arrangement involving a large casing housing several valves, Smaller, simpler steel castings or forgings of smaller wall thickness can satisfactorily be machined on all sides, so that surface layers with high residual stresses or casting flaws can be eliminated. Such casings must be provided with one or two endoscope sockets to make it possible to reveal all faults mentioned above by visually checking critical locations without dismantling (Figure 13).

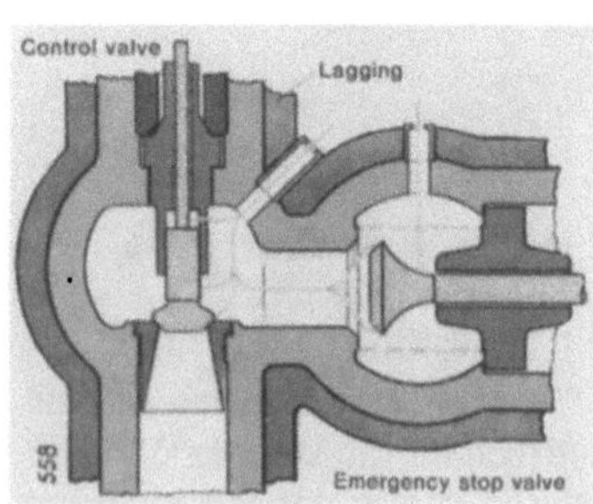

Figure 13. Endoscopic inspection of valve seats, steam strainers and internal walls of fittings. According to [12].

From the design point of view, all fittings should be classified in terms of
- the steam-carrying working section,
- the hydraulic power section, and
- the hydraulic, electrical or electronic control section.

These three groups of fittings should be coupled by quick-acting interlocks, so that, during short shut-down periods under barring-gear operation, the power and control sections can be replaced, and the working section can be subjected to a visual and manual check of their operation. In this way, overhaul of the power and impulse sections, including functional tests, can be carried out under optimum conditions in the manufacturer's special workshop equipped with checking instruments, or in the user's central workshop, without being pressed for time and independent of shut-down of the set.

5.1.2.3. The Control and Protection System [5.3]

Faults in the control and protection system lead to unintentional shut-downs and difficulties in regulation and, on rare occasions, to particularly dangerous damage through overspeeding.
Today industrial turbines should also satisfy certain minimum requirements which are standard with large turbines of modern design and which are to a large extent responsible for the decline in the number of cases of damage caused by overspeeding. These requirements are as follows:
All governor and emergency stop valves should open against steam pressure and in hydraulic systems against spring action with increasing hydraulic pressure. All these valves should close with shut-down of steam pressure and through spring action with decreasing hydraulic pressure.
The speed governor, irrespective of its design, should limit the transient speed increase at full load to a maximum of 108.5% and the permanent overspeed to 107% of the nominal speed. Special speed governing for saturated-steam turbines for nuclear power stations must be agreed by contract between manufacturer and operator. An increase of the usual tripping speed of 110% to 115% is possible.
The overspeed protection must act, independently of the governor system, through two individually operating speed sensors via two separate impulse lines, on the emergency stop and governor valves. In the case of an emergency stop, the governor valves should not remain open and certainly not completely open.
With reheat installations, the overspeed protection should not only trigger the stop valves and governor valves ahead of the HP and IP sections without delay, but should also effect the following operations:
- Closure of all pass-out quick-acting valves.
- Closure of shut-off or non-return valves in the cold section of the reheat line.
- Opening of the HP transfer station.
- Opening of the IP/LP cross-over station.

If these starting and shut-down stations, which are normally found in German power stations, are missing, the following governing operations must be carried out:
- Opening of the HP blow-off control valve.
- Opening of the IP blow-off control valve.

By quick regulation of live and reheat steam it is possible to prevent the damage which might result from exceeding the speed corresponding to the ultimate limit of loading, i.e., about 125% of the nominal speed, even with faulty closure of the live-steam and reheat-steam stop valves of the turbine set.
The two speed controllers should, independently of each other, be checked for correct triggering, smoothness of operation and impulse initiation. When doing so, the initiation and transmission of impulses should be tested right through to the emergency stop valve, if necessary by closing one after the other of these valves under low-load operation. This test programme should include establishing the safe cut-off of important stop valves of economizers and water-trapping devices in the LP section, particularly in saturated steam turbine installations.

Prevention of the Entry of Water into the Steam Turbine from Economizers
([28] with Extensive References)

The pass-out quick-acting valves should be designed as self-acting non-return types. The closing capacity of these valves, particularly with high-pressure economizers, is not satisfactory because

of the possibility of water hammer resulting from a sudden drop in temperature. It is therefore recommended, depending on the response of the non-return valve and/or the water-level controller in the economizer, to cut off the water inlet and outlet of the economizer concerned by quick-acting valves, to open the feed-water bypass and to cut off the economizer on the steam side by a gate valve.
In many existing installations, it is not possible to bypass individual economizers because the appropriate valving has not been installed. In the interests of safety, it is desirable in such cases, especially with costly and important installations, to retrofit these valves as far as is possible from the design layout. In new installations, the valve should be provided from the outset.

Prevention of the Entry of Water into the Steam Turbine from the Cold Side of the Reheat Line

In the past, the entry of water from this source has led to the most extensive damage by radial and axial rubbing in the HP section of reheat turbines. The reason is that, if the automatic water-injection equipment fails or if water injection from the boiler side at the entrance to the reheater is incorrectly adjusted by manual control, large quantities of water will usually enter the hot HP cylinder via the cold branch of the reheat line during warm or hot starts or during run-down of the boiler. The use of stop valves which close automatically in the case of reverse flow or no flow is therefore recommended, as well as interlocking of the water injection, the stop valve, and the water-drainage devices in the cold branch of the reheat line with the HP governor valves, as follows:

- Water injection ahead of the reheater open:
  Drainage devices in the cold branch of the reheat line open.
  Stop valve in the closed branch of the reheat line closed.
- Stop valve closed:
  Drainage devices in the cold branch of the reheat line open.
- High-pressure governor valve closed:
  Stop valve closed.
  Drainage devices in the cold branch of the reheat line open.

Safeguarding HP Sections in Reheat Installations and Back-Pressure Steam Turbines in Industrial Plants against Excessive Pressure in the Casings

In reheat installations, it is permissible to provide the HP section of steam boilers with safety valves that do not blow off to atmosphere but into the reheater. As with back-pressure installations, the danger exists of a non-permissible pressure rise in the HP casings [32]—in the extreme case, attaining live-steam pressure.
In reheat steam turbines, a pressure rise above the maximum pressure in the cold branch of the reheat line should be prevented by proper location of the safety valves in the steam generator. The IP/LP cross-over station should be so arranged that, even with a non-permissible rise in the reheater exit pressure, it should open without fail as a controlled safety valve, the steam being led to the condenser or, in the event of failing water injection, to atmosphere.
With back-pressure steam turbines in industrial plants, a safety valve with a 100% blow-off capacity must be provided as a protection against unduly high steam outlet pressure.
The entire control and protection system should be tested in advance, adjusted, provided with test certificates and coupled to the turbine on site.
The hydraulic system should be so designed (see Section 8.6.2.2. on Prevention and Extinction of Oil Fires in this handbook) that the HP hydraulic system can be separated from the oil supply by a fire protection valve in the case of a fire, and when operating with the barring gear, so that overhaul of the control system can be carried out without loss of time.

5.1.2.4. Rotors, Rotor Discs

In these components, the predominating causes of damage are distortion, crack formation at notches, and thermal stresses following axial and radial bridging of clearances. If insufficient pressure is exerted by the shrink fits, fretting corrosion sets in under shrunk-on rotor discs, shaft sealing rings, couplings and thrust plates, as well as wearing away of the shaft material and crack formation originating from seizure. Some typical cases of damage of this kind are shown in Fig-

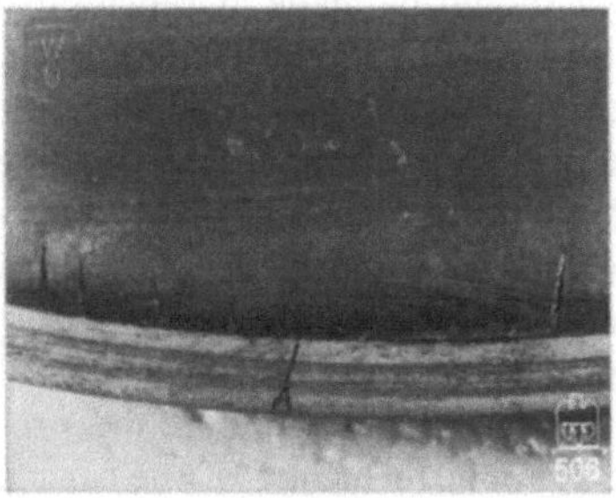

Figure 14. Shaft fracture area A is the origin of fracture at the root of the saw-tooth thread for the shaft nut.

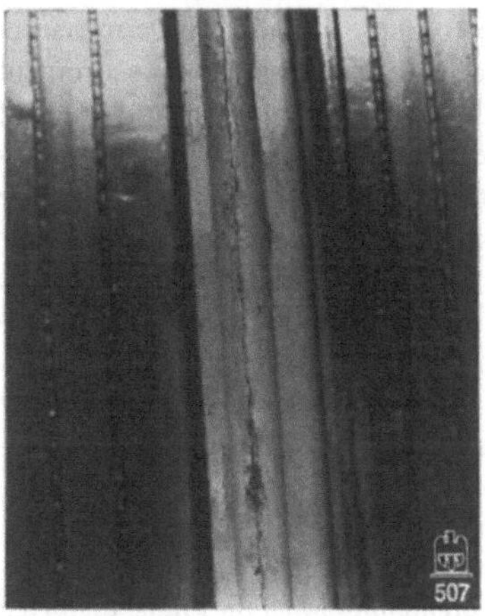

Figure 15. Circumferential crack in the blade root of a steam-turbine rotor.

ures 14–18. Foreign bodies at bearing locations can damage the shaft by scoring. Residues of soiling from dust and oil can cause deep grooves in the rotor body in the region of passing through the bearing pedestals.

To avoid *residual stresses from forging and heat treatment,* forgings should be ordered in accordance with reference [5.8]. Longitudinal, tangential and, in some cases, radial trepanned testpieces should be available. A concentricity test under hot running conditions at 100 °C above the operating temperature should be registered on a chart recorder and agreed upon by contract [21].

A second stress-relieved run should be carried out by the turbine manufacturer on the hot-testing bed after turning the rotor down to an allowance of 1 mm for the final cut. Deviations in

Figure 16. Shaft breakage in the region of the thrust bearing, originating from a turning-tool groove at the thrust-bearing plate/shaft transition.

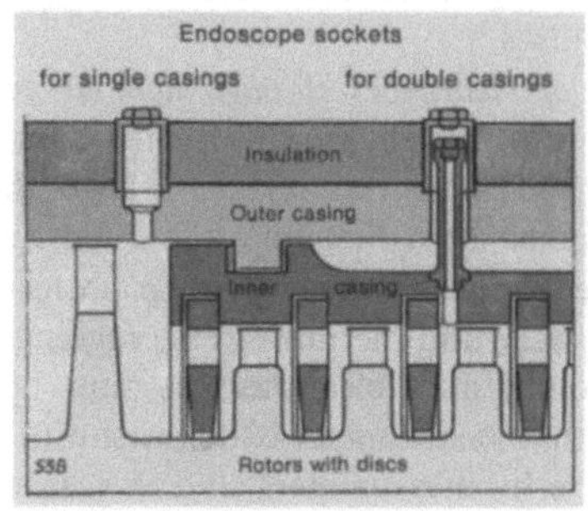

Figure 17. Disposition of endoscope sockets for the visual inspection of rotor and guide blades for checking radial and axial clearances while the casing is closed. According to [12].

Figure 18. Effect of rotor distortion on a diaphragm.

concentricity of the rotor under the influence of temperature should be less than 0.01 mm per metre of rotor length.

Axial bridging of clearances by exceeding relative expansions has become increasingly rare. The reason is partly due to improved methods of calculation, larger clearances, better matching of the masses of the rotor and casing, and sprayed casing insulation. However, the principal reason is an improved mode of operation during starting up and running down with reduced live and reheat steam temperatures and reliable supervision by means of contactless measuring instruments which indicate limiting values by giving an alarm and, if these limiting values are exceeded, trigger switching-off pulses.

Axial bridging of clearances through damage to thrust bearings is a frequent cause of rotor damage in reheat and industrial turbines. In both cases, axial contact must first be established by massive replaceable parts which serve to protect the rotor from wear and thermal-stress cracking. When dimensioning axial clearances, it must be borne in mind that axial contact is the least difficult to recognize by measurement and is easiest to remedy if it takes place in the region of the shaft seals. When deciding on the size of blading and shaft-seal clearances, a greater importance must be attached to the robustness of the steam turbine than to considerations of efficiency, which often cannot be verified. Subsequent to damage by axial contact and hence enforced increase in clearances, acceptance tests have established that changes in efficiency can often be within the measuring error of the test. This finding applies particularly to clearance increases between the rotor-blade outlet and guide-blade inlet. The clearance between rotor discs and diaphragms must be especially generous, since rotors of this design are particularly endangered by axial contact.

Radial bridging of clearances by distortion of the rotor, by uneven cooling while stationary, by water hammer from a pass-out or extraction point or from the cold branch of the reheat line, by casing distortion or displacement, by displacement or distortion of the inner casing, and by unevenly drying sprayed-on lagging are now the most fruitful starting points for applying loss-prevention measures. The following measures are recommended:

The first contact should be made at replaceable parts or parts that can subsequently be repaired, e.g., between rotor-blade tips and casing. Wear edges on shrouding, cover-plates or free-standing blades fulfil their purpose only if wear immediately causes an increase in clearance. Such initial bridging of clearance is rarely indicated by supervisory equipment if the rotor is not yet distorted. An endoscope socket should therefore be provided vertically at about the centre of the casing, thereby permitting, during short non-operational periods, a visual examination to be made to determine whether there are any signs of grazing which would indicate incipient danger to the rotor and casing. Using an inside micrometer, it is then possible to make a comparative measurement with respect to the original measurement from the outside in a simple manner and without the need for dismantling. These measurements check the position of the rotor in relation to the inner casing, of the rotor in relation to the outer casing, and of the inner casing in relation to the outer casing. The reference points for this radial-clearance control should be marked in such a manner on the endoscope sockets for the inner and outer casings that ambiguities in the interpretation of measurements are eliminated at the outset [25].

Taking account of critical speed, rotors must permit *repeated straightening when they have distorted* and *turning down to establish a new centre-line through the bearings.* This safety margin should be provided intentionally in the design of new rotors. It is recommended that minimum dimensions at the bearings, at the positions of the seals and of the rotor discs should be stated, so that, during overhaul, not all the considerations that have gone into the original design need to be repeated, often by different designers pressed for time.

Shaft sealing rings and shaft bushings must have freedom to expand on the shaft, since they attain a mean temperature different from that of the shaft. At the bearings and at locations of possible friction from a temporary shortage of lubricant or fault-induced clearance bridging, highly alloyed shaft materials must be protected by welding or spraying on materials which create less friction and have better sliding properties in the event of an emergency. Replacement bearings for the shaft must be stored in a premachined

Figure 19. Radial contact of the sheet-metal covering of a high-pressure shaft seal.

condition, i.e., left with a margin for finish-machining, because the diameter of the shaft at the bearings must be reduced after damage by distortion or scoring, and bearings finish-machined to the original shaft diameter would then no longer fit.

Keys, feather keys, radial bolts and taper pins for securing shrunk-on discs or sleeves to the rotor must be applied with rotational symmetry and must be fixed radially in such a way that, under the influence of non-steady thermal processes and of centrifugal force, there can be no distortion of the rotor, no shifting of discs or sleeves, and generally no deterioration of the state of balance.

Shaft nuts must, as a matter of principle, be mechanically secured positively, preferably by means of an axially or radially applied caulked set-screw. This locking arrangement must not make removal of the nut more difficult, and the thread on the rotor itself must not be damaged. Shaft threads must be cut so that the direction of rotation of the shaft opposes loosening of the nut.

Sheet-metal coverings on the rotor have the great advantage of easy reconditioning, which can now also be effected by machining. These covers are the best means of preventing severe rotor distortion in the event of radial clearance bridging. Sheet thicknesses of up to 0.4 mm have proved their value and, by placing sheets together at different heights, an adequate height of sheeting can be achieved. In this way, sufficient safety against break-outs through vibration is provided, without losing the advantage of instant increase in clearance in the event of damage by radial contact (Figure 19). The sheet material should not be too heat-resistant, e.g., a 13% chromium steel, and should in no circumstances be an austenitic steel. The sheet-metal covering should break or wear away in a brittle manner at the high temperature prevailing during radial contact, instead of producing wear-resistant smearing and large heat generation. This advantage is nullified with austenitic sheet.

On *each rotor component* a reference point should be clearly marked which should always be used by the manufacturer and user for checking concentric running, hot-running concentricity and balance. All phase reference points of the coupled-up rotor components should lie in the same longitudinal plane, thereby affording optimum possibilities for rebalancing in operation on the basis of rotor components individually balanced by the manufacturer and their balancing record giving angles and weights.

The balancing planes used by the manufacturer should be accessible for balancing on site without requiring a great deal of disassembly work, so that cooling down of the rotating parts is unnecessary before being able to attach balance weights. Experience has shown that this recommendation is fulfilled if one or more balancing planes are made accessible during run-down of the machine, i.e., in from 30 to 60 min, the rotor being positioned exactly by barring gear or manually, and the balance weights of each plane attached securely in about 5 min. In the intervals between weight attachment in different planes, the rotor can be turned over a few times by barring gear to ensure prevention of sagging or upward distortion. Closing up of the balancing planes can then be effected while the barring gear is in operation. If the steam turbine operates above the critical speed, three balancing planes per rotor component have proved useful. To be able also to use the middle balancing plane for balancing on site, it is recommended that the endoscope socket, which is in any case necessary for internal inspection and clearance checking, should be positioned above this balancing plane. For balancing, radial tapped holes on the circumference of the rotor, with uniform indication relative to the phase reference point of the rotor components and the complete rotor, should be provided. Prefabricated balance weights, made from materials of different density and adjustable simply by altering their effective length, facilitate quick balancing. Special tools, which grip the weights securely as they are being installed, must always be available to exclude the possibility of damage to the machine by foreign bodies. After the last test run to establish successful balancing, the weights must be secured finally by caulking. Since the balancing of large sets on site is, despite all precautions, indispensable because part of the blading can only be installed at the user's plant, the expense of providing facilities for proper balancing of the rotating assembly on site is certainly worthwhile.

The Position of Critical Speeds [1–3]

"Stiff" or subcritically running single rotors are today preferred in industrial turbines. The first critical speed in bending should be 5% above the highest possible transient speed and 20% above the highest permissible continuous speed.

"Soft" or overcritically running single rotors are now used almost exclusively in large steam turbines:

The first critical speed in bending should be removed from the lowest operating speed by at least 10 to 20%. However, if the first critical speed lies within the region of 40 to 70% of the permissible continuous speed, difficulties with self-induced vibrations must be reckoned with.

During recent years, problems in connection with quiet operation, with very large shaft vibration amplitudes at frequencies of 20 to 30 Hz, have been experienced on various occasions with steam turbines running at 3,000 rpm. With steam turbines running at 3,600 rpm, self-induced vibrations can attain frequencies up to 40 Hz.

Instability of the bearings is speed-dependent and occurs in the region of 70 to 110% of the nominal speed, whereas clearance is load-dependent in the region of 70 to 110% of nominal load [22–24]. Experience has shown that these difficulties can best be avoided by designing the rotor for a first critical speed in bending between 70 and 80% of the operating speed.

Removal of Rotors

The design must cater for clearance dimensions sufficient to permit the rotor to be removed from the casing without dismantling neighbouring turbine components. The screwed fasteners of the coupling must be capable of being removed and replaced with the rotor in its normal position in the bearing pedestals. Intermediate or centring discs in rigid couplings facilitate overhaul of turbine sections, partial revisions and readjustment of axial play of the different turbine components.

5.1.2.5. Turbine Casings with Fasteners, Baseplates, Bearing Pedestals

Castings should be ordered according to [5.8]. Cast-on test-pieces should be situated at various representative locations of the cast steel components. Repair welds and stress-annealing should be carried out at the foundry. The acceptance certificate, with values of materials properties, should also be given to the user. Today, damage by deformation and cracking in the hot region of cast steel casings predominates. The modern design trend in the last five years towards simple casing forms, without large differences in wall thickness and hence in rigidity, has clearly proved a remedy for such damage by deformation and cracking.

For loss prevention and speeding up disassembly work, recommendations include:

Slender joint flanges with the joint fasteners as close as possible to the inner wall. Support and packing flanges with leakage steam control of the space relieved by milling. Studs screwed into the flanges must not be used, but long through-bolts screwed together with extension sleeves. Threads should have increased flank play and unloading of the first three most highly loaded threads. The best method of assembling the joint fasteners must already be decided upon during the design stage [5.7] (see Section 5.1.3, Table 4). The use of hydraulic tightening devices is recommended. Low-pressure casings of welded construction deform measurably under the influence of a vacuum. The design should provide for clear separation of shaft seals and inner casing from the exhaust-steam hoods. The shaft seals can be flanged onto the lower parts of the bearing pedestals and elastically connected to the exhaust-steam hoods via compensators. Care should be taken to ensure that the compensators can be inspected for crack formation without extensive disassembly work. Split compensators are recommended because otherwise their replacement would be possible only after opening up the low-pressure section and removing the low-pressure rotor. The inner casings can be supported in the rigid central plane of the LP sections by carrying arms, which do not deform appreciably, on the

Figure 20. Broken-off disc of a steam-turbine rotor as a result of axial fouling of a low-pressure rotor caused by casing deformation.

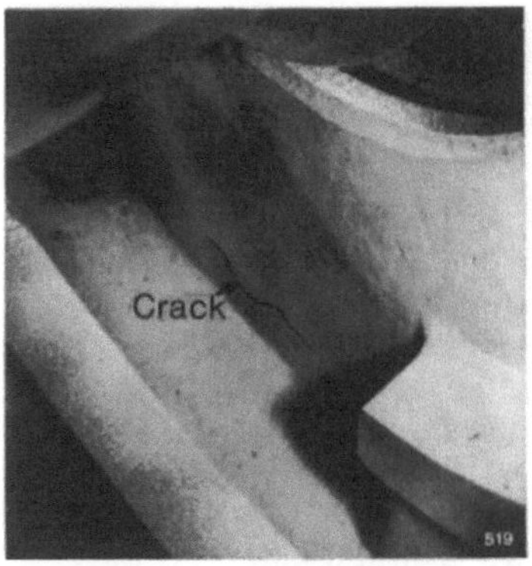

Figure 21. Crack in a bearing pedestal, caused by deformation of the grey cast-iron component through impeded sliding, as the sliding surfaces had not been lubricated.

foundation. Illustrated in Figure 20 is an example of severe rubbing damage, with loss of one rotor disc, through axial bridging of clearance as a result of deformation of the low-pressure casing.

Forces caused by thermal expansion should be transmitted from the casings along the central axis of the bearing pedestals or adjacent casings, so that only compressive and tensile forces, and no bending moments, can arise. Only force transmission without play via key connections in the casing supports or the joints of the thrust-rods will yield reliable measurements of expansion and dependable control of clearances during operation. Optimum pairing of the materials of highly loaded and thermally affected sliding surfaces, together with a reliable central lubrication system using high-temperature greases containing solid additives, should make it possible to convert the effects of expansions into casing displacements, without incurring non-permissible compressive and tensile stresses. Casing deformations, clearance bridging and cracks in foundations and bearing pedestals are the result of incorrect casing design (Figure 21) and the formation of rust during operation on the sliding surfaces between the pedestals and baseplates.

The baseplates must be securely and permanently attached to the foundation by anchoring under controlled pretension and using a grouting material free from swelling and shrinkage. Grouting materials sometimes contain iron filings, whose rust formation and hence increase in volume

Figure 22. Severe under-surface corrosion of the baseplate of a steam turbine.

counteracts the shrinkage of cement. Operational experience has shown that, through condensation or moisture as a result of leakage steam, rusting of the iron filings progresses during operation, so that the consequent swelling of the grouting material causes arching of the baseplates and hence deformation of the bearing pedestals, thereby constituting a serious threat to the condition of alignment of the steam turbine. Such under-surface corrosion (Figure 22) can also occur with too much water in the grouting compound. It is therefore recommended that baseplates should be designed in such a way as to avoid, with certainty, the formation of water bubbles underneath them during grouting. Since it is usually not possible, even when sealing adjoining surfaces, to exclude moisture during operation, only those grouting materials which have proved their worth in service should be employed for steam-turbine installations.

The surface of the baseplate should be machined in the direction of sliding of the bearing pedestal. These pedestals ahould be able to be lifted from the baseplates without having to dismantle the appropriate turbine casings, so as to be in a position to check the key guideways and sliding surfaces for proper functioning or to rectify any impediments to sliding. It is recommended that the longitudinal key should be made as long as possible and that it should be placed between the baseplate and bearing pedestal in such a way that it can be removed without having to dismantle the pedestal and turbine casing. Lubrication of the lateral faces of the longitudinal key is more important than lubrication of the sliding surfaces of the baseplate and bearing pedestal because of the high surface pressure caused by asymmetrical expansion of the casing.

Bearing pedestals should be so designed that their covers can be removed quickly for inspection of the bearings without extensive disassembly of measuring leads. All electrical and hydraulic connections should therefore be located in the side walls or, with free-standing bearing pedestals, in the front wall. Of proven value is a so-called "measuring insert", which accepts all pick-ups and their wiring and which can be removed separately without disturbing the wiring. In this way, recalibration of pick-ups after bearing inspection, etc., with its likelihood of errors, is largely avoided.

After removal of their covers, bearing pedestals should permit unimpeded alignment of rotors and casings. Couplings in bearing pedestals should operate in a separate chamber, so as to reduce ventilation heat and oil misting. For reasons of stiffness, the base of bearing pedestals is strongly ribbed. This ribbing should be so designed, in connection with oil drainage, that foreign particles can settle in an oil sump, even if it is shallow. Visual inspection of the sump during short stoppages offers the best guarantee for early detection of damage to bearings.

5.1.2.6. Radial and Axial Bearings with Oil Supply

In this section, consideration will be given only to specific problems of steam-turbine bearings. Radial and axial bearings are located in bearing pedestals on spherical supports to allow adjustment of the bearings to the bend line of the shaft. Each bearing is designed for vertical and horizontal adjustment to ensure perfect alignment, even with sub-assemblies.

In the case of bearings with shell inserts, three-material designs are to be preferred because of their superior running properties under emergency conditions. Even if a steam turbine is shut

Figure 23. Shell of a radial bearing with the thin white-metal layer partly broken away right down to the steel shell and partly melted away.

down in good time because of damage to a radial bearing, a period of up to 45 minutes will elapse before the rotor comes to a standstill. During this running-down stage, the bearing, which has already been damaged, will be even more highly stressed among other factors, because the disturbed running conditions when passing through the critical-speed region and the disturbed lubrication conditions engendered by the damage. In earlier days, the normal white-metal layer of from 1.5–5 mm in thickness had a running-down reserve and exhibited a remarkably small sinking effect on the shaft, despite often extensive damage, the reason being that, as damage was incurred, bearing play was immediately increased, with a consequent substantial increase in oil flow, which cooled the already softened white-metal and thereby exerted a self-healing effect on the white-metal bearing, stopping further aggravation of damage. With white-metal bearings of modern design, having a cast white-metal layer of 0.2 to 0.5 mm in thickness, the white-metal melts out as soon as damage occurs (Figure 23). The increase in play is, however, too small to augment appreciably the flow of oil, and the shaft can suffer severe damage by running in the steel shell of the bearing. The provision of an emergency bronze running layer prevents such severe results following damage to modern thin-layer white-metal bearings.

This consideration is even more important when using tilting-pad bearings on large turbines and when the present specific bearing loading of 25 kg/cm^2 is increased in the future up to 50 kg/cm^2 (Figure 24).

The problem of introducing shaft-jacking oil into radial bearings has been satisfactorily solved in recent years by improved design and manufacture. An important consideration is pressure relief of this critical location, to ensure that, in the event of leakages between the high-pressure oil nipple and the cast white-metal layer, leakage oil cannot build up pressure between the bearing shell and the white-metal layer. If there is no provision for drainage towards the outside, such leakage-oil flows lift the cast metal layer in the event of low lubricating-oil pressures, disturb the distribution of oil in the bearing and are therefore the cause of severe damage to the bearing. The shaft-jacking oil recess should be as shallow as possible, with gradual transitions, and should in no circumstances span the entire width of the

Figure 24. Because of dry running for a short time while switching on the emergency oil supply, the thin white-metal layer of a steam-turbine tilting-pad radial bearing was severely damaged. The lack of a bronze layer for emergency operation increased the extent of the ensuing damage.

bearing, because the hydrodynamic operation of the bearing is adversely affected by deep, sharply defined and wide shaft-jacking oil recesses. Moreover, at the transitions of such large shaft-jacking oil recesses, the bearing geometry can be damaged by cavitation. The pressure in the shaft-jacking oil recess can be measured as a function of bearing loading and the variation in bearing loading during operation by means of damped pressure gauges, even after the shaft-jacking oil pump has been switched off. This measurement is the more meaningful, the nearer the pressure in the shaft-jacking oil recess approaches the maximum hydrodynamic oil-film pressure. With tilting-pad radial bearings, resilient pressing-on of a statically unloaded tilting segment is strongly recommended because of the very small top clearance.

With steam turbines, the axial bearing is the rotor datum point. Michell and Kingsbury bearings predominate over simple thrust-ring types with milled-in wedge surfaces. With larger dimensions and more flexible plummer-block designs, fabricated instead of cast, the resilient support of the tilting segments in Michell bearings assumes greater importance than hitherto. The most frequent cause of damage is uneven loading through insufficient adaptation of the axial bearing to the thrust collar, i.e., oblique seating. For this reason, pressurized-oil lubrication of the spherical seating surfaces of axial bearings is strongly recommended. Rigidly built-in axial bearings without resilient tilting segments exert bending forces on the rotor. These forces can still lead to fatigue failures of turbine rotors, even after long periods in operation, particularly if they have been scored or grooved either through contamination of oil by foreign particles or while carrying out repair work on the rotor during revision (Figure 16). Every axial bearing must be so designed that adjustment of the rotor with respect to the casing and adjustment of play between tilting segments and thrust collar is possible without major remachining of the components.

5.1.2.7. Moving and Stationary Blading

As indicated earlier, the most critical component of a turbine has always been and still is the moving blade. After all, only the manufacturer can improve the quality of this component by superior conception, refined methods of calculation, precision production and trial on a test-bed. Moving-blade rows with partial admission, after discontinuities, before and after pass-out and extraction points, and in the transition region where steam becomes wet (Wilson zone) are the main contributors to blade damage. Moving- and stationary-blade rows should, as a matter of principle, be designed with the greatest possible damping. An opportune measure is lower stressing than in other turbine moving- and stationary-blade rows, and hence greater safety against unknown disturbing influences that are difficult to determine by calculation. Blades with approximately equal stressing in the critical root section and in the most highly loaded cross-section are preferable to the type with substantially higher stressing at the root than along its length. From many cases of damage to moving blades, we have compiled the stress values in the fracture cross-sections, i.e., in the most heavily loaded section of the root or along the blade length, and have compared them with the natural frequency of vibration.

Because of the variety of shapes of the blades and their roots and the location of the blades in the steam-expansion region of the turbine, all we would like to derive from this comparison are the following generally valid indications for loss prevention: The more unequal are the stress conditions in the fracture cross-section (e.g., between trailing and leading edges), the higher is the stressing in the root cross-section, as compared with a section along the length of the blade; also, the higher are the absolute stress values that have been selected, particularly in regard to static bending stresses by steam forces as a parameter for dynamically generated stresses, the more careful must be the attention given to the design, from the point of view of vibrations, of each individual blade in a moving-blade row. It is recommended that new blading offered by various manufacturers should be evaluated in accordance with these criteria.

To transfer weak points to the blade length proper, which can at all times be checked, the stresses in the blade root should be substantially lower than the highest stresses along the blade length, taking into account all influences such as locally high surface pressure and notch effects.

Only with this principle of blading design is it possible, with the present state of testing technology, to discover cracks with a degree of certainty during control checks and to exclude ensuing damage to the rotor or diaphragm through rotor groove break-outs.

In the regulating stage, it is preferable to install blade twins or triplets, welded together or milled out from the solid, or welded blade rows. In the HP and IP sections and in the first blade rows of the LP section of modern industrial turbines and all large steam turbines up to the highest outputs, moving blades with contiguous cover-plates or shrouding bands predominate. As in the past, blade roots are still designed with single or multiple inverted-T roots and fir-tree roots, and with inversion of both these designs as "straddle" roots, particularly in the U.S.A. Moreover, fork-root blading is commonly used from the regulating stage to the last stage. As the highest-quality solution, side-entry fir-tree and fork roots are employed for extreme stresses in the regulating and last stages. For large turbines in Germany, the longest blades (usually the last three blade rows in the LP section) are preferably of free-standing design without damping elements. In the U.S.A., the preferred design for large turbines and for industrial turbines, particularly those operating at variable speed, utilizes lacing wires for damping, including in the last stages, or even sometimes welded or riveted cover-plate elements.

Industrial turbines operating at variable speed are now almost exclusively manufactured with tied and hence strongly damped blading, the main exception being those turbines where the blading is generated electro-chemically from the rotor itself.

For loss prevention, it is strongly recommended that new blade designs should be thoroughly checked for vibrational behaviour at full speed—a proviso that involves test-bed operation or actual running tests with telemetric transmission of blade movements. Such measurements can also be carried out in a centrifugal test-pit under a vacuum, when vibration excitation can be simulated by magnets. The vibration pattern, entered into a *Campbell* diagram, should be supplied to the user. These calculated values must be corroborated by measurements. The scatter band, caused by manufacturing and installation tolerances, must also be entered, once the above-mentioned tests have been carried out. Today, certain minimum requirements for the vibration pattern of moving and stationary blades apply [14]. Generally, the following should be taken into account:

- The natural frequency and first harmonic in the direction of minimum and maximum moment of inertia and the torsional vibrations of individual blades and blade packets.
- A 10% difference from the exciting frequencies due to operational speed up to about the 5th harmonic and, in the last two or three LP blade rows, up to about the 8th harmonic:
  - by way of the moving or stationary blade rows ahead or behind,
  - the nozzle groups, particularly for tuning of the blading of the regulating wheel,
  - the gearing in transmissions,
  - symmetrically disposed locations of disturbances on the periphery ahead of and behind moving and stationary blades, such as flow guides, pass-out slots, steam inlets, etc.

Before commissioning the set, the manufacturer should establish, by a series of tests, that these requirements have been met. With blading of the LP section having a natural frequency from 50 to 250 Hz, it is absolutely essential, with turbines running at 1,500 or 1,800 rpm, to tune the moving blades individually, before installing them in the rotor, in a clamping fixture in which they are purposely held loosely, to ensure that the frequency scatter band of the moving blades resulting from manufacturing and installation conditions does not cause individual moving blades to be near resonance. In some cases, the 10% requirement cannot always be realized because of the small frequency band available. For industrial turbines operating at variable speed, certain speed ranges which endanger individual moving-blade rows by resonance must be prohibited for continuous operation, i.e., operation under load.

In impulse-type turbines, the design of the guide blades is similar to that of the moving blades. Of almost equal dimensions, these blades are not stressed by centrifugal forces and are therefore much more rarely subject to damage. The guide blades of turbines with a small degree of reaction

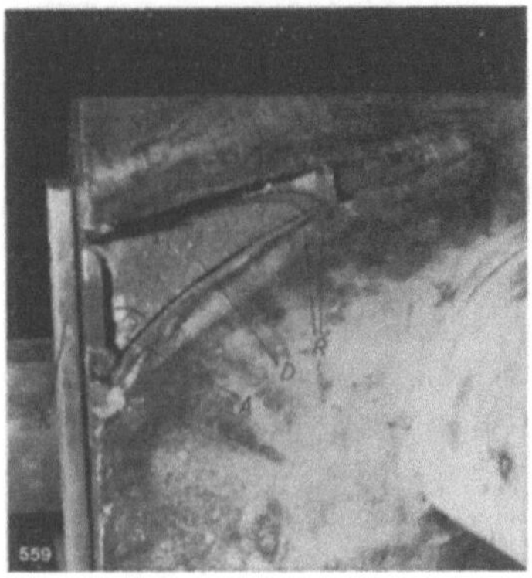

Figure 25. Fatigue failure, in the vicinity of the parting line, of a guide blade cast into a cast-steel diaphragm of a 250-MW steam turbine.

are built into the diaphragms or are welded or cast in. Welded designs have in the past created difficulties by plastic deformation after long periods in operation. When dimensioning the clearances of a set and carrying out strength calculations of diaphragms, care must be taken to ensure without fail that clearance bridging, even under conditions of temperature-induced relaxations, cannot occur during the normal operating life of up to 25,000 hours. Under-dimensioned diaphragms require time-consuming and very difficult repairs since, under the given space conditions available, stiffer designs of diaphragm can usually be fitted only at considerable expense. Cast diaphragms are mainly applied in the LP and IP sections. Grey cast-iron diaphragms and cast-steel diaphragms (widely used in the U.S.A.) both have their weakest points in the parting line through cutting in of the guide blades (Figure 25). It is recommended that guide blades should not cut into the parting line and that the whole diaphragm assembly should be offset to ensure avoidance of these critical cross-sections.

5.1.3. Loss Prevention during Assembly and Commissioning

Of very great importance are accurately drawn up measurement records containing all adjustments relative to safety in operation, together with indications of the measuring points and the method of measurement. The following measurement records should be compiled by the fitter, and their correctness confirmed in writing by the commissioning engineer or specialized engineer of the user before the plant is started up:

- Amount of play between radial and axial bearings and rotor (see Section 8.3. Bearings).
- Data on adjustment of oil supply (see 8.6. Oil Supply).
- Alignment of all couplings (see 8.4. Couplings).
- Radial and axial clearances between rotor, casing, moving and guide blades.
- Radial (and axial) clearances at all entry points of bearing pedestals and casings, shaft seals.
- Radial and axial clearances at endoscope sockets for subsequent inspection of the radial and axial positions of the rotor and inner casing in relation to the outer casing.
- Record of all balancing arrangements in all balancing planes of all rotors, with indications of weights and their positions.
- Record of elongations of all joint, live-steam flange, and coupling flange bolts.
- Values of settings for alarm and switch-off impulses for all supervisory and protective equipment.
- Values of settings for the control and safety system of the high-pressure oil or control-fluid circuit, particularly those for actuation of the auxiliary and emergency oil pumps.

The assembly of joint and coupling screw fasteners should here be examined in detail: Studs with waisted shanks and their nuts must be allocated to the appropriate holes in the flanges. During assembly, the effective elongation at room temperature as a measure of screw tension must be entered into the record. An indication of the torsional angle, of the expansion of the screw immediately after heating and tightening the nut, or of the hydraulically generated amount of elongation is not sufficient, because no account is taken of the plastic compression of the flanges in the region of nut support and of possible additional stresses resulting from the longitudinal stiffness of the flanges. This requirement can be catered for only if the flange fasteners, designed as stud-bolts with waisted shanks, afford the possibility, without reservation, of measuring their

Table 4. This record is applicable to all studs with waisted shanks, which permit measurements of length after tensioning. In [5.7], further information can be found for screws in the high-temperature region.

Location of installation	Condition on delevery	*Assembly*			*Disassembly*			Non-destructive tests					
Number of screw	Unstressed length [mm]	Length under stress [mm]	Elastic extension [mm]	Elastic strain [%]	Length after stress removal [mm]	Residual extension [mm]	Residual strain [%]	Ultrasonic tests		Hardness tests		Checking for cracks in threads	
								Accep-table	Reject	Hard-ness	Strength [kg/mm^2]	Accep-table	Reject
a	b	c	d=c−b	e=100 d/b	f	g=f−b	h=100 g/b						

length in their hollow bore or from the outside, as in the case of coupling flange fasteners. The record of screw pretensioning is useful, during revisions, for deciding with certainty whether a fastener can be used again, since the residual extra length of the screw after stress removal is an indication of the appearance of relaxation. Moreover, screwed fasteners should be checked in accordance with [5, 7]. An example of such a screw record is given in Table 4.

The procedure for taking the installation into operation commences with flushing the entire hydraulic system with oil (see 8.6. Oil Supply). After cleaning the hydraulic system before starting up for the first time, all supervisory and protective equipment must be checked for correct adjustment and functioning. Starting up is the most critical operational stage as, during this stage, there occur most oil fires, axial and radial bearing damage, damage through overspeeding, and particularly damage by clearance bridging through exceeding permissible relative elongations and by the ingress of water into hot turbine components. The majority of these cases of damage during start-up have led to severe damage to materials and to losses by outages, because of inadequate adjustment or even a complete lack of supervisory and protective equipment. For this reason, it is recommended that adjustment of all set-points for alarms should be carried out with circumspection and on the low side and that, during this stage, it is preferable to risk a few unintentional switch-offs than to do without any switching-off signals at all. Particularly critical is the oil supply to the bearings when switching over pumps (see 8.6. Oil Supply) and the relative elongation during the first cold and hot starts, since no experience is available in this respect with new turbine designs. It is recommended that, during start-up tests, the turbine set should be under constant supervision on site by experienced personnel, as this is the safest way of becoming aware of irregularities. Such personnel must be shown clearly how to recognize possible situations demanding immediate switch-off and must be given the authority to initiate emergency stoppages. Partly or fully automatic, as well as remotely controlled, operation of a steam turbine when first starting up or restarting after repairs or revisions must not be permitted without permanent on-site control.

5.1.4. Directives for Loss Prevention during Operation

In the following, the recommendations made in [5.9] and in the operating instructions of the turbine manufacturer will be supplemented only in regard to a few points.

5.1.4.1. Condition of Steam

To prevent damage it has proved useful to impose heavy demands on the condition of the steam when starting up a steam turbine. As an example, the steam purity required by a user for live and reheat steam is given as:

pH	$\geqq 7.5$
SiO_2	$\leqq 0.03$ mg/kg
Fe	$\leqq 0.05$ mg/kg
Conductivity	$\leqq 0.5$ µS/cm
Turbidity	$\leqq 2$ mg kieselguhr/kg

Difficulties with the steam purity required are particularly likely to arise with newly installed plant or after long periods of repairs and revisions.

Apart from operating instructions and steam purity, care should be taken of supervisory and protective equipment. With complicated modern installations, operation with incomplete or not fully connected supervisory and protective equipment is inexcusable.

5.1.4.2. Supervisory and Protective Equipment [5.1, 5.2]

This section surveys supervisory, safety and protective equipment for the steam-turbine plant. The limiting values for alarm or switch-off pulses are based on experience gained from cases of damage and should contribute to loss prevention. We shall be pleased to answer individual enquiries regarding the determination of limiting values. The manufacturer of the steam-turbine installation will also make these values available.

5.1.4.2.1. Protection from Underspeed, Overspeed and Reverse Running

Supervision: *Speed-Measuring Instruments (Tachometers)*

Measuring range from 0 to 130% of nominal speed without the need for switched selection of measuring ranges.

Alarm by limit contacts:

- at the upper speed limit of 105%,
- at the lowest permissible speed,
- when operating within a prohibited speed range.

Permissible speed ranges for continuous operation are marked in *green* and non-permissible ranges in *red*.

The switch-off pulse should be adjusted to the normal emergency trip speed of 110 to 115%.

Steam turbines driving pumps and compressors subjected to *danger from reverse running* must be provided with tachometers which indicate reverse rotation and, in such an event, actuate an alarm.

Control pulses are recommended for making possible automatic closure of stop valves in the pressure lines of pumps and compressors and for switching on the auxiliary oil pump to prevent damage to the bearings in the whole set when running in reverse.

Tachometers with measuring-range selectors can lead to severe damage from overspeeding by misinterpretation of information by operating personnel. Although tachometers without alarm contacts correctly indicate dangerous conditions of a machine, nobody's attention is drawn to them. Tachometers which, despite reverse running, indicate zero speed or even, as has occurred in a case of severe damage, a normal speed increase, prevent countermeasures by even the most careful and qualified operating personnel. From such cases of damage, it is clear that unsuitable speed-control equipment of this kind must in all circumstances be replaced, particularly when changing over to remote control.

Safety Equipment: *Speed Governor and Underfrequency Relay* (see also [5.3] and page 147, Section 5.1.2.3. Control and Protection System)

Limitation of transient overspeeding to 1.5% below the trip speed, with continuous overspeed to 3% below the trip speed.

Limitation of the lowest speed under load.
Alarm actuation if the speed limit is raised, e.g., for testing the trip gear.
Alarm actuation when reaching the lowest permissible speed under load. This latter alarm should be incapable of being cancelled.
Pulse for unloading the driven machine. As a rule, automatic switching off of the driven machine is not recommended, since operating personnel can protect the steam turbine equally well by unloading the driven machine and disconnecting it from the grid without the risk of a sudden load release.

Protection: *Speed Controller*
(see also [5.3] and page 147, Section 5.1.2.3)

Alarm and switch-off pulse at from 110 to 115% of the normal operating speed. For saturated-steam turbines in nuclear power stations, the manufacturer provides special circuits to reduce the risk of overspeeding through evaporation of water from the turbines in the event of a switch-off.

5.1.4.2.2. Protection from Non-Permissible Steam Pressures

Supervision: *Pressure-Measuring Instruments*

Live- and reheat-steam pressures,
casing pressure,
back pressures,
extraction and pass-out pressures,
pressures in the shaft-sealing system,
condenser pressure.
Supervision is best carried out by recording the measured values of pressure, in the form of curves, at certain load points. By comparing these curves, indications are given of changes in internal efficiency.

Safety: *Limiting Regulators and Safety Valves*
(see [5.3])

Limiting regulators serve to throttle the flow of live steam so that there can be no pronounced pressure drop, with the danger of a sudden temperature drop, and not too large a supply of steam, with the risk of excess pressure in the condenser.

Safety valves prevent pressure from rising to live-steam pressure in the high-pressure cylinders of reheat installations and in back-pressure installations.

Protection: *Pressure Controllers, Rupture Discs*

Condenser protection:
Alarm at 80 to 90% vacuum,
switch-off pulse at 50 to 80% vacuum,
and simultaneous pulse for closing cross-over stations.
Condenser rupture discs burst at from 0.1 to 0.3 kg/cm^2 above ambient pressure.

5.1.4.2.3. Protection from Inadmissibly High and Low Steam Temperatures and Non-Permissible Steam-Temperature Changes

Supervision: *Steam- and Wall-Temperature Measurements*

Live- and Reheat-Steam Temperature:
Alarm at
10 °C excess temperature in continuous operation,
20 °C insufficient temperature in continuous operation.
Switch-off at
20 °C excess temperature in continuous operation,
50 °C insufficient temperature in continuous operation.

Exhaust-Steam Temperature:
High-pressure section, back-pressure section:
Alarm at 50 °C above the normal value.
Switch-off at 100 °C above the normal value.
Low-pressure section, condenser:
Alarm at 60 °C, switch-off at 120 °C.
Closure pulse for cross-over valves at 100 °C.

Safety: *Temperature Regulation on the Boiler Side*

Protection: *Wall-Temperature Measurements of those Parts of Casings in Contact with Live and Reheat Steam;*
Reverse-Power Protection of the Generator

Wall-temperature measurements: In modern large steam turbines and automated steam turbo-sets, wall-temperature measuring devices have proved to be the best protection of the installation as a whole against non-permissible steam-temperature changes and their damaging effects on the life of casing and shaft materials [32, 33].

Reverse-power protection: Reverse-power protection should prevent dangerous motoring of the generator and hence driving the turbine from the generator. Under such conditions, it is possible that, because of the lack of a through-flow of steam in the low-pressure sections of the turbine, inadmissible temperature rises will be caused.

Reverse-power protection equipment is generally adjusted to respond at half the tractive power. The following recommendations have been derived from cases of damage:

Assurance of a supply of electric current for actuation of the generator main switch, even during outage of the station's own supply.

Possibility of manual release of the generator main switch from the steam-turbine control station if the reverse-power protection equipment fails.

5.1.4.2.4. Protection from Inadmissible Loads

Supervision: *Power Measurement of the Generator or Driven Machine*

Safety: *Controller*

Opening limitation on the controller: Alarm initiation when the limiting position is reached. In this way, it is possible, particularly in the case of pump and compressor sets, to give an indication of internal damage to driven machines.

Switch-off pulse when reaching the maximum permissible opening limit of the controller, adjusted above normal power consumption.

5.1.4.2.5. Protection from Axial-Clearance Bridging through Damage to Axial Bearings (see 8.3. Bearings) and through Inadmissible Relative Elongation

Relative measurements between rotors and casings,

alarm on 20% approach to the limiting value, switch-off on 5% approach to the limiting value.

5.1.4.2.6. Protection from Radial-Clearance Bridging

Measurement of wall-temperature differences in the centre of the casing between the highest and lowest points of the casing.

Alarm at 30 °C, switch-off at 50 °C.

For measurement of shaft vibration, see 5.1.4.2.7.

5.1.4.2.7. Protection from Inadmissibly Rough Running, with Consequent Damage to Bearings and Damage by Clearance Bridging

Measurement of bearing-pedestal vibration [7.1],
Measurement of shaft vibration [7.2, 28–30].

Whilst limiting values, subdivided into quality grades, are available [7.1] for measurement of bearing-pedestal vibration, only danger limits are given in [7.2], at which, according to previous experience of cases of damage, bearing damage occurs. It is strongly recommended that, although measurement of bearing-pedestal vibration should not be neglected, preference should be given to measurement of shaft vibration as a means of supervising and protecting against rough running, with consequent damage because of its direct connection with the rotor.

Shaft-Vibration Measurements Indicate with a High Degree of Certainty the Following Processes, Problems and Cases of Damage in Steam-Turbine Installations:

—Passage through the region of critical speed in bending, by a rise and fall in measured values in a certain speed range, with a constant speed increase in a relatively short time of minutes. This increase should, if at all, be used only for alarms that can be cancelled when exceeding the normal value during passage through the critical speed in bending. It serves as a guide for permissible shaft vibrations, without clearance bridging, in steam turbines.

—Changes in the state of balance through various causes, for example:

Moving-blade fracture: Sudden, immediate change, but not necessarily deterioration, of the measured values, depending on the state of balance.

Loosening of shaft-sealing rings, generally shrunk-on rings, bushes and discs: Slow increase in the measured values of shaft vibrations, sometimes only during processes of thermal instability during start-up and run-down, provided that loosening or shaft distortion by shrink-fit friction is only temporary.

Rotor crack: Rapid, but not sudden, increase in the measured values of shaft vibration. This deterioration of running behaviour sets in only with a far-advanced fatigue crack in the shaft. Depending on the kind of fatigue crack, i.e., rotating bending or one-sided bending fatigue crack, the change in the measured value for shaft vibration, depending on changes in the bending line of the rotor, can appear very late.

Rotor-disc crack: Very rapid increase in shaft vibrations, manifesting itself very late, i.e., when the crack has already led to a noticeable increase in imbalance through displacement of part of the disc.

Rotor distortion: Gradual deterioration of measured values, if distortion of the rotor is caused during measurement (i.e., during operation) by slight radial clearance bridging through, for example, slow or only temporary displacements of inner casings, shaft-sealing housings or also outer casings.

Poor measured values beginning with the lowest speed, with pronounced deterioration when passing through the critical-speed region and increasing deterioration on approaching the operating speed, are signs of marked shaft distortion through one-sided cooling during stand-still or the ingress of water after opening the valves.

A sudden deterioration of up to 500%, i.e., much more pronounced than with damage to the moving blades, occurs with water ingress during operation at normal speed.

—Rough running due to causes other than changes in the state of balance:

Oil whip: Deterioration of running smoothness in a scatter band in the upper *speed region.*

Gap excitation: Deterioration of running smoothness in a scatter band in the upper *load region.*

These two disturbances do not occur with the prevailing frequency of rotation, but with the natural frequency of the rotor concerned in the region of about half the frequency of rotation. Nowadays, these disturbances can be determined satisfactorily by frequency analysis, supplemented by the usual vibration measurements of shafts or bearing pedestals.

Because of the variety of criteria for supervision, the following well-proven procedure is recommended for establishing the limiting values for alarm and switch-off pulses:

Determine the greatest shaft-vibration amplitudes at each measuring point when passing through the critical speed, during normal load application, and during continuous operation.

Alarm release: 20% higher than the highest value of shaft vibration when applying load, releasing load, and during continuous operation.

Switch-off: 50% higher than the highest value when passing through the critical speed or the highest value when applying load, releasing load, and during continuous operation.

The alarm limit should be between 100 and 200 μm, and the switch-off limit between 150 and 360 μm.

A condition is determination of the maximum shaft vibration from two individual measurements displaced by 90 deg. In addition to these limiting values, deviations from normal values must be assessed by the operating personnel with the aid of the criteria given. The percentage values must be chosen in accordance with the characteristics of the installation.

5.1.4.2.8. Protection from Disturbances Emanating from the Oil Supply and the Radial and Axial Bearings

(see 8.3. Bearings and 8.6. Oil Supply)

5.1.5. Loss Prevention by Inspection and Revision

Inspection and revision are the classic means of loss prevention and have gained in importance as a result of the increasing potentialities of non-destructive testing and the complexity of steam-turbine installations. Attempts to increasingly lengthen the intervals between inspections and revisions have been instigated by a number of isolated reports of steam turbines being operated without revisions. However, neither in Germany nor abroad, particularly in the U.S.A., France and Great Britain, has steam-turbine operation without revisions proved useful or able to gain a foothold. Today, periods between major revisions range from two to five years, depending on the different degrees of technological maturity of installations, which, in the case of new large turbines with many prototype components, necessitates opening up individual casings every one to two years. Regarding experiences at the revision of individual turbine casings and the inspection of individual important turbine components during the available stoppage periods without dismantling the whole steam-turbine installations, thus far, no disadvantageous accumulation of cases of damage have been found after such *partial revisions*. However, the suitability of the various types of steam turbines for the complete and independent revision of individual components can be quite different. For instance, the disposition of valves at the side or below the steam turbine, with lateral inlet and outlet pipes, permits dismantling of valves without the risk of foreign bodies entering and damaging the turbine blading, whereas, with valve groups on top of the turbine casing, dismantling can cause damage to casing insulation, as well as to measuring lines, and foreign bodies can drop into the turbine. The same applies to inlet and cross-over piping which, when placed laterally in relation to the turbine, does not impede dismantling of the top halves of casings, so that a visual inspection and a check of measurements of a casing can be carried out in from 10 to 14 days, provided that efficient aids for flange fastener assembly are available. In nuclear power stations with ordinary-water reactors, the sectionalized partial revision of the turbo-set while fuel elements are being changed will have to prove its value fully in the future.

Check-List for Inspection and Revision of Steam-Turbine Components by Non-Destructive Testing

In the event of damage to turbine components, the checks suggested below can usefully be employed to complement the usual repairs of the damaged components.

1. Checks during Operation

1.1. Measurement of Steam Consumption
Comparison with the result of a previous measurement makes it possible to assess whether there has been a change in the efficiency of the turbo-set.

1.2. Determination of Internal Efficiency
If measurement of steam consumption is too costly, it is sufficient for the determination of internal efficiency η_i of individual turbine casings to measure the prevailing inlet, intermediate and final steam temperatures and pressures.

1.3. Measurement of Stage Pressures
Stage pressures as a function of steam flow and in comparison with those obtained with a "clean" machine give an indication of blade deposits and blade deformations.

1.4. Comparison of Casing Temperatures
The wall temperatures of casings should be compared with the associated steam temperatures. An assessment of the differential casing temperature provides information as to whether there exist internal leakages (e.g., at the inner casing or at the housings of shaft seals), steam leakage at the seals between outer and inner casings or distortion of the outer casing causing such leakages.

1.5. Vibration Measurements
Measurements of the vibration of bearing pedestals and shafts should be compared with previous measurements. If there is any suspicion of changes in the state of balance, manifested by a deterioration of bearing or shaft vibration measurements, the matter can be cleared up by measurement of the imbalance vector with the aid of suitable equipment.

1.6. Relative Shaft and Casing Elongations
A comparison of the relative elongations of the casing and shaft indicates whether the anchorage of the turbine installation to the foundation is satisfactory with regard to the sliding surfaces of bearing pedestals and baseplates.

1.7. Support of the Turbine Casing by the Bearing Pedestals
This support is controlled by the movability of shims or packing bushes at the bolting of the casing feet. If there are signs of jamming, a revision of the turbo-set should be arranged for.

1.8. Checks on Bearings
Comparison of oil inlet pressures to the bearings with previous measurements: a reduction in inlet pressures indicates increased bearing play.
Comparison of the temperatures of the bearing metal with previous measurements: a rise in bearing-metal temperature indicates deterioration of bearing geometry.
Comparison of shaft-jacking oil pressures with previous measurements: a pronounced variation in shaft-jacking oil pressure indicates a change in the loading of a bearing through shaft misalignment or internal bearing damage.
Whenever bearing checks yield negative results, the first step should be to arrange for a revision of the bearing, not revision of the machine.

2. Checks on a Machine during Run-Down and Cooling and while Operating with Barring Gear

2.1. Noises in the Region of Shaft Penetrations
Clinking or scraping noises indicate that the alignment of the rotor with respect to the shaft has altered during operation. Immediate opening of the casings is necessary.

2.2. Check on the Power Consumption of the Barring Gear
The power consumption of the barring gear or of the speed under constant oil pressure to the barring gear should be checked by comparison with previous measurements. An increased power consumption indicates clearance bridging in the turbo-set. Immediate opening up of the casings is necessary.

3. Checks on a Stationary and Cooled-Down Machine

3.1. Check of the Concentric Position of the Rotor with Respect to Penetrations through Casings and Bearing Pedestals
This check is only possible with modern steam turbines by using measuring points provided for in the design. With older machines, it should prove easy, during repairs or revisions, to retrofit the appropriate data points for feeler or tracer-stylus measurements. This clearance check yields the most accurate information on the respective positions of the rotor and casing, particularly if it is supplemented by a tracer-stylus measurement via endoscope sockets.

3.2. Check of the Longitudinal Displacement of the Rotor in the Casing
This check can be carried out only if the axial bearing is removed or the coupling is disconnected. To be able to determine changes, it is necessary to make comparisons with a previous revision or repair.

3.3. Check for Concentric Running
(see 8.4. Couplings)
A check for concentric running gives information on possible thermal distortions after water hammer or damage by clearance bridging. To be able to come to any conclusions regarding distortions between the bearings in the interior of the casing, it is recommended that the dial gauges used for the concentric-running check should be located as far away as possible from the bearings (e.g., on the coupling flange) or, better still, through an endoscope socket.

3.4. Checks on Moving and Stationary Blading
If no endoscope sockets (as shown in Figure 17) are provided, it is at least possible, with the aid of flexible endoscopes, to view those moving and stationary blade rows which are accessible through pass-out connections and exhaust and inlet branches without opening up the casing.

3.5. Check on Couplings for Eccentricity
There is no need to carry out this check unless there is a deterioration in quietness of operation. With the bearing removed, a concentric-running test is performed in the region of the bearing, and this test should confirm the proper assembly of the coupling. This check is possible only if the coupling is supported on both sides by radial bearings.

3.6. Prestress Check on Coupling Flange Fasteners by Length Measurements after their Removal.

3.7. Alignment Check after Removal of Coupling Fasteners (see 8.4. Couplings).

3.8. Check on Casing Keys in the Guide Planes Relative to the Bearing Pedestals and on the Bearing-Pedestal Guideway between the Pedestal and Baseplate.

3.9. Check on Bearings, with Measurement of Bearing Play and Inspection for Possible Damage to the Bearing Metal (see 8.3. Bearings).

4. Checks with Extensive Disassembly Work

4.1. Checking of Quick-Acting, Control, Pass-Out Quick-Acting, and Non-Return Valves, Steam Strainers, and Fittings in General

4.1.1. Testing casings for cracks by the dye-penetrant method, particularly on internal separating walls, using mirrors and endoscopes on casing intersections of difficult access, and at valve seats by the dye-penetrant method.

4.1.2. Testing of the spindle for roundness and visual checking of its surface.

4.1.3. Checking of the surface of the valve body for pitting in the valve guide and of the valve seat by the dye-penetrant method (Figures 12 and 26).

4.1.4. Checking of valve-seating with engineers' marking ink, and possible regrinding if any suspicion of inadequate valve tightness is confirmed during run-down.

4.2. Checks on Hot Screwed Fasteners according to [5.7], and on Coupling Flange Fasteners

4.2.1. Length measurement after unloading and comparison with the original length (see page 159).

4.2.2. Ultrasonic testing to discover cracks in the first three stressed screw threads.

4.2.3. Hardness testing of the screwed-fastener material for embrittlement.

4.3. Checks on Heat-Resistant Cast-Steel Casings

4.3.1. Visual checks with mirror and endoscope in live-steam, steam-inlet and nozzle-group channels for crack formation, blowholes, erosion, cracking of chaplets, in some circumstances by means of the die-penetrant or magnetic-powder methods (Figures 27 and 28).

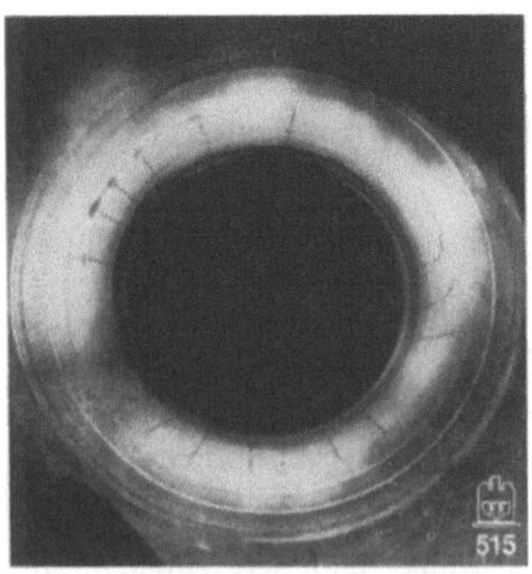

Figure 26. Cracks in a valve seat, rendered visible by the dye-penetrant method.

Figure 27. Crack in a cast-steel casing.

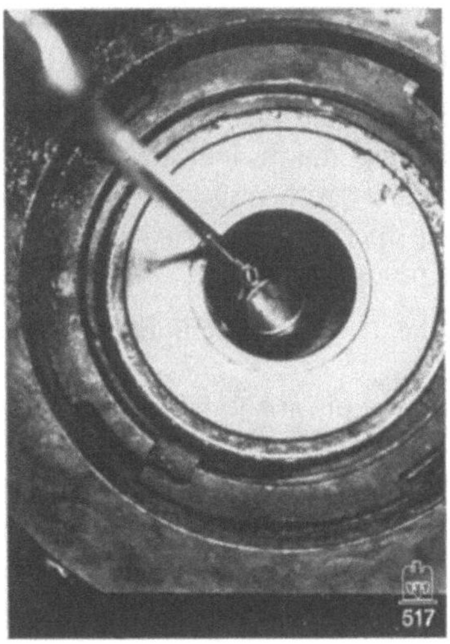

Figure 28. Examination of a steam inlet chest by means of a mirror.

4.3.2. On-site examination of structure to determine the influence of long-term exposure to temperature on casings that have been in service for about 100,000 hours.

4.3.3. Radiographic examination of cast-steel components containing cracks, so as to be able to assess whether they are suitable for further use.

4.4. Checks on Moving and Stationary Blades

4.4.1. Visual check for damage by foreign bodies and by radial and axial clearance bridging.

4.4.2. Checking for cracks after cleaning, particularly if there is any suspicion of damage, as follows:

4.4.2.1. On the blade, particularly at the entrance and exit edges, in the hardness zone of brazed-on or welded-on stellite wear-resistant shields; the roots of blades in the wet-steam region, as well as the damping-wire holes and the brazing of the lacing wire: Checking for cracks by the dye-penetrant method, by the magnetic-powder method (using d.c. yoke magnetization or a.c. coil magnetization), by ultrasonic testing with surface-wave propagation or by eddy-current testing (possible only at blade entrance and exit edges).

4.4.2.2. On rivet shanks: Ultrasonic testing, using a specially shaped probe for applying ultrasonic radiation to the flattened shank.

4.4.2.3. On shrouding and on damping and lacing wire: Dye-penetrant or magnetic-powder methods.

4.4.2.4. On blade roots of the fork, "straddle", large inverted-T, and large fir-tree types: Ultrasonic testing of the free blade length, of the transition of the root with oblique propagation of waves, and from the lateral faces of the root with surface-wave propagation (Figure 29).

4.4.2.5. On the fork pin: Ultrasonic testing.

4.4.3. Check of the natural frequency of moving and guide blades if, for example, there is a suspicion of resonance when damage has occurred on similar blading, and of strongly eroded blading in the wet-steam region because of the increased natural frequency arising from erosion: Contactless measurements of vibration in the assembled condition, also without removing the rotor from the casing.

4.4.4. Investigation of deposits: Laboratory analysis, looking particularly for chloride content.

4.4.5. Documentation of the progress of erosion by photographic or replica methods.

4.5. Checks on Casings and Rotors before or after Removal of the Rotor

4.5.1. Horizontal, axial and vertical measurement of play at the narrow locations of shaft seals, moving and stationary blades, diaphragm seals, rotor discs, and diaphragms.

4.5.2. Axial displacement of rotor with determination of the narrowest clearance when stationary, determination of the limiting values for supervision of axial position and relative elongation, taking account of thermal expansion during operation.

4.5.3. Deformation checking of diaphragms by measurements of clearance between the diaphragm and rotor at different points radially and on the circumference between diaphragm seal and diaphragm fit in the casing.

4.6. Checks on Rotors

4.6.1. Centricity check.

4.6.2. Checking for cracks at notches and at changes in cross-section, using the dye-penetrant

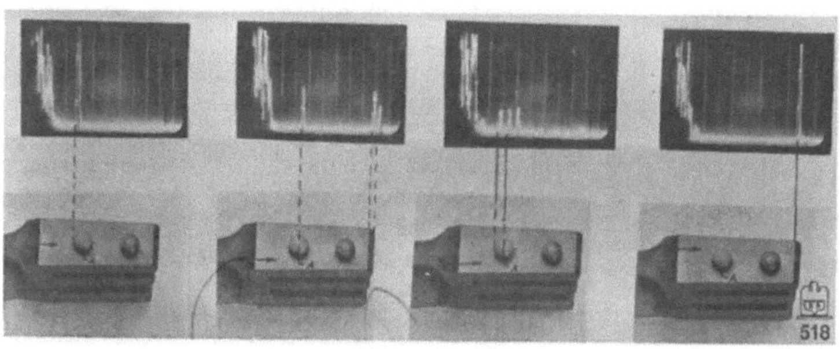

Figure 29. Series of ultrasonic tests of a fork root after formation of a crack at A. Wave propagation in the direction of the arrow through the blade/root transition. The ultrasonic trace in the second picture from the left clearly shows the fault. This procedure made it possible to test ultrasonically all blades of this fork-root type when the low-pressure casing was opened up, but without having to dismantle the moving blades.

method or ultrasonic testing with wave propagation through the end-faces of the rotor.

4.6.3. Checking of surfaces for the appearance of products of corrosion at shrink fits of couplings, shaft-sealing rings, bushings and rotor discs.

4.7. Checks on Casings

4.7.1. Crack formation at intersections.

4.7.2. Leakiness of internal sealing locations between outer, inner and shaft-sealing housings, guide-blade carriers, diaphragms, and all parting lines.

4.7.3. Obstruction of water drainage from casings.

4.8. Checks on Bearing Pedestals

4.8.1. Ability of the bearing pedestals to slide on the baseplate or of the casings on the pedestals.

4.8.2. Attachment of the bearing pedestals on the baseplates.

4.8.3. Mobility of the keys between bearing pedestals and casings and between pedestals and baseplates.

4.8.4. Examination of the oil sump in the bearing pedestals for signs of white-metal, bronze and steel abrasion.

4.9. Checks on Bearings (see 8.3. Bearings).

4.10. Checking of the Oil Supply (see 8.6. Oil Supply).

4.11. Checks on Couplings (see 8.4. Couplings).

4.12. Checks on Transmissions (see 8.5. Stationary *Gearing*).

4.13. Checks on Condensers (see 5.2. Condensing Plant).

5.1.6. Guidelines for Loss Prevention during Repairs

Every case of damage is preceded by a fault. Periods of non-operation during repairs should therefore be used for the elimination of faults, with special emphasis on the improvement of supervisory and protective equipment, which often show their shortcomings only when damage occurs. Provisional repairs with imperfect replacement parts require especially careful supervision. Every steam turbine can be operated with some blade rows missing if the load limitation thereby imposed is conscientiously observed, bearing in mind that, because of disruption of the process of uniform expansion, the power loss is greater than that corresponding to the missing blade rows. The question of removing the associated rows of guide blades must in each case be decided in consultation with the manufacturer. Operation of the plant, even as a temporary expedient, without the equipment necessary for its safety, such as auxiliary oil pumps, shaft-jacking oil pumps, and supervisory systems, must be strongly discouraged.

Bibliography to 5.1

[1] Dampfturbinen für allgemeine und besondere Aufgaben in der chemischen Industrie. VDMA 4310 sheet 1, 2. Draft 1972.

[2] General and Special Purpose Steam Turbines for Refinery Services. API-Standard 611, 612, 1969.

[3] Special Purpose Steam Turbines. ESSO Basic Practice 10-7-2, 1968.

[4] Empfehlungen und Richtlinien für Dampfturbinenanalagen der VIK. Essen, Richard-Wagner-Str. 41.

[5] Empfehlungen und technische Richtlinien für die Auslegung von Dampfkraftwerken und für den Turbinenbetrieb der VDEW. Published by VWEW, Frankfurt/Main, especially:

[5.1] Technische Richtlinien für Überwachungs-, Sicherheits- und Schutzeinrichtungen an Dampfturbinenanlagen. 1967.

[5.2] Empfehlungen für die Betriebsüberwachung von Dampfturbinen, insbesondere zum Bestimmen des Revisionszeitpunktes. 1966.

[5.3] Technische Richtlinien für Dampfturbinenregelungen. 1963.

[5.4] Ölbuch. 5th edition 1972.

[5.5] Technische Richtlinien für Ölversorgungsanlagen von Dampfturbinen. 1969.

[5.6] Lieferbedingungen für Dampfturbinen. Parts A, B, C. 1965/1969.

[5.7] Empfehlungen für Schrauben im Bereich hoher Temperaturen. 1964.

[5.8] Richtlinien für die Prüfung von großen Schmiede- und Gußstücken für Dampfturbosätze. 1965.

[5.9] Technische Richtlinien für den Turbinenbetrieb. 1972.

[6] VGB-Richtwerte für das Speise- und Kesselwasser von Wasserrohrkesseln. Essen: VGB-Dampftechnik 1968.

[7.1] VDI-Richtlinien für Lagergehäuseschwingungen No. 2056. 1964.

[7.2] VDI-Richtlinien-Entwurf für Wellenschwingungen No. 2059. 1971.

[7.3] VDI/VDE-Richtlinien No. 3521.

[7.4] VDI-Richtlinien No. 2060.

[8] DIN 4305 sheet 1,2. Dampfturbinen: Benennungen. 1968.

[9] Huppmann, H.: Häufigkeit und Ursachen von Schäden an Bauteilen großer Dampfturbinen. Der Maschinenschaden 43 (1970), No. 1, pp. 1/16.

[10] Leopold, J.: Laufschaufelschäden an axialen Dampfturbinen. Der Maschinenschaden 44 (1971), No. 3, pp. 2/12.

[11] Huppmann, H.: Schäden an Gleit- und Wälzlagern. VDI-Berichte No. 141, pp. 97/105.

[12] Braun, H.: Die Bedeutung der Schadenforschung für Konstruktion und Fertigung. Der Maschinenschaden 44 (1971), No. 5, pp. 175/180.

[13] Buchwald, K.: Neuere Erkenntnisse auf dem Gebiet der zulässigen Anfahr- und Laständerungsgeschwindigkeiten von Dampfturbinen. VGB-Conference Dampfturbinen und -betrieb. 1972.

[14] Mönch, R.: Mindestanforderungen and die Hersteller für die schwingungstechnische Auslegung der Beschaufelungen. Lecture „Erfahrungsaustausch der Energiebetriebe im Verband der chemischen Industrie". Oct. 1967, BASF, Ludwigshafen.

[15] Müller, H.: Das Anfahren von Blockanlagen im Gleitdruck-Gleittemperaturverfahren. Elektrizitätswirtschaft 63 (1964), No. 23, pp. 797/807.

[16] Meiners, K.: Automatisches Anfahren von Dampfturbinen. Escher-Wyss-Mitteilungen 40 (1967), pp. 3/16.

[17] Müller, H.: Einsatz moderner Grundlastblöcke zur Spitzenstromerzeugung. VGB-Mitteilungen 48 (1968), No. 5, pp. 306/321.

[18] Busch, H.: Schnelles Abkühlen von Dampfturbinen. Energie (1968), No. 1, pp. 8/20.

[19] Hennig, Th.: Kaltfahren von großen Dampfturbinenanlagen. Elektrizitätswirtschaft (1972).

[20] Höxtermann, W.: Technischer Bericht KWU-Nr. 71030: Überblick über häufige Ursachen für Schäden und Störungen an Dampfturbinen.

[21] Goedecke, R.: Beitrag zur Frage des Verhaltens von Turbinen- und Generatorwellen im Warmrundlaufversuch. Arch. Eisenhüttenw. 39 (1968), No. 3, pp. 201/211.

[22] Gasch, R.: Stabiler Lauf von Turbinenrotoren. Konstruktion 17 (1965), No. 11, pp. 447/452.

[23] Krämer, E.: Selbsterregte Schwingungen von Wellen infolge von Querkräften. BWK 20 (1968), No. 7, pp. 307/312.

[24] Thomas, H.J.: Instabile Eigenschwingungen von Turbinenläufern, angefacht durch die Spaltströmung in Stopfbuchsen und Beschaufelung. AEG Special publication 1958.

[25] Ringeis, W., et al.: Das 670-MW-Kernkraftwerk Würgassen. Atom und Strom 17 (1971), No. 11/12, pp. 173/192.

[26] Schinn, R.: Qualitätskontrolle großer Stahlguß- und Schmiedestücke für Dampfturbinen. Lecture at American Power Conference, 1971.

[27] Giesecke, D.: Sicherheitseinrichtungen an Vorwärmern des Wärmekreislaufs in Kraftwerken. Energietechnik 21 (1971), No. 7, pp. 293/297.

[28] Peters, G.: Wellenschwingungsmessungen an Turbomaschinen. Erfahrungen und Entwicklungstendenzen. Der Maschinenschaden 44 (1971), No. 5, pp. 157/163.

[29] Schöllhammer, F.: Beiträge zur Deutung von Relativ-Wellenschwingungsmessungen an großen Turboaggregaten. Der Maschinenschaden 44 (1971), No. 1, pp. 1/11; No. 5, pp. 164/166.

[30] Huppmann, H., Bohnstedt, J.: Beurteilung der Wellenschwingungen von Turbomaschinen. Der Maschinenschaden 44 (1971), No. 5, pp. 164.

[31] Schem, O.: Ermittlung und Anzeige der zulässigen Betriebsweise von Dampfturbosätzen. Siemens-Z. 37 (1963), pp. 126/131.

[32] Krüger, W.: Schaden an einem 180-MW-Block durch Nichtöffnen der gesteuerten Sicherheitsventile; Klärung des Schadens sowie Vorschläge für die Änderung der „SR-Sicherheitsventile". VGB-Mitt. 52 (1972), No. 1, pp. 57/68.

5.2. Condensing Plant

Surface condensers are generally cooled by means of fresh or recirculating water, though increasing use has been made in recent years of air-cooled installations. Experience of damage by Allianz is mainly concerned with water-cooled surface condensers.

Statistics of damage to condensers have not been prepared because difficulties arise preponderantly in the form of corrosion of tubes or tube plates, which generally are not covered by insurance and hence are neither reported nor recorded. However, within the framework of the technical service of Allianz, numerous examinations of these cases have been carried out. Table 1 shows the results of 46 examinations of damaged condenser tubes over the past seven years in regard to the types, locations and causes of damage involved.

In the foreground of these examinations are cases of damage through stress-corrosion cracking and local corrosion, each of which accounts for about one-third of the total number of cases of damage, because their causes, in both instances, are usually very complex. In operation, loss of material from the outside of tubes in the region of air extraction is also of major importance, though these cases are now investigated to a lesser degree as their causes are largely known [1, 2]. The situation is similar in regard to galvanic corrosion of tube plates, against which special protective measures are required.

5.2.1. Manufacture of Condensing Plant

Operating conditions for condensers are prescribed by their location and the layout of the power plant.

Important details for loss prevention:

Most cases of damage to condenser tubes emanate from the cooling-water side, so that attention should primarily be given to selection of the most suitable tube material related to the cooling water to be used [20]. Table 2 lists the most commonly used copper and wrought copper-alloy tube materials for condensers and heat exchangers, according to DIN 1785 (December 1967).

In addition, the following austenitic steels have established their value as material for condenser tubes:

For fresh-water installations:
X 5 CrNi 18 9 (ASTM A. Type 304)
Material No. 4301

For sea-water and brackish-water installations:
X 5 CrNiMo 18 12 (ASTM A. Type 316)
Material No. 4436

In experimental tubing, the following materials have been tested:

Material No. 4573, X 10 CrNiMoTi 18 12
longitudinally welded

Material No. 4401, X 5 CrNiMo 18 10
longitudinally welded

Material No. 4523, X 8 CrMoTi 17
seamless drawn

Experience in the U.S.A. where, in 1966, 50% of all new installations were already equipped with tubing made from austenitic materials [3, 4], and experimental tubing in the German Federal Republic, principally with air-cooler tube bundles, have shown the following advantages:

- High resistance to erosion and erosion-corrosion; higher alkalization of the water and steam circuit for the attainment of a low level of iron is therefore made possible, as well as
- higher cooling-water velocities;
- higher corrosion resistance, even with cooling water of very unfavourable quality.

An operational prerequisite is that deposits should be largely avoided, since pitting corrosion can occur on unevenly covered steel tubes (Figure 1). The use of austenitic steel tubing is not recommended with cooling water having a substantial content of manganese and, in the case of low manganese contents, only if the water is adequately conditioned ("shock" chlorination with a maximum excess of 1 mg Cl_2/l [5]).

The most important requirement for the selection of condenser-tube materials is an exact knowledge of the condition of the cooling water. In land-based installations, water condition is often subject to seasonal fluctuations and to changes over the years of operation of the plant. It is

Table 1. Survey of types, locations and causes of damage to condenser tubes

Type of damage	Number	Location of damage	Number	Prevailing cause of damage	Number	Material
Stress-corrosion cracking	15	Rolling-in, cooling-water side	7	Internal stresses from rolling-in + products of decomposition of algoid growths	7	CuZn 28 Sn
		Free tube length, cooling-water side	5	Residual stresses in the tube + products of decomposition of algoid growths	1 4	K-Ms 63 CuZn 28 Sn
		Free tube length, condensate side	2	Static operational stresses + NH_3 from excess hydrazine	1 1	CuZn 28 Sn X 10 CrNiTi 18 9
		Free tube length, condensate side	1	Residual stresses + mercury from damaged pressure gauges	1	CuZn 28 Sn
Local corrosion and pitting	15	Cooling-water side	3	Local galvanic action through inadequate protective layer	1 1 1	St 35.8 CuZn 28 Sn X 5 CrNi 18 9
		Cooling-water side	8	Local galvanic action through uneven internal deposits	1 2 1 2 2	CuZn 30 CuZn 28 Sn CuZn 20 Al SnBz 2 CuNi 30 Fe
		Cooling-water side	4	Dezincification through uneven deposits	3 1	K-Ms 63 CuZn 28 Sn
Loss of material on the outside of tubes	11	Air extraction at support plates	10	NH_3 from excess hydrazine + traces of O_2 in the drained condensate	10	CuZn 28 Sn
		Top tubes	1	Corrosion through impinging wet steam	1	CuZn 28 Sn
Mechanical circumferential cracks and locations of abrasive wear	3	Lane tubes between tube support plates	3	Forced vibrations caused by the flow of steam (partly excess steam generation)	2 1	CuZn 28 Sn CuZn 20 Al
Erosion-corrosion	2	Cooling-water inlet side	2	High cooling-water velocity + aggressive water (mainly brackish or sea-water)	2	CuZn 20 Al

Table 2. Most commonly used tube materials for condensers and heat exchangers according to DIN 1785 (December 1967)

Symbol new	Symbol old	Material number	Guidelines for use in condensers	Permissible cooling-water velocity
SB-Cu	ditto	2.0150	Good resistance in fresh water	≈1.5 m/sec
SF-Cu	ditto	2.0090		
CuZn 30	K-Ms 72	2.0265	Preferred for use in the food (sugar) industry	≈1.5 m/sec
CuZn 28 Sn	SoMs 71	2.0470	Mainly for land-based installations, with river water, recirculating-water-cooling plant, partly brackish water	≈2 m/sec
CuZn 20 Al	SoMs 76	2.0460	Land-based installations under arduous conditions, using brackish water or sea-water	≈3 m/sec
CuAl 15 As	AlBz 5	2.0918	Mainly in the potash industry	–
CuNi 10 Fe	ditto	2.0872	Under very severe corrosive and erosive conditions	≈4 m/sec
CuNi 30 Fe	ditto	2.0882	Preferred for sea-water	≈4 m/sec

Table 3. Main fields of application of commonly used condenser-tube materials

Material:	SB-Cu	CuZn 28 Sn (SoMs 71)	CuZn 20 Al (SoMs 76)	CuNi 10 Fe CuNi 30 Fe
Main field of application:	Fresh water	Fresh water, brackish water, water from a river mouth	Sea-water, brackish water, water from a river mouth	Sea-water, polluted sea-water, polluted brackish water and river-mouth water
pH-value:	5 to 9	6 to 9	6 to 9	6 to 9
Exhaust-steam residue	up to 500 mg/l	up to 500 mg/l	up to 30,000 mg/l	up to about 30,000 mg/l
Carbonate hardness	over 3° dH	over 3° dH	not applicable	not applicable
Chloride ions	up to 70 mg/l	up to 300 mg/l	up to 23,000 mg/l	≈25,000 mg/l
Sulphate ions	up to 500 mg/l	up to 500 mg/l	up to ≈ 3,000 mg/l	≈3,000 mg/l
Nitrate ions	up to ≈10 mg/l	up to ≈10 mg/l	up to ≈50 mg/l	≈100 mg/l
Oxygen	unimportant	unimportant	unimportant	unimportant
Iron and manganese	≦0.2 mg/l	≦0.2 mg/l	≦0.2 mg/l	≦0.2 mg/l
Sulphides	none	none	none	trace
Ammonia	≦5 mg/l	≦10 mg/l	≦10 mg/l	≦15 mg/l

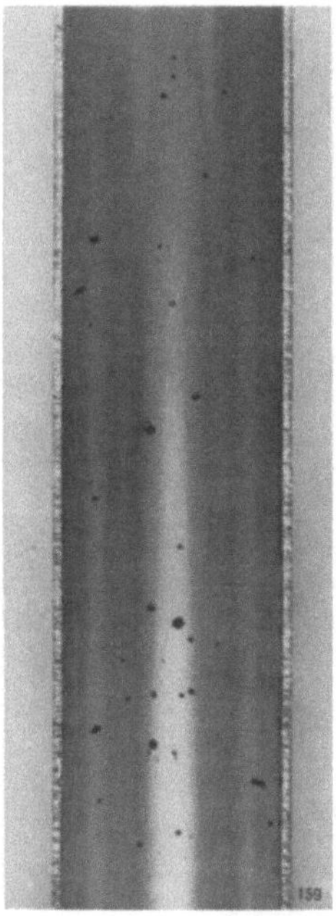

Figure 1. Damage by pitting corrosion to a condenser tube made of rust- and acid-resistant steel X 5 CrNi 18 9 (Material No. 4301, ASTM A. Type 304). Cooling water: sea-water from a tropical seaport.

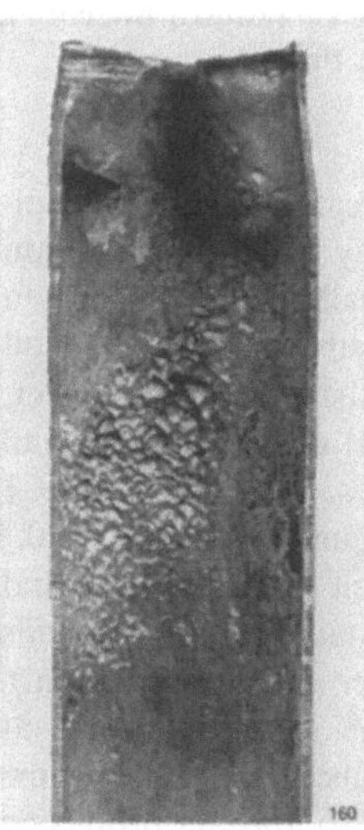

Figure 2. Damage by erosion-corrosion at the cooling-water inlet of a condenser tube made of CuZn 20 Al (SoMs 76) containing insufficient aluminium.

therefore necessary to know, at least, the analyses of the water over a full year with their extreme values. Furthermore, account should be taken of the trend towards increasing water pollution.

A survey of the types of cooling water for which the most commonly used condenser materials can be employed is given in Table 3 [6].

This classification can be regarded only as a preliminary step since, in particular, chemical impurities and the content of matter in suspension can have a decisive influence on materials selection. With difficult and unknown cooling-water conditions, it is recommended that, during the planning stage, a pilot plant should be installed at the projected site with tubes made of different materials and that a selection should then be made in accordance with the results obtained. As in the past, there still remains a wide range of materials from which to make a choice, because the reciprocal influence of the numerous parameters is largely unknown.

The selection of materials and the cooling-water velocity have a definite relationship. The values of velocity given in DIN 1785 (see Table 2) should not be exceeded. In fact, with polluted sea-water and brackish water, especially where higher cooling-water temperatures are involved (in the tropics), these velocities should rather be reduced by 0.5 to 0.75 m/sec; otherwise, the possibility of erosion-corrosion through turbulence in the region of the cooling-water inlet of the tubes (Figure 2) must be taken into account [2]. In such a case, the protective layer on the tubes is constantly carried away by the flowing medium, and the bare metal surface is subjected to the aggressive cooling water, though this condition can be remedied by the use of plastic sleeves or coatings [6]. An important point in this connection is that the cooling-water supply should be planned in such a way that the least possible amount of foreign particles, such as sand, suspended matter and air is sucked in.

The advantage of the use of *pickled new tubes* is essentially that they are free from drawing lubricant and, if they are taken into operation in the proper way, a protective film can be built up evenly (see Section 5.2.2, p. 177). On the other hand, an oxide film on new tubes, originating from the final heat treatment, can, under unfavourable commissioning conditions, take on the function of a first protective layer. For this rea-

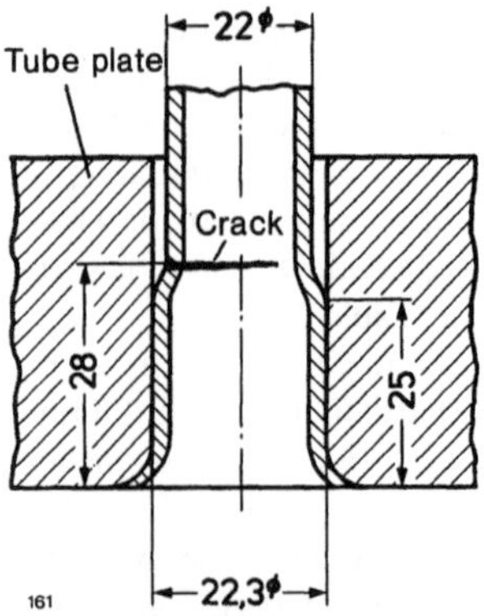

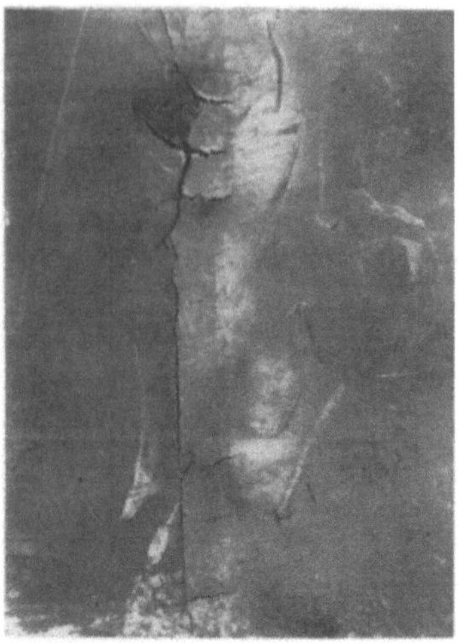

Figure 3. Damage by stress-corrosion cracking in the rolling-in region to a condenser tube made of CuZn 28 Sn (SoMs 71) F 38 (0.02% As, 0.01% P).

Figure 4. Stress-corrosion cracking caused by scoring on the outside of a condenser tube.

son, new tubes in a pickled condition and with an oxide film from the final heat treatment are employed to an equal extent.

A frequent site of damage is the attachment of tubes in the tube plates. In the German Federal Republic, attachment of the tubes is nowadays effected almost exclusively by rolling them securely into both tube plates. During the normal rolling-in process, a cold deformation in tension, far above the yield point, is applied to the rolled-in ends. In this way, high longitudinal stresses are set up in the transition to the free tube length and low tangential stresses in the tapered section in accordance with the mechanism of stress-corrosion cracking (Figure 3). With increased expansion of the tube, the sensitivity to stress-corrosion cracking also increases and, with pronounced expansion for greater adhesion, additional transverse cracks may occur in the region of rolling pressure. Actually, tube expansion should be as small as possible (about 2%) and flaring should not exceed about 15%. Moreover, rolling-in should not extend over the whole wall thickness of the tube plates, but should stop 1 or 2 mm from the end (Figure 3), so as to make sure of preventing expansion of the tubes behind the tube plate. By slightly tapering the roll ends (taper angle of 177 deg. over a length of 15 mm), it is possible to reduce the induced stresses at the transition to the free tube length. Short-time annealing after rolling-in also appears promising for the reduction of stresses.

Condenser tubes must todays be supplied free from internal stresses in accordance with DIN 1785.

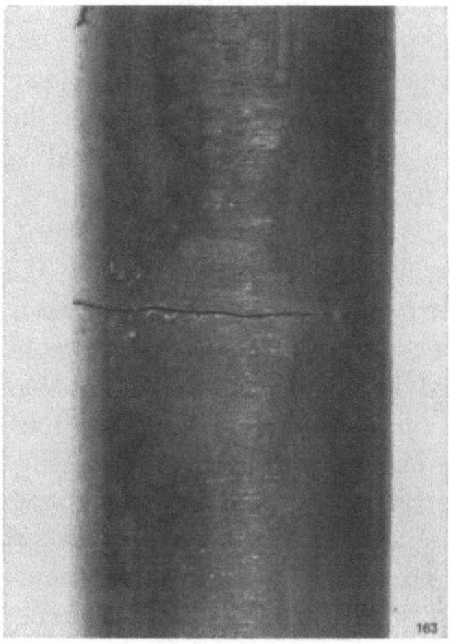

Figure 5. Fatigue fracture and areas of abrasion on a condenser tube as a result of vibrations excited by the flow of steam.

Stress-corrosion cracks in a longitudinal direction as a result of tangential stresses induced during manufacture therefore no longer present a problem. However, when drawing in the tubes, deformation of the tubes and damage to their surface must be avoided; otherwise, internal stresses can be created (Figure 4).

Among chemical reagents triggering stress-corrosion cracking from the cooling-water side, consideration should be given in the first place to ammonia, which is preferentially formed as a product of decomposition of organic coatings when the plant is not in operation.

Special attention must be given to even infeed of cross-over steam, i.e., over-production steam. High rates of steam flow often excite vibrations in top and lane tubes, resulting in abrasive wear and fatigue failures (Figure 5). A remedy can often be effected by installing tubes having greater wall thicknesses, in some circumstances of massive construction, at locations where such a risk exists. Erosion of the tubes by inflowing wet steam is also possible (Figure 6).

Selection of Materials for Tube Plates

With slightly aggressive fresh water: steel with a protective coating.

With aggressive fresh water, sea-water and brackish water: CuZn 40 (Muntz metal DIN 17660), CuZn 40 Sn (DIN 17661).

Steel tube plates, which, because of their simpler (welded) construction, are preferred for land installations, are endangered by galvanic corrosion in aggressive water, owing to their contact with the brass tubes (Figure 7). They can be protected from corrosion by painting and coating in the same way as steel or cast-iron water boxes which, however, are endangered to a lesser degree because of their greater distance from the brass tubes. The following protective procedures, under various trade names, are currently being used:

Coating with paints based on rubber, synthetic resins or bitumen.

Coating with synthetic resins or mixtures of synthetic resins and bitumen.

Priming with combinations of metal powders and synthetic resins.

Metal spraying, often as a base for painting and coating.

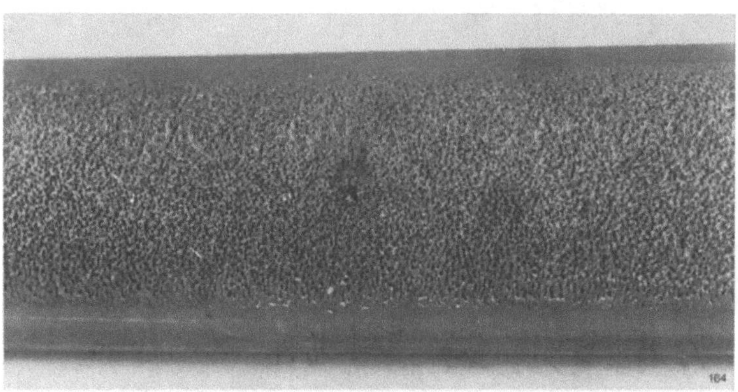

Figure 6. Erosion of a top tube through heavy impingement of steam.

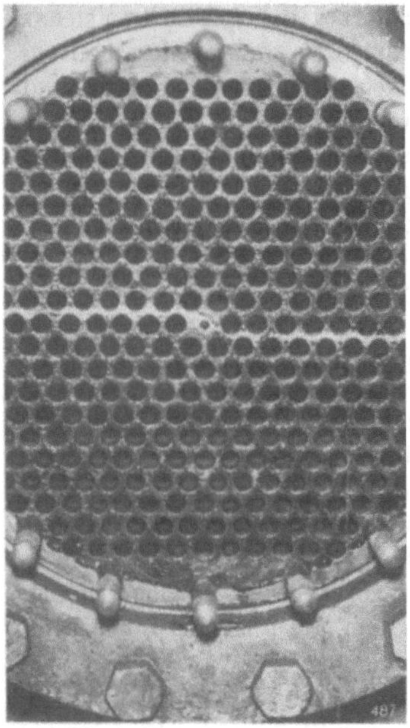

Figure 7. Concentrated tube-plate corrosion at locations of faulty protective layers.

The most important condition for all coatings is complete coverage of all steel surfaces and freedom from porosity achieved by multiple coats; otherwise, there is a risk of increased local corrosion and flaking off of the coating. The processing instructions (consistency, mixture ratios, air humidity, surface pretreatment by sand-blasting, temperature and dryness of the base material, as well as hardening times) must therefore be adhered to strictly. References should be obtained from specialist firms. If, on the basis of known water conditions, the use of primer coatings has already been decided upon in the planning stage, the condenser tubes should protrude from the tube plate by an amount equal to the coating thickness, i.e., 2 to 3 mm.

Experience has shown that absolute imperviousness of a coating is hardly possible. During the first two years of operation after its application, checks must be carried out on several occasions and repairs must be made where damage has occurred. Omission of these inspections and repairs to damage can lead to deterioration (Figure 8) which, starting from the point of damage, often progresses so rapidly that complete renewal of the coating and repairs to steel components become necessary. With local repair work, particular attention must be paid to careful preparation of the transition zone in accordance with the manufacturer's instructions. Supervision of this repair work is strongly recommended [19].

Good results in the prevention of corrosion of tube plates have been obtained by cathodic protection, which can be applied on its own or in conjunction with protective coatings. In principle, there are two methods of cathodic protection:

The installation of sacrificial anodes made from zinc or magnesium (Figure 9). These anodes must be fitted in such a way as to ensure good conductivity and hence an adequate protecting current at all locations [8]. Sacrificial anodes are consumed very quickly and must therefore be checked and renewed regularly.

The use of direct current, introduced via elec-

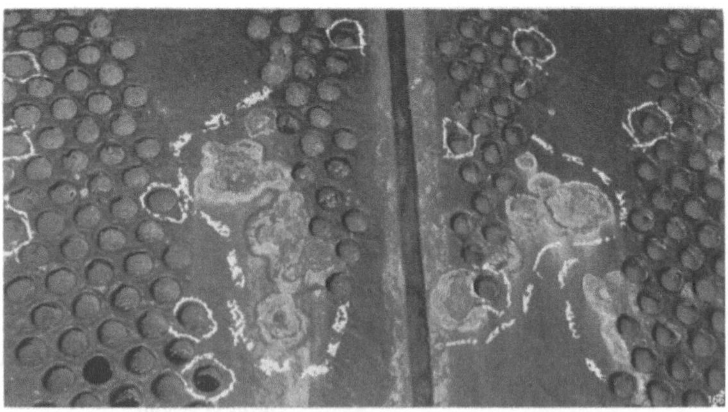

Figure 8. Severe corrosion of a tube plate made of steel.

trodes in a direction opposing that of the corrosion current. These electrodes last longer and, depending on operational conditions, need renewing on average after two to four years.

Cathodic protection is also often employed for protecting steel or cast-iron water boxes and covers if the tube plates are made from Muntz metal or brass.

Apart from concern about corrosion problems with condenser tubes and tube plates, the manufacturer must provide proper supports for the condenser. In German condenser designs, rigid connections between the exhaust-steam chest of the steam turbine and condenser predominate. Consequently, there is a need for resilient support of the condenser on the foundation.

For loss prevention, the spring assemblies should satisfy the following requirements:

Indication on a scale on each spring assembly of the limiting values as prescribed by the manufacturer and checked during commissioning for all operating conditions of the condenser.

Possibility of arresting the travel of the spring assemblies, so as to prevent thrust forces from being transmitted to the low-pressure cylinder of the steam turbine when the condenser is being emptied.

5.2.2. Operation of Condensing Plant

The weeks following the first filling with water, i.e., after taking the condenser into operation for the first time, are of decisive importance for the life of the condenser tubes. During this period, care must be taken that strongly adherent impervious protective films are able to form on metal

Figure 9. Sacrificial magnesium anodes in a condenser cover.

surfaces on the cooling-water side [4]. Slowly formed thin layers are denser than quickly formed thick layers and hence offer better corrosion protection.

To achieve the formation of a good protective layer, fresh water with as high an oxygen content as possible and a minimum amount of matter in suspension, preferably drinking water, should be used during the first period of operation. By dosing the cooling water with ferric sulphate (about 5 mg/l over one hour per day), the formation of a protective layer can be improved substantially [10]. Our own investigations at the Allianz Centre for Technology (AZT) have shown that a protective layer of $CuFeO_2$ is formed in this way.

Figure 10. Condenser tubes with unevenly distributed deposits that can lead to local corrosion.

With all tube materials, local corrosion is caused to a very large extent by uneven layer formation in the condenser tubes (Figure 10). For this reason, tube cleaning assumes a special significance in the prevention of damage by corrosion, apart from its purpose of maintaining good heat transfer.

In the presence of clean cooling water and in peak-load installations with relatively frequent shut-downs, periodic manual cleaning with brushes or rubber elements, shot through the tubes by pressurized water, often suffices. However, the degree of fouling, whose progress can vary greatly in accordance with the season and changing climatic conditions, must be supervised by control of the vacuum and of the exhaust-steam and cooling-water temperatures.

With twin shell condensers, each condenser can be cleaned manually during periods of part load. In this case, the spring supports must be arrested to avoid thrust forces on the exhaust-steam branch and displacements as a result of the missing cooling-water weight on one side.

With increasing fouling of cooling water and unit powers, continuously operating (i.e., semi-automatic or fully automatic) self-cleaning installations are finding ever-increasing acceptance. In practice, two systems have proved their value:

Self-Cleaning Installations with Foam-Rubber Balls, Taprogge System

In this system (Figure 11), foam-rubber balls, with a diameter slightly greater than the internal diameter of the condenser tubes, are introduced into the cooling-water inlet. The balls pass through the tubes, are caught in a screen, and are again introduced into the cooling-water stream by a pump.

With these installations, it is important that the distribution of the balls over all the tubes is as even as possible. Ball loss and wear must be checked constantly. Dead pockets, in which balls can accumulate, should be covered with perforated sheet-metal; also, in some instances, it may be necessary to provide baffle plates for improving the supply to certain tube sections. With very dirty cooling water (e.g., flood water containing hay and tree branches) the increase in pressure loss can cause deformation of the catching screen. In all circumstances, therefore, operation without

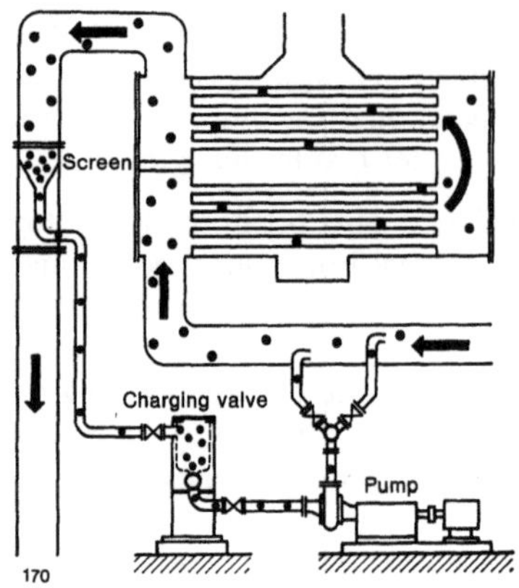

Figure 11. Schematic circuit diagram of a Taprogge-system self-cleaning condenser installation, according to manufacturer's data.

precleaning of the cooling water must be avoided. Coarse deposits, which may have formed despite cleaning, can be responsible for heavy ball wear. This contingency can be avoided by the use of foam balls with a corundum covering, by manual cleaning and also by pickling; however, only experienced specialist firms should be entrusted with condenser pickling. If necessary, the pickling process should first be checked by consultation with Allianz.

Condenser-Cleaning Installations, M.A.N. System

In this system, each condenser tube is provided with a plastic brush and a catching basket at either end (Figure 12). Cleaning is effected by a short-time reversal of water flow (Figure 13), thereby impelling the brushes from their rest position (in the catching basket on the water-outlet side) through the tubes and then back again. Depending on the degree of fouling, this cleaning cycle can be repeated automatically at predetermined intervals.

To prevent the brushes, at the moment of their reversal, from being shot through the tubes at high velocity by expanding air and thereby damaging the catching baskets, care must be taken to ensure that there is always adequate de-aeration (air suction) on the cooling-water side. Brushes must be inspected for wear, and brushes which have become stuck in the tubes as a result of large foreign bodies must be removed.

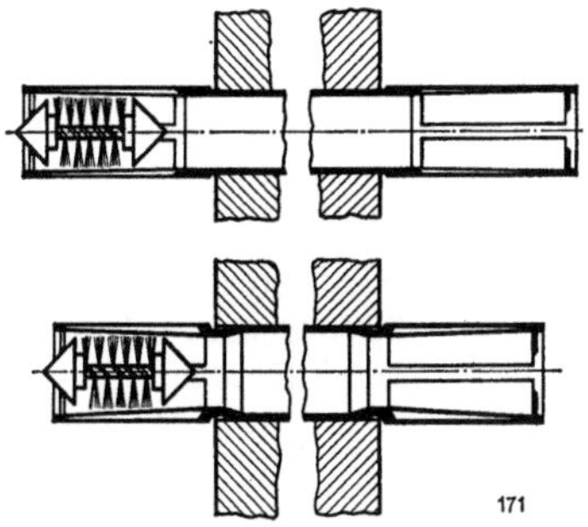

Figure 12. Disposition of catching baskets and brushes in the M.A.N.-system condenser-tube cleaning installation, according to manufacturer's data.

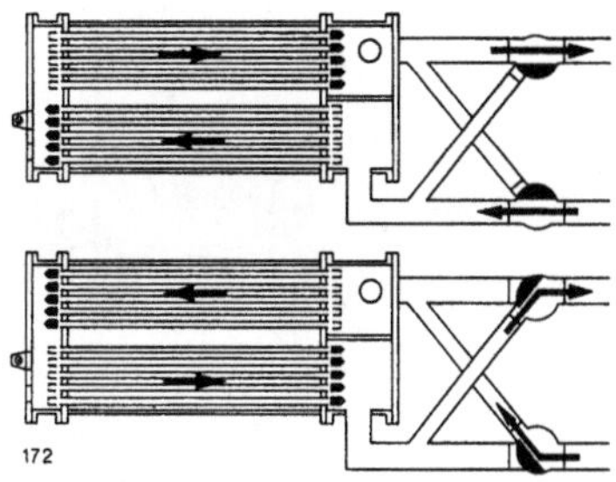

Figure 13. Circuit diagram for reversing the flow of cooling water in the M.A.N.-system condenser-tube cleaning installation, according to manufacturer's data.

Cooling-Tower Water

With cooling-tower circuits, conditioning of the cooling water and treatment of the make-up water assume great importance. Permanent control is necessary, since conditions can change easily within a short time through external influences [11, 12]. Essentially, treatment comprises

adjustment of the hardness of the cooling water (pH-value),
addition of a corrosion inhibitor,
addition of a biocide,
addition of a dispersion agent,
addition of a stabilizer,
treatment, in some cases, of the fresh water.

An automatic installation for monitoring pH-value and dosing with chemicals is recommended [13, 14]. Table 4 gives guidelines of the VGB for make-up and circulating water [15].

Corrosion on the Outside of Tubes

The most frequent cause of tube damage emanating from the steam-condensation side occurs through erosion-corrosion of material, mainly in the region of air extraction immediately adjacent to the support plates (Figure 14). This type of corrosion results from ammonia enrichment in the condensate, mainly in quiet areas next to the support plates, in conjunction with traces of oxygen [1, 2]. To reduce erosion-corrosion, it is recommended that alkalization of the feed-water with ammonia or hydrazine excess should be so chosen that the content of ammonia does not

Table 4. VGB guidelines for make-up and circulating water [15]

Make-up (fresh) water for cooling-tower circuit:	
Carbonate hardness (m-value)	~0.5 mval/l
Polyphosphate (P_2O_5)	$\leqq$0.2 mg/l
Cooling-tower circulating water:	
Chloride (Cl^-)	$\leqq$ 600 mg/l
Sulphate (SO_4^{2-})	$\leqq$ 400 mg/l
Salt content	$\leqq$3000 mg/l
Carbonate hardness (m-value)	
(a) in the presence of polyphosphates	$\leqq$ 4 mval/l
(b) without the application of polyphosphates	$\leqq$ 2 mval/l
Matter in suspension	< 10 mg/l
Oil	< 1 mg/l
Free chlorine (Cl_2)	4 mg/l (shortly after chlorination)
pH-value, 20 °C	$\geqq$ 7

Figure 14. Erosion-corrosion on the condensate side of condenser tubes through ammonia-containing condensate.

exceed from 0.4 to 0.7 mg/kg. Furthermore, a remedy can be effected by increased condensate impingement (installation of baffle plates) or by the use of tubes made from CuNi 30 Fe or stainless steel in the air-cooling zone.

If the conditions for corrosion by ammonia are present in a condenser, it is possible that components of air-extraction devices made of copper alloys, such as the impellers and protective sleeves of liquid-sealed pumps, will also suffer corrosion. In such a case, it is recommended that these components should be made of stainless steel.

Preventive Measures during Non-Operational Periods

During *short non-operational periods,* the condenser should remain filled and the cooling water should be circulated at a velocity of not less than 0.7 m/sec. This minimum through-flow velocity, as experience has shown, prevents the deposition of suspended matter in the tubes; for this reason, the velocity should also never be less than 0.7 m/sec during operation.

During *longer non-operational periods,* the cooling water must be drained completely and, if necessary, the condenser flushed with water having a low salt content, followed by drying by blowing fresh air through the manholes. Another possibility of protection during non-operational periods consists of draining the cooling water and then refilling the condenser completely with drinking water, perhaps after flushing it. Such measures are of decisive importance when interrupting operation during periods when protective films are formed at the time of first commissioning or after installing new tubes.

5.2.3. Inspection and Revision of Condensing Plant

The most important measures for loss prevention are regular inspections and revisions, since they can be carried out on condensers without great expense, even during short interruptions in operation.

Before taking a condenser out of operation, a vacuum test should be carried out [16]. With the air-suction gate-valve closed, the pressure rise (torr/min) in the condenser is recorded, e.g., every minute over a period of 10 min. During this period, the flow of exhaust steam from the turbine must remain unchanged, since vaccum tests can be compared one with another only if the condenser loading remains constant.

With loading, the size of the vacuum space also varies in accordance with the development of pressure in the turbine. If, apart from the condenser pressure and the exhaust-steam temperature, all pressures below 1 atmosphere absolute at bleed points and low-pressure preheaters are recorded and the vacuum test is carried out at different partial loads, it is possible to establish the locations of leakages [16].

During Short Non-Operational Periods the Following Inspections and Checks Must Be Carried out:

Checking of the condition of protective coatings or layers on tube plates and water boxes. Places where bulging or blistering has occurred can be an indication of corrosion underneath as a result of porosity or bad adhesion to the base. By opening up one or several blisters, it is possible to check for the presence of watery black to brown corrosion products. Such areas, as well as flaked-off layers, must be repaired as soon as possible—at the latest during the next longer period out of operation (see p. 175).

Figure 15. A spent sacrificial anode in an oil cooler.

Inspection and, if necessary, removal of sacrificial anodes or electrodes for cathodic protection (Figure 15).

Examination of tubes for freedom from obstructions, removal of trapped foam-rubber balls or brushes and, if necessary, manual cleaning of tubes. Also, analysis of deposits, comparison with a previous analysis and, if required, a change of cooling-water conditions.

Checking of all fittings, vacuum-breaking devices, slides, valves, non-return flaps, and water injection of the cross-over station for ease of operation and leak-tightness. Non-return flaps behind the condensate pumps are often neglected. If they fail, the condensate pump runs in reverse, leading to bearing damage and rubbing of the rotor in the casing.

Examination and checking of the functioning of the condenser protection equipment, which should, as a matter of principle, be in duplicate and should be connected to the emergency-stop gear of the turbine. Also, checking and adjustment of the limiting pressures and temperatures prescribed by the manufacturer for alarm and emergency-stop pulses.

Checking of the air-extraction equipment and, if necessary, replacement of worn parts.

During Major Revisions, the Following Additional Work Is Required:

Repair or renewal of damaged protective paints or layers.

Checking of condenser tubes for freedom from leaks. In this regard, the following procedures have proved their value:

Filling of the condenser on the steam-condensate side and adding a fluorescent medium, e.g., sodium fluorescein, to the water. Then, by illuminating the tubes on the cooling-water side with a quartz lamp, even small leakages are made visible by fluorescence.

Twin or double condensers can be partially emptied on the cooling-water side and operated under partial load, the vacuum thereby being maintained on the steam-condensate side. In this case, leaking tubes can be detected by sticking to both tube plates parchment paper, which ruptures where tubes are leaking because a vacuum is also generated in such tubes.

Examination of internal tube surfaces for local corrosion and cracks in the rolling-in zone. This examination can be carried out visually with an endoscope (Figure 16) or by the eddy-current method [17].

Checking of the outside of the tubes in the air-cooling region for corrosion by ammonia. This

Figure 16. Endoscopic inspection of condenser tubes by a member of a measuring team from the Allianz Centre for Technology (AZT).

Table 5. Application of supervisory, safety and protective equipment for condensers

Measured parameter	Type of supervision			Safety equipment	Protective equipment	Danger
	Indication	Recording	Alarm			
Condenser pressure	×	×	×	Turbine controller	Bursting discs 100%	Excess pressure
Exhaust-steam flange temperature	×	–	×	–	Emergency shut-down at 120 °C	Overheating by cross-over steam
Exhaust-steam flange temperature	×	–	×	–	Reverse-current protection, emergency shut-down at 120 °C	Overheating by motoring
Conductivity	×	×	×	–	Emergency shut-down with all ×	Corrosion

check can be carried out with the eddy-current method by highly skilled personnel (the influence of the support plates must be taken into account); generally, however, some plates will have to be removed for a meticulous check. Revision of the air-extraction equipment, and also of the condensate and cooling-water pumps, including their drives (see 5.3 and 5.4) in the fully opened-up condition.

5.2.4. Optimum Supervision

Supervisory, safety and protective equipment for condensers is discussed in [18]. Allianz recommendations are given in Table 5.

Bibliography to 5.2

[1] Splittgerber, E., Börsig, F.: Schäden an Kondensatoren und Oberflächenkühlern. Der Maschinenschaden 37 (1964), No. 11/12, pp. 213/226.

[2] Börsig, F.: Das Bild der Erosion und der Erosionskorrosion (Teil 2). Der Maschinenschaden 41 (1968), No. 2, pp. 51/56.

[3] Long, N.A.: Recent operating Experiences with Stainless Steel Condenser Tubes. American Power Conference 1966.

[4] Hoskinson, D.W., Kuester: Operating Experiences with Stainless Steel Condenser Tubes in Central Stations. ASME 1964, pp. 1/4.

[5] Held, H.-D.: Moderne Behandlungsmethoden in Kühlwassersystemen (Teil 2). Technische Überwachung 7 (1966), No. 7, pp. 226.

[6] Eichhorn, K.: Kondensatorrohre aus Kupferwerkstoffen – Richtlinien für Werkstoffauswahl und Betriebsbedingungen. Werkstoffe und Korrosion 21 (1970), No. 7, pp. 535/553.

[7] DIN 1785, Rohre aus Kupfer und Kupfer-Knetlegierungen für Kondensatoren und Wärmeaustauscher, No. 4.5.3.3.

[8] Schwenkhagen, H.F.: Schäden durch elektrolytische Korrosion. Der Maschinenschaden 29 (1956), pp. 161/168.

[9] Große Dampfkraftwerke. Edited by K. Schröder. Vol. III: Die Kraftwerksausrüstung. Part B: Dampf- und Gasturbinen, Generatoren, Leittechnik etc. pp. 197/198. Berlin, Heidelberg, New York: Springer-Verlag 1968.

[10] Corrosion Prevention and Control, May 1965 X.

[11] Held, H.-D.: Ausgewählte Kapitel der Chemie der Kühlwässer im Kraftwerksbetrieb. Mitteilungen der VGB, No. 96 (1965), pp. 161/174.

[12] Held, H.-D.: Moderne Behandlungsmethoden in Kühlwassersystemen (Teil 1). Technische Überwachung 7 (1966), No. 6, pp. 190/195.

[13] Weber, J.: Korrosionen und Ablagerungen in Kühlsystemen – ihre Ursachen und Bekämpfung. Schweizer Archiv 36 (1970), pp. 389/401.

[14] Richtlinien für die Aufbereitung von Kesselspeisewasser und Kühlwasser. Essen, VGB, 5th edition.

[15] Analysenverfahren für den Kraftwerksbetrieb. VGB, p. 59. Essen: Vulkan-Verlag 1962.

[16] Empfehlungen für Auslegung und Betrieb von Vakuumpumpen bei wassergekühlten Dampfturbinen-Kondensatoren. Verlags- und Wirt-

schaftsgesellschaft der Elektrizitätswerke. 1st edition 1970.
[17] Müller, E.A.W.: Handbuch der zerstörungsfreien Materialprüfung, C. 64, München: R. Oldenbourg 1971.
[18] Westhoff, G.: Einige Hinweise zur Überwachung von Turbinenkondensatoren. Energie 21 (1969), No. 3, pp. 89/93.
[19] Engelbach, W.: Erfahrungen mit der Korrosionsschutzbeschichtung von Kondensator-Rohrböden. Techn. Mitt. 63 (1970), No. 10, pp. 507/508.
[20] Hübner, H.: Werkstoffe für Kondensatoren und Kühler von Dampfturboaggregaten. Techn. Mitt. 63 (1970), No. 10, pp. 499/507.

5.3. Centrifugal Pumps

The following survey of causes and distribution of damage to centrifugal pumps is based on an evaluation of 400 clarified cases of damage. This evaluation spans a period of three years and includes the following types of pumps:

Single-stage standard pumps, multi-stage high-pressure boiler feed pumps, circulating pumps in industry and nuclear power stations, cooling-water pumps with or without variable-pitch impeller blades, large pumps and pump/turbines in pumped-storage installations.

In Sections 5.3.1 and 5.3.2, guidelines will be given for loss prevention in boiler feed pumps and cooling-water pumps.

Table 1 summarizes the percentage distributions of the numbers of cases of damage to centrifugal pumps in terms of their causes and locations.

Discussion of the Causes of Damage

The major portion of all cases of damage to centrifugal pumps is caused by product faults involving errors in planning and manufacture. Of operational faults, 18% can be traced to foreign bodies, this value referring to the 400 cases of damage to centrifugal pumps of all types evaluated. The same applies to damage caused by extraneous influences. If various types of pumps are regarded separately, these values are sometimes shifted appreciably.

Experience available indicates that faults due to mishandling generally occur during starting up or connecting the pumps.

Consideration of the signs and nature of damage occurring during operation (Table 2) permits conclusions to be drawn regarding successful loss prevention and the logical selection of protective and supervisory equipment. For example, supervision of quietness of operation, which today has been accepted only for large sets, would make it possible, on all pump sets, to indicate damage at the earliest possible moment. Abnormal power consumption of the prime movers should trigger alarm signals and effect shut-down of the set.

Furthermore, the frequency distribution of the types of damage which occur again and again (Table 2) is instructive. Thus, rubbing damage or clearance bridging (25%) predominates, while incipient cracks, fatigue failures and damage to bearings, i.e., damage that develops over a period of time, together account for 23% of all types of damage. Therefore, almost half the cases of damage occurring to pumps are typical in that they develop slowly in the course of operation, so that there is a high probability of discovering them by appropriate inspections and revisions before they result in major damage. Discovered in time, these distinctive types of damage often per-

Table 1. Distributions of the causes and locations of cases of damage to centrifugal pumps

Cause of damage	Distribution of the number of damage cases [%]	Location of damage	Distribution of the number of damage cases [%]
Product faults	*52*	Rotors	23
Errors in planning, design and calculation	18	Radial and axial bearings	22
		Balance arrangements	19
Manufacturing faults	17	Impellers, impeller blades	16
Faults in assembly and repair	10	Casings, fasteners	11
Materials faults	7	Fitting, strainers, transmissions	4
Operational faults	*44*	Guide blades	3
Maintenance faults	24	Others	2
Mishandling faults	20		
Extraneous influences	*4*		

Table 2. Results of the evaluation of the 400 cases of damage to centrifugal pumps with respect to signs and types of damage.

Sign of damage	Distribution of the number of damage cases [%]	Type of damage	Distribution of the number of damage cases [%]
Rough running	24	Rubbing damage	25
Sudden stoppage through jamming	23	Forced fractures	18
		Seizure	18
Fracture of components through mechanical overload	13	Incipient cracks	8
		Fatigue failures	8
Lack of lubricant	10	Wear of bearing metal	7
Distortions	8	Others	16
Others	22		

mit a clearer recognition of the true cause of damage and hence enable the defects in the pump to be rectified more successfully than would be possible after extensive major damage has been incurrred.

Particularly during planning and construction of new pump installations, manufacturers and users should follow VDMA guidelines:

No. 24296 *Centrifugal Pumps: Acquisition, Testing, Supply and Shipment;*
No. 24297 *Centrifugal Pumps: Technical Requirements.*

Understanding between the parties concerned is greatly enhanced by the use of uniform terminology; any subsequent deficiencies or faults in the installation, resulting from the use of non-uniform terms and definitions, can therefore be prevented right from the start.

Because of their special importance, cases of damage to boiler feed pumps and to axial-flow cooling-water pumps with variable-pitch impeller blades have been evaluated separately in the following.

5.3.1. Boiler Feed Pumps

Table 3 shows part of an evaluation of damage to boiler feed pumps according to reference [1] in the Bibliography.

Discussion of Damage Evaluation

The likelihood of *boiler feed pumps* components sustaining damage is in the following order of precedence:

1. Shafts with sleeves and impellers
2. Balance arrangements
3. Diffusers
4. Bearings
5. Shaft seals
6. Casings with screw fasteners

An evaluation has shown that about 23% of all cases of damage to boiler feed pumps can be traced to the failure of protective equipment and fittings. Particularly frequent are troubles associated with removing by-pass and balancing water, such as non-closure or jamming of the

Table 3. Distributions of the percentage number and costs of cases of damage to boiler feed pumps in relation to the causes of damage, according to [1].

Cause of damage	Distribution of the number of damage cases [%]	Cause of damage	Distribution of costs [%]
Product faults	49	Product faults	57
Operational faults	27	Operational faults	30
Extraneous influences	10	Extraneous influences	9
Not clarified	14	Not clarified	4

by-pass valve. Failure of non-return valves and non-closure of stop valves in the pressure line of the pump are other frequent causes of damage.

5.3.1.1. Planning and Design of Boiler Feed Pumps

The large proportion of product faults as the primary causes of damage has, in recent years, led to improved pump designs. The introduction of a new concept in pump design has been facilitated by increasing unit powers at about the same discharge pressures and hence substantially increased throughputs [2]. The use of higher driving speeds was a further step in the realization of this modern pump concept. Apart from improvements in the balance arrangements, this development of boiler feed pumps for large power stations proceeded from designs with many stages and a thin shaft to a design with fewer stages and a thick shaft, as shown in Table 4.

It should be noted that Table 4 and the numerical values given therein serve only as a rough guide. In practice, overlaps and exceptions are inevitable in view of special applications and constant progress in development.

When planning and ordering boiler feed pumps, a type of pump according to group (3) in Table 4 is to be preferred. The user can classify boiler feed pumps in terms of the three groups and can select the optimum means of loss prevention accordingly.

The new contact-free design of boiler feed pump has the following advantages:

- It is more robust and consequently less liable to trouble;
- starting up can always be contact-free;
- it ensures safety in operation, even with a high starting-up frequency and high peripheral speeds of the balancing gear and at the throttle gaps.

Even with this type of pump, care should be taken during planning, operation and maintenance to ensure that troubles can be recognized in good time or, better still, obviated by the provision of suitable protective and supervisory equipment and by checking its proper functioning. In the future, whenever operational parameters permit or require it, the new design of boiler feed pumps should be installed. The user will have to become

Table 4. Stages in the development of boiler feed pumps for discharge pressures higher than 100 atm gauge.

1 Original type of pump	2 Improved design	3 New contact-free pump [2, 3]
Balance arrangement with balance disc	Balance arrangement with balance disc and additional throttle sleeve; also as single- or double-disc design	Balance arrangement as single- or double-piston design
No axial bearing	Axial bearing with jacking device	Fixed axial bearing; hence adjusted minimum gap between balance disc and counter-disc
No hydrostatic intermediate bearing	With or without hydrostatic intermediate bearing	No hydrostatic intermediate bearing
Many stages, often more than 10	Many stages, often more than 10	Fewer stages, usually less than 10
Large distance between bearings	Large distance between bearings	Small distance between bearings
Thin shaft	Thin shaft	Thick shaft
Stuffing boxes	Stuffing boxes or mechanical seals	Mechanical seals
Low flow rates (up to about 300–400 tons/hr)	Low flow rates (up to about 300–400 tons/hr)	Higher flow rates (usually more than 400 tons/hr)

involved with the problems of both types. When planning a new installation, experience with both concepts will be a subject for discussion.

Axially Fixed Balance Arrangement

By securing the balance arrangement against axial movement on the pump shaft, as in the upper half of Figure 1, bridging of the gap between the balance disc and stationary disc can be prevented, e.g., with an empty pump shortly before restarting, if balance water acts incorrectly on the balance disc in the direction of the stationary disc.
If, however, the balance disc is axially movable (lower half of Figure 1), axial rubbing can occur and, in fact, in an actual case, resulted in considerable damage that could have been prevented by axial fixing of the balance arrangement. Thus, balance water from the balance-water line was admitted to the pump, which was in an empty condition for a short period before restarting (Figure 1). Since the balance disc could move on the shaft axially by 2 mm, the balance disc contacted the stationary disc during the starting process, leading to heavy scuffing. The results of such a rubbing action between the balance disc and stationary disc are shown in Figure 2.
The fixed balancing arrangement, particularly one with double-piston balancing (upper half of Figure 1), prevents metallic contact between the balance disc and stationary disc if the clearance S_{ax} is definitely fixed, i.e., with an immovable axial bearing. Such an arrangement is always preferred if disturbances in normal pump operation can occur as a result, for example, of evaporation, heavy overload or shock loading, or if, by an unfavourable mode of operation, such as frequent starting or operation at low speeds (e.g., turbine drive), no proper equalization of axial thrust can take place because of operation at less than minimum pumping pressure.

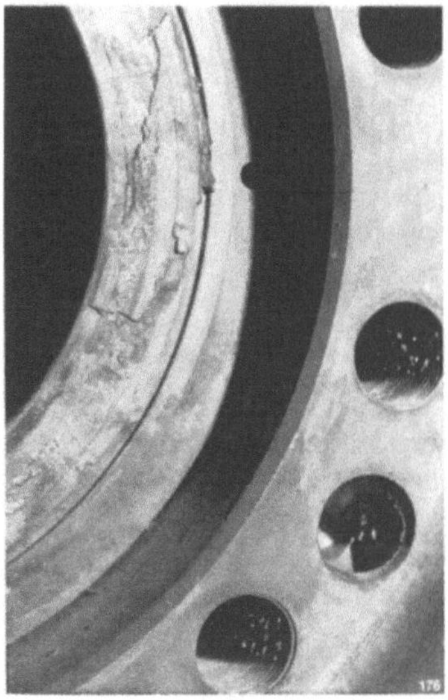

Figure 2. Stationary balance disc of a boiler feed pump ruined by scuffing.

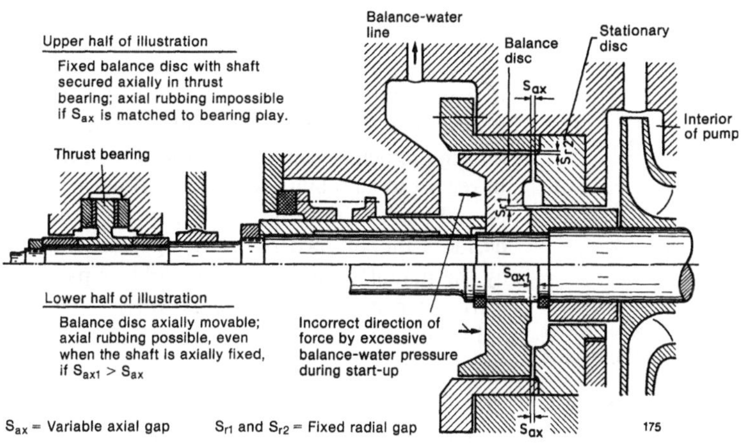

Figure 1. Axial fixing of the balance disc on the shaft, as in the double-piston balance arrangement shown in the upper half of the illustration, prevents rubbing on the stationary disc if, through a shortage of balance water, the balance disc is acted upon in the direction of the arrow by the balance-water pressure.

With a pump of the new design, the axial gap S_{ax} is so adjusted that it is about 0.2 mm when the bearing thrust collar makes contact. With this design, overloading of the thrust bearing, which is subjected only to residual thrust, can be recognized in time by the provision of an axial-force measuring device capable of triggering an alarm.

Hydrostatic Radial Bearing

The installation of a hydrostatic radial bearing [3] in a blind stage prevents instability of the rotor in cases where there is a long relative distance between the bearings, i.e., in a long pump with a thin shaft (original pump design), and is a safeguard against damage through radial clearance bridging. The vibrations produced, which are usually hydraulically excited, frequently occur when operating at low load. With seriously endangered pumps, the manufacturer should be consulted in good time regarding changes in design, so that, before major damage can occur, the necessary alterations can be made properly during a promptly scheduled non-operational period.

Throttling Elements

Shock loading and pressure fluctuations are often caused by instabilities in the balance-water line. In most cases, contact between the balancing elements results. By installing throttling elements or throttling lengths in the balance-water line (Figure 3), such instabilities can be rendered harmless.

In some cases, adjustable throttle elements are already installed in the balance-water line. Then, if instabilities do occur, it is usually possible to eliminate them by readjusting the throttles.

Balance-Water System

In order to be able to dismantle a pump without difficulty, e.g., during revisions or repairs, stop valves should always be installed not only in the discharge line, but also in the lines for balance water and bypass water to the feed-water tank. These valves must be secured against inadvertent closure, so that the pump cannot be run against a closed valve. In addition, a safety valve should be installed in the balance-water line between the pump and stop valve. The opening pressure of this valve should be about 1 to 2 kg/cm^2 above the maximum balance-water pressure, so as to ensure thrust equalization of the balance arrangement in every case of a disturbance.

The schematic arrangement of a balance-water system, in which the balance water and also the bypass water are returned to the feed-water tank, is shown in Figure 3.

It is recommended that each pump should be provided with a list similar to that in Figure 4, to indicate the positions of fittings during standstill, starting and operation.

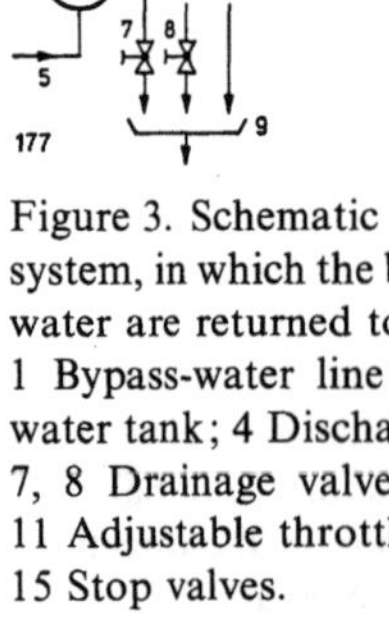

Figure 3. Schematic arrangement of a balance-water system, in which the balance water and also the bypass water are returned to the feed-water tank.
1 Bypass-water line; 2 Balance-water line; 3 Feed-water tank; 4 Discharge line; 5 Suction line; 6 Pump; 7, 8 Drainage valves; 9 Tundish; 10 Safety valve; 11 Adjustable throttle; 12, 13 Non-return valves; 14, 15 Stop valves.

Cold Starting

Not every pump is suitable for cold starting, but it is possible to design it for this purpose. The project and plant engineer should therefore state his requirements clearly right from the beginning in regard to suitability for cold starting, perhaps on full load, so that the pump manufacturer can decide on the correct design. Every guarantee given by the manufacturer respecting cold or partially warm starting presupposes, however, that the valves in the pressure lines of the pump close tightly.

Barrel-Casing or Ring-Section Pumps

With regard to the problem of choosing between barrel-casing and ring-section pumps [17, 19], all

	Item No.	Standstill						Start-up		Operation	
		Cold		Revision		Warm, ready for operation					
		open	closed	open	closed	open	closed	open	closed	open	closed
Stop valves											
suction line			×		×	×		×		×	
discharge line			×		×	×		×		×	
balance-water line	14		×		×	×		×		×	
bypass-water line	15		×		×	×		×		×	
heating-up line			×		×	×		×		×	
Drainage valves											
balance-water line	7	×		×			×		×		×
bypass-water line	8	×		×			×		×		×

Figure 4. Example of a proposal for a check list of fittings. In this example, the item numbers given correspond to those indicated in Figure 3.

that can be said within the scope of this handbook is that, with large unit powers, the barrel-casing pump is to be preferred if a 100% reserve has not been catered for. During an outage of a pump of the barrel-casing type, the possibility exists of restricting the time out of operation to a minimum, because it is feasible to dismantle and reassemble such pumps much more quickly.

5.3.1.2. Experience during the Operation of Boiler Feed Pumps

A cause of damage frequently observed with boiler feed pumps is evaporation of the liquid in the suction line or the interior of the pump. The reasons for this evaporation are given in references [4–7, 16, and 22].

Prevention of Evaporation: Pressure Reduction

The maximum rate of pressure reduction calculated by the pump manufacturer (from about 2 to 3 kg/cm^2 per minute) can be obtained in pumps without disturbances, provided that there is no evaporation in the suction line to the pump. In general, evaporation does not occur if:

- pressure reduction is continuous;
- the feed-water tank has an effective volume sufficiently large to permit good mixing of the water, and the take-off branch to the pump is so arranged in the feed-water tank that the coolest possible water is always supplied to the pump; the hot feed water from the deaerator must therefore not reach the suction line of the pump too quickly;
- the branch connection in the feed-water tank for the supply line to the pump must be so designed as to be conducive to favourable flow conditions, with an adequate radius and provided with flow guides, so that entry losses are kept small and inlet eddying is avoided;
- a separate supply line from the feed-water tank is provided for each pump;
- the supply line leads directly to the pump without horizontal sections and loops;
- the flow velocity in the supply line does not exceed 2 m/sec.

It is known that, even with low rates of pressure reduction in the suction line of non-operating pumps, evaporation can take place [6]. If a pump is connected up when such a condition of evaporation exists, it can lead to cavitation, resulting, in most cases, to damage by rubbing action. If no special countermeasures can be adopted, e.g., constant flushing of the suction line, the rate of pressure drop should not exceed 3 kg/cm^2 per minute. Evaporation in the suction line almost always leads to pronounced running disturbances of the pump and shock loading. Damage by rubbing action is the usual consequence (Figure 5).

Figure 5. A pump impeller destroyed by rubbing action. The cause of damage was evaporation in the suction line.

Leaking Stop Valves

Practical experience has shown that stop valves and other types of shut-off elements in the pipelines connected to the pump sooner or later start to leak. The effects of leaking stop valves on stand-by-pumps are known [7], results usually are distortion of the casing or the rotor during non-operative periods. When starting up such a pump, radial contact between rotating and stationary pump components often occurs, leading to extensive damage. This is particularly true with the chromium steels generally employed today which possess unfavourable emergency running properties in comparison with those of the previously used alloyed cast iron. The cause of the distortion produced is the uneven temperature distribution in the pump when not in operation. Such differential-temperature regions are created if, for example, small amounts of hot water enter via leaking stop valves in the discharge or tapping line and flow through the stand-by pump towards the suction branch. This possibility is twice as great with pumps having a tapping connection than those without tapping, because of the additional branch. Back-flow through leaking stop elements in tapping lines and the resulting distortion of casings are a more frequent cause of damage than had hitherto been assumed [1].

Evaporation in a pump standing by in readiness can occur if a stop element, such as a non-return valve of the pump, is leaking and, for example, the water entering via a common discharge line from another pump in operation is depressurized in the stand-by pump. If such a pump must suddenly be taken into operation, extensive damage generally ensues as a result of rubbing at the balance arrangement and the rotor. Moreover, leaking stop elements harbour the inherent risk that pressurized water will flow back and cause the pump to rotate backwards. Such a condition often leads to damage to the balance arrangement and to the bearings (see 8.6. Oil Supply Installations). Stop elements can be considered sufficiently tight if, after taking a pump out of operation, its temperature decreases steadily during the first hours.

Temperature Supervision

By measuring the temperature at the suction and discharge branches, it is possible to recognize evaporation in the interior of a pump. Also, temperature supervision of the suction line of a pump in operation is strongly recommended, if both pumps are connected to a common suction line, as in Figure 6. The suction temperature should not be higher than the temperature in the feed-water tank. Such a temperature differential arises, particularly in the case of a sudden pronounced drop in load, when large amounts of cold condensate flow into the feed-water tank. A higher temperature level ahead of the pump is permissible only to the extent that the resulting pressure drop does not conflict with the minimum pressure head stipulated by the manufacturer.

An Example of Damage

Evaporation in a pump can occur, in the same way as described above, by leaking or jamming stop elements in the balance-water and bypass lines if there are common connecting lines to the pump in operation or to the feed-water tank,

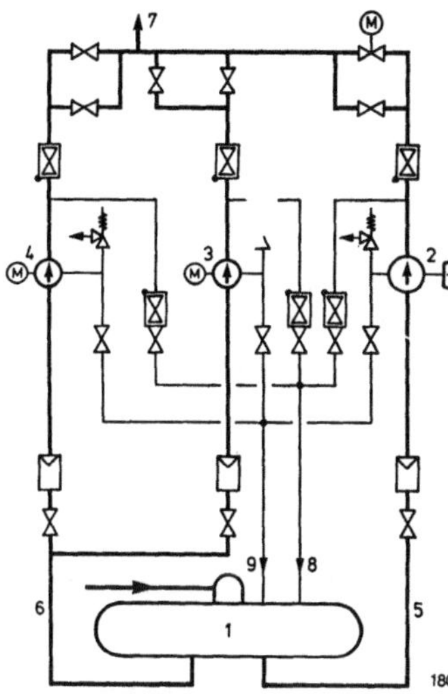

Figure 6. Simplified pipeline circuit of a group of feed pumps, with common suction and return lines for two or more feed pumps. The use of common lines 6, 8, and 9 can be conducive to steam formation in the pump and should therefore be avoided. Separate lines, as in Figure 7, are preferable.
1 Feed-water tank; 2 Full-load feed pump; 3, 4 Half-load feed pumps; 5 Suction line for pump 2; 6 Common suction line for pumps 3 and 4; 7 Discharge line; 8 Common bypass line for pumps 2, 3, and 4; 9 Common balance-water line for pumps 2, 3, and 4.

as in Figure 6. Incorrectly installed elements can also cause evaporation in the pump and consequent severe damage. An example is given in the following:
During assembly, the non-return valve (item 13 in Figure 3) behind the automatic bypass was installed the wrong way around, with the result that the requisite minimum flow quantitiy of fluid (bypass) could not be carried away. Moreover, this minimum quantity could not be removed, even by opening a manually operated valve, because the valve cone, which was provided with an internal thread, unscrewed itself out of the retaining nut. This nut was pressed by the stream of water against the outlet, and the water, heated up by the pump rotor, evaporated when the stop valve was opened. Heavy damage to sealing rings, distance sleeve, impellers, high-pressure cover and balance device ensued through rubbing action and thermal stresses.
If such difficulties are experienced when trying to start up, operation must be stopped immediately and all fittings must be checked for faultless functioning.

Combined Pipelines

By combining the pipelines of several pumps, there is always the risk that, in the event of leakage through stop elements, particularly non-return valves, liquid from an operating pump may reach a standing pump. Common balance-water and bypass-water lines in the immediate vicinity of the pumps has proved to be extremely unfavourable, as evidenced by several causes of damage brought about by leaking fittings. These lines, therefore, should in all cases be connected separately to the feed-water tank, as in Figure 7, or, if this is not possible, should only be combined close to the feed-water tank. In the latter case, the common inlet branch should have a cross-section at least 1.5 times the sum of the individual cross-sections of the pipes that have been combined. This loss-prevention measure can also be retrofitted to existing installations during overhaul.

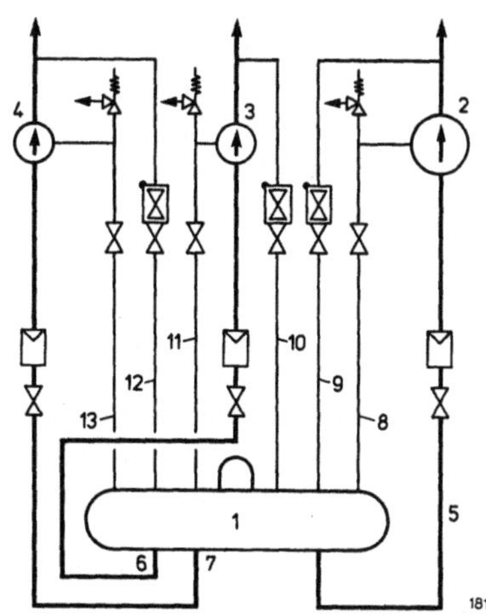

Figure 7. Group of pumps, as in Figure 6, but with separate suction, bypass and balance-water lines for each pump.
1 Feed-water tank; 2 Full-load feed pump; 3, 4 Half-load feed pumps; 5 Suction line for pump 2; 6 Suction line for pump 3; 7 Suction line for pump 4; 8 Balance-water line for pump 2; 9 Bypass line for pump 2; 10 Bypass line for pump 3; 11 Balance-water line for pump 3; 12 Bypass line for pump 4; 13 Balance-water line for pump 4.

Dangerous Differential-Temperature Zones

When connecting up boiler feed pumps, damage which occurs through distortions of the casing or rotor as a result of differential-temperature zones (compare also Leaking Stop Valves, p. 190) can be avoided by supervision of differential temperature Δt during start-up. For this purpose, the supervisory equipment utilizes, for example, the temperature difference between the water in the suction branch of the pump and the lowest point of the inner pump casing as a controlling parameter and prevents connecting up the pump if the temperature difference is inadmissibly high. Such supervisory equipment should be installed on all important pumps—particularly those which must be frequently switched into the circuit.

There are three different methods of combating dangerous temperature zones directly:

- Turning over the contents of the pump by means of a separate circulating pump.
- Continuous operation at low speed.
- Constant supply of feed water at operating temperature.

With the first two of these methods, the formation of temperature zones is prevented during the cooling process. With the third method, cooling is prevented by a constant heat input. One of these three methods should be employed, especially for multi-stage pumps in installations where frequent starting is required.

Heating up and Cooling down

A very effective means of heating up (or steady cooling down) is achieved if warm water is injected at the bottom of the pressure and suction casing (or cold water is withdrawn) and if, in a similar way, temperature zones are at the same time avoided in the region of the stuffing-box cover on the suction side. In effect, it is immaterial whether feed water is injected into these spaces or whether water is withdrawn into a container at a lower pressure. If the suction branch points downwards, injection or withdrawal is unnecessary because cold water can flow downwards and does not remain in the pump casing.

A common method of heating up by injection (Figure 8) consists of warming up the non-operating pump from the discharge of the booster pump of the unit in operation. In the circuit shown, feed water from the suction branch of whichever pump is in operation is conducted to a common line as a result of the pressure increase through the booster pump, this line connecting all three feed pumps. From this line water reaches the admission lines to individual injection points, such as pressure and suction casings, or only the pressure casing and the stuffing-box housing on the suction side. Regulation of flow quantity is effected by orifices. To prevent return flow and hence an inadmissible pressure rise in the pipeline circuit when the feed pump comes into operation, two non-return valves are provided in the injection line on the pressure side. In addition, a pressure-operated switch actuates an alarm should there still be an inadmissible pressure rise. The common line is protected by a safety valve and can also be connected to a separate heating pump, so as to be able, in the event of all pumps standing idle, to heat up the particular pump which is to be taken into operation.

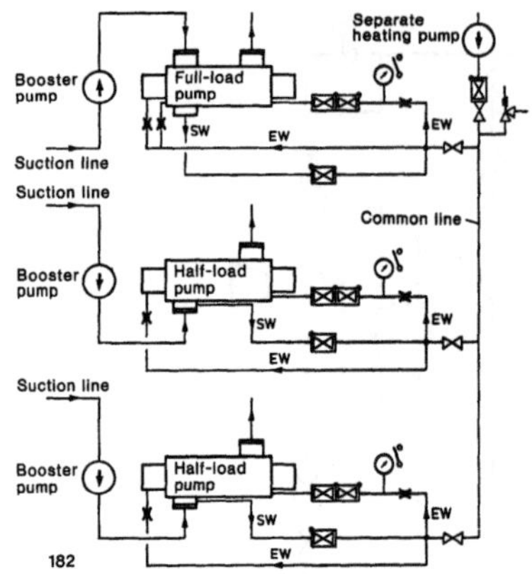

Figure 8. Injection-heating circuit for a group of feed pumps comprising one full-load and two half-load pumps, each with one booster pump.
SW = warm feed water to the common line from the suction branch of the pump which is running at the time.
EW = injection water from the common line to the points of injection of the particular non-operating pump to be heated up.

Protection from Foreign Bodies

Numerous cases of damage to pumps are caused by foreign bodies in the feed water. The theoretical requirement would be to install, ahead of the

Figure 9. Embedding of foreign matter into the throttle bush of the balance arrangement of a boiler feed pump.

Figure 10. Axial bearing of a boiler feed pump overloaded and destroyed by disturbance of thrust equalization.

pump inlet, a strainer having a mesh size smaller than the internal clearances of the pump. However, practical experience has shown that a mesh size of 0.5 mm is generally quite sufficient to eliminate the risk of damage. For pumps of corrosion-resistant construction, such fine-mesh strainers are an absolute necessity. Results of tests [1] have established that a fine strainer should even be provided for pumps of conventional design. The strainer and strainer casing should be matched so as to expose the strainer fully to flow.

Fine-mesh strainers must be located as near as possible to the suction branch of the pump. They should be capable of retaining welding beads, broken-off platelets of scale and other foreign particles that are always present in the circuit as a whole, particularly during the commissioning period. To prevent clogging of fine-mesh strainers, a differential-pressure supervisory installation is recommended, with switch-off or alarm facilities.

In cases where larger foreign bodies must be reckoned with, a coarser strainer with a mesh size of about 4 mm must always be provided in the suction line ahead of the fine-mesh strainer. In general, booster pumps should be protected by a coarse-mesh strainer on the intake side. The free cross-section of the strainer should correspond to at least three times the cross-sectional area of the suction line for both coarse and fine strainers.

In reference [1], examples were given of several cases of damage which can be traced back to a complete lack of or inadequate strainers. Figure 9 shows the throttle bush of a balance device, into which has been embedded foreign matter that has passed through the pump. The shaft sleeve and the balance piston were badly damaged by this foreign matter. Generally, damage caused by such foreign matter is responsible for a marked disturbance of axial thrust equalization and, in such cases, the result can easily be overloading and destruction of the axial bearing, as in Figure 10, or of the entire balance installation.

Guidelines for the Handling and Maintenance of Boiler Feed Pumps

As can be seen in Table 3, 27% of all cases of damage to boiler feed pumps are caused by operational faults, i.e., faults due to mishandling by operating personnel and to inadequate maintenance. Some of these cases of damage can be avoided if the following guidelines are followed:

If automatic pump protection is not provided, pump characteristics should be displayed clearly in suitable form next to the control installation. Figure 11 shows the characteristics of a pump, with limiting curves for maximum and minimum deliveries. In the example illustrated, with the pump is operating at a speed n_1, the opening point for the bypass valve is at $Q_{1\,min}$, while loading of the pump above $Q_{1\,max}$ is inadmissible.

Maintenance of pumps, fittings and strainers should be ensured by maintenance and reconditioning schedules, for which check lists, matched to the characteristics of operation and of the pump, are most useful.

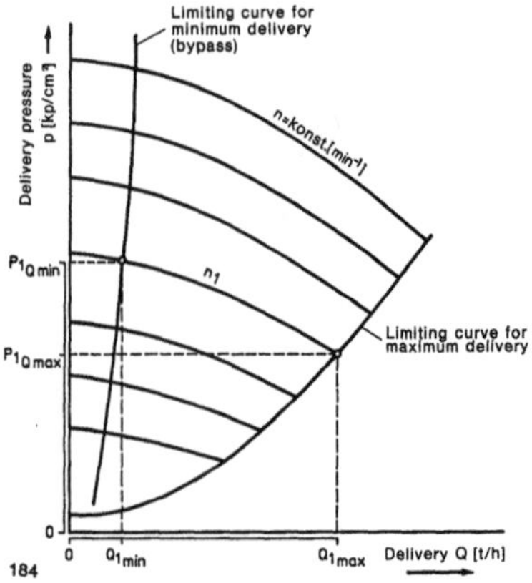

184

Figure 11. Characteristics of a boiler feed pump with limiting curves for maximum and minimum delivery.

As a matter of principle, the condition of boiler feed pumps, stop elements and strainers should be checked whenever revisions of boilers and turbines are being carried out. Quite independently, pumps of conventional design should be inspected after 250 starts at the latest with respect to internal clearances and the condition of the balance arrangement. A general overhaul becomes necessary if internal diametral clearances have increased to the maximum permissible amount stipulated by the manufacturer [2]. Further operation above this limit is unjustifiable because, in such circumstances, the natural frequency of the pump approaches the operating frequency.

Automatic Operating, Supervisory and Protective Equipment for Boiler Feed Pumps

Since a large proportion of operational faults is rooted in human fallibility, the maximum possible number of functions required in the course of operation of a pump should be carried out, and also monitored, by automatic control equipment. All actual operational data must constantly be compared with the desired values and, in the event of inadmissible deviations, the automatic control system must actuate accoustic or visual alarms, or take the group of pumps out of operation and, if necessary, bring stand-by pumps into operation. For this purpose, it has been found useful to group functions together and to automate them individually.

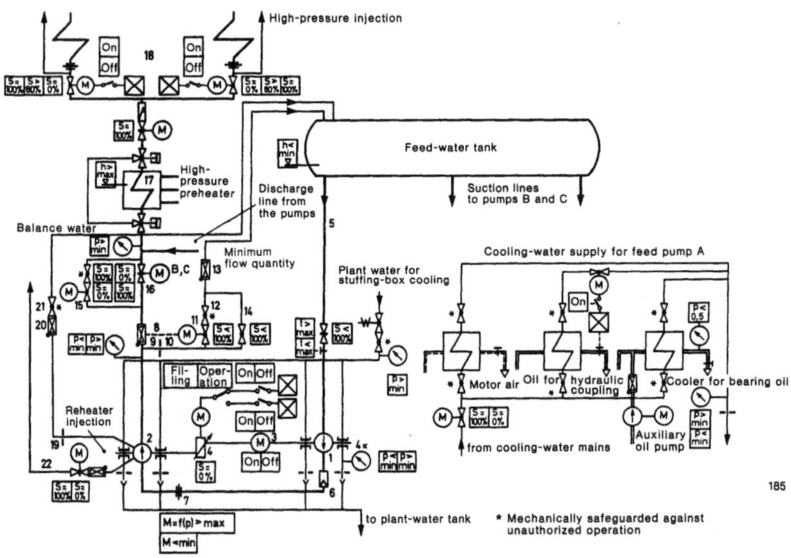

Figure 12. Schematic arrangement of a group of feed pumps with full protection and separate suction and return lines [13].
1 Booster pump; 2 Main pump; 3 Drive motor; 4 Hydraulic coupling; 5 Suction line; 6 Strainer; 7 Measuring orifice; 8 Non-return valve; 9 Bypass line; 10 Swirl nozzle in bypass line; 11 Electrically operated bypass valve; 12 Stop valve; 13 Non-return valve; 14 Manually operated control valve; 15 Bypass valve for filling the discharge line and the boiler; 16 Discharge stop valve; 17 High-pressure preheater; 18 Boiler; 19 Throttle in balance-water line; 20 Non-return valve; 21 Manually operated stop valve; 22 Water-injection line to reheater.

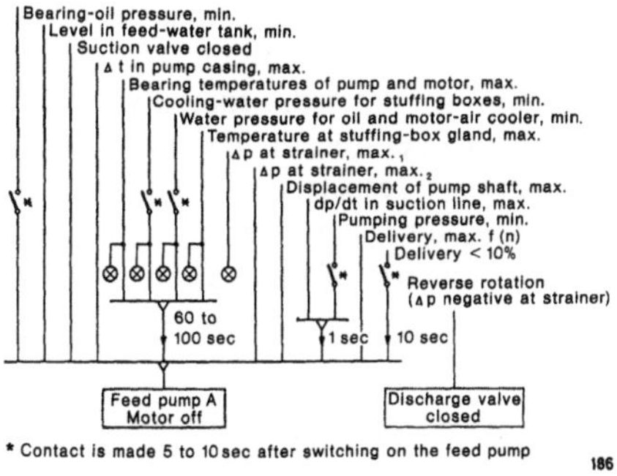

Figure 13. Full pump protection for the function group "feed-water supply" [13].

Industrially tested systems for the automatic control and complete protection of entire installations and of individual groups of functions have been developed and are in use [8–13]. Figure 12 shows schematically, according to [13], a pumping unit from the function group "feed-water supply", consisting of three identical half-load pumps A, B, C, equipped with the control, supervisory and protective equipment necessary for the prevention of disturbances and damage. In addition, Figure 13 shows schematically the mode of operation of the equipment for the full protection of such a pumping unit. In Figure 12, the demands made earlier for separate suction and return lines for bypass and balance water have already been met.

Pump groups must be protected from reverse rotation either by monitoring of the differential pressure Δp of the strainer or by a tachometer which indicates reverse rotation. In both cases, reverse rotation should trigger a pulse for closing the discharge valve and, for the prevention of damage to bearings, should automatically switch on the auxiliary oil pump.

The safety and interlocking system which has been installed for the protection of a group of three feed pumps in series (low, medium and high pressure) is described in [14, page 684].

5.3.2. Axial-Flow Cooling-Water Pumps with Adjustable Blades

In terms of their causes and locations, the evaluation of the distribution of cases of damage to cooling-water pumps of axial-flow design with adjustable blades yielded the results given in Table 5.

Discussion of the Causes of Damage

Of the total of 71% of product faults, 40% have their origin in incorrect constructional planning, incorrect information on operating data and inadequate matching of the prime mover and pump, while 60% are caused during manufacture and assembly. With cooling-water pumps, damage caused by *extraneous influences* is primarily attributable to sucked-in foreign bodies.

Table 5. Distributions of the causes and locations of cases of damage.

Cause of damage	Distribution of the number of damage cases [%]	Location of damage	Distribution of the number of damage cases [%]
Product faults	*71*	Rotor blades	40
a) constructional planning, incorrect operating data, inadequate matching of prime mover and pump	28	Screwed fasteners	20
		Shaft, rotor	10
		Casing, standpipe suspended	10
b) during manufacture and assembly	43	Bearings	10
		Guide vanes	6
Mishandling faults	13	Protective equipment	4
Extraneous influences	16		

5.3.2.1. Planning and Design of Cooling-Water Pumps

When planning intake structural work, it is imperative to ensure, particularly with an arrangement of several pumps situated side by side, that proper intake flow conditions to the pumps will be achieved with every possible combination of operating and standing pumps. Because of the higher building costs associated with the realization of these requirements, pumps are frequently expected to operate under inadequate intake conditions. Faulty intake flow conditions have on a number of occasions led to extensive damage to pumps or have, in conjunction with weaknesses in materials and design, encouraged the likelihood of damage.

Planning Requirements

- Each pump should, if possible, have its own intake supply basin with the same intake flow direction.
- Mixed operation should be avoided.
- Separate groups of pumps should be provided for operation with fresh water and with water from cooling towers.

The intake structure shown in Figure 14 does not fulfil the requirements of the pump manufacturer

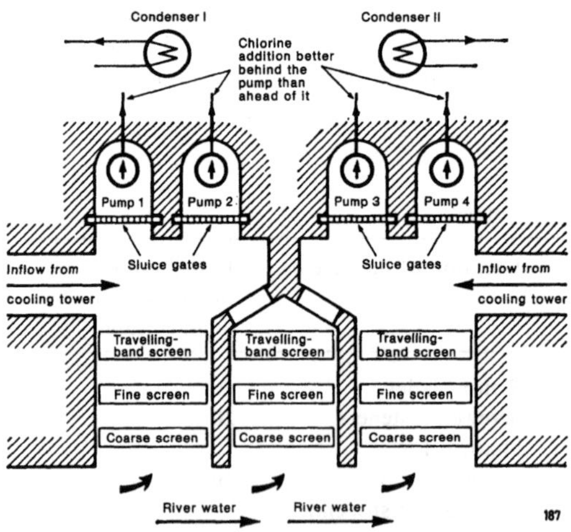

Figure 14. Very unfavourable intake conditions for the four cooling-water pumps (1 to 4) are caused by the fact that cross currents are created by the central intake and the lateral inflow from the cooling towers.

in respect of equal inflow conditions for each pump. Very complex flow conditions ahead of individual pumps are created, especially if only one pump each is operating in the two groups of pumps 1–2 and 3–4 respectively. Figure 14 also indicates how difficult it is to remedy shortcomings which cause damage, if the reasons for such damage are inherent in the intake structure.

Asymmetrical Inflow Conditions can:

- reduce the efficiency of the pumps;
- cause cavitation at the rotor blades;
- give rise, as a result of eddy formation, to vibrations which can lead to fatigue failures of pump components.

Since, as exemplified in Figure 14, the most propitious technical concept and the recommendations of the pump manufacturer are not always realized during planning and construction, it is considered worthwhile to mention the most frequent faults in the design of intake structures:

- Abrupt enlargements;
- insufficiently long intake channels;
- source of disturbances, such as projections or pipes at the bottom of the channel a short distance ahead of the pump;
- several asymmetrically located pumps in one pumping chamber;
- insufficient distance of the suction bell-mouth from the bottom of the pump basin or sump;
- water from pipes or conduits falling into the suction basin; a supply pipe reaching below the water level eliminates the otherwise turbulent flow conditions at the pump inlet and the entrainment of air;
- eccentric positioning of inflow with respect to the suction basin can, particularly with circular or semicircular chambers, produce dangerous swirling at the pump entry.

Recommendations regarding the most favourable dimensioning of pump basins as a function of pump delivery have been made and published [15] by the Hydraulic Institute, New York.

The inflow channel must be so dimensioned that, with full pump delivery, the water level cannot fall below the minimum depth of pump submergence stipulated by the manufacturer. A drop below this minimum depth must be prevented by a level-control system and/or dry-running protection equipment, to give a switch-off pulse

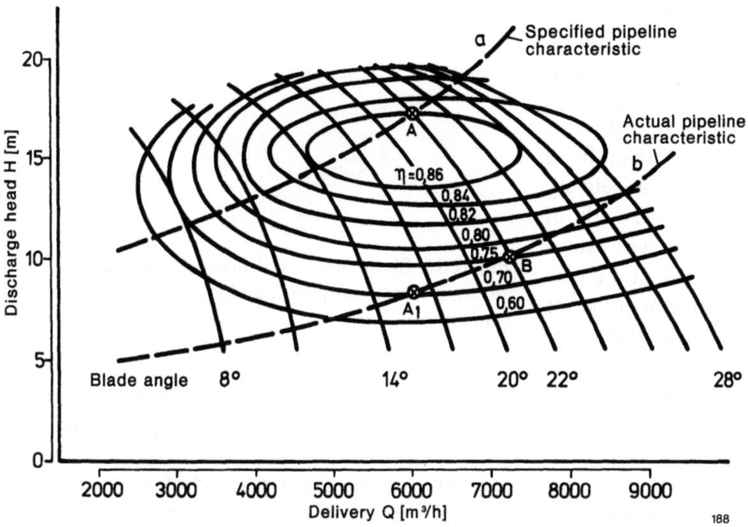

Figure 15. Characteristics of a single-stage axial-flow cooling-water pump with adjustable blades. η = Pump efficiency.

to the drive motor and to actuate an alarm, so that cavitation at the rotor blades, the formation of air "tubes", vibration and dry-running can be avoided. A level-control system is especially necessary in those cases where the amount of clogging of the intake screen varies very irregularly with time or where the pump intake is associated with a weir installation, which is also used for other consumers who have nothing to do with the pumps.

Damage to cooling-water pumps by foreign bodies indicates that, despite the cleaning of fresh water in the intake structure by coarse, fine and travelling-band screens, as is customary with condenser installations, larger foreign bodies still manage to get into the pumps.

The travelling-band screen installation should be self-cleaning and should be combined with a level switch. For the continuous control of screen clogging, the use of a differential-pressure monitor is recommended. The cross-sectional area of the mesh aperture of the band screen, as stipulated by the pump manufacturer, should be between 2 and 4 mm^2. The total cross-sectional area should be large enough to ensure that, if at all possible, a flow velocity at the fine strainer of 0.6 m/sec is not exceeded.

Planning Faults

A frequent planning fault is specification of too high a discharge head. Without preliminary correction of the blade angle, the pump will then operate in the region of higher, perhaps even inadmissibly high, deliveries and, moreover, lower efficiencies. These remarks can easily be corroborated by reference to the curves in Figure 15, in which it is assumed that the specification given to the pump manufacturer refers to the upper pipeline characteristic "a", with the operating point at A, corresponding to a delivery of 6,000 m^3/hr. If, however, the actual pipeline characteristic follows curve "b", the pump would no longer operate at point A but at point B, provided that there has been no change in the blade angle. In this case, the delivery is increased to about 7,250 m^3/hr, and the pump efficiency is decreased from 86% to 75%. Consequently, to reduce this higher delivery to its original required value of 6,000 m^3/hr, corresponding to point A_1 on curve "b", the blade angle must be reduced from 22 deg. to about 16 deg., involving a further reduction in efficiency to 70%.

Owing to such planning faults, vibrations and blade cavitation can occur as a result of exceeding the limits set by the manufacturer for delivery and blade angle. In addition, a substantial power loss is engendered by continuous operation of the pump in an inefficient region of the characteristics. Several cases of damage, in which flange fasteners fractured through fatigue at various locations in the pumps, have been caused by these faults in planning.

In one instance, where an installation had to be checked because of a case of damage, the pump did not deliver, as was specified in the order, against a head of 16 to 18 m w.g. but only against 8 m, but it had been found possible, with this inadequate discharge pressure, to obtain, by a slight adjustment of the blade angle, a delivery 50% higher than the specified delivery, without overloading the motor. However, as was established later, this delivery was higher than the permissible maximum for this particular type of pump, despite the fact that the blade angle was within acceptable limits. Operation of the pump under these conditions led to disturbed flow and vibrations, inducing stresses that exceeded the fatigue strength of the screwed fasteners (in another similar case, the fatigue strength of the rotor blades was exceeded). To prevent damage of the kind just described, it is imperative to check the discharge pressure during initial commissioning, so that blade angles can be adjusted to an optimum value and the permissible limits of blade adjustment established.

Harmful Vibrations

Damage to cooling-water pumps by vibration is often attributable to the fact that the drive unit, consisting of motor and motor bracket with interposed gearing, is frequently not supplied by the pump manufacturer and is inadequately matched with the pump in regard to vibrational characteristics. The customer should therefore order this unit immediately after ordering the pump and should ensure that the various suppliers match the vibrational characteristics of the different components with each other. If the data of the drive unit are made known to the pump manufacturer about 8 weeks after it has been ordered, optimum matching of the pump to the given vibrational conditions can usually be effected without major changes in design.

Since about 98% of all cooling water pumps are driven by electric motors, these guidelines for loss prevention apply to practically all pumps. With turbine drives, which are very rare, great importance should be attached to matching the critical speed n_{crit} of the pump, which is generally in the region of $n_{100\%} < n_{crit} \leqq n_{110\%}$, i.e., within the permissible overspeed region of the emergency stop gear of the steam turbine. The trip speed of the turbine must therefore be made known to the pump manufacturer promptly, so that he can raise the critical speed of the pump, by suitable design measures, above the trip speed of the turbine.

Figure 16. Blade of an axial-flow cooling-water pump, showing a crack originating from the stub-shaft/blade transition radius.

As indicated in Table 5, most cases of damage occur at the rotor blades. Although the cause of damage does not always lie in the rotor blade itself, it is still necessary to insist that this component should be as reliable as possible in operation, bearing in mind that inflow and water conditions are not always ideal, because damage to rotor blades usually results in severe damage elsewhere. A typical case of blade damage is illustrated in Figure 16. Those parts of the rotor blade which, on the leading and trailing side, project far beyond the stub shaft are relatively easy to excite by vibrations. The alternating loading thereby induced can cause high stresses at the transition of stub shaft to blade, with a particularly detrimental effect at these regions of stress concentration, because the transition and rounding-off radii cannot be given optimum dimensions. It is for this reason that the stub-shaft/blade transition radius must be executed with the utmost care.

Blade Materials

A further possibility of preventing such blade damage is by selection of the most suitable blade material with as high an alternating bending strength as possible for a known cooling-water analysis under the most unfavourable conditions. Alternating bending tests in air and river water on specimens of different blade materials have yielded markedly different values of alternating bending strength. As an example, Figures 17 and 18 show two results from our own series of tests,

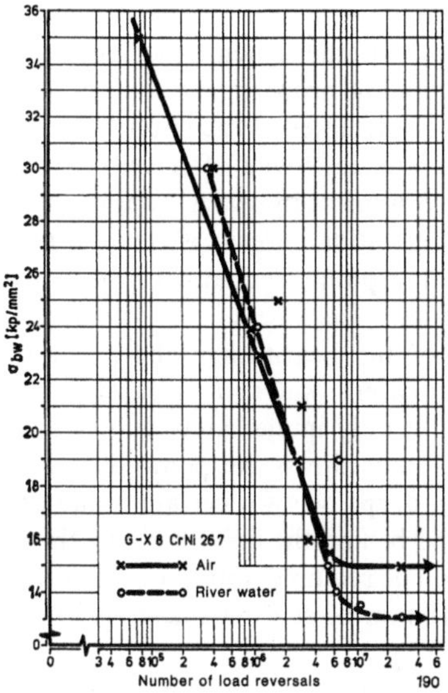

Fig. 17. Alternating bending strength of blade material G-X8 CrNi26 7 in air and river water.

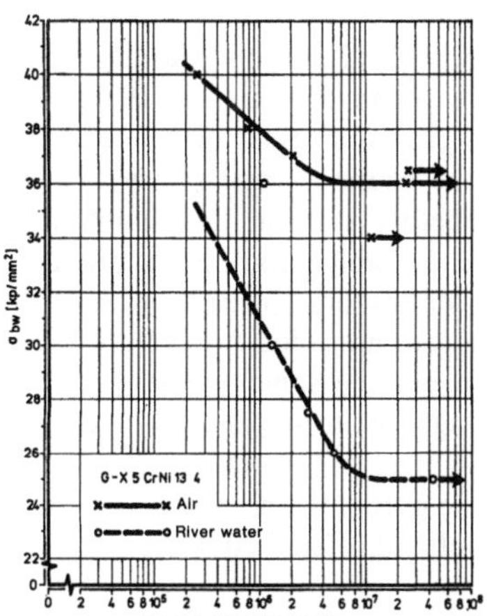

Figure 18. Alternating bending strength of blade material G-X 5 CrNi 13 4 in air and river water.

from which it can be seen that the material G-X 5 CrNi 13 4 possesses much more favourable strength properties than material G-X 8 CrNi 26 7. When selecting blade materials, the designer should have precise information about chlorination of the cooling water. If chlorine is added ahead of the pump, a further drop in alternating bending strength to below the values established in normal river water must be accepted and taken into account when designing rotor blades. For this reason, it is recommended that, as indicated in Figure 14, chlorine should only be added to the cooling water behind the pump.

5.3.2.2. Assembly and Commissioning of Cooling-Water Pumps

Damage to rotor blades has sometimes been caused by reverse rotation of the drive motor. In every assembly operation, particularly after repairs and revisions, the motor, when it has to be electrically reconnected, must first be uncoupled and checked for correct direction of rotation. During commissioning, the correct direction of rotation of the motor and pump must be indicated permanently and in a clearly visible manner in the vicinity of the coupling flange.

5.3.2.3. Experience in the Operation of Axial-Flow Cooling-Water Pumps

If an automatic control installation is not provided, the pump characteristics must be displayed clearly at the control console to enable the machine operator to adjust the blade angle to its optimum value in relation to the intake water level and the required delivery of cooling water or to take into operation an additional pump in good time when approaching the permissible limits.

In [14 (pp. 1046 and 1047)], in which the characteristics are given for the primary axial-flow cooling-water pumps for a particular power station, the author comments: "The delivery/head diagram indicates how the blade angles must be adjusted in dependence of the water level in the lower and upper basins to obtain the desired delivery. Before starting up the pumps, the blades must be swivelled by the blade-angle control gear to the smallest possible angle. Electrical interlocks prevent the drive motors from being switched on at other blade angles. In this way, motors and pumps are protected from inadmissibly high starting loads. Subsequently, the blade angles can be set, as required, within the region of adjustment, for the requisite delivery of water."

This starting instruction is the best means of loss prevention when starting these pumps.

Damage to Screwed Fasteners

As shown in Table 5, the second highest number of cases of damage occurs to screwed fasteners. Apart from those causes of damage to screwed fasteners that can be traced back to *planning faults,* loosening of unsecured set-screws has on several occasions led to fracture by overloading of other fasteners which have not become loose, with consequent severe rubbing and bearing damage to the pump. Since, during operation of the pump, shocks and vibrations are always present, all screwed fasteners must be locked. Set-screws can be locked effectively by the application of suitable self-hardening one-component adhesives. When carrying out revisions and repairs, care should be taken to ensure that screwed fasteners which have been disassembled and reassembled are provided with the same means of locking. It is recommended that installations should carry a notice indicating the means last used for locking the fasteners.

Damage to Bearings

Cases of damage to bearings included within the framework of damage evaluation always refer, with the exception of one case of damage by wear to the upper axial bearing, to the radial guide bearings. The principal causes of damage were found to be insufficient lubrication and wear caused by dirty lubricating water or by shock.

If sufficient clean water for lubrication is available from an extraneous source or if the water being pumped can itself be used after passing through a separate filtration system, water-lubricated rubber bearings are preferable for the guide bearings located in the stream of water. Such bearings can be applied successfully at peripheral speeds up to 6 m/sec and for bearing loads of 2 kg/cm^2. For higher bearing loads, rubber bearings can still be used but plastic bearings have recently proved successful competitors. Before entering the bearings, the lubricating water must pass through a fine filter, mainly for the removal of sandy contaminants. The great advantage of water lubrication is that grease or oil cannot penetrate into the cooling-water circuit from the lubricating system. On the other hand, the lack of emergency running properties is a disadvantage that must be accepted. When running dry, rubber bearings are destroyed after only a few revolutions. The lubricating-water supply must therefore be in operation at least 30 sec to 1 min before starting the pump and also during rundown of the pump. Because of the rapid rate of destruction of the bearings in the event of insufficient lubricating water, a separate flow control which not only immediately actuates an alarm, but also shuts down the pump automatically if the flow drops below a certain minimum must be provided behind the fine filter and ahead of each bearing. The required amount of lubricating water per 10 mm of shaft diameter is about 1 litre/min.

If it is necessary to support bearing loads lying above the permissible limits for rubber or plastic bearings, grease-lubricated leaded-bronze bearings are usually installed. These bearings are suitable for operation under water at peripheral speeds up to 12 m/sec and are continuously lubricated by a grease pump driven from the pump shaft. Here, too, provision must be made in all cases to supply the bearings with grease at least 30 sec to 1 min before starting the pump and, for this purpose, a prelubricating installation, driven by a separate motor, is necessary. A separate grease line is recommended for each lubrication point, to ensure the same grease pressure for each bearing. Water-repellent lime-soap greases are well suited as lubricants for pump bearings inside the pumping chamber, since they are not washed out and offer adequate protection against corrosion.

Supervisory Equipment

Apart from *supervisory equipment* for minimum water level in the pump sump, for differential pressure Δp at the travelling-band screen, and for the supply of lubricant, it is essential to install alarm-actuating bearing-temperature supervisory equipment for the upper, usually oil-lubricated, axial bearing, with provision for delivering an automatic shut-down pulse if the limiting values are exceeded.

So far, continuous supervision of the vibrational behaviour of cooling-water pumps in power stations has in most cases been dispensed with, and only occasionally have random test measurements been carried out. The assessment of these measurements follows VDI guidelines

2056 *Measures for the Assessment of Mechanical Vibrations of Machines*

Since most cases of damage in this evaluation manifest themselves by rough running, it is recommended, in the interests of attaining greater reliability in operation and early recognition of incipient damage, that the vibrational behaviour of pumps should be continuously monitored by permanently installed probes. This supervisory equipment must signal any excess of the vibration amplitudes normally permissible during operation and, better still, effect shut-down of the pump. Very simple supervisory equipment, without recording of values, is perfectly adequate for this purpose. To make it impossible to overload a pump during operation, the maximum permissible blade adjustment angles should be limited in both directions by automatically responding limit switches. The adjustment of these limit switches must be carried out by the pump manufacturer during the first commissioning phase after determining the operating point of the pump in each particular case.

5.3.2.4. Maintenance and Inspection of Axial-Flow Cooling-Water Pumps

Generally, fixed intervals between revisions are not customary with axial-flow cooling-water pumps. Proper maintenance of the installation (e.g., cleaning or exchanging filters for the lubricating medium; checking the operation of the lubricating pumps, the blade-adjustment gear, and the through-flow and pressure-measuring instruments; checking of the limit switches for blade adjustment) should be carried out at regular intervals in accordance with the recommendations of the manufacturer.

During longer periods out of operation, it is recommended that, if the pump as a whole is not being overhauled, an inspection should be made of the complete external guide and suspension pipe, particularly the individual screwed flange connections. A further desirable measure for the early recognition of damage would be checking the rotor blades for freedom from cracks with the aid of a flexible endoscope. However, a prerequisite for this latter check is that the pump basin or sump can be partitioned off, as indicated in Figure 14, and can easily be emptied. This proviso should be catered for when planning the intake structure.

Guidelines for Repairs

In conclusion, a few observations should be made concerning the repair of damage through cavitation, with special emphasis on rotor blades:
Hitherto, this type of damage was rectified by deposition welding, using a suitable filler material, followed by grinding. Such repairs are expensive, as the blade to be repaired has to be heated, prior to welding, to about 300 to 400 °C, depending on its material of construction, and then has to be stress-annealed after welding. However, a compound, consisting of a metallic material and a special plastic, has recently been used with great success for carrying out such repairs.

Bibliography to 5.3

[1] Horenburg, O.: Schäden an Kesselspeisepumpen. Hinweis zur Schadenverhütung durch Auswertung von Schadenstatistiken. Der Maschinenschaden 43 (1970), No. 4, pp. 135/147.

[2] Richter, H.: Entwicklungsstand der Kesselspeisepumpen. VDI-Z. 111 (1969), No. 3, pp. 147/152.

[3] Honold, E.: Die Entwicklung von Kesselspeisepumpen bei KSB zwischen 1935 und 1968. KSB Technische Berichte 15 (Feb. 1971), pp. 12/15.

[4] Strub, R.A.: Abfall des Saugdruckes von Speisewasserpumpen bei starken Lastschwankungen. Technische Rundschau Sulzer 42 (1960), No. 3, pp. 41/44.

[5] Wollschlaeger, K.: Maßnahmen zur Verbesserung der zulässigen Druckabsenkungsgeschwindigkeit bei Kesselspeisepumpen. Energie 18 (1966), No. 1, pp. 16/18.

[6] Rahlwes, H.: Untersuchung zur Klärung von Zulaufstörungen an Speisepumpen bei Gleitdruckentgasung. Mitt. VGB (Feb. 1967), No. 106, pp. 61/67.

[7] Stönner, A.: Ein Beitrag zur Schadenverhütung an Hochdruckkesselspeisepumpen. Energie 18 (1966), p. 360.

[8] Freymeyer, P.: Ein Konzept zur Automatisierung von Wärmekraftwerken, Electrotechnik 46 (1968), pp. 115/120.

[9] Ramsauer, H., Stoll, A.: Schutzeinrichtungen für Kesselspeisepumpen. Siemens-Z. 39 (1965), pp. 341/342.
[10] Freymeyer, P., Jung, M.: Simatic-Steuersystem P für die Betätigungsebene in Wärmekraftwerken. Siemens-Z. 42 (1968), pp. 436/442.
[11] Jung, M., Schmidt, R.: Simatic-Steuersystem P für die Funktionsgruppenautomatik in Wärmekraftwerken. Siemens-Z. 41 (1967), pp. 410/416.
[12] Friedewald, W., Zwetz, H.: Regelung der Temperaturen im Wasser-Dampfsystem von Bensonkesseln. Regelungstechnik 13 (1965), pp. 62/68.
[13] Handschuh, H., Jung, M., Sterff, J.: Der automatische Betrieb von Kesselspeisepumpen im Dampfkraftwerk. Siemens-Z. 39 (1965), No. 9, pp. 1039/1045.
[14] Große Dampfkraftwerke. Edited by K. Schröder. Vol. I: Kraftwerksatlas mit Kennwerten von 220 Kraftwerken, 98 Kraftwerksbeschreibungen und 6 Ausführungsbeispielen. Berlin, Heidelberg, New York: Springer-Verlag 1959.
[15] Recommendations for Sump Dimensions. Power, October 1960, p. 112.
[16] Honold, E.: Kesselspeisepumpen. KSB Technische Mitteilungen 59 (1966), No. 8, pp. 401/407.
[17] Honold, E.: Vergleich von Kesselspeisepumpen für hohe Enddrücke in Glieder- und Topfbauweise im Hinblick auf die Entwicklung großer Kraftwerksblöcke. Mitt. der VGB (June 1966), No. 102, pp. 149/152.
[18] Strub, R.A., Rymann, W.: Neueste Entwicklung auf dem Gebiet der Speisepumpen. Sulzer Research Reports 1961.
[19] Honold, E.: Ein kritischer Vergleich von Topf- und Gliederpumpe am Beispiel der Entwicklung der Höchstdruckkesselspeisepumpen in Deutschland und den USA. KSB Technische Berichte, No. 1 (October 1960).
[20] Ratzeburg, W.: Aktuelle Probleme bei Kesselspeisepumpen. BWK 11 (1959), No. 11, pp. 528/532.
[21] Odendahl, W.: Mindestmengen-Steuerung für Hochdruck-Kreiselpumpen. Mitt. der VGB (Oct. 63), No. 86, pp. 317/322.
[22] Lange, F.: Kesselspeisepumpenzulaufsystem bei gleitendem Entgaserdruck. BWK 23 (1971), No. 7, pp. 321/328.

5.4. Turbo-Compressors and Blowers

Under this heading, discussion is mainly limited to those compressores which are used in the mining and chemical industries. Logically, however, the statements made also apply to the fuel gas and combustion air of stationary gas turbines and jet engines, the problems of which are touched upon in Section *5.5 Gas Turbines*.

Large axial and centrifugal blowers for fresh air, coal dust and flue gases are also included in the following survey (Table 1) of cases of damage.

Discussion of Survey of Damage

In 85% of the clarified cases of damage, no part was played by extraneous influences. Of the 15% caused by these influences, 6% were instigated by foreign bodies, and 3% each by climatic conditions, by disruption of electricity supplies, and by fire.

Origin of Damage

Of the cases of damage, 24% were accompanied by a deterioration in running smoothness, 21% were attributable to a lack of lubricant, and 18% were caused by mechanical overloading of a component. The remaining 37% were distributed more or less evenly among reverse rotation, loosening, wear, erosion, corrosion, and the failure of protective devices. The most effective means of combating the most costly cases of damage to compressors and blowers would be careful attention to the quietness of operation and painstaking supervision of the oil supply and the bearings themselves.

Overall Pattern of Damage

Typically, the overall pattern of damage takes the following form:

30% damage to bearings—induced mechanically and by lack of lubricant,
24% fatigue and long-term failures, particularly at rotor discs and blades,
18% forced failures,
14% damage by radial and axial rubbing,
8% rotor distortions,
6% incipient cracks and deformations.

These manifestations of damage also support the demand for improved supervision of the bearings and the running smoothness of turbo-compressors and blowers. Here, too, the high proportion of slowly developing damage points to the need for routine inspections.

Table 1. Distributions of causes and locations of cases of damage to turbo-compressors and blowers.

Cause of damage	Distribution of the number of damage cases [%]	Location of damage	Distribution of the number of damage cases [%]
Product faults	*49*	Bearings with oil supply	37
Errors in planning, design and calculation; technological imperfections	18	Rotors, rotor discs	27
		Rotor blades	13
		Gear wheels, gear transmissions	6
Manufacturing faults	12		
Faults in assembly and repair	10	Guide vanes	5
Materials faults	9	Fittings, control elements	5
		Casings, screwed fasteners	4
Operational faults	*36*	Shaft seals	3
Maintenance faults	26		
Mishandling faults	10		
Extraneous influences	*15*		

Table 2. Criteria for the selection of turbo-compressors and blowers.

Type	Medium to be handled	Delivery range [m^3/hr] relative to condition at inlet	Pressure ratio
Centrifugal design			
Compressor with multiple interstage cooling (external cooling)	air or gases of similar properties	5,000 to 30,000	6 to 12
Compressor with multiple interstage cooling (integral cooler)	air or gases of similar properties	30,000 to 200,000	6 to 10
Mixed-flow compressor	air or gases of similar properties	200,000 to 360,000	6 to 10
With intermediate branches for inlets and outlets or external intercooling (cooling less intensive)	air and other gases	5,000 to 200,000	6 to 25 per casing and depending on the type of gas
Without intermediate branches, without intercooling	air, gases and vapours	2,000 to 200,000	4 to 6 per casing and depending on the type of gas
High-pressure compressor, barrel-casing design	light gases	1,000 to 20,000	6 to 25
Special designs			
Pipeline blowers	natural gas, hydrocarbons	5,000 to 50,000	1.2 to 1.9
Exhausters	air and gases	6,000 to 72,000	2 to 3.5
Reactor blowers (centrifugal or axial)	CO_2/He	25,000	1.02
Axial design			
With fixed guide blades	air, gases and vapours	from 50,000	1.2 to 8 per casing
With adjustable guide blades		from 50,000 but extended operating range	

Final pressure range [kg/cm^2] abs.	Application	
6 to 12	general compressed-air and process-air supply	
6 to 10	mining, metallurgical engineering, automobile industry: compressed-air supply	
higher pressures with booster	air-decomposition plant:	compression of process air, nitrogen and oxygen
6 to 10	process engineering:	process air for ammonia synthesis and production of nitric acid
50 to 80 depending on the type of gas	air-decomposition plant:	compression of oxygen and nitrogen
	petrochemical industry:	compression and circulation of gases and refrigerants of all kinds
	nitric-acid industry:	compression of process air
50 to 80 depending on the type of gas	petrochemical industry:	compression and circulation of gases of all kinds
	inorganic chemistry:	manufacture of soda, nitric acid, chlorine
	metallurgical engineering:	blowers for roasting plant, charging of shaft furnaces
	gasworks:	extraction of gases from coke ovens, gas-distribution technology
	sewage plant:	aeration of waste water
	refrigeration plant:	compression of refrigerants
50 to 450	chemical industry:	gas compressor for the synthesis of methanol and ammonia, compression of CO_2 for the production of urea
	refineries:	platforming plant and circulators
	petrochemical industry:	distribution and liquefaction of natural gas
40 to 100	compression of natural gas or circulation of hydrocarbons at high pressures	
6 to 7 m w.g. vacuum	paper industry:	extraction of water from paper-making machines
	chemical industry:	extraction of waste gases from reaction chambers
10 to 25	circulation of carbon dioxide or helium in reactor circuits	
1.2 to 12 max.	metallurgical industry:	blowers for blast furnaces and roasting plant
	chemical industry:	compression of process air
	petrochemical industry:	compression of process air
	refineries:	compression of process air
	air-decomposition plant:	compression of air
	natural-gas liquefaction:	refrigeration compressor
	research centres:	operation of wind tunnels
	gas-turbine installations:	compressors for combustion air or air circulation

5.4.1. Planning, Design and Manufacture of Turbo-Compressors and Blowers

Environmental Conditions

The decision as to whether the installation should be erected in the open air or under cover in a hall is largely influenced by environmental conditions. Troubles with open-air installations have not been reported in cases where the manufacturer has been properly informed of climatic conditions when the order was placed. The design of open-air installations must be carefully considered, particularly with regard to vertical displacements through thermal effects on the various foundations of the driven machine, gearing and prime mover. It is recommended that the whole set should be erected on a coherent base made from one material, so as to minimize changes in alignment at the couplings of machines in the event of displacements of the base as a result of exposure to sun or wind, rain or snow. Unstable steel structures for equalization of the various centre-heights of machines should be avoided, especially with gear transmissions.
For repair work, protection from the weather and dirt must be provided, as well as proper lifting tackle with slow and rapid hoisting and lowering facilities.

Selection of Compressor Type

Table 2 and references [1] and [2] are an aid to the selection of compressor type, while Table 3 indicates the possibilities for the regulation of compressors. Because machine types partially overlap, additional criteria must be considered in the selection. For example, air compressors with deliveries of about 100,000 m^3/hr and pressure ratios up to 8 are available in both centrifugal and axial designs. If an adequate supply of cooling medium is available, the logical choice is the centrifugal compressor, with its suitability for intensive interstage cooling, so as to save driving power. If, however, preheated air is desired or if cooling water is not available, there is a

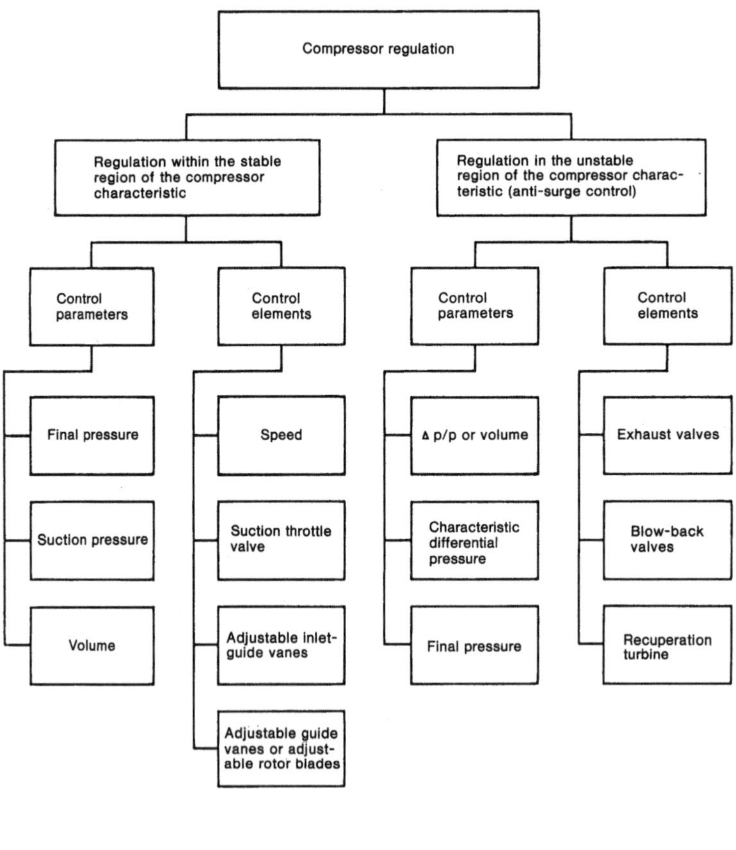

Table 3. Possibilities for the regulation of compressors.

choice between the uncooled centrifugal compressor or the axial compressor, which is, under the same conditions of operation, of smaller dimensions and has a higher efficiency, but costs more. For economic reasons, the use of axial compressors is recommended only for deliveries above 70,000 m^3/hr. By comparison with centrifugal compressors, axial compressors are more sensitive to dirt and erosion, so that precautions must be taken if the medium delivered contains moist or dry abrasive particles. These precautions are necessary not only for air in the vicinity of marine or industrial areas, but particularly also for process and flue gases.

The following tabulated data compare the advantages and disadvantages of the different types of compressor drives:

Electric Motor

Advantages

1. Simple construction.
2. Boiler plant not required.
3. Easy maintenance.
4. Non-pollution of the environment since there are no emissions.
5. Possibility of matching the most favourable compressor speed through appropriate gearing.
6. Immediate starting.

Disadvantages

1. Constant speed (no speed reserve).
2. Efficient regulation and favourable characteristics under partial load are generally possible only with adjustable blades.
3. Gearing is nearly always necessary.
4. With asynchronous motors, heavy load on the grid during starting.
5. For high powers, motors are available only as expensive individually manufactured units; increased cost of switchgear.

Steam Turbine

Advantages

1. Variable speed, economical regulation at part load.
2. In conjunction with adjustable blades, large stable operating region.
3. Mostly without gearing.
4. Particularly economical if steam is generated from process heat or supplied from a back-pressure, extraction back-pressure, or extraction condensing plant.

Disadvantages

1. Costly if an outside source of steam or process steam is not available, as boiler plant will then be required.
2. More maintenance required than with an electric motor.
3. Longer starting-up times.
4. With condensing turbines, additional cooling water required.

Gas Turbine

Advantages

1. Variable speed.
2. Mostly without gearing.
3. Economical if a cheap fuel, such as blast-furnace or natural gas, is available, particularly if the exhaust gases are utilized in a waste-heat boiler.
4. No cooling water for a condenser.

Disadvantages

1. More maintenance required than with an electric motor.
2. More sensitive than an electric motor or steam turbine.

For a plant which either produces or requires process steam, a back-pressure or extraction condensing turbine is preferable. If steam is not generated, the advantages of the electric motor predominate. If neither electric current nor steam is available or if the fuel is cheap, the gas turbine is useful and, in fact, gas turbines are often found in the compressor stations of natural-gas pipelines, in blast-furnace plant for blast generation and in oil and natural-gas fields for driving compressors.

The required operating region of a compressor is a factor which speaks for or against a particular drive. All turbo-compressors have a characteristic which is divided into a stable and an unstable region. If a compressor is wrongly operated in its unstable region, pulsating flow (compressor "surge") takes place [3–8]. The dividing line between the two regions is the surge limit. By using fully effective anti-surge control, it is possible to operate a compressor in the endangered region because part of the delivery is exhausted or is blown back, i.e., the actually compressed amount of gas is greater than that delivered. It is for this reason that, when operating below the surge limit, there is no saving in driving power, despite the fact that the amount delivered is further reduced.

With a parabolic pipeline characteristic, a variable-speed drive is more favourable from the point of view of efficiency than a fixed-speed drive. On the other hand, with a constant delivery pressure over the whole delivery range, adjustable blades [3, 4, 9–14] are more favourable than variable speed. Moreover, the combination of both these possibilities can be economical if the plant is frequently operated at part load, despite the higher initial costs.

Protection from Corrosion and Erosion

During planning, particular attention must be paid to protection from corrosion and erosion. In the chemical and metallurgical industries, as well as in coastal areas, atmospheric conditions prevailing today can cause heavy corrosion of *air* compressors. Compressors with interstage cooling are especially at risk. Impellers and diffusers (blading after each cooling stage), as well as the first stage, must be made of corrosion-resistant steels. Corrosion of impellers can be prevented by:

- Installing water separators behind the cooling elements. Even so, corrosion-resistant materials should be employed for the following impellers and diffusers (e.g., 13% chromium steels).
- Operating with a gas temperature behind the cooler above the dew-point.

Effective separation of water, within a limited volume of separator construction, can be obtained only at the expense of additional pressure losses. Moreover, the air emerging from the separator is never entirely free from entrained water. The finest droplets are difficult to separate, though they evaporate quickly in dry air.

The preventive measure of *operation above the dew-point* involves a reduction in internal efficiency and requires a series arrangement of cooling-water flow or regulation of the flow through individual coolers. In cases where atmospheric conditions are unfavourable, it is essential to plan for *hot operation* of the machine at the time when it is being ordered. In this way, the risk of major damage and breakdown of efficiency can be safely avoided. The extra energy requirements of from 0.5 to 1.5% are within tolerable limits. The construction of a suction chimney is recommended to enable the intake air to be drawn from a higher level.

The majority of damage cases observed, mainly to cooled but also to uncooled compressors, can be traced back to *corrosion while out of operation* [15, 41]. Such machines are often taken out of commission in a damp condition and containing highly corrosive condensate residues, and are then left, without being dried out, to themselves for months under changing atmospheric conditions. When taken back into service, the resulting one-sided corrosion leads to rough running. To avoid corrosion while out of operation, a fan-type drier should be provided to blow warm air into the machine after it has stopped running. In humid atmospheres and where there are large temperature fluctuations, this drier should be kept running permanently to prevent condensation.

Compression of Chemically Aggressive Gases

When compressing chemically aggressive gases, the choice of appropriate protective measures and materials will depend on whether chemical attack on the components takes place while the machine is out of operation or during operation with or without the presence of moisture. With explosive or readily combustible gases, note should be taken of the appropriate safety regulations for the selection of materials and design. Recommendations that have proved their value in practice are given in API-617 [16].

Because of the attendant fire hazard, grease and oil must be kept away from the gas-carrying components of oxygen compressors.

Casings, diffusers and return passages made of grey cast iron have proved to be largely insensitive to chemical and mechanical attack. Corrosion-resistant materials must be used for the oxides of nitrogen. For impellers, steels with chromium contents from 12 to 18% offer good protection not only from weak acids, but also from nitric acid. Information on the use of turbo-compressors in various chemical and process industries, and the associated special problems, will be found in [17–20].

The blading of *turbo-compressors of axial design* can be effectively protected from *corrosive attack* by chromizing. However, like protective layers of non-metallic materials, the relatively thin surface layers produced by this treatment do not offer long-lasting protection against *erosion*. If erosive conditions exist, these protective layers are removed locally and, at these locations, corrosion is intensified. Under even more severe corrosive conditions in conjunction with erosion, austenitic steels with a high content of chromium and nickel and additions of molybdenum must be used. Although these special steels are highly corrosion-resistant, they usually exhibit lower strength properties, thereby necessitating a reduction in peripheral speed and hence an increase in the number of stages for an equal increase in pressure. If the yield point drops to half the value associated with normal steels, the peripheral speed, with the same safety factor, must be reduced to about 70%. For an equal increase in pressure, the number of stages must be doubled. In addition, the impeller diameter must be made 40% larger to achieve the same volume flow. Such special materials therefore make the compressor more expensive because of

high-quality materials,
a larger number of stages,
a larger volume.

A compressor with conventional materials, but requiring periodic replacement of endangered components, can certainly constitute a more economical solution.

Fine dust in the air causes clogging, particularly in combination with moisture [21]. Coarse dust from the air or from process or flue gases is responsible for erosion damage, mainly to impellers, rotating and guide blades. For this reason, dust-laden suction air must be filtered. Dust-carrying gases must also pass through filter or gas-scrubber installations.

Filters

Whilst, at one time, oil-wetted box-type filters or oil-circulation filters were mainly used to remove dust from air, *dry filters* are usually employed nowadays. Although oil-circulation filters involve small operating costs and also retain coarse dust particles, they generally release oil which, in conjunction with fine dust or soot that has not been separated, tends to clog flow paths and coolers. The result can be severe reductions in efficiency and volume flow, sometimes leading even to complete functional incapacity of the compressor.

Dry filters are manifestly superior to oil-circulation filters in respect of their dust-removal capabilities but entail higher operating costs, particularly with high dust concentrations. To reduce operating costs, drum-type filters with air-scrubbing facilities connected in series ahead of the filters are often provided, but these facilities, in turn, cause an increase in the water content of the air. The first step to take is to locate the air suction inlet far enough above ground level to ensure that the coarser particles of dust are

Figure 1. An impeller "sawn through" by water erosion after 4,000 operating hours. The cause of damage was insufficient removal of water behind the intercooler.

Figure 2. Impeller of a process-gas compressor nearly "sawn through" by water erosion. The cause of damage was insufficient removal of water behind the intercooler.

eliminated in advance. Apart from erosion by coarse dust, it is possible that water which has not been entirely removed behind the coolers, or water droplets formed as a result of operation below the dew-point ahead of the first compressor stage, will completely "saw through" exposed components by erosion (Figures 1 and 2) or cause severe damage by erosion/corrosion (Figure 3).

Protection from Noise

When planning compressor installations, acoustic problems, including measures for noise abatement, should be considered right from the start. Retrofitted silencing measures are expensive. The demand for more efficient compressors leads to higher flow velocities, i.e., to greater sonic energies. Sound radiation takes place directly via the medium being carried in the suction and pressure lines and is then emitted from the compressor casings and pipelines into the machine room and its surroundings [22–25].

For noise abatement in machine rooms, the following points merit attention:

1. Small rooms with smooth sound-reflecting walls should be avoided.
2. In medium-sized rooms, noise levels can be reduced by about 5 to 7 dB by sound-absorbent wall linings.
3. The machine-room basement, which is generally much noisier, must be carefully isolated from the machine room. Break-throughs for ventilation, pipelines and stairways must be properly sealed. Machine-room floors of latticed construction are unacceptable for acoustic reasons.
4. Suction, pressure and exhaust lines must be provided with sound-absorbent insulation. For the assessment of permissible noise emission to the surroundings, the noise level of the envi-

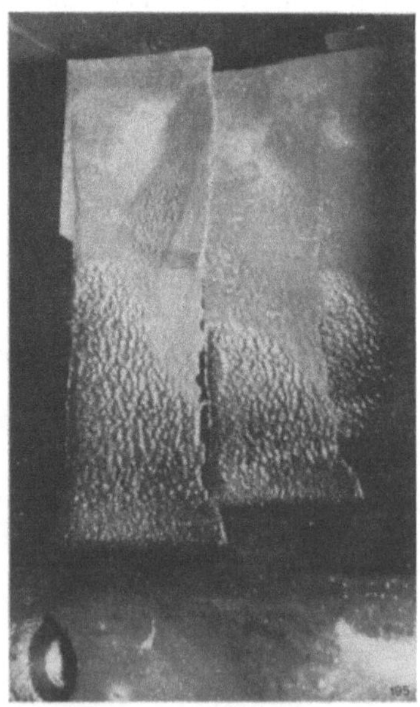

Figure 3. Erosion/corrosion at the first row of guide vanes of an axial compressor after 9,000 hours in operation.

ronment and the associated frequency spectrum must be known. This knowledge will then make it possible to estimate how much it will be necessary to spend on noise abatement outside the machine room.
5. A silencer should be installed in the suction line.
6. A silencer should be installed in the exhaust line. As a rule, this measure is necessary, since it is virtually certain that, during exhausting, the noise level, even for an industrial area, will be exceeded.

Regulation

Depending on the application of a compressor, different demands are made in regard to constant delivery pressure or constant volume flow. The different operating points are attained either by throttling the air flow at the compressor inlet, by changing the speed, or by adjusting the blades. Table 3 gives an indication of the possibilities for the regulation of compressors; further information will be found in the literature [3, 4, 26].

Endangering of Compressors by Surging

Operation of turbo-compressors at partial load is limited. By favourable selection of design parameters, e.g., [27], and by precision construction, it is possible to improve the surge limit substantially, particularly with axial compressors. When the surge limit is reached, flow in the compressor reverses, and the final pressure is greatly reduced. As a result, gas flows back and hence the contents of the air receiver and the pipeline on the pressure side are evacuated. Depending on the consumer characteristic, the compressor can then again attain an operating point in the stable region, and the final pressure rises again to the surge limit, this process being repeated periodically. The frequency of surging depends, among other factors, on the volume flow, on the volume of the pipelines ahead of the compressor or the pipelines or receivers behind the compressor, and on the shape of the compressor characteristic. With uncooled compressors particularly, this surging leads to marked heating-up of the whole compressor because flow through the blading, i.e., dissipation of the heat losses through the medium being transported, is lacking. The resulting rise in temperature usually leads to extensive damage through clearance bridging, but especially to severe blade damage (Figure 4). In this respect, axial compressors are more liable to damage than centrifugal compressors.

Anti-Surge Control

Shortly before reaching the surge limit, the anti-surge control [3, 4, 28] should prevent a further reduction in volume flow through the compressor

Figure 4. Damage to the blading of the rotor of an axial compressor as a result of flow disruption and surging.

by opening an exhaust or blow-back valve. Anti-surge controllers utilize delivery pressure and volume for control pulses. As far as possible, the surge limit should be determined on the test bed and the anti-surge controller adjusted provisionally. When commissioning, the surge limit should again be approached carefully at the required points and the anti-surge control finally adjusted. This work should not be carried out without fully effective surge protection.

Surge Protection

The surge-protection equipment should automatically switch off the compressor immediately before entering the surge region or at the latest after a few surges. Pressure loss in the suction or delivery branch or temperature rise in the blading can be utilized for triggering the pulses for surge protection. With both methods, it is necessary to use rapid-response measuring instruments which, without an intermediate alarm, immediately protect the compressor from damage by opening an exhaust valve or, should this fail, by switching off the drive.

Rapid automatic release is essential because, according to experience, damage by overheating of the blades and clearance bridging can occur within a few seconds after the inception of surging.

Rotating Stall

Apart from the appearance, as here described, of surge, which is clearly audible, local break-away of flow can also take place. This phenomenon, which is generally rotating, is known as "rotating stall" [3–5, 29, 30] and, with axial compressors, it can lead to increased vibrational stresses, even before the surge limit is reached.

These stresses are caused by regions of flow break-away. Up to seven such regions have already been observed and, depending on their number, a rotating frequency of about 30 to 70% of the operating frequency results.

As tests have shown, blades can be excited to vibrate by these regions of flow break-away if the exciting frequency or an integral multiple thereof coincides with the natural frequency of the blades. Such cases of resonance lead to the highest stresses. The lower the order of the harmonics, the higher the stresses will be. The random appearance of resonance is enhanced by the fact that the true natural frequency of the blades is subject to considerable scatter, owing to the differing conditions for securing the blades in the rotor arising from manufacturing tolerances, discontinuities such as filling-up pieces, and manual assembly. Experience during the last few years has shown that, with ratios of natural blade frequency to maximum rotational frequency of 4.5:1 or higher, blade breakage will not normally take place as a result of excited vibrations.

Shaft Seals

Special requirements regarding shaft seals of compressors arise when handling noxious, flammable, costly, or (mixed with air) explosive media, or if, because of the desired purity, air must not enter the system [16, 31, 32].

Labyrinth Shaft Seals

Although labyrinth shaft seals have the merit of simple construction and freedom from any friction, they have the disadvantage of a certain amount of leakage. Their application is preferred for the compression of air, nitrogen, carbon dioxide, and oxygen. In the event of clearance bridging, sparking can be prevented only by the appropriate choice of materials for the packing rings and bush. In an atmosphere of oxygen, sparking can cause complete destruction of the compressor set by burning up. For pairing seal materials, the noble metals or their alloys are suitable because, apart from their freedom from sparking, they cater for rapid heat dissipation in the event of clearance bridging.

Oil-Barrier Shaft Seals

For other media to be compressed, particularly those which, because of their toxicity or danger of reaction with the surrounding air, must be prevented from reaching the outside, oil-barrier sliding-ring or floating-ring shaft seals are employed; if there is a risk of explosion, water is used as the sealing medium. Not all liquid-barrier shaft seals remain effective during relatively long non-operational periods, and leakage of gas can then be avoided only by separate seals for stationary conditions (e.g., gas locks, positive seals, etc.).

Sliding-ring shaft seals have a low oil requirement but are not free from wear, so that they can be recommended for peripheral speeds up to only 80 m/sec approximately.

Floating-ring shaft seals have a higher oil requirement but, provided that the entire sealing-oil circuit is properly maintained, are not subject to wear and to limitations on peripheral speed.

Both these designs of shaft seal hermetically isolate the compressor by means of a film of sealing oil at a pressure higher than that of the medium to be sealed. The sealing oil separates the rotating and stationary sealing elements and exits partly to the gas space and partly to atmosphere.

The oil flowing into the gas space is mixed with gas and must generally be degasified in gas separators, the gas being led back to the process and the sealing oil being reused. This advantage applies mainly to floating-ring shaft seals.

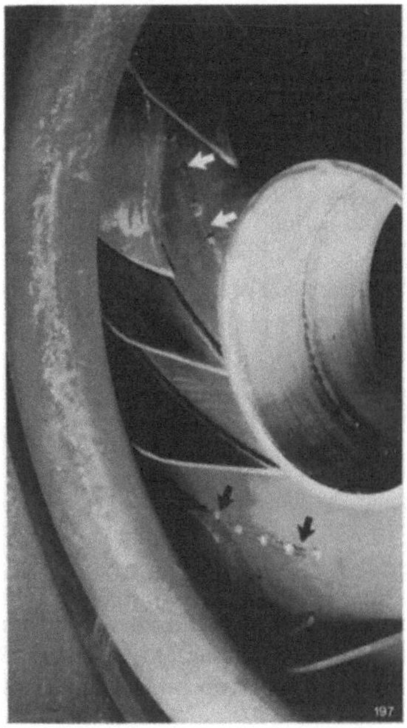

Figure 5. Fatigue failures at the rivet shanks of an impeller of a centrifugal compressor.

Impellers and Impeller Blades

Impellers of Centrifugal Compressors

Riveted Impellers

Rivet locations are critical. Impellers with Z- or U-shaped blades should be made with a rivet pitch as close as possible and with short shanks on the rotating blades, so that bursting off of the rivet head because of corrosion underneath is obviated. The rotating blades themselves are usually manufactured by cold-bending appropriately pre-cut sheet material. Since the fastening strips must be bent at an angle of 90 deg., stressing at the bent corner is relatively high. Faults in the sheet material, such as laminations and segregations, can then lead to incipient cracks at the bent corners and subsequently, in operation, to blade fracture. Hot-forming of the rotating-blade sheet material and testing for cracks after bending the blades can exclude this source of faults during manufacture.

Riveted blades with a twisted profile require the greatest care during riveting. Over-long rivet shanks or excessively hard riveting lead to the generation under the rivet head of cracks which cannot be detected visually. Even ultrasonic tests do not reveal these cracks during manufacture with any degree of certainty. It is only during operation that such incipient cracks can develop into fatigue failures of the rivet shanks, resulting in loosening of the rotating blades between the shrouds (Figure 5). Moreover, faulty contact of the rotating blades with the two shrouds can leave local gaps which can also lead to interfacial corrosion with this design of rotating blade. Faulty fitting of the rotating blades to the shrouds, possibly with resilient riveting, is also the cause of vibration fractures of the cover or hub shrouds.

With riveted impellers, damage by vibration at the rotating blades themselves is very rare, whereas fractures at the cover or hub shrouds, particularly by reason of faulty riveting of shrouds and blades, occur more frequently. The most extensive damage to centrifugal compressors is caused by parts expelled from the shrouds of such impellers.

Welded-in Rotating Blades

Impellers with welded-in rotating blades must be tested to destruction during the manufacturing stage, so that the optimum welding technique for final production can be established. Depending on the width of the impeller passage, a distinction

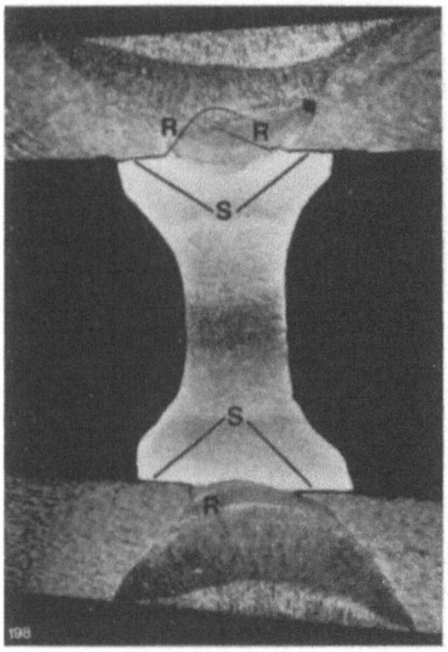

Figure 6. Gaps S between shrouds and rotating blades, as well as insufficient root-welding, led to cracks R and fractures (Figure 7) under operating conditions.

Figure 7. Broken impeller of a centrifugal compressor. The damage was caused by unwelded root gaps (Figure 6) which led to incipient cracks due to notch effects.

can be made between internally welded blades and blades that are welded from the outside through the shrouds. With both procedures, the greatest care should be taken to ensure perfect welding of the blades to the cover and hub shrouds. Unfavourable weld roots, gaps, local hardening or stresses resulting from the welding operation are sources of crack formation that particularly endanger the impeller shrouds themselves (Figures 6, 7 and 8). Such cases of damage can be avoided by favourable welding design, as exemplified in Figure 9.

Figure 8. Complete destruction of the impeller of a centrifugal compressor after a few hours in operation, with substantial ensuing damage. The cause was unrecognized hardening and residual stresses as a consequence of welding.

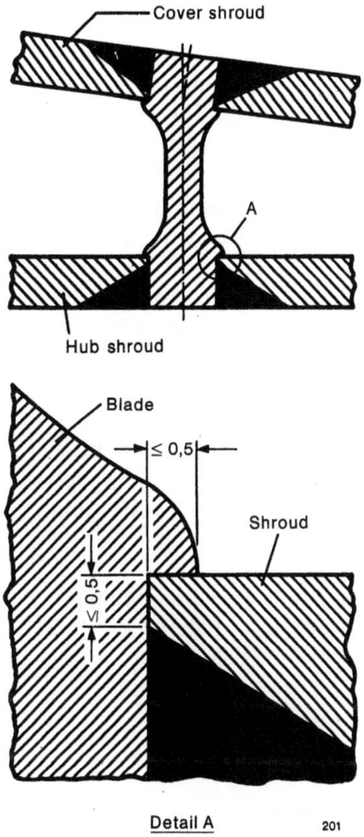

Figure 9. Proposal for a favourable welding design for the impeller of a centrifugal compressor.

Figure 10. An impeller of a centrifugal compressor completely stripped of blading during operation. The damage was caused by a substantial reduction in the strength of the sintered connection between the blades and shroud through bonding faults.

Brazed Impellers

Brazed impellers have recently gained in importance for impellers having extremely narrow passages. Statistics of damage with these modern brazed impellers are not yet available to us. Since the shrouds are brazed to the rotating blades fully automatically, without manual intervention, great importance attaches to the determination of optimum brazing conditions, fluxes, and brazing materials. Also, examination of these brazing processes can be carried out only by extensively pretesting impellers so produced to destruction over a long period because, with non-destructive test methods, only a very limited examination of brazed connections can be performed over the whole impeller. Damage due to inadequate bond strength, such as that shown in Figure 10, which refers to an impeller manufactured by another modern bonding technique, can in most cases be severe.

Cast Impellers

Cast impellers require careful radiographic testing after casting, so as to discover blowholes in these very difficult castings before further manufacture or before the centrifugal test. In general, impellers with blowholes must be discarded.

Axial Rotating Blades

Of particular importance is the level of the natural frequency of the blades in relation to the existing frequencies present in the compressor [3–5, 33, 34]. Vibration of the rotating blades is excited by the frequency of the guide-vane cascade and by possible flow-separation phenomena in the compressor (see also under "Rotating Stall"). Moreover, there should be no chance of exciting the blades by low harmonics of the frequency of rotation. Parting lines, reinforcing ribs or blow-off valves must also be taken critically into account in considerations of vibrational behaviour. Whether made from the solid or from a rough forging by a chip-producing machining process, as well as by finish-forging and electrochemical forming, the rotating blades of a compressor exhibit a certain amount of scatter of their natural frequencies, depending on the method of manufacture.

The attachment of the rotating blades to the rotor can cause additional scatter of the natural frequency of a wheel. This scatter can only be revealed with any degree of assurance if periodic measurements are made of vibration of the rotating blades following a centrifugal test of the completely bladed compressor rotor. The centrifugal force generated during this test, carried out at 115% of the maximum operating speed, has the effect of settling the blade roots in their grooves and hence of taking up in the rotor the final position determining their vibration characteristics. The centrifugal test is also essential for radial impellers. To check the operational safety of radial impellers, it is recommended that they should be tested for cracks at critical locations of the shrouds and at the blade ends.

Guide Rings and Guide Vanes

Whilst there are no major difficulties with the guide rings of centrifugal compressors, axial-flow compressors, particularly if they are provided with adjustable guide vanes, require the greatest care in the design of these items. If there is insufficient play at the pins of the guide vanes, jamming through soiling or corrosion often takes place. Corrosion-resistant materials and appropriately larger clearances make it possible to eliminate these difficulties with adjustable guide vanes. Guide vanes in axial compressors must be dimensioned and checked according to the same vibrational considerations as their rotating blades, even though they are as a rule not as sensitive to additional dynamic stresses as the rotating blades, because no influence is exerted by centrifugal force.

The designer should, by the provision of inspection apertures which also permit the introduction of endoscopes, enable the condition of the interior of the compressor to be checked visually, without having to dismantle the machine in every case. In this connection, the reader is referred to the corresponding sections in the chapters on steam and gas turbines. If these inspection apertures are supplemented by a few measuring holes, the possibility of checking soiling and wear in the interior of a compressor without disassembly during non-operative periods is greatly simplified (see also 5.1. Steam Turbines, Figure 17).

Recommendations are given in reference [16] for the position of the critical speed and for permissible values of vibration in operation. In compressors with stiff rotors, the first critical speed should be at least 20% above, and in those with elastic shafts not more than 60% of, the maximum continuous speed. Moreover, these critical speeds must differ by at least 10% from any possible operating speed. A vibration analysis of the compressor, transmission and prime mover must be agreed upon in the contract. The vibration analysis permits determination of coupled critical speeds in bending and torsion of the entire shaft line [35–37]. It is recommended that this vibration analysis be carried out during commissioning as it enables dangerous operational conditions to be recognized clearly and promptly. Extensive damage to bearings and couplings, with possible ensuing damage to rotating parts, in addition to losses arising from outage, can thus be avoided.

The machines should be so designed that on-site bearing pedestal and shaft vibration measurements can be carried out [38]. According to [16], the following measured values of shaft vibration are currently permissible:

Maximum continuous speed [rpm]	Vibration amplitude A [μm]
up to 4,000	32
4,001 to 6,000	26
over 6,000	19

Allianz has experienced that these values can also be exceeded in continuous operation, since the "danger" values for sliding bearings are very much higher. In this respect, there are at this time* no agreed rules because the draft for VDI guideline 2059 applies only to turbo-machines in the speed range from 3,000 to 3,600 rpm. If shaft-vibration measurements on turbo-compressor sets are difficult to interpret, the Technical Insurance Department of Allianz will gladly furnish data on experience gained so far with permissible values of shaft vibration. Regarding problems concerning the bearings of turbo-compressors, the couplings of such sets and the oil supply, the corresponding sections should be consulted. With a combined lubricating and sealing-oil supply, attention should be paid to particularly careful degasification of the sealing oil before re-entry into the oil tank. Separate lubricating and sealing-oil circuits are preferable because of the decreased risk of damaging the bearings through dirty oil.

* 1.5.1972

5.4.2. Assembly and Commissioning of Turbo-Compressors and Blowers

Not all installations can be tried out under realistic operating conditions over the entire load range on the manufacturer's test bed [39]. In each case, however, at least a mechnical test run should be carried out, during which there should be functional tests and adjustments that are indispensable for the safety of the set during the first commissioning at the user's works. On the other hand, insistence must be made on running prototype machines under load on the manufacturer's test bed, unless the medium being compressed makes it impossible to carry out such runs on the test bed.

When prototypes are being put into operation at the user's works, a prolonged commissioning period is necessary. For commissioning, the recommendations given in the preceding sections on *Anti-Surge Control* and *Surge Protection* should be carefully observed. The turbo-compressor protective and control equipment responding to temperature, delivery pressure, flow rate, and rotational speed must be adjusted to full effect *before the first commissioning* or, as in the case of surge limit control, pre-adjusted.

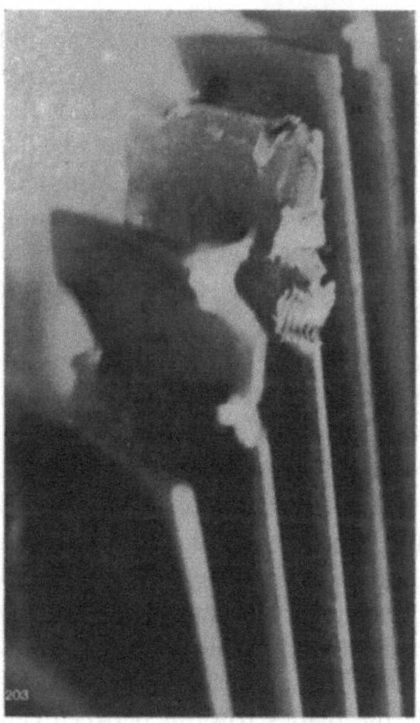

Figure 11. Damage to the blading of an axial compressor through foreign bodies, i.e., by loosened screws.

Hot Alignment

Where applicable, checking of alignment after the first heating-up of the prime mover or after the first normal operating run of the turbo-compressor and gear transmission is indispensable. This so-called *"hot-alignment"* check provides protection from damage to bearings and couplings of the machines concerned and can, as a rule, be carried out without great expense, since gear-type couplings are used (see also 8.4).

Securing of Pipelines

During assembly and commissioning, particular attention should be paid to the lay-out of pipelines, which should be clamped at short distances in such a way that vibrations in the pipeline system are avoided. All pipelines on the compressor should, in accordance with [16], be provided with flanges and not with screwed connections.

Foreign Bodies

Commissioning personnel should take care to ensure that foreign bodies, such as washers, screws, tools and cleaning rags, are not left in the suction region of the machine or in the machine itself. Particular attention should be directed to making certain that there are no inadequately fastened and secured components that can become loose under the suction effect of the compressor and can thereby cause damage in the interior of the machine. Such damage by foreign bodies, as, for example, that to the compressor blading in Figure 11, can be avoided. All accessible openings in the suction region of the compressor must not be covered up, even temporarily, with cleaning rags, carboard or similar materials.

Measurements According to DIN 2045

All measurements and tests associated with proving the performance of the compressor must, unless otherwise agreed, be carried out in accordance with VDI Compressor Regulations DIN 2045 [39].

5.4.3. Optimum Supervisory and Protective Equipment on Compressors and Blowers

Apart from the usual supervision of the bearings (see 8.3. Bearings) and of the oil and cooling-medium temperatures, the thrust bearings must be monitored and protected in both directions by limit-value alarm and switch-off pulses. The same facilities apply to supervision of the suction filter.
Moreover, the supervision of vibrations by means of bearing-pedestal and shaft vibration transducers is strongly recommended, as only these two procedures indicate changes in the state of balance and give protection from ensuing damage. Proven values for bearing-pedestal vibration measurements, alarm release and switch-off pulse should be taken from VDI guideline 2056. For measurements of shaft vibration, the recommendation is to determine the normal level in the range of operating speeds at different loads and to adjust the protective equipment to emit a switch-off pulse if this normal level is exceeded by from 50 to 100%. It also may be desirable to couple this protective equipment with the bearing-pedestal vibration equipment in a dual-circuit arrangement.
With compressor sets that have been dirtied by the medium being compressed, the measurement of axial thrust by means of load cells is a useful method of enabling the optimum intervals for cleaning to be determined and of making sure of preventing damage by overloading the axial bearing.
An advisable safeguard is the ability to check all supervisory and protective equipment manually in such a way that the measuring transducers, the corresponding signal transmission, and the trip mechanisms can be tested for their functional capability during operation without stopping the set. The measured values should be registered on a dot or line recorder with a chart speed of 60 mm/hr. Alarm release by limiting values is recommended, because there is a tendency to neglect changes, brought about by damage, at measuring positions which are only recorded. By the appropriate choice of measuring positions on a recording instrument, it should be possible to ensure simultaneous indications of important measured values at the correct time. An automatic supervisory, protective and control installation for a motor-driven compressor is described in [40].
The usual hourly notes made by the machine operator are suitable only for the determination of changes that occur slowly. These observations should therefore be concentrated more on checking the condition of those aspects of the installation as a whole which cannot be controlled by measuring instruments. In this connection, it is recommended that checks should be made of the noise of the whole set, especially if it incorporates gearing; of the entire oil pipeline system in respect of vibrations and oil leakages; and of possible fume or smoke generation in the region of the set. In modern installations, these functions of the machine operator can be replaced by observation of the set by television and by transmission of noise level by a microphone to a loudspeaker installation at the control station. Such existing supervisory equipment is complemented by periodic checking of the stage pressures of the compressor, making it possible to draw conclusions regarding soiling, the wear of guide vanes and rotating blades, and the condition of the seals.

5.4.4. Maintenance of Compressors and Blowers

The compressor set must be given a special check for soiling if irregularities have occurred in those parts of the installation preceding the compressor which could be responsible for the presence of dirt in the compressor itself. Maintenance work should be concentrated on checking the filtration equipment, the cooling system, the water separator, the oil supply, and the safety and control equipment. Also, about once a year, a routine check should be made of alignment.
When there is a deterioration in running smoothness or unusual occurrences of surging, an additional check in the nature of an inspection must be carried out. The inspection of compressors can be markedly improved by the use of endoscopes, since these instruments permit a visual assessment of the condition of the surface of the blading. Inspection covers enable an examination to be made of the first and last blade rows at little cost, while the blades rows in between can be inspected very well with an endoscope if sockets are available.

5.4.5. Revision of Compressors

Of particular importance is checking of the impellers and impeller blading and the guide rings and guide vanes of compressors. In this connection, non-destructive tests based on magnetic-powder or dye-penetrant methods are useful for detecting cracks. Apart from this revision work, which is specific to compressors, all checks must be carried out on couplings, bearings, coolers, associated fittings, etc., as in the case of steam-turbine installations (see 5.1. Steam Turbines).

The impellers of centrifugal compressors must be checked with great care, especially in the region of impeller inlets, since erosion at the blade entry edges and loosening of the rivets have frequently been observed. With welded impellers of centrifugal compressors, the shrouds must also be examined very carefully for crack formation along the run of blades. The blade ends at the inlet and outlet must be examined in detail by non-destructive crack-detection methods.

Bibliography to 5.4

[1] Breshr, S.A., Smith, J.H.: Anleitung zum Entwurf und Betrieb sowie zur Wahl ströungsfreier Verdichter. Chemical Engineering 77 (1970), No. 12, pp. 161/170.

[2] Moody, J.F.: Wahl einer geeigneten Verdichterbauart. Plant Engineering 24 (1970), No. 14, pp. 66/69.

[3] Eckert, B., Schnell, E.: Axial- und Radialkompressoren. Berlin, Göttingen, Heidelberg: Springer-Verlag 1961.

[4] Traupel, W.: Thermische Turbomaschinen, Vol. 1. 2nd edition. Berlin, Heidelberg, New York: Springer-Verlag 1966.

[5] Horlock, J.H.: Axialkompressoren. Karlsruhe: Verlag G. Braun 1967.

[6] Kellermann, H.: Über die „Pump"-Grenze vielstufiger Verdichter. Zeitschrift für Flugwissenschaften 12 (1964), No. 12, pp. 425/435.

[7] Strub, R.A., Suter, P.: Das Pumpen von Verdichtern in Gasturbinen- und Hochofenanlagen. Technische Rundschau Sulzer 2 (1966), pp. 80/83.

[8] Hemingway, E.W.: Surge and its implications in gas turbines. The Oil Engine and Gas Turbine, Sept. 1961, p. 186/188.

[9] Benvenuti, E.: Flexibility of axial compressors with adjustable guide vane control. Quaderni Pignone 7 (1967), No. 4, pp. 5/16.

[10] Rupp, A.: Gebläse mit Laufschaufelverstellung. VDI-Nachrichten 19 (1965), No. 25, pp. 6/7.

[11] Meienberg, M., Keck, F.: Hochofen-Axialverdichter mit im Betrieb verstellbaren Laufschaufeln sämtlicher Stufen. Escher-Wyss Reprint 230 57d–10/65250.

[12] Canonica, H., Feer, K.: Sulzer-Axialverdichter "Varax" mit verstellbaren Leitschaufeln. Sulzer-Research Reports 1961, p. 6.

[13] Wachter, M.: Luftverdichter für Anlagen großer Leistung und hohen Förderdruckes. Technische Rundschau Sulzer 49 (1967), No. 3, pp. 163/174.

[14] Hoppe, H.: Die Leitradverstellung in Axialverdichtern. Year book 1962 of WGLR, pp. 340/348.

[15] Leopold, J.: Laufschaufelschäden an axialen Dampfturbinen. Der Maschinenschaden 44 (1971), No. 3, pp. 81/91.

[16] Centrifugal compressors for general refinery services. American Petroleum Institute (API), Standard 617, April 1963.

[17] Bauermeister, K.J.: Einsatz von Kreiselverdichtern in Prozeßanlagen. Chemical & Process Engineering 50 (1969), No. 9, pp. 79/81.

[18] Opitz, W.: Turboverdichter in der chemischen Industrie. Verfahrenstechnik 4 (1970), No. 8, pp. 358/365.

[19] Meienberg, M.: Die Verwendung von Turbomaschinen in Anlagen zur Salpetersäure-Herstellung. Escher-Wyss-Mitt. (1958), No. 3.

[20] Abel, H., Özvegyi, F.: Chlorverdichtung und Chlorverflüssigung. Technische Rundschau Sulzer 50 (1968), No. 3, pp. 123/132.

[21] Wellauer, M.: Verschmutzung von Axialkompressoren. Technische Rundschau Sulzer 47 (1965), No. 2, p. 97.

[22] Morfey, C.L.: Sound generation in subsonic turbomachinery. Transactions of the ASME. Journal of Basic Engineering Sept. 1970, pp. 450/458.

[23] Morfey, C.L.: Axial-compressor noise, some results from zero-engine research. Lecture held at the Conference on Gas Turbines of ASME, Zürich, March 1966.

[24] Pfau, H.: Über einige neue Bauelemente ortsfester Turboverdichter. VDI-2101 (1959), No. 13, pp. 521/526.

[25] Sobel, J.A., Schalldämpfung für Axial- und Schraubendichter. Noise Control 7 (1961), No. 5, pp. 9/11.

[26] Stoeckicht, A.: Die Entwicklung der Axialgebläse und ihrer Regelverfahren in der Industrie. Technische Rundschau Sulzer 1 (1961), No. 1, pp. 17/28.

[27] Hoheisel, M.: Über den Einfluß von Belastung, Lieferzahl und Reaktionsgrad auf das Abreißverhalten von Axialverdichterstufen. Deutsche For-

schungsanstalt für Luft- und Raumfahrt (DFL), Research Report 69–17.
[28] Filippini, U.: Surge condition and antisurge control of a centrifugal compressor. Quaderni Pignone 5 (1966), pp. 1/9.
[29] Simon, M.A.: Schaufelschwingungen in mehrstufigen Axialverdichtern infolge Abreißens der Strömung. BBC-Nachrichten 51 (1969), No. 2, pp. 100/107.
[30] Janson, W.: Rotating stall in a radial vaneless diffusor. Journal of Basic Engineering, Transactions of the ASME, Series D, Dec. 1964, pp. 750/758.
[31] Loch, E.: Hydrodynamische Gleitringdichtung. Konstruktion 20 (1968), No. 9, pp. 364/367.
[32] Sangiovanni, L.: Research and tests of high pressure centrifugal compressors. Quaderni Pignone 5 (1966), pp. 1/12.
[33] Traupel, W.: Die Beanspruchung schwingender Schaufeln in Resonanz. BBC-Mitt. 58 (1971), No. 4/5, pp. 138/147.
[34] Armstrong, E.K.: Recent blade vibrations techniques. Transactions of the ASME. Journal of Engineering for Power, July 1967, pp. 437/444.
[35] Shiraki, K. et al.: Über die Schwingungen von zwei durch eine Zahnkupplung verbundenen Rotoren. Technical Review Mitsubishi 7 (1970), No. 1, pp. 22/23.
[36] Thomas, H.J.: Möglichkeiten und Grenzen der Schwingungsanalyse an großen Turbomaschinen. Der Maschinenschaden 42 (1969), No. 4, pp. 113/120.
[37] Vogel, D.H.: Das Schwingungs- und Stabilitätsverhalten unwuchtbehafteter, mehrfeldriger Wellen auf Gleitlagern. Konstruktion 22 (1970), No. 12, pp. 461/466.
[38] Bauermeister, K.J.: Über die Erhöhung der Betriebssicherheit von Turbokompressoren durch direkte Messung der Wellenbewegung. DEMAG-Nachrichten 195, Dec. 1970, pp. 19501/19507.
[39] Abnahme- und Leistungsversuche an Verdichtern. DIN 2044, DIN 2045, sheets 1–3.
[40] Koch, E.: Automatische Steuerung von Verdichteraggregaten. Siemens-Z. 41 (1967), No. 6, pp. 504/506.
[41] Stephan, G.: Empfehlungen zur Verhütung von Stillstandskorrosion an Turbomaschinen. BBC Nachr. 47 (1965), No. 4, pp. 199/208.

5.5. Gas Turbines

Experience available with gas turbines refers mainly to installations which are operated with blast-furnace gas as the fuel [1–4, 18, 19]. Because of these difficult operating conditions, they assume a special position, characterized by many set-backs. In this connection, high-temperature corrosion [5] of the rotating blades and guide vanes is particularly worthy of mention. Moreover, the most recent cases of damage to machines, which occurred to compressor blades through *surging* and at the first rotating-blade rows by *insufficient cooling* [6], prove that the problems of the development period still exist. Difficulties were also experienced with the first peak-load gas-turbine installations with jet-type fuel-gas generators. These difficulties occurred at the well-proven jets and at the newly designed quick-start casings for prime-mover turbines [7–11].
Development is currently characterized by single-shaft turbines with outputs from about 80 to 100 MW.
An evaluation of the damage that has so far occurred to gas turbines is given in Table 1.

Discussion of the Locations of Damage

The primary locations of weakness in gas turbines are the hot first two rows of rotating blades. Improvement of these components, and timely inspection and replacement are now subjects of discussion with new designs, particularly in view of the higher gas inlet temperatures to the blading of the turbine.
In general, guide vanes are considered to be safer components than the rotating blades by reason of their lower mechanical and dynamic stressing and their simpler and hence more reliable cooling.
The combustion chamber with its liner or with its built-in components, e.g., flame tubes, turbulators and thermometer sockets, occupies second place in the damage-frequency scale. Irrespective of their design, combustion chambers can be inspected economically, particularly by using endoscopes. Damage by foreign bodies (i.e., by combustion-chamber parts) to the blading of turbines amounts to about 20% of all damage to the moving blades. For this reason, the need for

Table 1. Survey of the distributions of causes and main locations of cases of damage to gas turbines.

Cause of damage	Distribution of the number of damage cases [%]	Location of damage	Distribution of the number of damage cases [%]
Product faults	*69*	Rotating blades	42
Materials faults	17	Combustion chambers	25
Faults in planning, calculation and design	20	Hot casing parts	17
		Rotor and wheel discs	9
Manufacturing faults	19	*Others*	7
Assembly, commissioning, repair	13		
Operating faults	*15*		
Maintenance faults	8		
Mishandling faults	7		
Extraneous influences	*16*		
Foreign bodies	8		
Mains supplies	8		

Table 2. Technical data of the 18 gas-turbine installations examined.

No.	Type		Driven machine	Fuel	Gas-turbine inlet temp. [°C]
1	2 shafts, open circuit	HP LP	– generator, BF blower*	blast-furnace gas, oil since 1965 oil	730/750 550/525
2	1 shaft, open circuit		generator, BF blower	blast-furnace gas, gas oil	750
3	1 shaft, open circuit		generator	blast-furnace gas	750
4	1 shaft, open circuit		BF blower	blast-furnace gas, coke gas	730
5	1 shaft, closed hot-air circuit		generator	coal dust	680
6	1 shaft, open circuit		BF blower	blast-furnace gas, coke gas	750
7	1 shaft, open circuit		BF blower	blast-furnace gas, coke gas	750
8	1 shaft, open circuit		generator	blast-furnace gas	750
9	2 shafts, open circuit dual combustion	HP LP	– generator	fuel oil fuel oil	680 630
10	1shaft, open circuit		BF blower	blast-furnace gas, coke gas	680
11	2 shafts, open circuit dual combustion	HP LP	– generator	fuel oil fuel oil	680 630
12	1 shaft, open circuit		generator, BF blower	blast-furnace gas, oil	750
13	1 shaft, open circuit		generator	fuel oil	720
14	1 shaft, closed hot-air circuit		generator, compressor	coal dust	680
15	1 shaft, open circuit		generator	natural gas, fuel oil	650
16	1 shaft, open circuit		generator	fuel oil	750
17	1 shaft, open circuit		BF blower	blast-furnace gas, coke gas	710
18	1 shaft, open circuit		generator	fuel oil	750

* BF = blast-furnace

Table 2 (continued)

Turbine speed [rpm]	Effective power [MW]	In operation since	Observation period	Mean time in operation during period of observation [hr/year]	Remarks
4,280 to 4,750 3,000	6	(5. 5.) '57	1957 to 1966	7,000	taken out of service in 1966 after an operational life of 55,000 hr.
3,060	16	(30. 7.) '58	1958 to 1968	6,000	
3,000	16	20.9.'58	1958 to 1966	4,700	
3,580 to 4,250	5	1.10.'58	from 1968	4,300	non-operational from 1963 to 1966
8,217	6.4	13.10.'58	from 1964	7,350	
3,580 to 4,250	5	10.3.'59	from 1968	6,100	
3,580 to 4,250	5	19.1.'60	from 1968	6,500	
3,000	15.5	11.2.'60	1960 to 1964	4.000	
4,500 3,000	25	Sept. '60	from 1960	800	
3,750 to 3,960	6	5.12.'60	from 1968	4,000	non-operational from 1963 to 1965
4,500 3,000	25	Jan. '61	from 1961	800	
4,750	5.8	2.8.'61	1961 to 1966	4,400	taken out of service in 1966 after an operational life of 22,000 hr.
3,000	24	2.9.'61	from 1964	5,300	
8,224	6.4	25.4.'63	1963 to 1968	7,800	
4,760	4.5	Oct. '63	from 1964	4,000	
3,000	25	9.4.'63	from 1963	3,400	
2,380 to 3,670	5.8	9.7.'64	from 1964	5,000	taken out of service in 1968 after an operational life of 19,400 hr.
3,000	25.9	20.11.'64	from 1965	5,000	

combustion-chamber inspections to be carried out strictly in accordance with the rota basis prescribed by all manufacturers is emphatically confirmed.

Discussion of the Causes and Manifestations of Damage

Over 30% of all cases of damage were caused by thermal overloading and 25% by mechanical overloading, while 17% of the damage cases were due to distortion of rotors and casings and 8% to lack of lubricant.

With regard to the manifestation of damage, thermal stress cracking and rubbing damage were predominant, amounting to 30% of all damage cases. All other characteristic manifestations of damage, such as fatigue failure and violent rupture, deformations and bulging, were evenly distributed, and this is explained by the relatively small number of cases of damage included in the data evaluation. Thermal stress cracking, as well as rubbing through clearance bridging, constitutes damage that develops slowly, underlining the usefulness of inspection and particularly revision with extensive measurements.

Since an electronic data-processing (EDP) evaluation cannot answer to all interesting questions because of the relatively small number of damage cases and the whirlwind development of the gas turbine, the operational lives of 18 gas-turbine installations were evaluated, taking into account not only material damage in the sense of machine insurance, but also revision reports and known assembly damage and damage covered by guarantee. Table 2 shows the most important technical data of these installations.

Discussion of the Locations of Damage to the 18 Gas Turbines in Table 2

These installations, from four manufacturers, have effective outputs from 4.5 to 26 MW, with inlet temperatures up to 750 °C.

Figure 1 shows the operational times of these installations, with indications of the type of damage, its cause and location. Included in the evaluation are all cases of damage known to Allianz in respect of the following components:

- Combustion chamber and air heater, including hot-gas duct. (BK)
- Gas turbine. (GT)
- Combustion-air compressor. (BL)
- Blast-furnace gas compressor, if any. (BG)

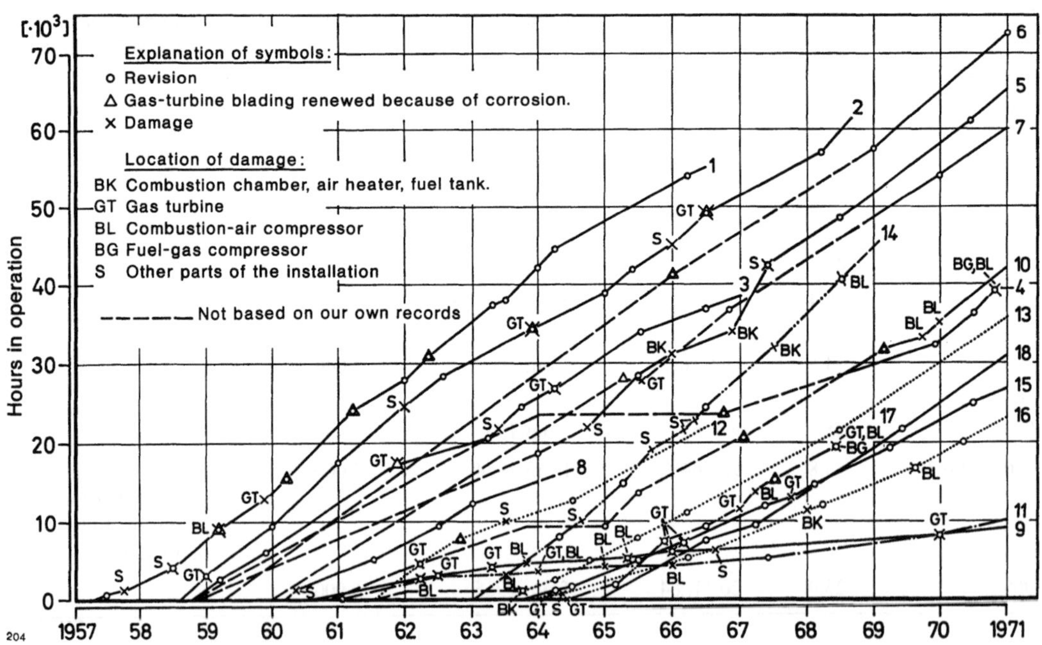

204

Figure 1. Operational times of and chronological manifestations of damage to the 18 installations in Table 2.

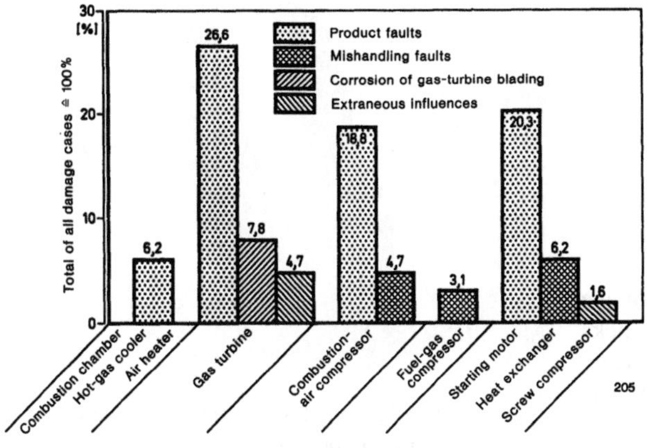

Figure 2. Frequency distribution of the causes of damage to the 18 gas-turbine installations in Table 2.

- Other parts of the installation (gearing, oil supply, heat exchangers, starting motors, etc.). (S)

Studies were made of 64 cases of damage and 83 reports of revisions.

Damage to the driven generators or compressors has not been taken into account.

In addition to the evaluations so far mentioned, corrosion of gas-turbine blading was determined separately. The results in Figure 2 show, as also do the EDP statistics, the eminence of product faults as the cause of damage in 71.9% of the cases of damage investigated. Of these, the major part consists of faults in planning, design and materials, brought about by the high demands imposed by new developments in gas-turbine installations.

From the operational times in Figure 1, it can be established that, of damage by product faults,

53.3% occurs between
0 and 10,000 hours in operation,
11.1% occurs between
10,000 and 20,000 hours in operation,
11.1% occurs between
20,000 and 30,000 hours in operation,
17.8% occurs between
30,000 and 40,000 hours in operation,
6.7% occurs after more than
40,000 hours in operation.

The first operational phase, up to 10,000 hours, merits the greatest outlay in terms of inspection and revision. Only in this way is it possible to gain knowledge which can be applied for the repair of these installations and to make the manufacture of subsequent units less liable to damage.

After 30,000 hours in operation, the effects of ageing processes manifest themselves. The relatively low percentage of damage cases after 40,000 operational hours can be explained by the small number of installations which have so far exceeded this operational period.

Discussion of the Results of the 83 Revisions

Since the current demand for definite intervals between revisions often meets with little appreciation, it is appropriate to discuss here the result of the 83 revisions (Figure 3):

Only those deficiencies that gave rise to unscheduled repairs and examinations by the user and manufacturer have been taken into account. Of these, 35 revisions resulted in objections, usually to several parts of the installation. In the case of 16 revisions, extensive and unexpected damage was discovered. Revisions which were

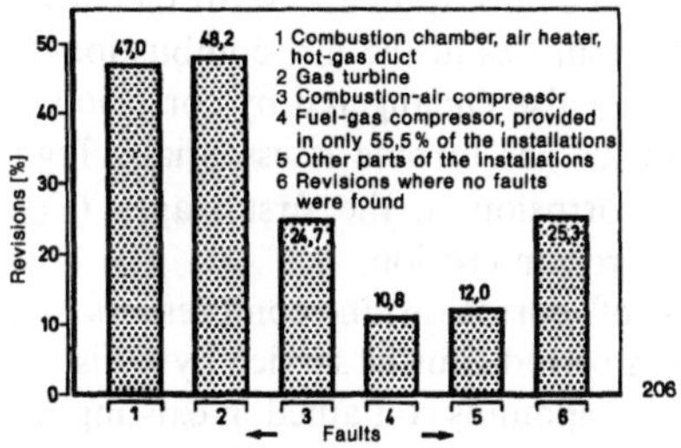

Figure 3. Faults found during revisions which initiated measures for repairs and examinations. Number of revisions 100%; number of faults 100% because *one* revision could reveal *several* faults.

carried out following damage are excluded from these considerations.
With 74.7% of revisions, there were substantial, partly known, deficiencies whose rectification and inspection exceeded the normal cost of revisions.
In the hot zone, i.e., the combustion and air heater, and in the gas turbine, unacceptable conditions were found in 50% of revisions.
In combustion chambers, the majority of damage was in the form of distortions, scaling or cracks in the lining and mountings. With mixed firing, burners and linings were damaged by crack formation.
In gas turbines, more complaints were made about blading and casings, in approximately equal proportions, than about all other components.
In 55.5% of installations with fuel-gas compressors, from 20 to 25% of the revisions of these installations resulted in finding deficiencies in these compressors, about 50% of which were concerned with the blading.

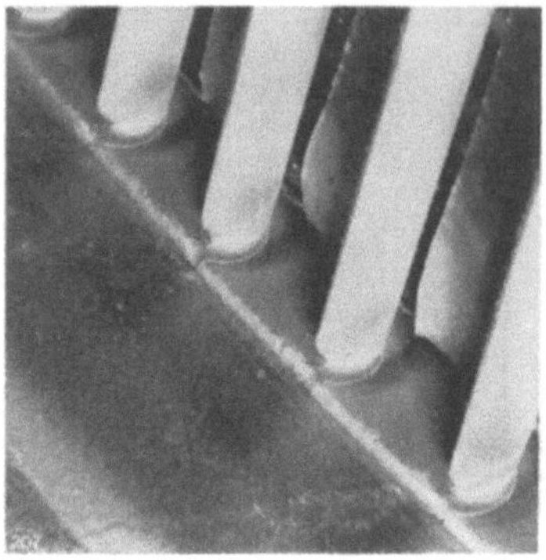

Figure 4. Erosion by dust at the guide vanes of a combustion-air compressor.

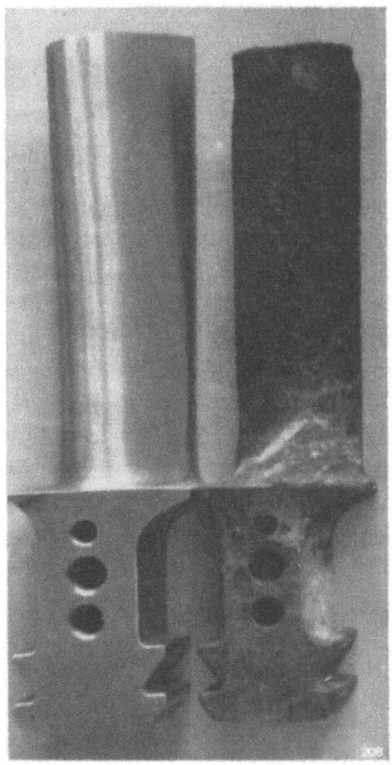

Figure 5. High-temperature corrosion of a gas-turbine blade. Left: a new blade; right: after 31,300 hours in operation with blast-furnace gas.

5.5.1. Guidelines for Prevention of Damage to Gas Turbines during Manufacture

5.5.1.1. Planning of Gas Turbines

The life of gas-turbine blading and of combustion-air or fuel-gas compressor blading, properly designed from the point of view of stressing, is essentially determined by erosion and corrosion [2, 3, 5].
So far, pure erosion has been found only in compressors (Figure 4), the cause being dust in blast-furnace gas and combustion air. With increasing amounts of impurities in the air, combustion-air compressors are also endangered by corrosion.
Corrosion in the gas turbine is essentially high-temperature corrosion in the first stages (Figure 5) and corrosion-erosion.
Some cases of corrosion in compressors are caused during periods out of service by moisture and sulphur compounds contained in existing deposits from contaminants in the fuel gas or combustion air.
Particularly responsible for promoting corrosion are the effects of alkalis in conjunction with sulphur and vanadium [31]. The corrosion mechanism can vary in accordance with the compounds present and the hot-gas temperatures. At temperatures above 620 °C, if heavy fuel oil is used, measures must be taken to remove the above mentioned constituents or to cater for their bonding

Figure 6. Effect of sulphate on gas-turbine blading at high temperatures.

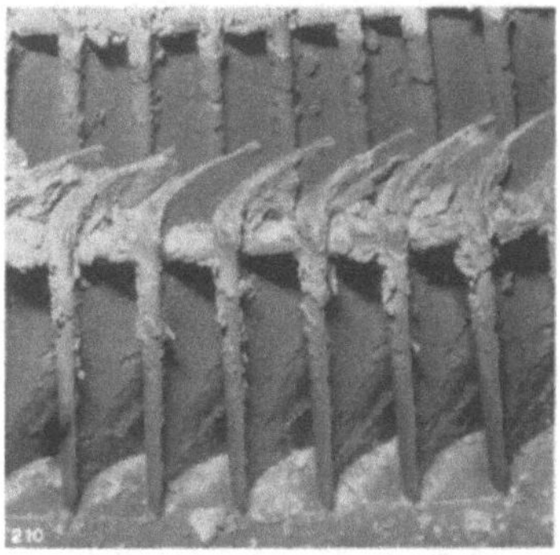

Figure 7. Deposits on the guide vanes of a gas turbine.

with appropriate additives. The effect of sulphate on gas-turbine blading at high temperatures is exemplified in Figure 6.

With blast-furnace gas, the high proportion of alkali chlorides in the combustion chamber leads to complex alkali sulphates whose eutectic melting points can be lowered to less than the operating temperature of 720 to 750 °C by the dissolution of heavy-metal oxides and sulphates.

The results of research in connection with deposits and their origin on the blading of blast-furnace gas turbines are available in [1, 2].

According to these references, corrosion-promoting substances in the combustion air generally have particle sizes less than 5 µm, making filtration more difficult. As a result, filtration of the air can cause a relative enrichment in these substances. The output and efficiency of an installation are greatly influenced by deposits, quite apart from the corrosion and erosion mentioned earlier.

Open-circuit gas turbines exhibit blade-profile weakening by corrosion in the first stages, while profile thickening by deposits occurs in the last stages (Figure 7). The effects of such changes in profile are reported in [17].

Combustion-air compressors are in a large measure endangered by dirt, since the conditions for flow disruption and surging can be displaced into the permissible operating range.

Careful planning of combustion-air intake and filtration can overcome these dangers. With a given position and height of the exhaust-gas stack, the air intake must be laid out in such a way that sucking in of exhaust gases under the usually prevailing climatic conditions is effectively prevented. In this connection, account must be taken not only of the main wind directions, but also of the location of buildings in the vicinity.

The following types of filter installations have proved successful:

- *Rotating dry filters, possibly with air prewashing*
  (Advantages: Freedom from maintenance, automatic operation.
  Disadvantages: Wear of filter cloth, replacement)
- *Drum filters with water cleaning*
  (Advantages: Freedom from maintenance, automatic operation.
  Disadvantages: Wear of filter cloth, replacement)
- *Oil-wetted filters*
  (Advantages: Low operating costs, since oil-changing only is necessary.
  Disadvantages: Entrainment of oil)

Supervision of filters by measurement of pressure difference, automatic advance of the filter cloth and warning in good time before the cloth comes to an end are recommended, since this procedure is the only way of safely avoiding damage to combustion-air compressors by sucking in filter elements that are soiled and overloaded by differential pressure. Filter installations must be provided with safety bypass flaps, particularly

against snow-packing, so as to prevent danger to the compressor.
The second revision of an installation after 17,000 hours in operation showed incipient cracks on vanes of the inlet guide-vane row of the combustion-air compressor. Examination at the Allianz Centre for Technology revealed fatigue corrosion cracking caused by strongly aggressive deposits through sucking in chloride-containing turbine exhaust gases.
By *matching the characteristics of the units of an installation*, especially the combustion-air compressor, a stable operating mode must be ensured over the whole load range which, in addition, must not be endangered by such influences as corrosion and soiling, since cleaning of the compressor blading is possible only at definite intervals.
Generally, an *analysis of the contemplated fuels*, together with details of tolerances, should be given to the manufacturer before placing the order, to take account of unfavourable developments and enable him to determine, by deciding on the temperature and choosing the materials, the safety of the blading and its desired operational life. The manufacturer will then also give information from his own experience on the use of additives.
Natural gas is considered to be the ideal fuel, followed by light fuel oil with a low sulphur content.
If hot-gas temperatures are above 620 °C, *heavy fuel oils* should, for reasons of corrosion, be used only in special circumstances. Residual oils require special investigations.
With *blast-furnace gas*, dust extraction by mechanical pre-separation and electrostatic precipitation, as well as drying, is necessary.
The admixture of silica additives is also advantageous here, since liquid slag components are transformed into the solid state and the corrosive action of molten slag on the first gas-turbine stages is alleviated. In addition, the early solidification of slag prevents the formation of deposits.
Difficulties are often experienced with the control characteristics of installations when changing from one fuel to another with a different calorific value.

5.5.1.2. Design of Gas Turbines

Design measures for loss prevention are concentrated on components exposed to the hot gas stream:
With the higher gas temperatures of 750 to 900 °C, the stresses in the *moving blades* are controllable by means of an amply proportioned supply of cooling air in the region of the blade roots. The cooling-air supply must be so designed that all moving blades are cooled evenly on their periphery. Cooling-air quantities of about 2 to 4% of the combustion-air throughput have proved successful in practice. The allotted cross-sections of the cooling air must not suffer one-sided reduction during operation. The cooling-air ducting must not cause excessively high pressure losses, so as to ensure the flow of cooling air in the desired direction under all operating conditions. Through-flow and pressure measuring points in the cooling-air supply system should already be catered for in the design. A check on the effectiveness of cooling can be executed by the inspection of inserted fusible plugs.
The *rotors* of gas turbines currently on the market are in most cases assembled from self-centring discs made radially movable by Hirth-type gearing, with a central hub or with several axial tie-rods on their periphery; alternatively, the discs can be made into a rigid rotor by welding. In each case, the discs are protected from inadmissibly high heating by a through-flow of cooling air or by heat shields—a particularly important requirement for the roots of the moving blades and the root grooves in the discs. Here, too, a check on proper cooling can be carried out by the inspection of inserted fusible plugs.
The *casings* are protected from local overheating and particularly from locally occurring large temperature differences by cooling and by external and internal insulation. Since relatively short periods for starting and application of load are demanded from all gas turbines, accumulations of material at horizontal and vertical flanges must be avoided. Light guide-vane carriers, suspended in the outer casing and centrally slidable under the influence of heat, can be used to shield the outer pressure-sustaining casing from high temperatures and have proved to be the most satisfactory for this purpose. Deformations of the outer and inner casings constitute the most costly dam-

age to gas turbines and are the most difficult to repair. For this reason, it is recommended that the cooling-air supply for the casing, the rotor and the blades should be reliably supervised by measuring pressure differences, by measuring the cooling-air temperature at the cold end of the gas-turbine rotor and by measuring the temperature of the casing, at least with new designs until they have been proved satisfactory.

The blading of *stationary gas turbines* is normally designed now with 50% reaction, so that particular care must be exercised in the cooling of the casing because of the small clearances.

Prime-mover gas turbines are also produced as impulse turbines, since this type of design permits large radial clearances without detrimental effects, and hence casing deformations are not so critical.

Examples of Cases of Damage

The following few examples will serve to show how practical these recommendations are:

On two gas-turbine rotors, cracks at the shoulders of the blade grooves of the first row of moving blades were found after 7,000 and 20,000 hours in operation respectively. The cause was temporary failure of the *cooling-air supply* and unfavourable stressing through *notch effects*.

Blade fracture and further *incipient cracks* were found in a combustion-air compressor. Examination by the Allianz Centre for Technology (ATZ) revealed fatigue cracks, abetted by slight pitting. The manufacturer had based his blade calculations on limiting values which had to be reduced as a result of subsequent experience.

Displacements of components in their guides under the influence of temperature differences must be allowed to take place unhindered. The guides must have sufficient play to permit displacements without serious obstruction, even under conditions of uneven heating. In the hot region, a stellite-on-stellite pairing has proved its value.

Multiple *incipient cracks* in the *fastenings* for the lining of the inlet duct of a gas turbine occurred. The welds had to be renewed at almost every revision. Cracks in the clamping rings of the combustion-chamber lining enforced changes in the design of these rings.

Rewelding of welding-seam cracks at the *linings* of combustion chambers, hot-gas ducts and gas-turbine inlets is part of the normal programme of revision. However, the possibility of thinning of the walls by corrosion must be borne in mind to ensure that the break-away of whole surface areas cannot occur.

Stresses through temperature differences in a component can lead to plastic deformation, particularly under non-steady operating conditions. The frequent occurrence of such temperature differences (e.g., with fluctuating hot-gas temperatures, frequent starting and stopping) already represents low-cycle alternating stressing, with a corresponding reduction in permissible stresses.

With *thick-walled components,* checking by calculation and deciding upon a permissible temperature gradient are indispensable. From the design angle, it is possible to cater for their effects by even mass distribution and elastic configuration.

Uneven mass distribution and ribbing that imparted excessive stiffness led to wall fractures in two gas turbines after 3,000 and 4,000 hours in operation respectively and in two compressor casings after 6,000 and 4,500 hours in operation respectively.

For the *selection of materials,* a knowledge is required of creep strength and long-time rupture strength, as well as of resistance to scaling and corrosion at the required temperature level. Although this knowledge has been greatly expanded by manufacturers and users since the inception of gas-turbine development, its present level cannot be regarded as the ultimate. Further basic research into permissible creep-strength values at limiting temperatures, and also into the cause and effect of disturbing operational influences, is necessary to attain the desired better economic utilization by a reduction of outages resulting from damage, and not merely by improved efficiency [19, 30].

Revision-Conscious Design

A *reduction in outage times* can be achieved by a design which caters for ease of inspection and revision [20]. Visual observation still provides the safest information about the progress of corrosion, erosion, and the degree of soiling.

Access to the interior of an installation should, if possible, be provided by manholes. Welded or flanged inspection ports should make it possible to carry out measurements of clearances and visual examinations at less accessible locations. With the aid of suitable endoscopes from their many variants, it is possible, for example, to inspect the moving-blade rows of compressors and gas turbines [21, 22] by using special sockets. If an installation suffers a loss in output, it can therefore be established quickly whether the installation and which part thereof should be opened up for cleaning or repair.

Nowadays, it must be practicable to check the functional capabilities of the control, supervisory and protective equipment without having to take the installation out of operation.

Such a check is fully effective if the entire functional chain, from pulse emitter to actuating element, is submitted to tests on several components. In this connection, the manufacturer must supply unambiguous instructions and outline the facilities, since he is best acquainted with the components that are important for the safety of the set.

Recommendations for Supervisory and Protective Equipment

Hot-gas temperature control should give the minimum of permissible deviation from the rated temperature.

Exhaust-gas temperature control is often safer since it operates at a lower temperature level and does not require any facilities at the gas-turbine inlet.

In each case, control of the propellant-gas temperature must always be complemented independently by two co-ordinated *limit-temperature monitors* with alarm and switch-off facilities at the combustion-chamber exit. The combustion chamber is usually provided with *flame monitors,* whose purpose is to give protection from explosion via interlocking devices.

The *cooling-air supply* of the gas-turbine section must be supervised by an alarm system.

Combustion-air compressors must be protected from overheating and blade damage by an anti-surge device via automatic actuation of blow-off valves in the bladed section and at the compressor exit. These blow-off valves must be open during starting up to a partial speed determined on the test bed, in cases of emergency shut-off or reverse current, and during run-down from the above partial speed. The valves can be closed automatically by simple circuits embodying switches actuated tachometrically and by final compressor pressure.

Overspeed protection should be provided by two pulse emitters independently of each other. With two-shaft installations, the separate prime-mover turbine must be protected individually from overspeeding (e.g., during sudden load-shedding).

Bearing-oil pressure protection should be provided to switch off the set if the required bearing-oil pressure is not attained and to switch in the a.c. or d.c. (or diesel) auxiliary oil pump.

Protection of axial bearings (two-channelled) should be effective in both directions of thrust and should be derived from the bearing-metal temperatures in at least two bearing pads displaced by 180 °C, with alarm actuation at 80 °C and switch-off at 100 °C.

Supervision of operational quietness should be effected by measurements of bearing-pedestal and shaft vibrations, with alarm and switch-off facilities [23, 24].

Generator protection should be as with other turbo-generators (see 2.1. Turbogenerators).

5.5.1.3. Guidelines for Loss Prevention during the Manufacture and Assembly of Gas Turbines

The prescribed condition of materials must be assured by thorough quality control. For example, the following cases of damage occurred through the faulty condition of materials:

- During a check about 2,000 operating hours after reblading a gas turbine, it was found that a moving blade had elongated because of a mix-up of materials (Figure 8).
- Cracks appeared on the blades of the third moving-blade row of a gas turbine after 7,000 hours in operation. The cracks were caused by grain coarsening during heat treatment in association with inclusions.

The supervision of manufacture and the purposeful inspection of components ready for assembly give information regarding the proper condition of all components.

During *assembly work,* the prescribed axial and radial clearances must be measured once again in the presence of a representative of the manufacturer. These clearances should be ascertained

Figure 8. Because of a mix-up of materials, one of these moving blades of a gas turbine elongated during operation and resulted in radial rubbing.

at each revision, at least on the basis of random tests—and *also before lifting out the rotor*. In this way, possible displacements and casing deformations will be disclosed.

The alignment of shafts with respect to each other, axially and radially, must, even with the use of flexible couplings, lie within the framework of tolerances prescribed by the manufacturer (see 8.4. Couplings).

The facilities provided for expansion must be measured again. Impeded expansion can lead to deformations and clearance bridging.

Built-in components, as well as all screws, must be secured against loosening or dropping out:

A cover ring at the inlet of a gas turbine became loose after 500 operational hours, with consequent damage to the blading, because the cover plate had only been tacked in place, instead of welded, as required.

It is strongly recommended that all measurement records during assembly, revision, repairs and commissioning should be countersigned by a representative of the user. In addition, a register should be kept of all adjusted settings and characteristics that are important for assembly and operation, and a copy of this register should remain with the plant management, so as to make available information on the history of the machine in the event of disturbances, damage and revisions.

5.5.1.4. Requirements for the Commissioning of Gas-Turbine Installations

Before commissioning, proper functioning of all measuring and supervisory equipment should be assured.

Test runs should be made for adjusting safety elements under mild operating conditions, e.g., for checking the safety governor under excess speed at a reduced hot-gas temperature.

Prototype sets should in all circumstances undergo extensive test-bed operation. The characteristic of the compressor and the surge limit must be determined, so that adjustment of protective devices for the prevention of damage by surging can still be carried out under the manufacturer's responsibility.

The behaviour of the gas turbine must be tested under partial and over-load, and slow, normal and quick starting, including changing over to different fuels.

Even today, the variations in flame configuration with different fuels can lead to local overheating, particularly of the combustion-chamber lining. The design of the burner and the guidance of cooling air are here of primary importance. Every new burner design and cooling-air duct must be tested during commissioning over the whole measuring range by measuring the temperature in the combustion-chamber lining.

The efficiency curve of the compressor and gas turbine must be plotted. The Bacharach-Ringelmann anti-smoke test and a check on dust emission must be carried out.

Baseplate vibration, torsional frequency and coupled critical-speed calculations must be available *before commissioning,* so as not to remain in the critical-speed region when putting the set into operation for the first time. With prototypes, checking of the calculations by measurements is recommended.

Tooth breakages in the main transmission of a gas-turbine installation were caused by resonance of torsional vibration and operating speed. By uncoupling the starting motor, hitherto running idle with the set, the critical torsional vibration was raised to a safe region.

It is recommended that commissioning should be initiated with excess cooling air and that the safe and necessary quantity of cooling air should be adjusted by slow throttling in accordance with the results of pressure and temperature measurements.

After revisions or repairs to components which affect the quantity of cooling air, the correct supply of air must be re-established by measurements when putting the set into operation again.

5.5.2. Guidelines for Loss Prevention during the Operation of Gas Turbines

Automatic Supervision

Faulty measuring points must be clearly marked as soon as they are perceived. Damage to the thrust bearing of a gas-turbine blower was indicated by a rise in the oil-outlet temperature on both sides of the bearing. Machine personnel instigated a "hot-spot" check of the bearing because thermocouples installed shortly before in the white metal were known to be faulty and did not respond properly. Thrust-bearing protective equipment triggered emergency stoppage at an axial displacement of 3.9 mm (adjusted value 0.9 mm).

The cooling-air supply must be continuously checked for its effectiveness during operation by suitable measuring instruments.

Cleaning

Combustion-air compressors can be cleaned during operation by wetting with solvents, followed by flushing with a copious supply of water delivered through injection nozzles. They should then be dried with rice or coarsely ground nutshells (caution: there is a risk of damaging corrosion-protection layers that may be present).

Combustion chambers can be cleaned with a jet of water or mechanically by brushing.

Gas turbines can be cleaned by means of solvents and water.

Inspection

An inspection comprises a survey of the combustion chamber, the hot-gas duct, the gas-turbine inlet and outlet branches, and also the inlet and outlet branches of the combustion-air compressor.

Its purpose is to assess the progress of corrosion, soiling and deformations, and also planning of the next revision with respect to the time required and spare parts.

The interval between inspections depends on the mode of operation, the fuel and the materials used. With installations operating in an open-circuit mode, an interval of 6,000 to 8,000 operational hours, with a maximum interval of one year, should not be exceeded.

Inspection is facilitated by the use of endoscopes, for which sockets or other suitable apertures should be provided in the hot-gas region. A premise for the use of the most common types of endoscope currently available is a temperature of less than 100 °C during inspection.

5.5.3. Revision of Gas-Turbine Installations

The periods between revisions for hot parts (so-called minor revisions) could, depending on the number of starts and the fuel used, and particularly by improvements in design and materials, be lengthened from an initial 2,000 to 8,000 hours up to as high as 16,000 hours in operation. During the course of further developments and the experience thus gained, even longer periods could be considered. Intervals of from 8,000 to 16,000 hours in operation are also appropriate for the other parts of the installation in light of the present state of technology. As can be seen in Figure 3, faulty compressors in about 25% of revisions constitute a substantial proportion.

Experience with installations in operation today clearly argues against the trend emanating from foreign countries of doubling the interval between major revisions, bearing in mind that details of operating experience in Germany with sets produced in series and using the certainly more favourable natural-gas fuel (but involving higher inlet temperatures) are not yet available.

The yardstick of "equivalent operating hours", partly introduced by manufacturers for determining revisions, is an attempt to include additional stressing from starting procedures and excesses of temperature as a criterion. Undoubtedly, this requirement will become increasingly obligatory as stresses or temperatures approach limiting values.

Bibliography to 5.5

[1] Umland, F.: Chemische Untersuchung an Beschaufelungen von Gichtgasturbinen. V.I.K.-Bericht No. 164 (September 1965).

[2] Umland, F.: Physikalische und chemische Untersuchung von Brennluft- und Gichtgasstäuben bei Gasturbinen. Part I: MTZ 30 (1969), No. 10, pp. 367/372. Part II: MTZ 30 (1969), No. 11, pp. 407/411.

[3] Möller, W.: Fortschritte bei der Bekämpfung der Hochtemperaturkorrosion an Gasturbinenschaufeln. BBC-Nachrichten 48 (1966), No. 12, pp. 669/678.

[4] Condorf, E., Wehe, H.: Arten und Eigenschaften von Schutzschichten gegen Hochtemperaturkorrosion unter besonderer Berücksichtigung der Anwendung im Gasturbinenbau. Stahl und Eisen 91 (1971), No. 14, pp. 839/841.

[5] Keller, A., Stauffer, W., Arnet, F.: Werkstoffprobleme bei Hochtemperaturturbinen. Escher Wyss Mitt. 40 (1967), No. 3, pp. 11/21.

[6] May, H.: Neuere Untersuchung über die Kühlung an Gasturbinenschaufeln mit Luft und Flüssigkeiten. BWK 19 (1967), No. 5, pp. 260/265.

[7] Gasturbinenanlagen mit Flugtriebwerken. BWK 17 (1965), p. 440.

[8] Kunze, N.: Erstes Gasturbinenkraftwerk mit Strahltriebwerken in Deutschland. BBC-Nachrichten 48 (1966), No. 6, pp. 374/378.

[9] Ellwanger, E.: Strahltriebwerke in der Elektrizitätsversorgung für Spitzenlast und steile Lastanstiege. Elektrizitätswirtschaft 68 (1969), No. 15, pp. 523/531.

[10] Gasturbinenanlage für 100 MW mit 10 Triebwerken. BWK 18 (1966), No. 12, pp. 621/622.

[11] Jacobsen, G.: Gasturbinenspitzenkraftwerke mit Strahltriebwerken der NWK. Aufbau der Anlage, Wirtschaftlichkeitsvergleich mit einer normalen Gasturbine, Betriebserfahrungen. VGB-Fachtagung „Gasturbine und Gasturbinenbetrieb" 1971.

[12] Schmoch, O.: Große einwellige Gasturbinen in der öffentlichen Energieversorgung. MTZ 31 (1970), No. 5, pp. 198/203.

[13] Friedrich, R.: Stand und Entwicklungstendenzen im Gasturbinenbau. VGB-Fachtagung „Gasturbine und Gasturbinenbetrieb" 1971.

[14] Maikranz, F.: Vergleich zweier Heizkraftwerksplanungen auf der Basis Gasturbinen 1960 und 1970. VGB-Fachtagung „Gasturbine und Gasturbinenbetrieb" 1971.

[15] BBC-Nachrichten 48 (1966), No. 6. Special issue: Kraftwerke für Spitzendeckung. pp. 327/390.

[16] Pfenninger, H.: Vergangenheit, Gegenwart und Zukunft der BBC-Gasturbine. MTZ 27 (1966), No. 11, pp. 449/461.

[17] Bammert, K., Stalle, M.: Messungen an einer mehrstufigen Axialturbine mit normalen verdünnten und verdickten Schaufelprofilen. MTZ 31 (1970), No. 5, pp. 189/198.

[18] Hülig, M.: Gichtgasturbinen in Hüttenwerken. Stahl und Eisen 89 (1969), No. 6, pp. 118/124.

[19] Betriebszeiten von Gasturbinen. Energie 18 (1966), No. 10, p. 447.

[20] Braun, H.: Die Bedeutung der Schadenforschung für Konstruktion und Fertigung. Der Maschinenschaden 44 (1971), No. 5, pp. 175/180.

[21] Huppmann, H., Löbert, P.: Erfolgreiche Schadenverhütung in Kraftwerksbetrieben durch den Einsatz von Endoskopen. Der Maschinenschaden 44 (1971), No. 1, pp. 11/13.

[22] Splittgerber, E.: Probleme bei Schadenuntersuchung und -auswertung. Mitt. d. VGB 51 (1971), No. 4, pp. 301/11.

[23] VDI-Richtlinie 2056: Lagerschwingungsmessungen.

[24] VDI-Richtlinie 2059: Wellenschwingungsmessungen.

[25] Gašparović, N.: Gasturbinen, Probleme und Anwendungen. Düsseldorf: VDI-Verlag 1967.

[26] Kruschik, J.: Die Gasturbine. 2nd edition. Berlin, Göttingen, Heidelberg: Springer-Verlag 1960.

[27] Gerlach, H., Kautz, R.: Untersuchungen an Werkstoffen für den Gasturbinen- und Abgasturboladerbau. MTZ 29 (1968), No. 2, pp. 50/57.

[28] Bollinger, H.R.: Betriebserfahrungen mit einem Gasturbinen-Industriekraftwerk in Venezuela. BBC-Mitt. 55 (1968), No. 1/2, pp. 15/19.

[29] Hänschke, A.F.: Die Arbeitsturbinen der 121/139-MW-Gasturbinenanlage im Kraftwerk Sewaren. BWK 20 (1968), No. 3, pp. 122/125.

[30] Bammert, K.: Industriekraftwerke mit Gasturbinen. Atomenergie 13 (1968), No. 5, pp. 381/383.

[31] Schoberl, H.: Erfahrungen mit einer Gichtgasturbine und ihre energiewirtschaftliche Bedeutung. Stahl und Eisen 89 (1969), No. 6, pp. 281/289.

[32] Müller, E.O., Jakobeit, W.: Molybdän, ein Schaufelwerkstoff für Helium-Reaktor-Gasturbinen. BBC-Nachrichten 51 (1969), No. 7, pp. 392/396.

[33] Möller, W.: Zur Korrosion von Gasturbinenwerkstoffen in V_2O_5-reichen natriumhaltigen Salzschmelzen. BBC-Nachrichten 51 (1969), No. 7, pp. 382/391.

[34] Skowronek, M.: Schwedische Gasturbinen für den Einsatz im Spitzenlastbetrieb. BWK 21 (1969), No. 7, pp. 369/373.

[35] Nakao, K., Hirai, H.: Zweijährige Erfahrungen mit einer mit Heizöl S direkt befeuerten Gasturbine. Betriebsökonom 24 (1970), No. 3, pp. 42/45.

[36] Konings, J.A.K.: Einsatz von Gasturbinen verschiedener Bauweisen zur Wärme- und Krafter-

zeugung in der Industrie. Energie 22 (1970), No. 4, pp. 122/128.

[37] Bartelmann, W.: Gasturbinen für die Industrie. Siemens-Z. 44 (1970), No. 4, pp. 225/227.

[38] Bernhardt, S.: Planungsgrundlagen, technische Ausführung und vorgesehene Betriebsweise der Anlage mit kombiniertem Gas-Dampfprozeß der Neckarwerke AG. Mitt. der VGB 50 (1970), No. 3, pp. 153/161.

[39] Czermak, H.: Betriebserfahrungen mit der Kombination eines Gasturbinen-Dampfkraftprozesses in der Anlage Wärmekraftwerk „Hohe Wand". Mitt. der VGB 50 (1970), No. 3, pp. 161/166.

[40] Steam injection can improve gasturbines. Power 88 (1970), No. 6, pp. 82/85.

[41] Gašparović, N.: Die Problematik der „nuklearen" Gasturbinen. BWK 22 (1970), No. 7, pp. 334/339.

[42] Schultze, R., Bartz, W.J.: Versuche zum Korrosionsproblem der Verbrennung schwerer Heizöle. MTZ 31 (1970), No. 9, pp. 377/383.

[43] Hanisch, F.: Das Zweistromtriebwerk Rolls Royce RB 211. MTZ 32 (1971), No. 5, pp. 176/179.

[44] Keen, J.M.S.: Fortschrittliche Technologie bei Rolls-Royce-Flugtriebwerken mit hohem Bypassverhältnis. MTZ 32 (1971), No. 3, pp. 90/96.

[45] Deuster, G., Plür, H.: Betriebserfahrungen mit Heißluftkraftwerken und Folgerungen für die weitere Entwicklung von Heliumkraftwerken. Mitt. der VGB 51 (1971), No. 2, pp. 81/91.

[46] Russische 100-MW-Gasturbine. Energie 35 (1971), No. 5, p. 176.

[47] Aguet, E.: Technische und wirtschaftliche Vorteile kombinierter Gasturbinen/Dampfkraftwerke. Mitt. der VGB 51 (1971), pp. 402/409.

5.6. Water Turbines

Water turbines are required to deal with large quantities of liquid at relatively low flow velocities. Large flow cross-sections and large masses per unit of power output characterize their design.
An evaluation of 106 technically clarified cases of damage over the last three years is shown in Tables 1 and 2.

Discussion of the Causes and Locations of Damage

The most critical components of water turbines are the axial bearings, followed by radial bearings and the oil supply. As indicated in Table 1, 38% of all cases of damage must be apportioned to these components. Transmissions, notably bevel-gear drives, are in second place on the danger

Table 1. Distributions of the causes and locations of cases of damage to water turbines of all types and powers.

Cause of damage	Distribution of the number of damage cases [%]	Location of damage	Distribution of the number of damage cases [%]
Product faults	*52*	Bearings, especially axial bearings with oil supply	38
Faults in planning, design and calculation, technological shortcomings	21.6	Transmission	17
		Pipelines, seals, fittings	13
Manufacturing faults	16	Guide apparatus with guide vanes	9
Assembly and repair faults	10.4		
Materials faults	4	Runner with vanes and shaft	8
		Casing and fasteners	8
Operating faults	*28*	Regulation	7
Maintenance faults	23.2		
Mishandling faults	4.8		
Extraneous influences	*20*		
Foreign bodies	14		
Climatic conditions	6		

Table 2. Damage-initiating faults and manifestions of damage to water turbines of all types and powers.

Damage-initiating fault	Distribution of the number of damage cases [%]	Manifestation of damage	Distribution of the number of damage cases [%]
Lack of lubricant	20	Forced rupture	28
Loosening	19	Changes in load pattern and damage to bearing metal	23
Running roughness	18		
Jamming	13	Endurance and stress/time rupture	19
Overload	12	Abrasion	9
Distortions	7	Scoring	8
Wear	5	Deformations and incipient cracks	7
Erosion, cavitation	4	Rubbing	6
Ageing	2		

Table 3. Evaluation of cases of damage to water turbines in relation to their output power, i.e., over 2,000 hp and from 100 to 2,000 hp

Causes of damage to machines over 2,000 hp	Distribution of the number of damage cases [%]	Causes of damage to machines of from 100 to 2,000 hp	Distribution of the number of damage cases [%]
Product faults	*62*	*Product faults*	*37*
Manufacturing faults	19	Faults in assembly and commissioning	11
Faults in design and calculation, technological shortcomings	18	Faults in design and calculation, technological shortcomings	10
Faults in assembly and commissioning	17	Repair faults	8
Repair Faults	7	Manufacturing faults	6
Planning faults	1	Planning faults	2
Operating faults	*13*	*Operating faults*	*51*
Maintenance faults	8	Maintenance faults	45
Mishandling faults	5	Mishandling faults	6
Extraneous influences	*25*	*Extraneous influences* particularly foreign bodies, climatic conditions	*12*
Foreign bodies	13		
Climatic conditions	2		
From mains supplies	2		
Others	8		

list because of the often difficult alignment conditions in water-turbine installations.

Table 2 shows the distributions of damage-initiating faults and manifestations of damage to water turbines. A lack of lubricant is predominantly responsible for damage to bearings. Loosening particularly occurs on shrunk-on runners, sealing rings and bushings, and gearwheels, and also on the thrust discs of axial bearings. It is noteworthy in Table 2 that 18% of all cases of damage have been marked by a deterioration in running smoothness. Jamming is mainly due to broken-off pieces of runners or guide vanes and also to foreign bodies that have entered the turbine from the outside.

A more refined evaluation of damage distinguishes between water turbines used in public utilities for the generation of electricity and turbines of lower ratings in industrial use, especially in small industries. This evaluation is given in Table 3.

Turbines of Public Electric Utilities with Outputs of over 2,000 hp

Included are 200 cases of damage, each involving a rehabilitation cost of more than 5,000 DM. Of these, 30% are apportioned to cases of damage caused by exceeding the permissible period in operation of components and by wear, as well as by omitting to replace wearing parts in good time. The remaining 70% of the cases of damage can be allocated to the various causes shown in the left-hand column of Table 3.

Small Water Turbines, Mainly in Industrial Plant (Sawmills, Flour Mills, Paper Mills, etc.) and Small Electricity-Generating Plant down to 100 hp

Of 100 cases of damage, each involving a rehabilitation cost of more than 3,000 DM, 18% were due to exceeding the permissible period in operation of components and to wear. The causes of the remaining 82% of damage cases were sta-

tistically evaluated, as listed in the right-hand column of Table 3.
Comparison of these two evaluations demonstrates the dominant role played by product faults as the cause of damage to large water turbines (i.e., 62%), and by operating faults (51%), particularly maintenance faults, as the cause of damage to small turbines.

5.6.1. Guidelines for Loss Prevention by the Manufacturer

Examination of cases of damage for which responsibility was attributed to the manufacturer shows that designs aimed at saving materials and at cheaper production were the main causes of damage with large new types. For instance, the highest proportion of damage to screens, guide vanes, casings and draft tubes can be traced back to constructional weakness and cheap manufacture, i.e., to substantially increased stressing. A particularly striking example is the change-over from cast steel to welded fabrication. Moreover, assembly personnel are insufficiently familiar with innovations, resulting in faults during assembly and through commissioning.
A surprising indication in Table 3 is the small percentage of faults during planning, which is carried out by the user with his operational experience. However, such faults as do occur mainly have an effect on the appearance of wear, excluded from the foregoing considerations. In many cases, cavitation appears because, depending on planning, operation over a relatively long period per year may involve unfavourable values of cavitation coefficient σ due to too low a tailwater level with reduced availability of water.
To save costs with impulse turbines, the most favourable layout for the pressure line is often dispensed with, e.g., by the insertion of a number of elbows that cause disturbances at the inlet during operation; with extra expenditure, such disturbances could be avoided by using a different line layout.
In particular, damage to water turbines which has occurred during the assembly and warranty period can to a major extent be traced back to manufacturing faults related to components that have been newly designed or are much larger in their dimensions, to design faults with more highly stressed or simplified components, and to assembly faults with such innovations.
Based on this experience, it is strongly recommended that purposeful inspections should be made of new components by the manufacturer in conjunction with the insurer during the first period in operation after the first load and test runs. If no faults are found, these inspections can be repeated at longer intervals.

The Following Loss-Prevention Measures Have Individually Proved of Particular Value:

Guide-vane adjustment has time and again been endangered through jamming by foreign bodies. In new installations, safety devices are now incorporated in every second guide vane to reduce the possibility of damage during closure. During revisions or design conversions of water turbines, it is possible to fit these slide- or shear-type safety devices to guide vanes. Particularly well proven are stopper arms which, after a certain amount of rotation, lock the guide vane and thus prevent damage to neighbouring guide vanes. In each

Figure 1. If the slide safety device responds or the shear bolt fractures, the stopper arm on a guide-vane adjuster will limit the movement of the guide vane and particularly its tendency to flutter. A limit-switch signals response of the guide-vane safety device. Centralized lubrication of the guide-vane bearings was installed as a rationalization measure.

case, limit-switches must announce immediately the response of the slide safety devices, so that the installation can be restored to a proper condition by readjustments after the removal of foreign bodies. These slide safety devices (Figure 1) have proved to be excellent for loss prevention in all known water-turbine installations.

Because of the special function of axial bearings as weight-carrying elements in vertical water turbines, it is recommended that the oil supply of the axial bearing should be designed in such a way that the oil inlet and outlet lie above the Michell-bearing segments, so that, in the event of failure of the oil supply, there will be enough oil left in the bearing housing to safeguard rundown of the water turbine without damage to the axial bearing.

With regard to the oil supply to water turbines, only those guidelines that are specific to water turbines are given here; for other considerations, see Section 8.6. Oil Supply. The pressure-oil tank is protected from explosion by a safety valve adjusted above the maximum possible oil pressure. The pressure-oil supply to the guide-vane adjustment device must be safeguarded by a non-return valve in the pressure line behind the oil pump to prevent, in the event of pump failure, a flow-back of oil and hence pressure loss. Despite the low working temperature in water-turbine installations, the same recommendations apply to the components of the oil-supply system as given generally in Section 8.6. Oil Supply.

5.6.2. Guidelines for Prevention of Damage to Water Turbines during Operation

While damage from mishandling faults is low with large and small turbines because of the high standard of supervisory, safety, and protective equipment, damage arising from maintenance faults or lack of maintenance is disproportionately high. With large turbines, components subject to wear, such as bearings and runner vanes, are often inspected too rarely and, during revisions, are not evaluated sufficiently critically. A photographic documentation is recommended, so that the progress of cavitation and erosion can be evaluated reliably. With small turbines, revisions are usually carried out too late, e.g., after functional irregularities or because of a deterioration in running smoothness.

The user often realizes only after damage has occurred that an unscheduled outage has resulted in increased material damage and, in addition, substantial losses due to outage, all of which could have been avoided with a planned revision, especially in respect of the precautionary replacement of affected or worn components. The cause of damage as a result of inexpert or temporary repairs is often that the original manufacturing facilities are no longer available at the time of carrying out repairs which in the majority of cases are necessary because of the appearance of wear. For some inexplicable reason, the cast white-metal lining of large radial bearings is now not always of the same quality as the original production years ago—a finding which is borne out by extensive bonding faults in newly cast bearings shortly after installation.

5.6.2.1. Guidelines for the Supervision of Water Turbines

Practicable Supervisory and Protective Equipment common to all Water-Turbine Types (see Table 4, pp. 240–241)

Protection of Water-Turbine Installations against Fire

The fire risk of water-turbine installations is due to the oil-supply system, the seals and the generator. The use of built-in ionization-type alarms in the region of the stator windings of the generator, the oil supply and the gearing, if applicable, is recommended. In addition, it is recommended that this fire-alarm installation be complemented by a remote-controlled fire-extinguishing plant. For the generator winding, a CO_2 protection system has proved to be excellent.

For supervision of the oil supply and of radial and axial bearings, reference should be made to Sections 8.6. Oil Supply and 8.3. Bearings.

Television Monitoring of Water-Turbine Installations

Water-turbine installations include a dam, often far removed from a long machine hall, and

Figure 2. In the region of a gearbox, an ionization-type alarm takes over fire protection.

screening plant that stretches along the whole length of the hall. The headwater level and the tailwater are important for the safety of a hydraulic power station and also for the safety of the environs. Central television monitoring stations with remotely controlled panning and vertically tilting cameras are installed for improving safety and as a measure of rationalization. Each television camera is in circuit with a monitor unit in the control room and can, if necessary, be switched over to a large television receiver. To permit one operator to exploit this effective visual observation of the whole hydraulic power station in the headwater and tailwater areas, at intervals or continuously (e.g., in the event of a high water level or debris-laden water), not only by day but also by night, each television-camera unit is fitted with a remote-controlled searchlight.

Before starting control manoeuvres, every opening in *the dam* is checked for freedom from floating objects with the television installation. The television cameras then supervise raising or lowering procedures for all sluice gates.

The headwater is observed by television cameras a long way ahead of the dam or the screen in front of the machine hall, to detect swimmers, pleasure boats and water skiers quickly and warn them of the risk they run by means of loudspeakers. Particularly successful is supervision of the headwater for approaching oil slicks, whose removal before entry into the hydraulic power plant can be effected satisfactorily. Especially useful has been the supervision of the headwater, the dam, and also the machine-hall screen in the event of a substantial amount of drift ice and debris.

The machine-hall screen is also monitored by a television camera for fouling. In addition, the degree of fouling is determined by measurements of differential pressure, and cleaning at intervals is carried out accordingly.

The television camera for *the tailwater* takes over supervision of boat locks; also, shore control in the region of the dam is carried out by television cameras, particularly in the case of high water. Before actuation of all sluice gates in the dam installation, a check is made with television cameras in the vicinity of the tailwater to ascertain whether people are still present in the danger zone, despite the loudspeaker warnings.

A favourable disposition of cameras in *the machine hall* makes it possible to carry out an overall visual observation from the control room in the hall of all water-turbine installations, even in areas that are difficult of access. Since, with appropriate adjustment of the sensitivity of television cameras, even oil vapour, oil spray, and particularly the slightest signs of smoke can be detected perfectly, a television monitoring system for machine installations is an extremely effective precaution against fire.

5.6.3. Guidelines for Loss Prevention by Inspection and Revision

Valid for Water Turbines Generally and for Francis Turbines

Monthly:

- Samples of bearing oil and pressure oil from transmission and bearing housings and oil tanks should be taken from the lowest position and checked visually for signs of white-metal and steel abrasion and for water content.
- Oil filters in the bearing- and pressure-oil circuits must be cleaned. Abraded metallic particles must be examined so as to localize wear sites and damage. Abraded particles of white metal or bronze indicate damage to bearings,

Table 4. Recommendations for supervisory and protective equipment with indications, proved in practice, for

All radial bearings, also generator and gearing (guide bearings with vertical water turbines)	Oil-temperature measurement Bearing-metal temperature measurement Oil-level measurement with a float or immersion probe in the bearing housing Oil admission by flow controller
Axial bearings (weight-carrying bearings with vertical water turbines)	Oil-temperature measurement Bearing-metal temperature measurement in at least three tilting pads displaced by 120° in the region of the exit edge Oil-level measurement with a float or immersion probe in the bearing housing Oil admission by flow controller Shaft-jacking oil pressure during pump operation At normal speed between tilting pad and non-return valve
Gearing, if applicable	Oil inlet pressure Oil admission by flow controller
Oil supply	Bearing inlet pressure ahead of the highest or most distant bearing or gear transmission with pressure controller
Leakage oil	Float measurement Immersion-probe measurement
Oil tank	Oil-level measurement Oil-temperature measurement Oil-temperature measurement
Cooling water for the bearing-oil circuit	Flow controller
Pressure-oil tank	Magnetic (glass-free) oil-level measurement with indication Pressure controller Pressure measurement
Regulating-oil circuit	Pressure measurement
Runner	Excessive speed
Guide-vane adjustment	Position measurement "closed"
Each guide vane with shear or slide safety device	Change of position through limit-switch
Shaft seals	Pulse for switching on leakage-water pumps
Leakage water in the region of the shaft seal	Float indication and immersion-probe measurement
Relief discharge or jet deflector	Measurement of position

alarm, quick-closure and emergency shut-down action.

10° above normal	alarm	at 80 °C	quick closure
10° above normal	alarm	at 120 °C	quick closure
too high	alarm	too high	quick closure
too low	alarm	too low	quick closure
no circulation	alarm	no circulation	quick closure
10° above normal	alarm	at 80 °C	quick closure
10° above normal	alarm	at 120 °C	quick closure
too high	alarm	too high	quick closure
too low	alarm	too low	quick closure
no circulation	alarm	no circulation	quick closure
over 100 kg/cm^2	alarm		
above normal value	alarm		
10% too low	alarm	20% too low	quick closure
no circulation	alarm	no circulation	quick closure
10% too low	alarm	20% too low	emergency shut-down
too high	alarm		
too high	alarm		
too low	alarm		
under 30 °C	alarm		
over 60 °C	alarm		
no circulation	alarm		
oil level high	switching on air compressor and alarm	too high	quick closure
oil level low	switching off air compressor and alarm	too low	emergency shut-down emergency shut-down
oil pressure low	alarm	inadmissibly low	emergency shut-down
oil pressure high	switching off air compressor and alarm	too high	air release through safety valve
		inadmissibly low	emergency shut-down
10% above normal speed	alarm	depending on type	emergency shut-down
in closed position	alarm		
guide vane twisted with respect to pivot	alarm		
with long operating periods or too many starts	alarm		
too high	alarm	when exceeding the maximum permissible level, continuous alarm	
response	release of emergency shut-down alarm		
no response	locking of emergency shut-down		

while steel abrasions point to damage to gear teeth.

- An inspection should be made of all seals on shaft break-throughs, guide-vane bearings, servo-motors and fittings.
- Limit-contacts on pressure gauges and thermometers, flow-pressure monitors, immersion probes, floats and position indicators must be checked for correct adjustment and functioning without response by the quick closure or emergency shut-down.
- The pressure of the balance water for axial-thrust equalization should be compared with the rated value at a given tailwater level.

Quarterly:

- Checking of the satisfactory operation of all fittings important for shut-down.

Yearly:

- Examination of oil by oil suppliers.
- Cleaning of oil coolers.
- Checking of all screwed connections of the entire installation.
- Internal inspection of the pressure-oil tank and the oil tank.
- Checking of the whole control, supervisory and protective equipment for correct adjustment and functioning.
- Complete revision of all main, auxiliary, shaft-jacking oil, leakage oil, and leakage water pumps with their drives and automatic switchgear.
- Visual examination of the screening installation by divers.

Every Two Years:

- Visual inspection of the runner by the manufacturer, particularly for crack formation in the rim, hub and vanes. Also for damage by foreign bodies, corrosion and cavitation [1].
- Measurement of clearance on shaft seals.
- Visual examination of all guide vanes, protection rings on the regulating ring, and guide-apparatus cover for incipient cracks, damage by foreign bodies, corrosion and erosion.
- Checking of gaps at the guide vanes for jammed-in foreign bodies.
- Checking for faults and, if necessary, renewal, of the corrosion-protection coatings on all parts in contact with flowing water.
- Removal of radial-bearing shells, ultrasonic and tapping tests for loosening of bearing metal, and re-measurement of clearance, particularly in comparison with the clearance of the shaft seals. The clearance of the radial bearing must always be smaller than that of the shaft seals.
- Renewal of bearing-housing seals.
- Axial bearing:
  Removal of tilting pads,
  checking of the load-bearing pattern,
  pressure-testing at the shaft-jacking oil inlet,
  ultrasonic and tapping testing for loosening of the bearing metal.
- Alignment and centricity testing, only in the case of damage to the bearings or unsatisfactory running smoothness prior to revision.
- Centricity testing of vertical water turbines after each disassembly of the axial-bearing (load-bearing) disc from the runner, with the radial (guide) bearings removed.

Special Considerations for Kaplan Turbines

Every One to Two Years:

- Visual checking of the turbine runner for tightness of the blade fastenings, for tightness of the blade bushings, and for cavitation on the runner and runner blades [1].
- Checking of an oil sample from the lowest possible position in the runner hub for water content.
- Visual inspection of the runner body for cavitation and of the draft tube for cavitation and lifting up of concrete.
- Visual examination of the lower water-guide ring for abrasion, if there is an overlap when the runner blades are at maximum pitch.
- Visual checking for perfect lubrication of the lower and upper guide-vane pivots.
- Checking of the functioning of the whole guide apparatus.
- Visual inspection of the guide vanes.

Special Considerations for Pelton Turbines

Yearly Visual Inspections:

- Internal and external checking of nozzle distribution piping for cracks and corrosion, if necessary with special aids (endoscopes, mirrors).
- Examination of pit armour-plating, with all built-in components, for incipient cracks, loosening and corrosion.
- Checking of the runner for incipient cracks at the bucket roots, and for wear at the inlet edges.
  Repair work should be carried out by the manufacturer; alternatively, the entire welding procedure should be effected in accordance with his recommendations [1].
- Checking of nozzle spears and mouth rings and of the cutting-off edge of the deflector for wear.
- Testing of the overspeed-protection equipment by tripping it and checking that it functions.

5.6.4. Guidelines for Loss Prevention by Correct Repair (Examples)

An insured company provided a repair aperture in the runner of a Kaplan turbine. In this way, it is possible for repair welds to be made economically at the inside of the runner blade. The shaft-jacking oil pump permits positioning of each runner blade in front of this repair aperture. With the small radial clearance of 2 mm between the front of the runner blades and the hub, such repair work would otherwise be impossible.

With vertical water turbines, it is recommended that, in the event of unsatisfactory running smoothness, a careful check on concentricity should be carried out with the turbine emptied (Figure 3). With the radial bearings removed, the whole runner is suspended in the thrust bearing and, with the shaft-jacking oil pump switched on, i.e., with hydrostatic lubrication, even the heaviest runner can be rotated by simple means. During run-down, shaft eccentricity is measured at all radial-bearing positions, when the alignment of the turbine and generator can also be effected by relating eccentricity measurements to markings on the shaft periphery. By readjusting

Figure 3. With guide bearings removed, the vertical rotor, suspended only in the thrust bearing, is rotated while the shaft-jacking oil is turned on. During the drive-free run-down of the emptied turbine, shaft eccentricity is measured in the planes of the radial bearings. By adjusting the thrust-bearing disc relative to the runner, optimum quietness of operation and minimum loading on the radial bearings are achieved. In this particular example, a shaft eccentricity of 0.7 mm was reduced to 0.06 mm.

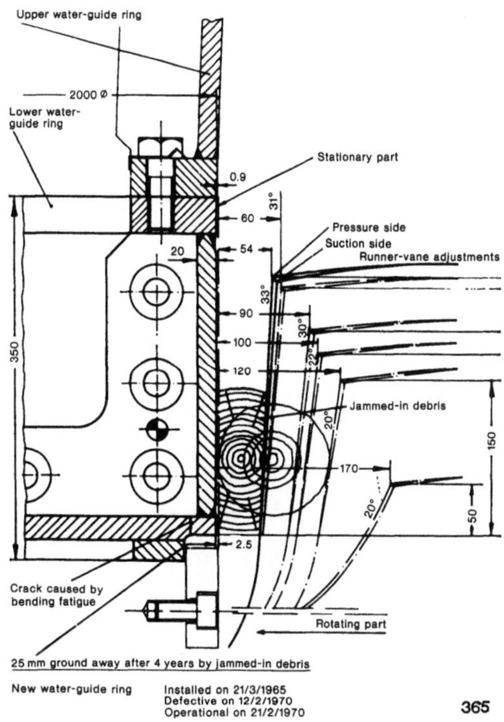

Figure 4. The cause of damage by abrasion and crack formation at the lower water-guide ring is overlap of the runner vanes at high flow rates. As the illustration shows, debris (pieces of wood) are in this way jammed in by the runner vanes, thereby overloading the water-guide ring.

the axial-bearing disc relative to the shaft, e.g., by remachining the thrust-bearing ring, optimum running of the lineshaft is achieved and hence the smallest bearing loads and minimum operational noise. In this example, it was possible, on a 40-MW turbine set, to reduce shaft eccentricity from 0.7 to 0.06 mm. Faultless running is attained with a shaft eccentricity in the plane of the guide bearings of from 0.05 to 0.1 mm.

The runner blades of Kaplan turbines, with large admission angles, can overlap the lower water-guide ring. Debris can then be wedged in and cannot get free, weakening the ring.

Long-term alternating bending stresses exerted on the internally ribbed water-guide ring ultimately result in cracks in the region of attachment of the water-guide ring, allowing water to penetrate the oil compartment of the turbine guide bearing. Such damage was discovered on the one hand by the large amount of leakage water and on the other hand by an oil check. It was found that 10% of the wall thickness had been worn away. Figure 4 shows the cause of damage, Figure 5 the additional internal ribbing used to reinforce the weakened ring, and Figure 6 a special improvement, based on weld-deposited strippers, for preventing further damage from this cause. These strippers are disposed at different heights around the whole circumference of the water-guide ring and serve to distintegrate the jammed-in debris (pieces of wood) to small splinters which are carried away from the gap. If necessary, the strippers can be reconditioned during revisions by further weld deposits.

Figure 6. Decisive for loss prevention was this improvement on the outside of the water-guide ring: Weld-deposited strippers at different heights completely disintegrate debris jammed in and dragged along by the runner blades.

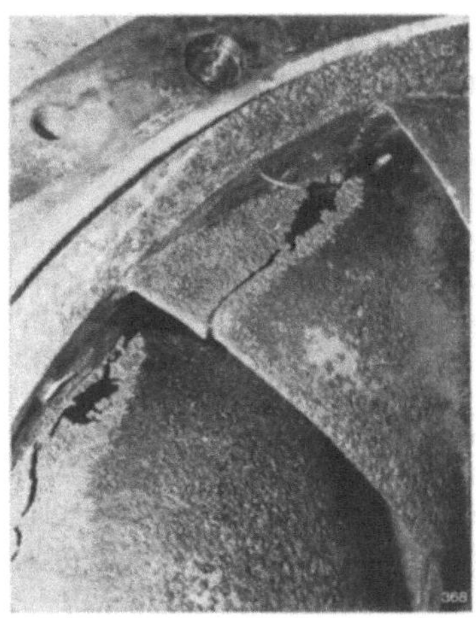

Figure 7. Cavitation and the resulting cracks produced in the runner vanes of a Francis turbine.

Figure 5. This ring, its wall thickness having been reduced by 10% by abrasion, was stiffened by closely spaced additional internal ribs. The crack responsible for the leakage that led to detection of the damage was welded over on both sides.

Cavitation can occur at runner vanes (Figure 7) and in the casings, particularly at the transition to the draft tube. Guidelines for the repair of runners by welding are given in [1]. It is re-

Figure 8. Eight regions of cavitation, 25 × 30 cm in extent, with pore depths of from 3 to 5 mm, were discovered during revision.

Figure 10. The V4A plate is set in position and is pressed against the base material by the two holding straps. Slots permit fusion of the base material with the 6-mm V4A plate.

Figure 9. The cavitation region was ground away to a depth of 3 mm over an area of 30 × 40 cm. Bolts for two holding straps were welded on.

Figure 11. The plate is welded in and ground over.

commended that welding repairs should be carried out in consultation with the manufacturer. Figure 8 shows the extent of cavitation discovered during a revision of a draft-tube liner, while Figure 9 shows the preparatory work involved for reconditioning the draft-tube liner. The armour plate was then carefully fitted to the base material and pressed against it before and during welding by two holding straps (Figure 10), and finally the completely welded-in plate was ground over (Figure 11).

If there are voids between the draft-tube liner and base, injection of concrete has proved useful. Figure 12 shows how simply this loss-preventive work can be effected during a revision. By ventilating the voids, it is possible to ensure that bulging of the draft-tube liner cannot occur inadver-

Figure 12. Injection of concrete between the base and draft-tube liner by machine 1.

tently during injection. The following injection composition has proved its value:

4 parts of cement, 1 part of sand, 2 parts of water; 0.03 parts of hardener, 0.05 parts of oil.

The timely filling up of voids and the adhesion of the draft-tube liner to the base attained by the additions prevent major damage by crack formation, as has been established by satisfactory experience with this technique over three years.

Bibliography to 5.6

[1] Technische Anweisung für die Unterhaltung von Laufrädern und Bauteilen von Wasserturbinen durch Schweißung. EDF working instruction.

[2] Schwierzer, T.: Bewertung und Überwachung des Schwingungszustands großer Wasserkraftmaschinen. Part 1. VDI-Z 114 (1972), No. 1, pp. 28/32.

6. Selected Lifting and Loading Equipment

The following recommendations are based on knowledge and experience of the insurer gained from numerous cases of damage. The guidelines given are primarily aimed at avoiding or removing those faults and deficiencies which are frequently repeated and lead to damage and interrupted operations.
Practice shows that a very high percentage of all damage can be traced to factors which can certainly be influenced. In this regard, technical shortcomings in design, faulty manufacture and assembly, unsuitable materials, faulty handling, inadequate maintenance and (of even more significance) ignorance, thoughtlessness and negligence, as well as insufficient co-operation, are the reasons for extensive and costly damage.

Regulations Serving Operational Safety

Existing building codes, operational instructions and supervision directives have, after all, been drafted with a view to loss prevention and hence to the protection of human beings and property.

Sources of information on the most important relevant technical regulations and directives are:

DIN leaflets, VDI guidelines and AWF leaflets:
Beuth-Vertrieb GmbH
Burggrafenstr. 4–7, 1000 Berlin 30,
or
Beuth-Vertrieb GmbH
Friesenplatz 16, 5000 Cologne,
or
Beuth-Vertrieb GmbH
Gutleutstr. 163, 6000 Frankfurt/Main.

VDE regulations:
VDE-Verlag GmbH
Bismarckstr. 33, 1000 Berlin 12 (Charlottenburg).

Regulations of the employers' liability insurance companies for the prevention of accidents:
Carl Heymanns-Verlag KG
Gereonstr. 18–33, 5000 Cologne.

6.1. Slewing Cranes and Loading Bridges

Representative of the large number of damage cases to slewing cranes and loading bridges reported to the insurer during recent years is the survey given in Table 1, in which 476 cases of damage to slewing cranes in shipyards and cargo-handling facilities are divided in accordance with the causes and primary locations of damage.

Product Faults

Apart from design faults, product faults include manufacturing and materials faults and the so-called *late assembly faults* (e.g., as a result of faulty alignment of machine components), as well as planning faults, the responsibility for which lies not only with the manufacturer but also with the user. In addition, product faults include all repair faults which later give rise to further damage.

Assembly Accidents

This group comprises all damage which has occurred during assembly (e.g., by crashing down as a result of rope breakage).

Operational Faults

Operational faults include all damage caused by mishandling and faulty maintenance. Also, damage arising from collisions, storms and broken ropes must be allocated to this group, since it can be traced almost exclusively to mishandling and inadequate maintenance.

Discussion of Damage Causes and Locations

First of all, the high proportion (42%) of all damage arising from product faults stands out in Table 1. This high proportion can be explained, on the one hand, by the fact that the cranes concerned are long-lived products which have partly been built in the post-war years, at the time when circumstances made it necessary to improvise and also on occasion to make use of less suitable materials. On the other hand (and this primarily concerns the high proportion of damage arising from design faults), the changes to which crane construction was subjected in the 1950's have contributed to the increase in the damage quota ascribed to product faults. The demands for maximum economy, new technical assumptions and changed conditions in important fields of application have resulted in converting the old design of slewing crane into a cargo-handling machine of high efficiency, but in the course of this conversion numerous problems of many different kinds had to be solved.
With respect to the primary damage locations, i.e., those from which damage (e.g., rupture, distortion and short-circuits) originates as a result of manufacturing faults, inadequate maintenance,

Table 1. Distribution of the causes and primary locations of 476 cases of damage to slewing cranes.

Cause of damage	Distribution of the number of damage cases [%]	Location of damage	Distribution of the number of damage cases [%]
Product faults	42	Structural steelwork (including booms)	26
Assembly accidents	2		
Operational faults	56	Mechanical equipment (travelling, slewing, lifting, luffing gear, etc.)	51
		Electrical equipment	7
		Wire ropes	3
		Various (through collision, storms and other causes)	13

etc., the relatively small proportion assigned to ropes is surprising. In this connection, it should be appreciated, however, that rope breakages almost always lead to extremely severe damage and that, precisely for this reason, ropes deserve particularly careful attention and maintenance.

6.1.1. Guidelines for the Planning of Lifting Equipment

Always Seek the Advice of the Manufacturer before Buying

The planning and lay-out of lifting equipment are primarily determined in accordance with the prevailing nature and environmental conditions of the application. If these factors are not clearly defined, damage is almost certain to occur. Close co-operation between the user and manufacturer of the lifting equipment is therefore essential in the planning stage.

Unfortunately, it is not at all rare that a crane is selected solely on the basis of its maximum lifting capacity, i.e., without taking essential factors into account. Only too soon does the equipment then prove unsuitable for the application concerned, as is often manifested by a continuous series of cases of damage. The blame is then usually attributed to the manufacturer—sometimes not altogether unreasonably. Quite apart from the fact that his reputation would certainly not be enhanced by such an "unlucky" crane, the manufacturer should, in every instance and on his own initiative, enquire about the conditions of the application and advise the customer accordingly. In the event that the purchase is arranged through an agent, the manufacturer should ensure that the agent has trained specialist personnel at his disposal and should make him responsible for properly advising the purchaser (e.g., by information sheets or questionnaires).

Purchase of a Second-Hand Crane

When acquiring a second-hand crane, the manufacturer should be consulted regarding an assessment as to whether the crane is suitable for use in its new location and also for the duties for which it is required.

Disassembly of a second-hand crane and its re-erection at the new site should be assigned only to the manufacturer or should, at least, be carried out under his supervision.

Influence of the Crane Track on Trouble-Free Operation

Track unevenness and crane construction are interdependent. The stiffer is a gantry, the more must unevennesses in the track be kept within limits, so as to

- ensure security against derailment,
- exclude the possibility of overloading individual gantry wheels,
- avoid poor stress conditions in the upper gantry girder or the platform.

Example of Damage

Because of their better controllability, purely series-wound motors were initially used to power the drive mechanism of the propulsion system of a luffing and slewing four-legged travelling gantry crane. Owing to local conditions—crossing tracks—the gantry was guided on one side only by two flanged wheels, the two wheels on the other side being flangeless. In addition, the two shafts for the driven wheels were separately powered. However, because of height differences in the crane track (up to ± 10 mm for a track gauge of 6,000 mm), the crane stood on three legs in some positions, as the gantry was relatively rigid. Under these conditions, it was possible for only one leg to be driven, in which case the drive mechanism of the unloaded leg speeded up temporarily as a result of the characteristic of its series-wound motor. Then, when the unloaded leg made contact again, the resulting restoring moment gave rise to

- exertion of torque on the gantry,
- shock-loading of the drive mechanism itself,
- displacement of the gantry and
- a tendency to skewing because of the different speeds of the gantry legs,
- high flange friction and
- high eccentric loading of the free-running flangeless wheels.

Although an auxiliary winding for the drive motors would not have removed the basic cause of trouble, it would have counteracted its effects.

Operation of Several Cranes on a Common Track

The operation of several cranes or loading bridges on a common track also merits attention. As a matter of course, the buffers should be of equal height and should not be offset with respect to each other.
The rear overhanging parts of cranes are often the cause of severe damage by collision with each other. For cargo-handling operations on ships, when two cranes often have to work side by side on neighbouring hatches, portal jib cranes having slewing rings with double-row ball bearings or columnar bearings are suitable because of their small rear overhang. With such cranes, the hollow gantry legs are filled with concrete to increase the stability of the crane.

Parallel Operation of Cranes

If two cranes must be used in parallel operation for the common handling of heavy loads, special measures are necessary to counteract the increased risk of accidents. Apart from equal characteristics of all travelling, slewing, luffing and hoisting actions, the two cranes should be equipped with synchronous control of all crane motions, whereby common control of both cranes is possible from one crane (the master crane).

Skewed Running of Loading Bridges

With increases in the track gauges of crane installations and particularly, therefore, with large-span loading bridges, the difficulty arises of moving both ends of the installation in parallel at the same rate. As a result of track imperfections, different friction values between rails and drive wheels, etc., it frequently happens that loading bridges run skewed, with consequent increased wear of the wheel flanges, rails and drive mechanism, and derailments and distortion of the bridge construction. This problem, which is well known, is dealt with in reference [1] in the Bibliography, together with straight-running devices intended to overcome it.

6.1.2. Design and Manufacture

Once the conditions of use of a lifting installation have been adequately clarified and its type decided upon, its design, dimensions and safety are to a large extent determined by the appropriate regulations and guidelines.
Because of the requirement for maximum safety, the following points should be given consideration in the design:

- Provision of self-adjusting and, if possible, wear-resistant brakes.
- Dimensioning of gearing according to the stresses to which it will be subjected.
- Distortion-free suspension of the gear-transmission casing.
- Easy access to mechanical and electrical equipment for maintenance.
- Buffer and track limitations matched to the design and rate of travel.
- Smooth and positive starting and braking of movements, i.e., with accelerations and decelerations of the crane movements largely remaining uninfluenced by the crane driver.
- Matching of slewing and luffing speeds to inhibit pendulum movements of the suspended load.
- Maintenance- and trouble-free operation of safety equipment.
- Unevenness in the track and its influence on design and the wheel-drive mechanism.
- Functionally safe protection against wind, especially with small crane installations and tower cranes.
- Drivers' cabins with as extensive a field of view as possible, i.e., ahead, behind, above and below, without the crane driver having to leave his position.

Execution of Welded Connections

Apart from considering future requirements in the design stage, some thought should also be given to eliminating a number of trivial but recurrent faults.
For example, the shape of welding seams often takes no account of the stresses to which the lifting equipment will be subjected during operation. Because of the numerous damage cases caused by unsuitable or inexpert welds, some

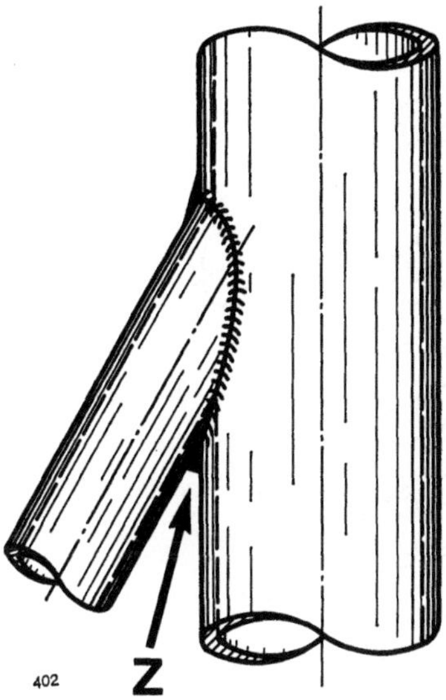

Figure 1. In view of the electrode thickness, too acute an angle between the two tubes can no longer guarantee a perfect weld in the corner Z.

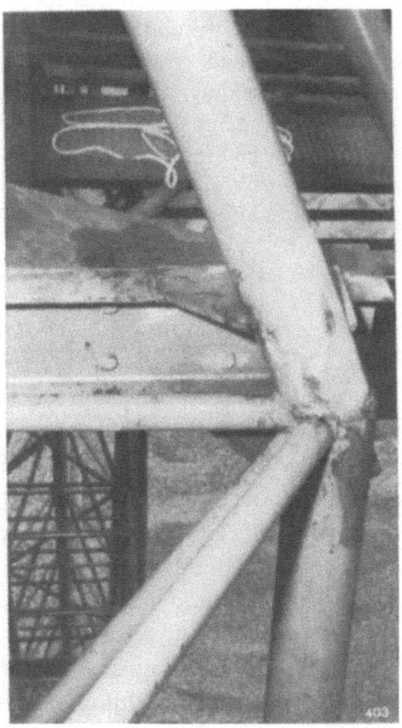

Figure 2. Welding-seam accumulations of this kind, particularly in the most highly stressed region, should be avoided. The welding seam in the knee of the corner tube of the support leg of a loading bridge exhibits a crack extending to about 25% of the circumference. The crack was discovered in the course of an inspection.

basic premises are given here for properly executed welding connections which take into consideration the stresses to which they will be subjected:
Attention should be paid to the feasibility of executing welds and to their accessibility (see examples in Figures 1, 11 and 12). This consideration applies especially to welds which must be carried out on site, where there often are more difficult conditions. Overhead welds should be avoided.
Welding-seam accumulations (Figure 2) should also be avoided, as well as concealed and uncontrollable welds on load-carrying components.
Dirty corners and pockets of water give rise to corrosion.
Inadequate weld preparation and unwelded root gaps, which therefore act as notches, repeatedly lead to welding-seam failures (Figure 3).

Figure 3. The joint of this rail-type slewing ring was welded only from the outside—and messily at that. Cracks emanated from the unwelded root on the inside of the rail and now extend to the whole cross-section.

Optimum matching of the filler material to the components to be welded is essential. In this regard, it should be mentioned that it is by no means rare that contact corrosion between the component and the filler metal contributes to failures of welding seams. With deficient or damaged paint coatings, it is possible in some cases under certain climatic conditions for galvanic corrosion to occur, leading to possible weakening and destruction of the welding seam.
The position of the weld should be such that, as far as possible, an undisturbed (i.e., a rectilinear) flow of force is ensured between the components to be connected.

Figure 4. The incipient crack, involving no deformation, in the tension chord from a jib of truss construction was caused by a substantial sudden change in stiffness, with a consequent extraordinarily high local notch effect.

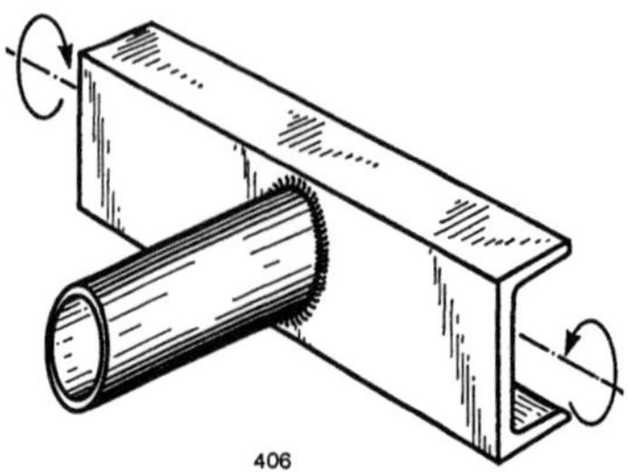

Figure 5. If this system is subjected to torsional stresses in operation, it is likely sooner or later that the welding seam between the torsionally weak U-profile and the torsionally stiff tube will fracture.

Figure 6. The fillet weld between the broken corner strut and the welded-on strut of a tower crane is excessively rounded in the region of the fracture. As a result, there was an increased notch effect in this region which, together with the operational stresses, led to stress peaks that exceeded the strength of the component.

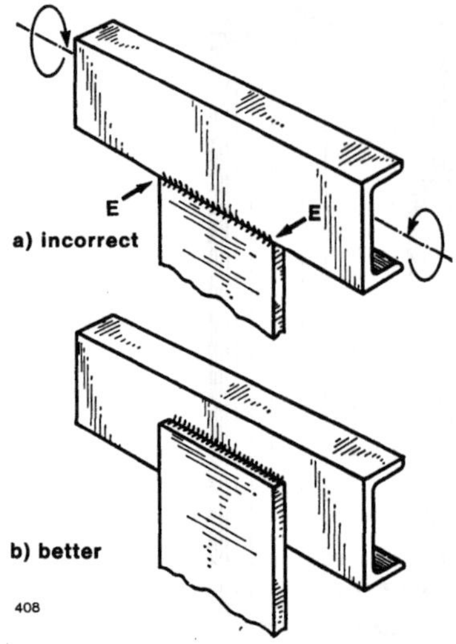

Figure 7. Welding of an iron plate to a U-girder stressed in bending. In the arrangement (a) elongation of the lower region of the U-girder, which is stressed in tension, is impeded by the weld. Moreover, the iron plate, which lies in the plane of bending, increases the stiffness of the girder, so that bending is locally inhibited. High stress peaks are generated at the corners E and, because of the end craters present, it is possible that cracks in the welding seam will develop at these locations. The arrangement (b) is more favourable, inasmuch as here the longitudinal seam lies in the neutral axis of the girder and therefore at least does not take part in elongations of the girder engendered by bending.

Sharp transitions in cross-section and sudden changes in stiffness should be avoided. Figure 4 clearly shows the reason for the incipient severe damage to a jib. By detecting the crack in time, a major accident could be averted. Above all, special attention should be paid to the joining of open to closed profiles, i.e., those that are weak and strong in torsion (Figure 5).

If possible, welding seams should not lie in the region of high stress peaks because, in general, welds have a lower strength than that of the base metal. In particular, the ends of welding seams should lie outside the most highly stressed zone. Penetration notches and excessively rounded welding seams cause stress peaks and should therefore be avoided (Figure 6).

The position of a welding seam should be matched to the mode of stressing of the components to be joined (Figure 7).

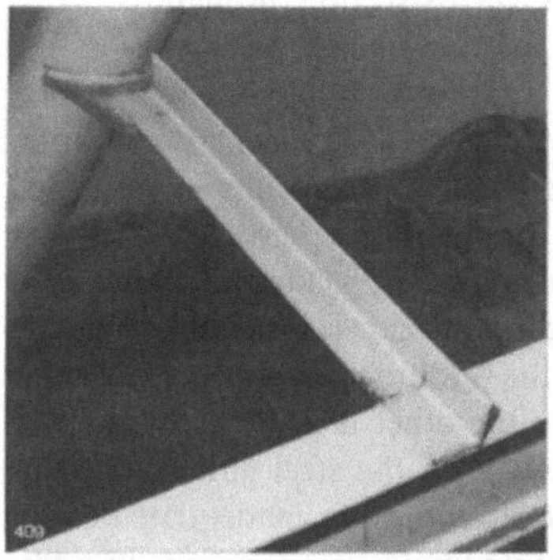

Figure 8. With this bracing arrangement between the corner tube of the fixed support and the lower chord of the main girder of a loading bridge, the forces arising from the traversing movements of the trolley were not taken sufficiently into account. The horizontal leg of the angle-iron broke next to the connection to the lower chord of the main girder as a result of relative movements between the fixed support and main girder. Chipping of paint is an indication of overstressing of the material.

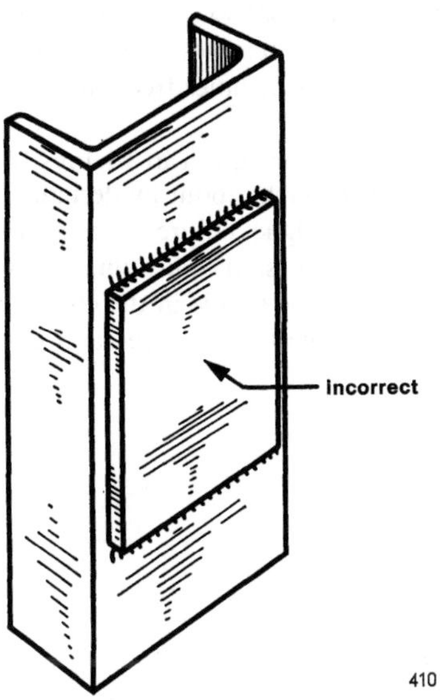

Figure 9. "Reinforcements" of this kind, as often found for attaching auxiliary equipment or minor building components to load-bearing elements in tension, are highly dangerous. Such fillet welds generate high stress peaks locally in the load-bearing elements stressed in tension, not infrequently leading to incipient cracks and eventually to fracture of the tension member.

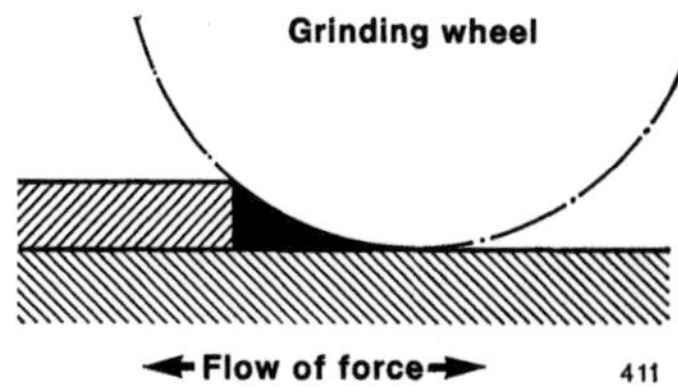

Figure 10. If a fillet weld at right-angles to the direction of stress is unavoidable from the design point of view, care should be taken to ensure that, at least, the weld is carried out neatly and without penetration notches and end craters. The surface of the welding seam should be cleaned up by grinding in the direction of stressing.

Figure 8 shows a joint in a loading bridge between the corner tube of the fixed support on the main load-carrying side and the lower chord of the main girder. This joint is unsuitable for the stress conditions imposed.

Fillet welds which run at right-angles to the direction of stress should, in all circumstances, be avoided in components stressed in tension (Figure 9). It is unfortunate, however, that neglect of this requirement is a common fault found when pipelines, etc. are fitted at a later date to load-carrying structural components.

If a fillet weld perpendicular to the direction of stress is unavoidable from the design point of view, care should be taken to produce an absolutely clean seam surface, as well as flat transitions between the seam and component. Penetration notches and craters at the ends of the seam must always be avoided. If necessary, the surface of the seam must be ground clean, with the plane of the grinding wheel at right-angles to the welding seam (Figure 10), so as to prevent grinding grooves that are perpendicular to the direction of stress.

An accumulation of welding and design faults is shown in Figures 11 and 12, which refer to the foot of the mast of a fixed slewing crane. After a relatively short period in operation, numerous cracks appeared at the foot, which essentially comprises two steel girders welded together crosswise and reinforced by a flanged plate located above. The cross-section of the square mast consists of four equal-sided angle-irons. The almost right-angled position of the triangular ribbing of the mast ends on the legs of the cross formed by the foot of the mast makes the purpose

Figure 11. Section from the foot of the mast of a fixed slewing crane. A combination of inadequate design, insufficient seam preparation and welding faults led to numerous cracks in the welding seams and also in the base material.

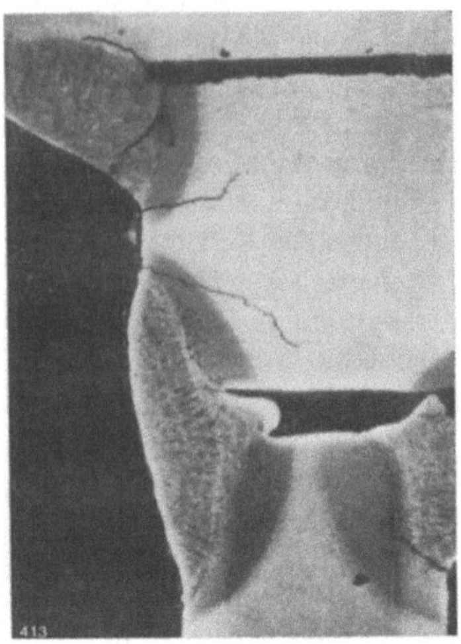

Figure 12. Section through the weldments between the cover plate, the flange of the girder and the additional welded-in web of the section at the foot of the mast shown in Figure 11.

of these ribs anything but clear. The question arises as to why the ribs were not arranged at an acute angle to each other and also in the direction of the hole for mounting the foot of the mast. Also the purpose of the screwed connections provided between the steel girders and the cover plate is not clear. In addition, there is the further question as to why the additional web plates welded to the steel girder were not extended to below the corner posts of the tower. It should also be noted that, apart from shortcomings in design, the weldments were carried out messily.

Figure 12 shows a cross-section through the weldments between the cover plate, the flange of the girder and the additional welded-in web. Here, it is possible to discern insufficient preparation of the components, as well as faults in the roots. Moreover, the designer has here made almost impossible demands on the welder. The fillet weld on the lower right-hand side is inside the rectangle covered up by the additional web, this rectangle being formed on the other three sides by the two flanges and the web of the steel girder. The numerous cracks bear witness to inadequate design, as well as to shortcomings in execution. In addition, analysis of samples of the materials revealed that the chord plates and also the webs had been made from steels which, according to DIN 120 and to DIN Draft Proposal 15018, are not acceptable for such components. The crane was repaired by using an improved design for the foot of the mast.

Avoidance of Notch Effects

The generally known problems of high notch effects at transitions of cross-sections, of stress concentrations at sharp recesses and of the weakening of components by holes frequently receive only scant attention.

Representative examples are given in Figures 13 to 17 of notch effects brought about by design.

Figure 13 shows the fatigue failure of a gear-transmission shaft in torsional bending. Numerous incipient cracks are visible in the circumferential region of the shaft, and these soon co-

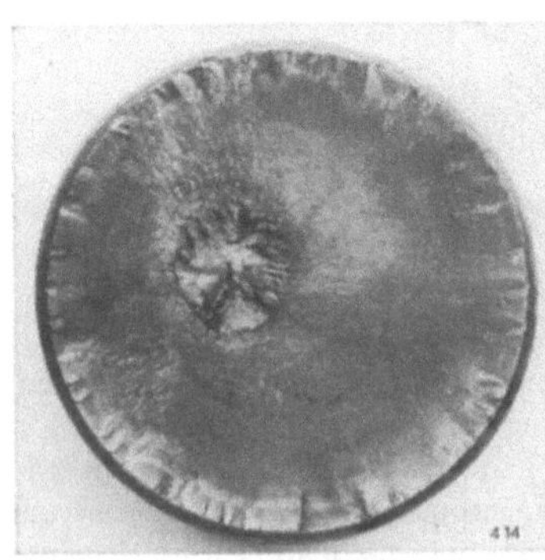

Figure 13. The fatigue failure in torsional bending of this gear-transmission shaft was caused by insufficient rounding of a shaft step.

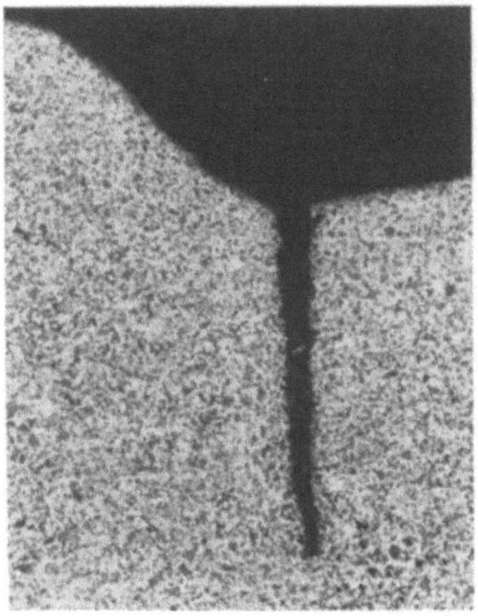

Figure 14. Section through a bending-fatigue crack (approximately 1 mm in depth) at an insufficiently rounded fillet of a shaft step.

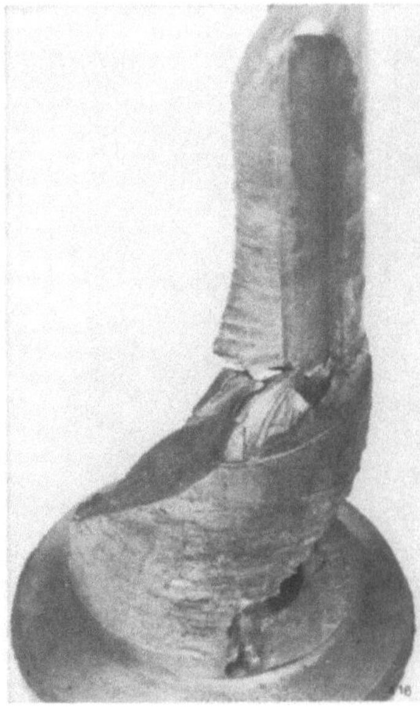

Figure 15. Fatigue failure under dynamic torsional stresses of this planet-carrier shaft of a hoisting gear emanated from the sharp-edged base of the slot for a feather key.

alesced to a common fracture front. Clearly recognizable are approximately circular shell-like marks which characterize the progress of the fracture. The relatively small residual fracture area points towards comparatively low nominal stressing. The cause of the cracks was insufficient rounding of a step in the shaft diameter, and the residual fracture, which does not lie quite in the centre, indicates a slight eccentricity of one of the two shafts or gearwheels in relation to each other.

Figure 14 shows a bending-fatigue crack caused by insufficient rounding of a step in a shaft.

The planet-carrier in Figure 15 also suffered a fatigue failure under dynamic torsional stressing, the cause being the sharp-edged base of a feather-key way.

In the chain link shown in Figure 16, the position of the fracture is determined by a step in the cross-section with a local notch effect.

With holes, particularly tapped holes, care should be taken to ensure that enough material remains around the holes (Figure 17).

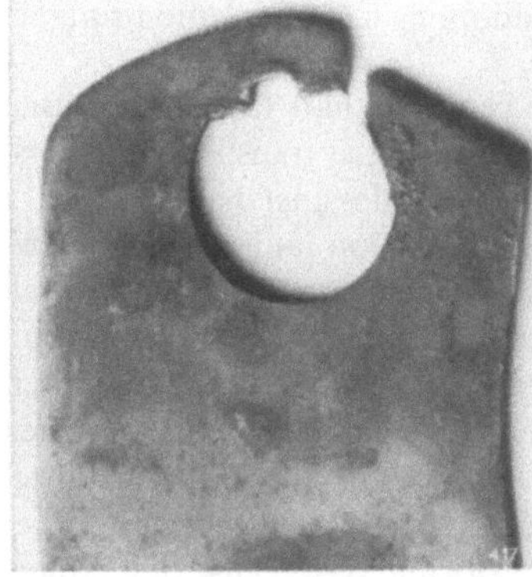

Figure 16. Forced fracture of one link of a chain for suspension of the movable counterweight of a luffing and slewing jib crane (laboratory tensile test). The link comes from a chain whose fracture resulted in the jib crashing down. Most of the remaining links of the chain exhibited incipient cracks which, without exception, emanated from the corners of a flattened hole or groove at places that, under operational loads, were exposed to locally increased stresses as a result of notch effects.

Correct Design of Shaft Steps with Regard to Stressing and Manufacture

When designing a component on the drawing-board, account should be taken of the fact that

Figure 17. Too little material was left around the blind tapped holes of this ball-bearing slewing ring in the region of the base of the internal gearing and the ball race, so that high stress concentrations occurred in this area under shock loading, leading to an incipient crack. Under operational stresses it was possible for this crack to develop quickly into a fatigue or endurance failure.

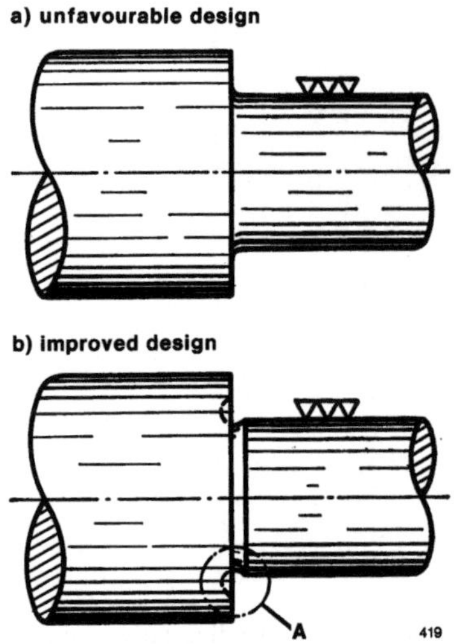

Figure 18. Designs of a relatively high shaft step. With design (a), a fine but sharp-edged step often remains as a result of grinding near the shoulder. With design (b), however, a free run-out (A) of the grinding wheel is ensured.

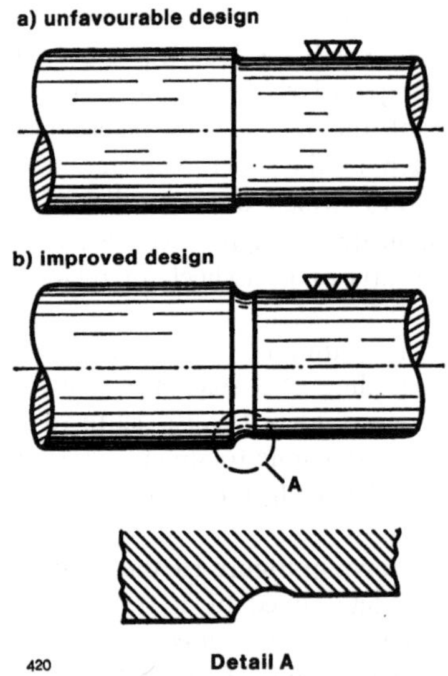

Figure 19. Designs of a shaft with a relatively small step in cross-section (see also Figure 18).

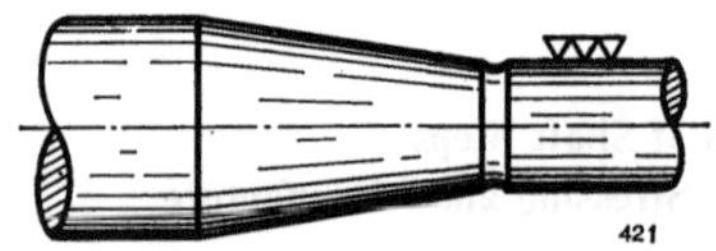

Figure 20. It is often possible to avoid an abrupt cross-sectional step of a shaft by a conical transition.

mechanical production can in some circumstances have an unfavourable effect on the strength of the component. This effect can often be prevented by adopting a configuration suitable for manufacture.

There are many cases of shaft failures resulting from insufficient rounding at transitions of cross-section. Even when the designer has specified a perfectly adequate transition radius on his drawing, failures still frequently occur and, in such an event, the cause is often that the transition radius demanded by the designer has not been adhered to for manufacturing reasons. When grinding an adjacent journal, the grinding wheel leaves a sharp-edged step, which makes the large-radiused fillet illusory and is the cause of a fatigue failure. Such shortcomings can be avoided by backing-off the shaft step. The designs shown in Figures 18 to 21 permit a groove-free run-out of the grinding wheel and have proved their value in practice on numerous occasions. However, it must be pointed out in this connection that all such measures fulfil no purpose if the surface quality of the fillets leaves much to be desired and if they contain score marks.

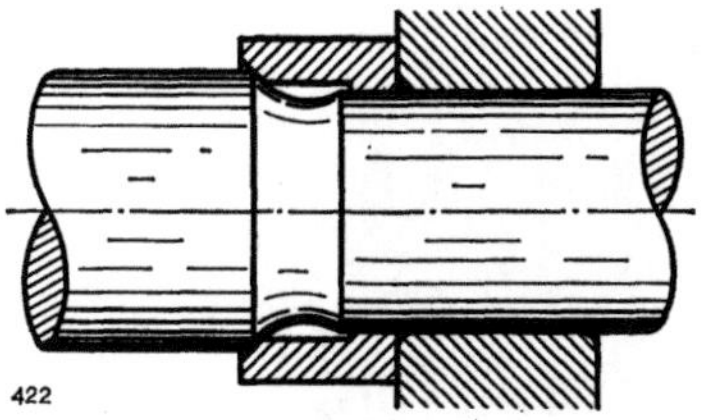

Figure 21. Excessively small shaft shoulders are frequently unsuitable for an axial precisely defined position of the neighbouring component seated on the journal (e.g., rolling bearings with their relatively large radii at the edges of the bore). In such cases, a perfect shaft shoulder can be created with the aid of a shoulder ring.

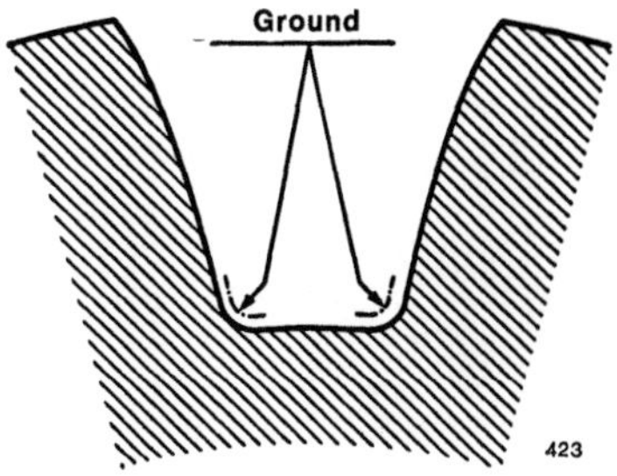

Figure 22. As a matter of principle, the transition radii at the root of the teeth of a slewing-gear pinion should be ground.

Damage to Gearwheels and Transmissions
(see also Section 8.5)

Slewing pinions are highly stressed at the roots of their teeth. The stress peaks caused by inertia forces and jerky modes of operation at the transition radii can easily lead to incipient cracks. Here, subsequent grinding of the transition can bring about a decisive improvement of conditions (Figure 22).

Similar damage to gear transmissions on crane installations occurs repeatedly and motivates some special remarks. Detailed description of damage to gears can be found in [2, 3].

Figure 23 shows a spur pinion that has been broken and bent outwards. The rim of this pinion was substantially weakened by two slots which, together with the transitions at the root of the teeth, were sharp-edged. It can be seen from Figure 24 that the initial fatigue crack emanated from the root of the tooth.

Unsuitable material and loose seating on the shaft were the reasons for the tooth breakages shown in Figure 25.

Gears with large tooth widths remain problematic. Flank damage often emanates from manufacturing faults. The one-sided wear, plastic flow and break-outs resulting from high local flank loading indicate inadvertent angular misalignment, i.e., inaccurate meshing conditions (Figure 26).

The mainly one-sided pitting of the tooth flanks of the helical gearwheel from the hoisting unit of a travelling crane (Figure 27) points to inadequate parallelism of the pinion shaft with respect to the associated shaft of the gearwheel, i.e., to angular differences between the gears in mesh and hence to manufacturing faults. If the gear transmission had remained in operation, its complete destruction would have been inevitable.

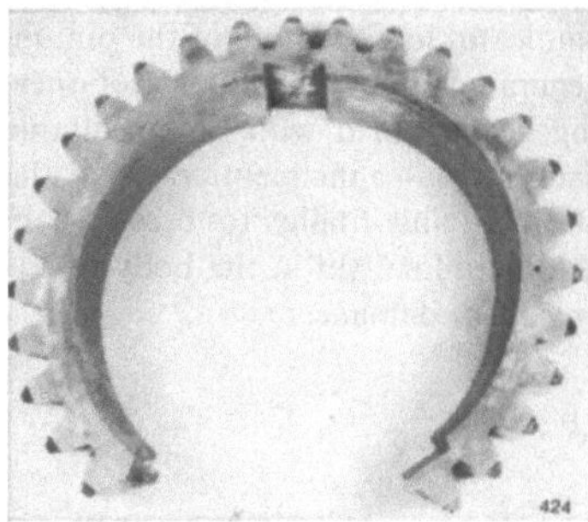

Figure 23. The marks of scoring, as well as of fretting corrosion in the bore of this pinion, indicate loose seating on the shaft. The resulting shock loads gave rise to such high stresses at the root of the tooth adjacent to the slot that the fatigue strength in this area was exceeded.

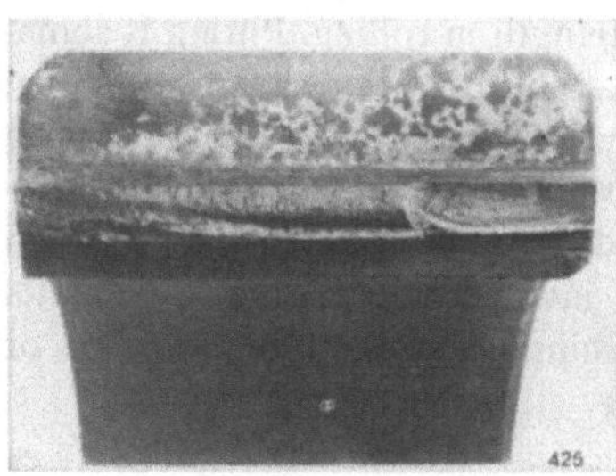

Figure 24. View of the fracture surface of the pinion shown in Figure 23. It can be seen clearly that the first incipient crack (right) emanated from the sharp-edged transition at the root of the tooth.

Figure 25. Fretting corrosion in the bore of the pinion of a slewing gear indicates loose seating of the pinion. The shocks thus generated under the influence of operational load changes, combined with an unsuitable notch-sensitive material, led to incipient cracks at the scored tooth-root fillets and finally to breakage of several teeth. The longitudinal slit in the body of the pinion is not the result of damage.

Pitting

Pitting is a phenomenon which occurs relatively frequently with new transmissions after even a short time in operation. In many cases, it is a question of so-called running-in pitting, which is particularly common with gears having a large tooth width. Its cause, among other factors, is a low standard of manufacturing accuracy and roughness of the flanks, leading to local exceeding of the fatigue strength in rolling. Pitting is sometimes arrested after further operation and then disappears in the course of time if manufacturing inaccuracies have been balanced out during operation. It should also be noted that the origin of pitting is very involved and pairing of the materials of the meshing gears, the lubricating oil or grease used and other factors all play a part. If pitting occurs, the manufacturer should be consulted in every case.

Figure 27. The mainly one-sided pitting of the tooth flanks of this gearwheel indicates installation or manufacturing faults.

Damage to Ball-Bearing Slewing Rings

Damage to the ball-bearing slewing rings of slewing cranes is not infrequent. Insufficient alternating shear strength of the material sometimes leads to premature wear. In such a case, surface hardening of the raceways often effects a remedy.

Damage to ball-bearing slewing rings as a result of an inadequate supporting structure is also not uncommon. In the design according to Figure 28, the horizontal plate ring of the L-shaped turntable frame for receiving the upper raceway on the upper carriage was insufficiently dimensioned and braced against the cylindrical part of the frame. As a result, the plate ring became arched upwards on its inner circumference under the operational load, in a manner similar to that of a Belleville washer. The ball raceway therefore could not find adequate support and deformed in the same manner, giving rise to a disturbance of the bearing geometry (Figures 29 and 30).

Figure 26. The clearly visible one-sided wear points towards faulty meshing conditions of the gears in engagement.

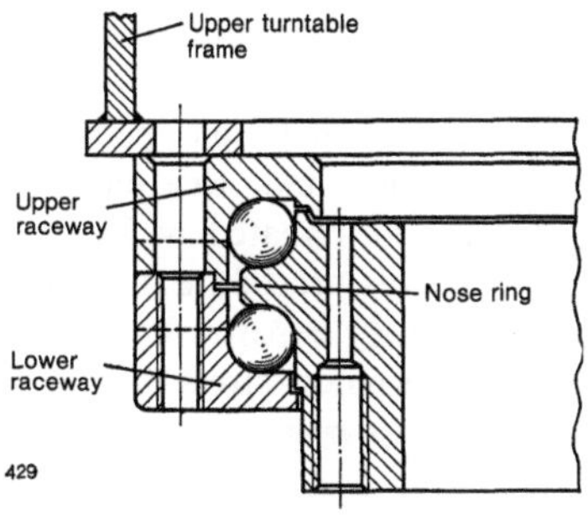

Figure 28. Insufficient support of the ball-bearing slewing ring of a slewing crane. The consequences of the lack of adequate supporting action for the upper raceway are shown in Figures 29 and 30.

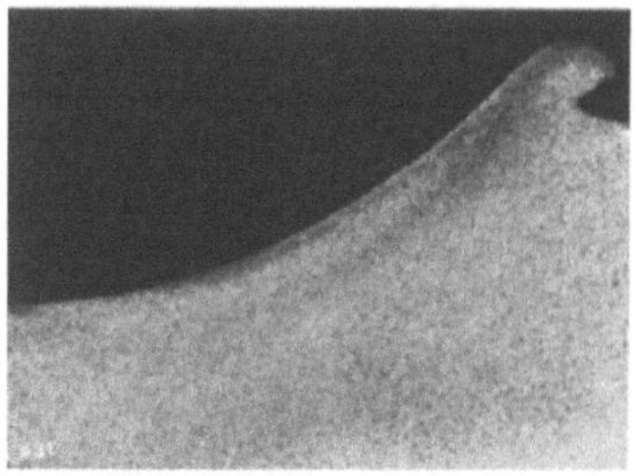

Figure 30. Greatly enlarged view of part of a section through the ball raceway shown in Figure 29. Plastic flow of material towards the right is clearly recognizable. In the present case, however, the pattern of damage was also engendered by locally inadequate hardening of the ball-raceway surface.

Figure 29. Damage to a ball raceway as a result of disturbed bearing geometry. Plastic flow of the material in a radial direction at the inner circumference of the ball raceway is clearly discernible. The cause of damage was insufficient support of the raceway (see Figure 28).

Further Design Guidelines for Loss Prevention

Good accessibility, means of facilitating inspection and maintenance (platforms, ladders) of the steel structure and the machine and electrical equipment should be a matter of course, and the possible need for repairs should be taken into account right from the beginning.

For example, experience shows that, if they are of difficult access, lubricating points are often neglected and screwed fasteners are often insufficiently tightened.

Provision should also be made for easy replacement of wear components, as well as for the possibility of removing entire machine units without extensive manipulations. Missing supports or mountings for jacking equipment can in some circumstances lead to disproportionately high repair costs. For instance, in one case involving replacement of a ball-bearing slewing ring, it was found necessary to dismantle the whole superstructure of the installation, working from ground level, since there was no possibility of the gantry sustaining the high jacking forces involved.

Hydraulic components of the installation (brake cylinders, pipelines, etc.) should not be positioned directly above brake discs, rails, travel wheels and similar items. Numerous cases of damage arising from the failure of braking equipment fouled by leaking oil argue against such arrangements.

Cables and pipelines should be laid under cover in such a way that they are afforded protection from damage during operation and inspection of the installation, as well as during repair operations.

Particular attention must here be drawn to the requirement for accuracy of central holes in the steel construction. Thus, the holes for the axles of travel wheels in the carriages, the fulcrum points for the jib system and similar components must be made in a precise geometrically defined position; otherwise, enforced fractures, increased flange friction, repeated derailments and similar occurrences are inevitable consequences. There should really be no need to mention that the holes must not be made until all welding and riveting work on the steel structure concerned has been completed.

An optimum field of vision for the crane driver is desirable. Although a completely unobstructed view all round hardly appears practicable, the driver should still be provided with as good a view as possible to all sides without having to subject himself to excessive contortions. In this regard, it is not at all rare that, because of the narrowness of the crane-driver's cabin, such contortions cause inadvertent actuation of control levers and trigger crane movements which lead to severe damage.

For the same reason, the crane-driver's cabin should be spacious enough to permit the presence, for instance, of a trainee next to the driver or to make it possible for the works electrician to carry out an operational check of the electrical installation without obstruction.

In addition, it would be desirable for the driver's compartment to be provided at the front with electrically operated windscreen wipers, so that a clear view of the working area—and the jib—is at all times ensured. A windscreen wiper on each of the side windows would also seem appropriate, manual operation of which would probably suffice.

6.1.3. Assembly and Commissioning

During the relatively short assembly period, there is a critical stage in the life of a crane installation which should not be underestimated. In this connection, it is necessary to prevent not only pure assembly accidents, such as crashing down or collision of components, but also subsequent damage resulting from basically incorrect execution of assembly itself.

Here, too, loss can be prevented if simple mistakes are avoided. Experience, know-how and reliability of the assembly personnel have a decisive influence on future operation. The careful selection of personnel is therefore an essential premise for faultless assembly.

Causes of damage which certainly could be avoided include air pockets in lubricating lines, spanners inadvertently left in gear transmissions, and missing warning notices when switches have been bypassed during testing under non-operative conditions. To cope with these and other pitfalls, internal memos should be circulated widely to the assembly personnel. This procedure may appear superfluous to assembly experts; however, numerous deplorable occurrences stress the need for such precautionary loss-prevention instructions.

Assembly of the Crane Track

Proper assembly of the crane track is of decisive importance. As indicated in Section 6.1.1, the tracks of slewing gantry cranes primarily merit special attention. The gantry and rotatable superstructure must be erected strictly vertically. Damage to the gantry structure as the result of an uneven track is by no means rare.

Moreover, it should be noted that the rails of the crane track must be properly earthed (copper or galvanized materials should be used). Also, apart from the fishplates, the rail joints must be interconnected by electrical conductors.

Adjustment of the Supporting Rollers for the Rotating Post of Jib Cranes

Special care is due to the supporting rollers for the rotating post of jib cranes. An imperfect adjustment of the rollers causes an off-centre position of the rotating post, so that its axis describes a circle during operation. By disturbing the meshing conditions of the slewing pinion, damage to the slewing gear is the unavoidable result. In such cases, the thrust bearing is also eccentrically loaded and is sooner or later destroyed.

Lubrication of the Post-Supporting Bearing

It should also be mentioned that the post-supporting bearing (thrust bearing) must be adequately lubricated before, and not only during or even after the first revolution. If necessary, the post must be lifted for this purpose and the supporting surface provided with the lubricating medium. A series of cases of damage to supporting bearings, which had hitherto remained inexplicable, was eventually traced to a lack of or inadequate lubrication of these bearings.

Alignment of Slewing Gearing

The slewing gearing requires precise alignment during installation. The slewing-gear pinion in Figure 31 clearly shows evidence of disturbed meshing conditions between the pinion and internal teeth of the rotating rim.

Figure 31. The different tapering of the teeth across the tooth width of this slewing-gear pinion indicates faulty alignment of the slewing gear during assembly.

Precautions When Making Safety Equipment Non-Operative

The bypassing of limit switches when carrying out final operations or when testing functions in preparation for commissioning is a constant source of danger, in that the unblocking of interlocks is not always immediately counteracted. Non-operative safety equipment is often forgotten, and severe damage with the control equipment concerned can then occur during subsequent manipulations. In such cases, it is essential to bring this changed and dangerous situation to the attention of personnel by providing clearly visible notices. In some circumstances, actuation of the control levers of a crane by unauthorized persons can be prevented by locking the driver's cabin.

Often the only missing precaution is a set of the necessary warning and notice boards and such a set should be supplied by the manufacturer as part of each crane installation. The driver's cabin or the crane house is a suitable location for storing these boards.

Overload Test

The procedure for acceptance of a crane installation is laid down precisely in the relevant regulations. In this regard, there is no need to go into details here, but merely to indicate that only the test prescribed with a stationary overload warrants special mention. Experience shows that a wire rope stretches primarily in the course of its first period in operation. The wires, as the rope is manufactured, lie relatively loosely in the strand and, as it were, initially fit themselves into a position in which they are in equilibrium with their adjacent wires. Only after a certain period has elapsed do they take up their final stable position in the rope, and this natural process should not be brought forcibly to a premature conclusion. It is therefore recommended that the test be carried out with a stationary so-called "large test load" only after four to six weeks in operation, according to DIN 15019, sheet 1 (Draft); otherwise, the individual wires in a new rope will be subjected locally to excessive stressing, which may be the cause of its premature unserviceability.

This postponed test has no effect on the stability of the crane because, in this respect, it is primarily the test with a "small test load" and with all operational movements that is decisive.

6.1.4. Operation of Slewing Cranes and Loading Bridges

Statistics show that mishandling—often involving complete disregard of the regulations drawn up for the protection of the installation and personnel—is to a large extent the cause of damage. The collision of cranes with neighbouring similar installations (see also Figure 33) or parts of buildings, damage arising from negligence, bad visibility, etc. are causes which pervade the whole damage statistics. Furthermore, damage frequently stems from overloading (Figure 32).

Here, the user has possibilities for loss prevention which are as simple as they are effective, inas-

Figure 32. The jib of this slewing gantry crane collapsed as a result of overloading. The overload safety device was incorrectly adjusted.

Figure 33. The breakage of this counterweight jib occurred after it was caught during slewing by the wire ropes of a neighbouring half-gantry crane.

much as they are aimed at ensuring that the operation of crane installations is entrusted only to specially trained personnel and also that the relevant regulations are adhered to [4, 5]. Regulations and operating instructions should be in the hands of the crane driver.

It is not without reason that employers' liability insurance companies stipulate certain conditions for the licence to drive a crane. Thus, for example, it is stated in the regulations for the prevention of accidents with tower cranes: "The tower crane must be handled by reliable personnel, over 18 years of age, who have been trained in its operation and are familiar with its mechanical and electrical installations". Similar obligatory regulations exist for the drivers of other crane installations.

Training of Drivers in Crane-Driver Schools

Every effort should be made to employ as crane drivers only those persons who have successfully passed a crane-driver's course. For years these courses have been organized especially for tower-crane operators by unions, the building industry, and also by the manufacturers of the corresponding equipment. These courses end with a practical and theoretical examination, in which employers' liability insurance companies play an active part. Information regarding where and when crane-driver courses will be held can be obtained from, among other sources, the crane manufacturer and the employers' liability insurance companies.

It is strongly recommended that, in addition to the prospective crane driver, an extra person should be sent to attend the course, so as to have available a useful stand-in in the event of unforeseen difficulties, such as illness of the driver.

Periodic Repetition of Training and Instruction

Every crane installation has its own peculiarities in regard to operation and maintenance. Experience indicates that, once mistakes or acts of negligence have been committed, they can easily become habitual. This tendency must be counteracted by repetition from time to time of the training and detailed instruction of crane drivers, mainly on the basis of personal discussions with the individual.

Prohibition of Forcible Pulling out, Dragging and Oblique Pulling of Loads

The pulling out of individual items from a large accumulation of material, e.g., when loading scrap metal, when transporting bent concrete-reinforcing bars or when lifting long baulks of timber, which are wholly or partly jammed in place by other pieces of timber, is a dangerous operation. Moreover, in the latter case, it is possible that the piled-up timber will start to move and cause more severe damage.

Just as the forcible pulling out of jammed load items is prohibited, so are dragging and oblique pulling of loads with a crane.

Attempts are made again and again to increase the reach of a crane by oblique pulling or by dragging or towing loads (e.g., wagons and boats).

Figure 34. This boom collapsed as a result of a whole series of mishandling faults on the part of operating personnel, inasmuch as the cause of trouble with the luffing gear, which was jammed in its position of minimum overhang, was not investigated. Instead, an attempt was made to move the boom out of this position by an inadmissible procedure involving oblique pulling.

When doing so, no thought is given to the fact that the overload device does not respond to horizontal forces acting at the end of the jib. As a result, the stability of the crane is highly endangered. Moreover, when an obliquely pulled load rises clear of the ground, it can develop a pendulum motion and generate uncontrolled forces which not infrequently result in the crane toppling over.

Figure 34 shows a slewing luffing-jib gantry crane whose cantilever boom system was wrecked by a whole series of mishandling faults, starting with prohibited oblique pulling. For reasons which could not later be established, the luffing gear of the boom jammed. The boom was in a position of the smallest overhang and could not be moved outwards, despite the fact that the electrical equipment was unimpaired. Then, a sudden decision, a girder weighing two tons was attached, the luffing gear was engaged and simultaneously the crane was propelled backwards, dragging the girder along the ground.

Unfortunately, during this operation, the girder got caught and, according to eye-witnesses, the result was a rain of rivets before the cantilever boom system collapsed completely. This damage could certainly have been avoided and, in any case, with a little thought and patience, it would have been possible to trace the reason for the failure of the luffing gear and rectify it.

Injunction against Filling Grabs Completely with Wet Materials

It should be borne in mind that, when operating with grabs, wet material is much heavier than dry material. Consequently, the lifting gear is almost always overloaded when filling the grab completely with wet materials. The crane driver must not rely on the load-limiting device, as it often does not respond quickly enough. In the past, several cases of damage have shown how the structural steelwork can buckle with the first attempt at lifting the excessive load before the switch of the overload device has had time to respond.

Taking Grab Weight into Account

When using a grab, it is important to note that the grab weight must always be added to the pay load. In other words, the known grab weight should be subtracted from the rated load capacity of the crane concerned to determine the actual pay load that can be carried.

Prevention of Rope Slackness

Load hooks must not rest on the ground while lowering; otherwise, the rope will become slack and there will be a danger that it will not coil uniformly on the winding drum during lifting operations. Damage to the rope and the lifting gear is the likely result.

Prohibition of Sudden Changes in the Direction of Motion

A serious warning concerns inadmissible reverse switching actions to speed up a change in the direction of motion (travelling, slewing) before the first motion has stopped. The field of damage so induced extends from broken gear-transmission housings to buckled jibs. By installing locking or delayed-action relays, etc., it is possible to prevent such damage from rough operation.

Co-operative Use of Two Cranes When Lifting a Common Load

The co-operative use of two neighbouring cranes for lifting large loads requires special measures and must be carried out only under expert supervision. A prerequisite is the provision of a suitable load-carrier. In no circumstances should be total load—including the weight of the load-carrier—amount to more than 80 to 85% of double the permissible load capacity of the smaller of the two cranes utilized. If this proviso is not adhered to, experience shows that the smaller crane can easily be overloaded and suffer damage as a result of differences (even relatively very small ones) in the motions of the two cranes and the associated pendulum movements of the gravity centre of the load.

Communication between the Crane Driver and Other Personnel

Faultless communication between the crane driver and other personnel is essential for operation of every crane. For this purpose, there should be a set of uniform and unambiguous signals (the use of "Hand Signals for the Operation of Cranes", according to Memorandum AWF 27, is recommended). Insofar as verbal instructions by direct calls or by radio are possible at all, consideration must be given to language difficulties and hence to misunderstandings that can arise if foreign workers are employed.

Some Further Reminders that are Frequently Necessary

Electrical resistances are not provided for the purpose of drying damp clothing. Also, cleaning rags do not form part of the covers of resistances. There is a risk of fire!

Crane tracks should be inspected once a day to discern incipient damage in good time. The tracks should be kept free from solid objects, accumulations of flying dust, etc.

It is also necessary to remember the regulation which stipulates that the effectiveness of brakes and emergency limit switches should be checked daily—in the case of rarely used cranes, each time before starting operations.

Defects in a crane must always be entered in the crane inspection book and reported to the supervisor concerned and also, if there is a change in the crane driver, to the man who takes over. If there are defects that pose a threat to operational safety (e.g., if brakes or emergency limit switches fail, or a wire rope gets damaged, runs off the winding drum, becomes snarled or kinked, etc.), operation must be stopped immediately.

Before leaving the control position, the operator must set down the load, hoist the load hook and turn off the crane switch.

6.1.5. Measures for the Prevention of Storm Damage to Crane Installations

6.1.5.1. Precautionary Measures

The operational directives laid down by the manufacturer and the regulations for the prevention of accidents issued by employers' liability insurance companies, as well as internal plant instructions, must be observed strictly.

The instructions given by representatives of employers' liability insurance companies, by technical supervisory associations and by supervising authorities must be carried out immediately.

Training of Personnel

Personnel must be given information about existing safety equipment for storm protection and must also be instructed in its use, particularly before any work especially endangered by wind is to be performed. The carrying out of storm-protection exercises at periodic intervals is recommended. The responsibility for the proper maintenance and constant readiness for use of the safety equipment must be entrusted to a reliable and proficient member of the operational staff (crane driver or attendant, deputy safety officer).

Desirability of Connection to a Weather Station

For the prompt transmission of storm warnings, connection to an official weather station is useful. Moreover, developments in the weather should be observed while a crane is in operation. If wind force is increasing or there are gusts, current work should be suspended and the appro-

priate safety measures instituted, even if there has been no storm warning from the weather station, as these disturbances could be merely local and not recorded by the station. Work must not be resumed until it is safe to assume that the abating storm will not be revived (if necessary, the weather station should be consulted).
Even if no agreement has been reached with the appropriate weather station for the transmission of storm warnings, information regarding likely weather conditions should be obtained before starting work that could be endangered by storms or if the weather prospects appear uncertain.

Wind-Warning Instrumentation

The installation of wind-warning instrumentation is recommended. In all cases there should be one instrument for indicating wind force and one for indicating wind direction, both of which should give an acoustic warning of changes in wind conditions beyond certain adjustable limits. The installation of the measuring instruments should be entrusted to an expert.

Safeguarding of Installations not in Use

If installations are not being used for some time, they must be secured particularly carefully against storm damage. The measures adopted must be checked from time to time (the dates should be noted in a special diary).
With older installations in particular, it is necessary to check whether the design and the brakes correspond to DIN specifications and to the accident-prevention regulations issued by employers' liability insurance companies, which go much further. In certain cases, reconstruction, which modernizes the installation may be necessary.

Brakes for all Travel Carriages

All travel carriages must be equipped with brakes; otherwise, there is a risk that, as a result of the tilting moment caused by wind forces, braked carriages can become unloaded and hence the braking resistance is lowered. Brakes which act directly on the carriage are more advantageous than a central motor brake, since here, in the event of danger, the effective transmission of braking forces to the wheels can be impaired by deformation of the bridge.

Unsuitability of Drive-Mechanism Brakes and Drag Shoes for Storm Protection

According to the relevant regulations issued by employers' liability insurance companies for the prevention of accidents, drive-mechanism brakes of rail-bound crane installations cannot be regarded as safety devices against the effects of wind. It is true that they are required to bring the installation to a halt against wind forces, provided that these forces are not above the limit at which operation is still permitted, but they are nevertheless not sufficiently dimensioned to hold the installation against more severe storm conditions. All cranes, loading bridges, etc. that are exposed to the wind and run on rails must be provided with additional mechanisms to prevent them from being driven along by the wind.

Suitable Safety Equipment for Wind Protection

Apart from automatically acting friction-type rail brakes, sprags, etc., fixed built-in rail clips (one per carriage) have proved their value more than other devices. However, practical experience shows that not all rail clips actually fulfil their purpose (Figure 35).
The most reliable equipment for protection against wind comprises positive anchoring devices between the crane installation and the rail foundation, always assuming, however, that these locking devices are made operative in good time. The well-known bolt-locking units have the disadvantage that, for securing a bridge, it is usually necessary to have an additional man who can guide the crane driver to the exact position. This disadvantage can be overcome by devices which allow the crane or loading bridge to be held on both rails by means of two pairs of movable tension members with associated tensioning elements. With this arrangement, the crane driver need only get the crane into approximately the right position and can then, with the aid of the tensioning elements, equalize any differences that may exist. An example of such a system is shown in Figure 36.
Devices to which the installation can be locked by means of bolts or shackles must be fixed to

Figure 35. A defective rail clip of a tower crane which toppled over from the effects of storm action. Although the clips were properly tightened before cessation of work, they were inadequately dimensioned and were not positively supported by the rails in the longitudinal direction. As a result, they were able to give way under wind pressure in the direction of travel and became distorted.

both tracks at regular, not excessively great distances apart. If possible, they should be visible to the crane driver, so as to enable a rapid correspondence between the holding devices of the foundation and the crane in the event of danger.
When using the positive anchoring method just described, the following warning notice at the control console should not be missing (unless the locking device is so designed that it automatically interrupts the electric current): *"Caution! The bridge is anchored. Do not start!"*
Positive anchoring has proved its value in numerous installations and should not contribute very much to the overall cost of new installations. For that matter, it can also often be applied to older installations without incurring excessively high expenditure.
Friction-type storm brakes must be so designed that the contact pressure is not reduced by changes in cross-section of the rail heads and that they will not collide and deform in the event of movement of the crane installation and thereby lose their braking action. Rail brakes and sprags, like drive-mechanism brakes, must not be impaired in their action by a tilting moment caused by wind forces.

Buffer-Stops

Securely anchored buffer-stops must be provided at the ends of both rails of crane and bridge runways. The buffers at each end of the track must be equidistant from the corresponding buffers on the crane or bridge. If possible, spring-type buffers should be used, so as to avoid severe impact shocks and consequent damage, as well as the risk of tilting. In no circumstances are baulks of timber, chains wrapped around rail ends, and similar expedients adequate as buffer-stops.

Consultation with the Manufacturer in Case of Doubt

For checking the installation, it is recommended that the user, particularly in case of doubt, should

Figure 36. In this system for anchoring a loading bridge against the effects of storm, two flat pieces of iron, formed at their ends into hooks, engage in slots in two steel girders installed later next to the travel rails. By manipulating the tensioning elements, small differences can be equalized and the loading bridge firmly fastened to the foundation.

get in touch with the manufacturer and seek his advice in connection with the fitting of safety devices for protection against storm.
From the beginning, the user should clarify the question as to when the installation should be taken out of operation and secured against storm. If no data are available concerning permissible wind forces during operation, information should be sought from the manufacturer.

Daily Inspections

The track and all safety equipment must be checked constantly, daily under arduous operating conditions, for serviceability and proper functioning. Particular note should be taken of bent brake rods, corroded or worn joints, worn brake linings and rail clips, etc. Brake shoes must make contact with their entire surface and must lift completely during operation of the installation; otherwise, they will be ground away and will hence become unusable. Defects must be rectified immediately. The travelling rails must be kept clean. Loose rust and residues of oil and grease must be removed.
Further measures and devices for the protection of crane installations from storm are reported in detail in reference [6] in the Bibliography.

6.1.5.2. Measures during Non-Operational Periods or Imminent Storm Conditions

Prompt Stoppage of Operation

When a storm rises, operation must be stopped in good time and the installation safeguarded—at the latest at a wind speed stipulated by the manufacturer or laid down in the operating instructions (often at a wind force of 6 on the Beaufort scale, corresponding to a wind speed of about 45 km/hr). The maximum permissible wind speed for operation by the manufacturer is, however, valid only if the drive-mechanism brakes and their linings are in perfect condition. In some circumstances, the installation should be taken out of operation at even lower wind strengths. Operation must never continue if slewing against the wind is no longer possible with a suspended load or if the slewing brake can no longer hold the boom or safe guiding of the load without pendulum movements is no longer possible.

Even with low wind strengths with gusty wind conditions, care is also necessary and, in such cases, maximum loads should not be lifted. Moreover, the transportation of large-area or bulky loads (e.g., shuttering and other casing components) should be avoided in such situations. A load forced into pendulum movements by wind gusts can prove disastrous to even heavy dockyard cranes.

Securing of Installations during Non-Operational Periods

An unexpected storm may arise during periods out of operation. For this reason, installations should be secured in position by the existing wind safety devices, even during short interruptions in operation and during repair, maintenance, and lubricating work. If the need arises, the anchoring devices must be moved into position and locked on the spot. For this purpose, it will often be necessary for a second man to guide the crane driver accordingly.
If no anchoring is provided, a loading bridge should be run up to those buffer blocks of the track which lie in the main direction of the prevailing wind (e.g., to the eastern buffers if the wind comes from the west).
Coal or ore dumps do not necessarily constitute a protection against wind. The slopes of the dumps create ascending air currents which, if the prevailing wind is in the direction of travel, can in some circumstances favour movement of a crane installation. Gaps between buildings and corners of buildings can, like lanes between ore or coal dumps, give rise to particularly violent turbulence and strong air currents.
Travel brakes must be engaged and assurance obtained that they grip sufficiently. All rail clips must be inserted and firmly tightened.
All control devices must be set to zero and the main switch turned off.
The stays or guy-ropes provided for storm protection (e.g., with tower cranes) must be secured. Hydrants, roof pillars, pylons, lighting masts, pipeline and pipe-supporting bridges are, however, unsuitable as anchoring points for stay wires.
The attachment and extension of wire stays must be effected only by splicing or with the aid of proper wire-rope clamps, the number and disposi-

tion of clamps being in accordance with DIN 4129. Spliced connections must be carried out only by experts. The tying of ropes is strictly prohibited.
Stays with kinks, pronounced corrosion or frayed wires must not be used. Ropes should never be led around sharp corners without being protected.
It is recommended that the wire ropes used as stays for safeguarding against storm should not be employed for other purposes but should always be left on the crane in an accessible position.

Booms of Tower Cranes

The booms of tower cranes should be brought into the position of *greatest overhang* and into the direction of the wind, i.e., the boom should be on the side of the tower turned away from the wind (Figure 37). The boom must not be locked in position, so as to ensure freedom of the crane to turn. However, if the wind veers and can cause the boom to come into contact with buildings, scaffolding or overhead electrical lines, it must be let down or firmly anchored in compression and tension against strong structural components.

Figure 37. When a wind sprang up, the boom was left in its steepest position and, as it was caught by a gust of wind, it buckled towards the rear.

Setting-Down of Loads

Loads and heavy load-carrying appliances (grabs, cross-beams, etc.) should, as a matter of course, be set down on the ground. However, this precaution should not be assumed to confer protection of the crane from movement.

Crane Hooks

The attachment of the crane hook to a fixed point as a safety measure against wind is a nonsensical procedure and actually results in additional hazards. The unloaded crane hook should be hoisted to the highest possible position after first removing wire ropes, chains, suspension attachments, etc.

Precautions during a Storm

During storms of long duration, brakes, anchoring and other safety devices should be checked from time to time. If necessary, they should be re-adjusted, as these devices can become loose under pulsating wind forces. This precaution applies to a large extent to rope-tensioning devices.

Precautions after a Storm

After the storm has abated, the crane installation, particularly the storm safety devices, should be inspected for possible damage. Also, the track must be cleared of coal dust, sand, solid objects, etc.

6.1.6. Protective Devices for Crane Installations

In accordance with relevant regulations, crane installations must be provided with a series of automatically operating safety devices to prevent the occurrence of dangerous situations and severe damage to the installation, possibly as a consequence of mishandling.

Overload Safety Devices

An important aspect of loss prevention is safeguarding the crane against overloads. Regulations for the prevention of accidents therefore demand that overload safety devices be installed not only on those cranes which may have to lift uncontrollable heavy loads, but also on all power-operated cantilever-type cranes, with fixed or variable reach, particularly tower cranes.

Cantilever cranes are specially endangered by exceeding the permissible load moment. In such a case, the crane can lose its stability and topple over. Hence the overload safety device primarily ensures the stability of the crane; in addition, it protects the structure and the drive mechanism from excessive stressing.

For cantilever cranes with an equal load-carrying capacity at all amounts of reach, a load switch suffices to prevent the lifting of loads beyond the permissible load-carrying capacity. In particular, the load switch is adequate for cranes with a fixed boom, as long as the load-supporting point always has the same distance from the tilting axis of the crane.

However, for cantilever cranes whose load capacity depends on the variable reach obtaining at the time, safety devices dependent on load moment (load-moment limitation) are necessary.

There are no generally valid rules concerning which safety device is required in individual cases for the particular crane involved. The type of overload protection is much more defined by the design and mode of operation of the crane.

The operating principle and design of most of the usual load switches and load-moment limiters are extraordinarily varied. For this reason, individual systems will not be described here in detail. Since the user's personnel do not themselves have the expertise to repair these often very complicated safety devices in the event of irregularities or defects, it is urgently recommended that, if malfunctioning occurs, they should not try to remedy the fault personally, but that recourse should be had to the manufacturer.

Generally speaking, the nearer to the load that measurements are taken, the greater is the sensitivity of these devices.

When a jib is in a steep position, a relatively small change in angle produces a large change in reach. The device must therefore be highly sensitive to changes in angle.

The influence exerted by frictional and accelerating forces, starting current, etc. must form part of the measuring parameters for overload protection.

Load Cells

Since, as a rule, inaccuracies are soon generated in mechanical measuring elements as a result of backlash and wear, and because load measurements based on motor current often become inaccurate owing to numerous extraneous influences, the use of the load cell is recommended as a measuring device. Apart from its high accuracy, a load cell offers the advantage of measurement without any moving parts.

If possible, however, a load cell should be chosen one size larger than necessary. A measuring cell which is too tightly dimensioned will, in some circumstances, lose its calibration if exposed to an overload and will have to be re-adjusted. Overload switch-off must lie within the permissible measuring range of the cell—at about 75% of the possible cell loading.

Release Setting for Overload Protection

Regulations regarding the release-value settings of overload-protection devices have not yet been issued. To exclude the possibility of a crane toppling over as a result of overloads, a release setting of 115% of the nominal load capacity is often recommended. This value is based on a 10% allowance for accelerating forces and on a $\pm 5\%$ allowance of the nominal load for inaccuracy of the measuring device. In the most unfavourable case, the switching point would therefore still be 5% under the small test load of 125% of the nominal load specified in DIN 120 and DIN 15019, sheet 1. For cantilever-type cranes, however, a higher sensitivity is required. We recommend a release limit equal to the nominal load for the angle of inclination plus 10%. With this adjustment, it should just be no longer possible to lift the load.

Other extremely important safety devices include emergency switches for

- the highest and lowest jib positions,
- the end positions of the boom-retraction equipment,
- the highest and lowest hook positions,
- the ends of the trolley and crane tracks.

Emergency Switch for the Lowest Hook Position

Emergency limit switches for the lowering movement of load hooks are necessary in cases where construction cranes are used, for example, for work below ground level, with the consequent risk of coiling the wire rope on the winding drum in the reverse direction.

Preliminary Limit Switches for High Travel Speeds

At high travel speeds, it is in some cases useful to install preliminary limit switches ahead of the emergency limit switches to effect a reduction in speed and thereby give a warning to the crane driver at the same time.

Adjustment of Emergency Limit Switches

When adjusting emergency limit switches, account should also be taken of the run-out distances from full speed.

Anti-Skewing Equipment

With gantry cranes and large-span loading bridges, there exists a risk of skewed running, even under normal operating conditions, either through uneven loading, uneven travel resistances or different wear of the travel wheels. Structural steelwork can then be overstressed and, in particularly difficult cases, the result can be derailments or even toppling over of a loading bridge.

This problem of skewing is generally known but, unfortunately, it has not yet been completely solved—at least in a form which can be justified economically. However, a range of simple and inexpensive equipment, also suitable for retrofitting, is now available as a means of enabling inadmissibly severe skewing to be avoided [1].

Equipment for Measuring Clearance between Crane Installations

In view of the present state of technology and the large number of damage cases caused by collisions of slewing crane jibs with neighboring crane installations or buildings, the desire for equipment which could prevent these collisions with absolute certainty is still unfulfilled. However, it should be noted that, for maintaining a safe distance between gantry cranes and loading bridges travelling on a common track, several reliable "collision monitors" are already on the market. These units frequently operate optically, but some recent designs utilize the ultrasonic-echo principle [7].

Maintenance and Attention

The maintenance of and attention to safety equipment in accordance with the instructions of the manufacturer should be carried out as a matter of course. In their own interest, crane drivers should ensure that the safety equipment is in a condition of maximum serviceability.

In the vast majority of cases of damage reputed to have occurred as a result of failure of a safety device, it has been found on closer examination that the device was either taken out of operation deliberately or was unable to function because of inadequate maintenance.

Prohibited Interference with Safety Equipment

It is an extremely dangerous and hence prohibited practice to take safety devices out of operation—even only temporarily. The blame and liability for any accidents or damage to equipment that result must attach exclusively to the person responsible for ordering or carrying out such prohibited actions, e.g., re-adjusting or rendering inoperative the overload safety device to permit the hoisting of an excessively heavy load.

Attention to "Unimportant" Safety Devices

The foregoing observations should be an incentive to pay more attention to the maintenance and care of safety equipment. In addition, apart from storm safety devices, it is also necessary to remember those safety devices which—quite unjustifiably—are all too often regarded as unimportant, such as

- track clearers,
- broken-wheel supports,
- load-hook safety catches,
- slack-rope safety devices on cranes which, under normal service conditions, are exposed to the risk of rope-slackening (grab or magnet operation).

Importance of Correctly Adjusted Safety Equipment

The importance of correct adjustment of safety equipment, e.g., overload safety devices, particularly with older crane installations, is exemplified in Figure 32.

6.1.7. Maintenance, Inspection and Revision of Crane Installations

The serviceability and safety of crane installations depend on regular and conscientious attention on the part of appropriate skilled personnel.

Maintenance by Operating Personnel

As regards the necessary attention and maintenance, as well as the required checks in the course of operations, these tasks can be carried out by properly trained personnel of the user. A prerequisite is, of course, that the personnel are fully familiar with the operating and maintenance instructions of the manufacturer.

Apart from the adequate supply of the appropriate lubricants to all bearings, joints, gearing, etc., regularly essential maintenance includes checking of all mechanical, electrical, hydraulic and pneumatic equipment, and especially checking of the dependable functioning of all brakes and of supervisory and protective equipment. It also includes a check on the wear of travel wheels, pulleys, wire ropes and other structural elements subject to wear.

To complement the maintenance instructions issued by the manufacturer, some suggestions follow which appear to be necessary on the basis of relevant damage cases, but which are not, or not always, included in the maintenance instructions.

Daily Visual Inspection of Crane Tracks

A careful inspection should be made daily—and also after every stormy wind—of the crane track, which must be kept clean and free from obstructions of all kinds. Wind-blown deposits of coal dust, sand, etc. must be removed, since the functioning of the rail brakes may otherwise be impaired and, in some circumstances, there is even a risk of derailment.

Loose or missing rail fastenings and joints must be tightened carefully or replaced. Settling of rails and foundations always constitutes a hazard to crane installations and must be reported immediately to the management. Particularly endangered are tower cranes. Even if only a small amount of rail subsidence appears, operation must not be resumed until the trouble has been rectified—if necessary, by filling in. Note should be taken of rail breakages.

Freedom of Sliding-Contact Channels from Ice and Snow

The covers of channels for underground sliding-contact conductors must always be kept free from ice, snow, etc. For instance, flap-type covers are rendered inoperative by packed ice and, as a result, the covers and current collectors will be deformed when the crane installation is moved, and there is also a danger in some cases of short-circuits during a thaw.

Freedom of Travel Emergency Switches

Drag cables and, of course, ramp rails, etc. for the actuation of travel emergency switches must be kept free from ice, snow and other impediments.

Dangers from Frost

Ice constitutes a hazard that should not be underestimated. Condensation or rain water that has penetrated leaking covers has often had the effect at low wintry temperatures of rendering limit switches, brake-lifting magnets and control systems unserviceable. The result has been jibs that have been pulled back too far and have buckled or even crashed down, and deformed track-limit buffers, carriages, gantries and bridge supports.

Pneumatic control equipment must, therefore, always be drained. For hydraulic installations, only hydraulic oils free from water, as stipulated by the manufacturer of the installation, must be used. Since hydraulic fluids have only a limited life, note should be taken of the prescribed periods for their renewal. In winter, hydraulic and pneumatic control installations must be prewarmed above the freezing point before commencing operation for the first time in the morning. (If necessary, the heating in the driver's cab should be switched on well before starting operations. Naked flames must never be used.)

Lubrication of Crane Installations

The lubrication of crane installations should be carried out in accordance with the lubrication schedule drawn up by the manufacturer. In addition, we give the following recommendations:

- Lubrication should always be ample and, under continuously high loading and particularly arduous and dusty operating conditions, it should also be effected at intervals shorter than those proposed by the manufacturer.
- Only high-quality lubricants, as specified by the manufacturer, should be employed. No experiments should ever be made with cheaper —or even more expensive—self-styled "wonder" lubricants.
- The oil levels of gear transmissions should be checked daily to detect leakages and irregularities in oil circulation.
- Oil changes should be made in accordance with the instructions of the manufacturer of the gear transmission, i.e., at the proper time. The old oil should be drained while it is still warm and therefore flows easily, i.e., immediately after a long period in operation. For flushing the transmission, use must be made of oil of the same quality and viscosity as that normally utilized in operation, heated if necessary to between 50 and 60 °C. Petrol, kerosene or similar fluids must never be used.
- So-called maintenance-free rolling bearings should also be included in the maintenance schedule because a grease filling is not likely to last for the estimated 10,000 hr of operation.

To facilitate maintenance, it is recommended that lubricating nipples should be marked clearly by surrounding them with a coloured rim.

When carrying out lubrication operations, the component concerned and its immediate surroundings should be inspected. The little effort involved does not take much time, but contributes to the prevention of extensive damage, inasmuch as loose screws, cotter pins that have fallen out, bent rods, etc. can be the first signs of incipient damage.

Daily Checking of Brakes and Emergency Limit Switches

It should be taken as a matter of course that brakes and emergency limit switches must be checked daily (or, with cranes which are seldomly used, each time before starting operations) for their effective performance, as prescribed rigorously in the regulations for accident prevention. If these safety devices fail, the crane installation must not be operated.

With brakes, freedom of movement of all moving parts, such as brake levers and rods, must be ensured. The prescribed clearance between brake discs and linings in the released condition of the brake (usually 0.5 to 1.0 mm) must in all cases be maintained. Grooves in a brake disc indicate rubbing rivets or worn linings.

Maintenance of Electrical Equipment

The electrical equipment also requires a certain amount of attention and should be inspected once a week. Before starting the inspection, the electrical system must be disconnected from the grid on all phases. Unauthorized switching on again must be made impossible by appropriate measures (padlocks or other safety locking devices).

The contact pieces of switches and contactors must be renewed if the contact coating has been almost burned away. A black discolouration of the contacts has no special significance, so that contacts should not be reconditioned. Contacts must not be greased. All junction and connection screws, particularly those on terminal blocks and resistances, must be tightened, as screws which have fallen out can cause serious electrical troubles and damage. Charring of screws and fuses indicates bad contact and they should be replaced. The prescribed fuses must be used. Spare parts must always be the same as those they replace.

With regard to motors, strict adherence to the maintenance instructions of the manufacturer should be observed. In the case of particularly dusty operational conditions, shorter maintenance and lubrication intervals than those stipulated are recommended. With slip-ring induction motors, attention should be paid to the free movement of the carbon brushes in their holders. Carbon brushes must be renewed at the right time, i.e., before they are completely worn away.

Particles of carbon and dirt must be removed carefully (perhaps by blowing out with dry air). Oily rags or cotton waste should never be used.

Checking of Ballast Arrangements

A check of the ballast or counterweight arrangements is also desirable. Cracked counterweights made of concrete endanger the installation and personnel, the highest potential hazard arising if pieces break away from the concrete. If a gravel filling is used as ballast, care should be taken that the container is leakproof. With tower cranes in particular, the completeness of central and counterweight ballast arrangements should be checked frequently.

Officially Prescribed Checks

Apart from continuous maintenance work, inspections and revisions which must be carried our periodically serve not only to safeguard personnel and machines, but also to prevent unforeseen operational outages. Such checks are therefore prescribed as a matter of principle by the relevant authorities.

According to the regulations of the employers' liability insurance companies for accident prevention, the condition of crane installations should be checked "as needed, but at least once a year, by an expert". Moreover, tower cranes must be checked every time they are erected and also every four years by an expert.

The obligatory checks are largely laid down in "Principles of Crane Testing", issued by the main association of employers' liability insurance companies (reprinted in reference [5]). It has, however, been shown that these regulations satisfy modern demands only with reservations. The mode of operation and technological considerations often enforce shorter intervals between revisions than those stipulated in the regulations for the prevention of accidents. Also, inspection of the installations by so-called "experts" who, according to these regulations, include machine or crane foremen, among others, does not always satisfy requirements.

Revisions by Expert Personnel of the Manufacturer

For a comprehensive inspection that is decisive for the operational safety of a modern crane installation, highly specialized expert knowledge is frequently essential. In this regard, consider as an example the often quite complex electrical and hydraulic equipment of modern installations. In view of the problems which arise with such equipment, even the best operator, who must, after all, be concerned with other matters during his daily work, is generally subjected to demands beyond his capacity. A main revision of such a crane should therefore not be carried out only by an expert of the user, but the user should utilize the experience and specialized knowledge of the manufacturer. For this reason, we strongly recommend that main revisions should be carried out by specialist engineers on the manufacturer's staff. Some users may consider this recommendation rather costly. However, only the wide range of experience at the disposal of the manufacturer quite generally and particularly with the type of crane concerned can give adequate assurance that the incipient sources of danger are recognized and that an early diagnosis can be made of impending damage.

Revisions prevent unforeseen operational outages. A revision carried out at the proper time by appropriate specialists is more economical than continuing operation until the next case of damage occurs. Furthermore, a scheduled revision can be arranged during a period when there is relatively little work and hence has an insignificant influence on the availability of the crane. It is advisable to agree on the time of such a main revision with the manufacturer. Also, it is recommended that a contract for regular revisions should be concluded with the manufacturer.

Determination of the Intervals between Revisions

In the regulations issued by the employers' liability insurance companies for the prevention of accidents, the intervals between main revisions are laid down in very general terms. When it is stated therein that the relevant checks should be carried out as the need arises, but at least once a year, this is understandable, as cranes that operate in several shifts and/or lift the maximum permissible load almost always require shorter intervals between revisions than less highly stressed installations. The user should also seek the advice of the manufacturer with regard to the time required for revisions.

Drawing-Up of Check Lists

The extent of a revision naturally depends on the type and design of the installation and can hardly be laid down in generally valid terms. The drawing-up of a check list or revision schedule is recommended.

With the aid of such a schedule, it is not only possible to deploy, the personnel available efficiently but also to substantiate that the work has been carried out. This measure serves to a large extent as a precaution against overlooking essential components during a revision.

As a rough preliminary basis for the preparation of such check lists, use can be made of "Principles of Crane Testing", reprinted in [5].

To what extent the constructional groups listed in the revision schedule should be subdivided into individual items is a question which depends on the reliability and expertise of the personnel carrying out the revision and on how comprehensive are the relevant instructions given by the manufacturer.

If these instructions are sufficiently detail, it should suffice to provide the constructional groups listed in the revision schedule with corresponding references to the pertinent sections in the operating instructions, etc. If they are not adequately detailed, an extensive subdivision of the components to be inspected and hence individual execution of revisions are recommended.

When planning such an inspection schedule, the manufacturer would certainly be pleased to be of assistance. In addition, the engineers of the Technical Insurance Department of the Allianz Insurance Co. are available for this purpose.

The following test and maintenance directives for several particularly important sections of a crane installation are reproduced here as an example of a suitable working basis. We recommend these directives as a complement to the crane test books currently in use, which, in our opinion, need revision as they deal with the subject in too general terms. The following directives have been subdivided into different fields of validity and determine when a revision is due, not only on the basis of the number of hours in operation, but also in the basis of time alone. These guidelines closely follow those of a large cargo-handling undertaking. They are the result of experience amassed over decades and play a permanent role in the crane-supervision procedures of the company.

Example of Test and Maintenance Directives (Abridged)

Field of validity:

Wire ropes, rope pulleys, fastenings, thimbles, cast capping, slides.

Test and maintenance work:

Ropes must be tested in accordance with DIN 15020, sheet 2. The operational safety of a rope must be assessed on the basis of the criteria given therein. When a rope comes to the end of its useful life, it must be replaced.

Depending on its external condition, the rope must be cleaned and greased. Old grease on rope pulleys and slides must be removed. The rope attachment to the drum, rope thimbles, including cast capping, and all pulleys must be checked.

Interval:

175 registered operational hours, but a maximum of 3 months.

Field of validity:

Load hooks and hook fittings.

Test and maintenance work:

Hook fittings must be completely dismantled, cleaned and checked for external damage and deformations. Defective parts must be renewed. Testing covers all individual parts, including shackles. The measured results must be entered on a test sheet, as exemplified in Table 2. If, at one measuring point, a reduction in diameter of about 10% is registered, the part concerned must be replaced. The measurements of all replacement parts must be entered in the measurement record. At the load hook, the distance (measuring point 10) between the centre-punch marks must be measured. If there are any deformations, the specialist must decide on the suitability of the hook fittings for further possible use. On the upper eyebolt, the thread and thread undercut must be checked. The components mentioned, especially the load hook, must be subjected to a crack test.

Interval:

12 months.

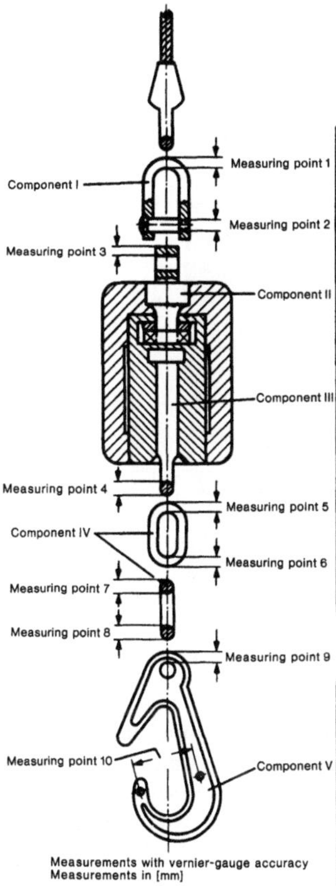

Hook fitting No.....

	1972	1973	19
Component I			
Measuring point 1			
Measuring point 2			
Component II			
Measuring point 3			
Component III			
Measuring point 4			
Component IV			
Measuring point 5			
Measuring point 6			
Measuring point 7			
Measuring point 8			
Component V			
Measuring point 9			
Measuring point 10			
Tested and measured on:			
Signature			

Component		
Designation		
No. removed		
No. installed		
Manufacturer		
Tested for freedom from cracks		
Signature		

471

Table 2. Example of a test record for the revision of hook fittings.

Field of validity:

Braking equipment, emergency load switches, starting-position restraint, limit switches and preliminary limit switches, alarm equipment, locking devices, automatic crane circuit-breakers.

Test and maintenance work:

Braking equipment and safety devices must be tested for proper functioning and condition. Any necessary repairs must be carried out immediately. All defective components must be replaced. Tests for the functioning of brakes must be conducted under nominal load.

Interval:

250 registered operational hours, but a maximum of 6 months.

Field of validity:

Control equipment, contactors, resistances, motors, slip rings, circuit-breakers, heating equipment, overcurrent protection, terminal connections.

Test and maintenance work:

The components mentioned must be tested and cleaned. They must be maintained in accordance with their condition. Defective and worn parts must be replaced.

Interval:

250 registered operational hours, but a maximum of 6 months.

Field of validity:

Load limitation.

Test and maintenance work:

Load-limiting devices must be adjusted so that they respond to the nominal load for the angle of inclination plus 10%. This test must be carried out in the presence of an expert.

Interval:

6 months.

Supervision of Weak Points

Certain weak points manifest themselves sooner or later in many installations, possibly by increased wear or by the frequent failure of certain components. Special attention should therefore be paid to such parts. If need be, their maintenance and inspection should be carried out at shorter intervals than those recommended by the manufacturer or stipulated in official regulations.

Checking of Newly Installed Components

With regard to maintenance and inspection intervals, it is also recommended that newly installed replacement parts or groups of parts should be given a further check after a short period in operation, so that possible faulty installation and other defects can be recognized quickly and suitably rectified.

Written Records of Defects and Repairs

Another strong recommendation is that all defects noted during maintenance, inspections and revisions, as well as on other occasions, should be recorded in writing, irrespective of whether they are being rectified immediately or whether their elimination—if it is a question of deficiencies that have no bearing on the safety of operation—has been postponed to a later date. In addition, a written record should be made of the repairs that have been carried out.

In this way, it is possible

- to recognize weak points early,
- to order in good time spare parts that are most likely to be required in the future and hence to build up the most useful stock of spares,
- to plan the execution of necessary repairs and maintenance work at a favourable period from the point of view of work load.
- to make optimum preparations, with respect to time and the organization of personnel, for effecting repairs and maintenance.

Apart from the above advantages, it is often possible, on the basis of such records, to clarify the causes of defects and damage that have occurred and hence to adopt the appropriate measures for their elimination [8].

Figure 38. This photgraph shows a number of defects in the track of a loading bridge. Among these are a broken rail joint, displaced or missing clamp and anchor plates, and loose clamping bolts, with the occurrence of local break-outs from the concrete beam.

Once Again: Crane Tracks

Crane-track deficiencies, such as loose or missing rail attachments, broken screws, gaps of several centimetres between rail joints, insufficient packing, settling of the foundation (particular attention is required in mining districts), can lead to derailments, rail and wheel fractures, etc.

Figure 38 clearly demonstrates the need for inspection of crane tracks. Actually, the track should never have been allowed to get into this condition, even after prolonged operation, and can only be explained by the apparent lack of a competent inspection. Any further deterioration of this track would undoubtedly have led to severe damage to the loading bridge. The strong recommendation in cases of this kind is that the installation should be inspected at shorter intervals, i.e., also between prescribed inspections, by an outside expert.

The fracture of the rail-type slewing ring shown in Figure 38 was obviously promoted by inadequate seating of the rail joint on the packing plate.

Damage to Ball-Bearing Slewing Rings

Pronounced local appearance of wear, break-out from the tracks, and destruction of rolling elements are often solely the consequences of constantly occurring shock loads. Such shocks can be caused by unevennesses in the track and are transmitted through the gantry to the slewing ring; also, they can arise from the sudden displacement of the centre of gravity through unloading when operating with a grab, when the counterweight is perhaps not optimally dimensioned.

Some breakages of slewing-ring bearing races can be traced to too soft a supporting structure or to inadequate seating of the ring mount on its support. The latter cause can often be detected early by loosened or even broken screw fasteners. If the support is not perfectly flat, the ring mount will deflect locally under load, so that the neigh-

bouring bolts become overloaded and, when these bolts fracture, the ring mount tends to lift itself up at these places.
Obviously, loose bolts on a slewing ring are not always signs of a non-rigid supporting structure or of inadequate seating. In many instances, unsecured bolts have simply been insufficiently tightened and have become loose solely by shocks during operation. When repeated loosening of the ring-mount bolts occurs, a search should, however, be made for its cause.

Loose Screwed Connections

Loose or completely missing bolts and nuts always indicate impending danger. Here are two examples from many similar cases of damage:
When an unsecured screw dropped out, the gib key of a bevel gear for the travel drive of a loading-bridge gantry leg came loose. The steel structure of the bridge was severely damaged by skewed running.
As a result of loose foundation bolts, the hoisting unit of a crane tilted to one side, causing the hoisting rope to spring off the drum and become coiled up on the drum shaft. The winding gear was badly damaged and the hoisting rope completely destroyed.

Excessive Bearing Play

Attention should be paid to excessive play of articulated joints and bearings. Heavy shocks are generated during operation by excessive bearing play. In addition to rapid destruction of the bearing itself, breakage of shafts, wheels, etc. and even damage to the steel structure often result. The journal bearings of pillar-type jib cranes, among others, merit special attention.

Loose Press-Fits, Fretting Corrosion

In the case of rolling bearings, gearwheels on shafts, etc., loose press-fits always lead to unforeseen outages if these defects are not detected and rectified in time. In the event of relative movements—even very slight ones—between press-fitted components, fretting corrosion takes place, further enlarging existing play. Load shocks contingent on the mode of operation then give rise to ever higher stress peaks in the structural components and finally lead to their failure.
As an example, reference should again be made to Figure 25, in which the slewing-gear pinion shown exhibits pronounced fretting corrosion in the pinion bore. Evidently, in the manufacturing stage, there must have been an insufficiently tight fit of the pinion (which was, incidentally, made of an unsuitable material) on the shaft, so that, under the conditions just described, the first cracks appeared at the root of the teeth and finally led to the complete fracture of whole teeth. In this case, the transition radius of 2 mm at the root of the teeth was not the cause of fracture, but merely determined its location.
A loose press-fit can often be recognized at an early stage by the exudation of fretting-corrosion products, usually of a brownish to reddish-brown colour, from the press-fit ("the steel is bleeding"). If the trouble is detected in time, it can sometimes be rectified by relatively simple means [9].
In this connection, breakages of drive-shaft pinions, ring gears, countershafts and the corresponding drives frequently occur as a result of inadmissibly large play between tooth flanks.

Cracked Brake Discs

Attention should be paid not only to brake linings, but also to brake discs or drums. Figure 39 shows a broken piece from a hoisting-unit brake drum, which fractured while braking a load dur-

Figure 39. A broken piece from the ruptured brake drum of a hoisting unit. Numerous thermal-fatigue cracks traverse the braking surface, one of which penetrated the drum over a greater length and thereby triggered the damage.

ing a lowering movement. Like the broken piece in Figure 39, the brake surfaces of all pieces exhibited numerous thermal-fatigue cracks running in an axial direction. The presence of temper colours signified very excessive operational temperatures.

Inspection of New Brake Linings

It has already been pointed out that newly installed components should be inspected at shorter intervals. This precaution applies quite specially to new brake linings, which—as is well known to every crane driver—are subject to relatively high wear during the initial running-in period. It will therefore be necessary to re-adjust the brakes frequently during the first period after the installation of new brake linings.

Figure 40. As a result of completely clogged oil-supply channels, the gearing for a luffing-mechanism limit switch ran dry for a long period. Because of the inability of the limit switch to function, the luffing-mechanism gearing and the spindle nut of the slewing crane were severely damaged.

Inspection of Equipment Rarely Required to Function

Precisely because safety and protective equipment (limit switches, devices for protection against overloads, safety devices for protection from storm effects and for preventing tilting over, broken-wheel supports, etc.) are seldom required to function. They merit special attention within the framework of an inspection. It is a completely wrong assumption to expect continuous serviceability from these devices on the grounds that they were new and free from defects when installed and since then have hardly been used. The effects of remaining idle, such as corrosion and gummed greases, can completely nullify the effectiveness of individual safety and protective devices. Care should also be taken to see that limit switches and overload protection equipment are appropriately secured to prevent them from being re-adjusted or bypassed by unauthorized personnel.

The more frequent failures concern emergency limit switches, which should therefore receive special care. Attention should be paid to the operational ease of all moving parts. Seals, particularly at cable, bolt, and shaft breakthroughs, must be checked for freedom from leakage, and seals that are not perfect must be replaced. Moisture in the interior of limit switches points to leaks and entails the risk of corrosion. The switches, especially their contacts, must always be kept clean. Oxidized switch contacts can, in an emergency, be reconditioned with fine emery paper. All screws must be firmly tightened. When closing up switch boxes, care should be taken to ensure that all cover screws are tightened securely and evenly.

An example of the effects of omitting to inspect a limit switch is shown in Figure 40. This example is concerned with the gearing for a luffing-mechanism limit switch, which was rendered inoperative because the gearing ran dry for a long period as a result of complete clogging of the oil-supply channels. Then, when the luffing mechanism was inadvertently run into its end position, its drive motor only cut out via the motor-protection device. Because of the failure of this limit switch, the luffing mechanism and spindle nut of the slewing crane were badly damaged.

Checking of Lubricating-Oil and Hydraulic Lines

Unfortunately, clogged lubricating-oil and hydraulic lines give rise to extensive damage. All pipelines must therefore be checked for freedom from obstructions on the occasion of major revisions. A check of all pipe connections is also strongly recommended.

Vibration-Free Attachment of Pipelines

Pipelines which tend to vibrate during operation must be secured carefully by additional clamping devices. These devices should not be welded onto load-bearing structural components but should, in case of need, be clipped or adhesive-bonded in place.

One incident of damage concerns breakage of a screwed connection of the hydraulic system for jib adjustment as a result of a vibrating pipeline. This breakage caused the pressure in one half of the system to drop suddenly, and the pantographic guide mechanism very rapidly moved to its end position. After repairing the fault in the hydraulic system, the crane was again put into operation but, within a short time, the jib could no longer be actuated: The cross-beam of the connecting guide for the transmission of the hydraulic actuating forces to the guide system fractured and buckled, while the lateral members of the system became severely deformed.

Figure 41 shows a section from the fractured area of the cross-beam near the guide pivot for the transmission of hydraulic forces. Incipient corrosion and the partly undeformed fracture surface suggested that the fracture must have started some time before buckling of the cross-beam, probably when the hydraulic system failed.

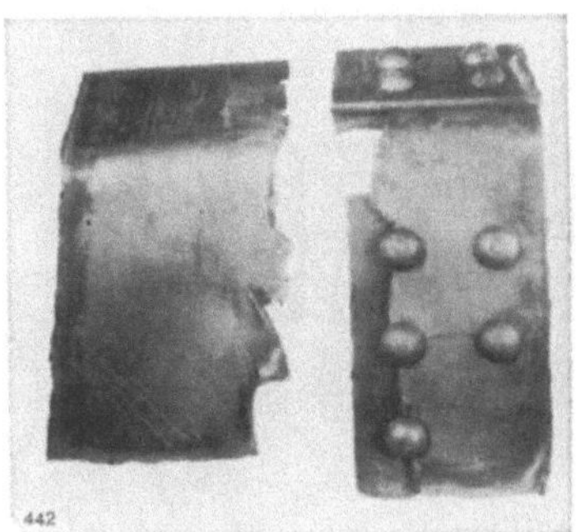

Figure 41. Associated fragments from the cross-beam of the connecting guide of a pantographic guide mechanism. The initial crack emanated from the rivet hole nearest to the plate edge (top) and, during normal operation, subsequently developed into a fracture and finally buckling of the cross-beam.

Non-Destructive Testing of Highly Stressed Components, Particularly after Exposure to Unusual Shock Loads

The damage described above indicates that, apart from the need for vibration-free attachment of pipelines, components of the installation which have been overstressed in one way or another by shock loads, as in the present case, must be subjected to a careful examination. This requirement applies, above all, to shock loads after breakage of ropes; after abnormal dropping or crashing down of heavy loads; when the crane strikes against a track end-stop, perhaps driven along by a storm; when slewing or travel brakes suddenly lock, e.g., as a result of a sudden electricity failure at full slewing speed; etc. (see also Figure 51).

If, following the damage to the hydraulic line described above, a conscientious examination had been made of the guide system, particularly the cross-beam, it would certainly have made it possible to detect the first incipient crack, and the resulting severe damage would have been prevented.

Furthermore, it is recommended that non-destructive tests should be carried out not only on structural steelwork, but also on other highly stressed components, such as slewing-gear shafts, king-pins, winches, travel-mechanism drives, etc., during long periods out of operation or on occasions of inspections. With structural steelwork, the points of force application, joints, nodal points and welding seams primarily merit attention.

To supplement visual inspections, dye-penetrant and magnetic-particle crack-detection equipment, ultrasonic testing equipment, and, for special cases, radiographic equipment are available for non-destructive testing. When carrying out such examinations, it has quite often happened that an incipient crack at a critical location has been discovered and a major mishap averted. In addition to the relevant institutes, many manufacturers also have trained test teams available. The test programme should be discussed by the user, the manufacturer, and the test institute and appropriately prepared in advance to include all essential structural components and to facilitate its rapid execution.

Figure 42. The fracture pattern of a diagonal strut from the truss construction of a loading bridge caused by fatigue failure which resulted from an initial crack in the flange. This crack would have been revealed by a non-destructive test.

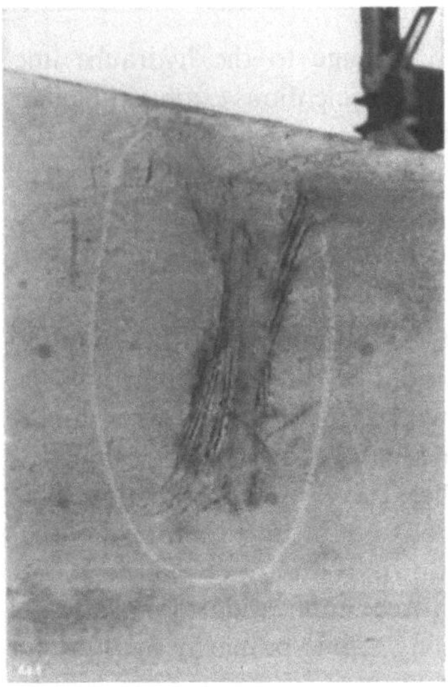

Figure 43. The incomplete nature of this paint coating on structural steelwork is clear evidence of previous overloading.

According to the fracture pattern shown in Figure 42, which refers to a diagonal strut from the riveted structural steelwork of a loading bridge, a fatigue failure developed over a long period of time from an initial crack in the flange. Under the stresses exerted during operation, a sudden complete separation of both flanges resulted. Decisive for the position of fracture was the notch effect of two rivet holes. A non-destructive test, carried out in good time, would in this case certainly have revealed the impending damage. During inspections, special attention should be paid to "bleeding" rivets, loose screws and chipped or swollen paint coatings (Figure 43). Such manifestations always indicate existing defects or previous overloading.

Critical Points of Portal-Type Gantries

In slewing cranes with portal-type gantries, the origins of incipient cracks are, among others, between the supports (legs) and the upper cross-beam. In particular, those cranes whose supports have been made from cold-worked sections, i.e., sections manufactured by cold deformation from flat material by rolling or press-brake methods, often exhibit cracks at these critical points or next to the welding seam. This observation applies especially to supports of half-open cold-formed sections (Figure 44), where the origins of the cracks must be sought in the region of the fold radii.

The cause of the cracks lies in the cold deformation of the plates. Thus, stretching of the material through folding gives rise to recrystallization of the structure, whereby notch toughness is substantially reduced by coarse-grain formation. Moreover, further embrittlement (artificial ageing) takes place in the heat-affected zones of the welding seam. In these zones, the material loses its capability of being plastically deformed and can no longer even out the stress peaks which are bound to occur in operation.

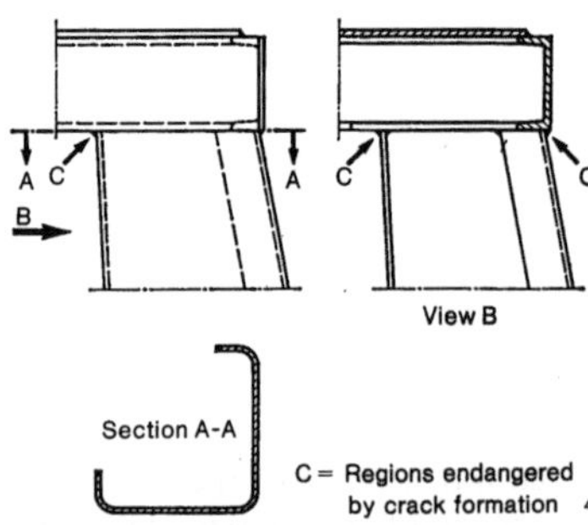

Figure 44. Regions endangered by crack formation between the supports (legs) made of half-open cold-formed sections and the upper cross-beam of slewing cranes with portal-type gantries.

In this instance, the cause of the cracks can no longer be eliminated, because fundamentally the material should have been subjected to a suitable heat treatment prior to welding. It is sometimes possible to remedy the trouble by providing stress-relief openings in the region of the fold radii, in which case the flame-cut edges must be carefully ground to make them notch-free. In these circumstances, it is strongly recommended to seek the manufacturer's advice.

Corrosion Damage

Great importance attaches to corrosion and to the damage it can cause to ferrous structures. Particularly at risk are nodal points and lap joints.

It is highly desirable to carry out inspections from time to time, especially with older cranes or loading bridges, and to check the structural steelwork for signs of bleeding and sub-surface corrosion, particularly in the region of rivet joints, so as to be able to take remedial measures promptly.

Rainwater that has penetrated into the joint between two riveted components can, in conjunction with ventilation conditions at these sites, lead in some circumstances to crevice corrosion. As a rule, under the influence of mechanical alternating stress, the result is fracture of the component in accordance with the mechanism of corrosion fatigue.

Figure 45 shows a section of a broken gusset-plate from the cantilever lattice-work of the boom of a slewing crane. Because of an inadequate protective paint coating, fracture occurred as a result of corrosion fatigue through the simultaneous effects of crevice corrosion and mechanical stressing.

The angle-iron (Figure 46) from the lattice-work boom of a slewing crane was ruptured by overloading in tension; weakening of the profile as a result of severe corrosion was also an influence. Rotting away of the paint coat in this manner should never be allowed to occur.

With regard to the protection of structural steelwork against corrosion, reference should be made to the guidelines given in Section 6.1.8.

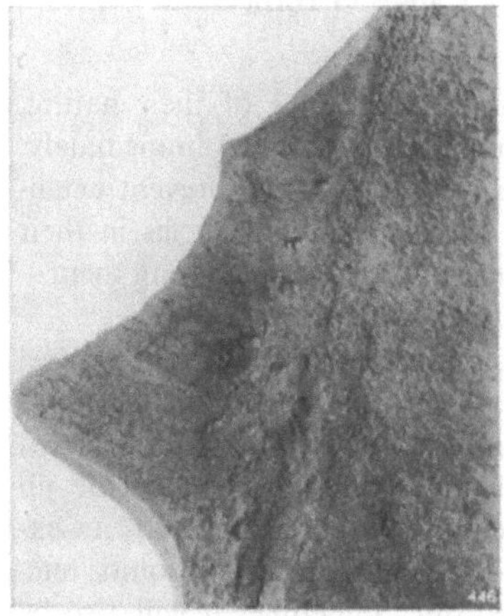

Figure 45. Section of a gusset-plate fractured by the simultaneous effects of crevice corrosion and mechanical stressing, in accordance with the mechanism of corrosion fatigue. From the corrosion groove it is possible to perceive the outline of a component which had once been riveted to the plate. The breakage of the gusset-plate was initiated at a source of corrosion above the illustration. In the lower part of the picture, a further incipient crack can clearly be seen inside the corrosion groove.

Figure 46. This angle-iron ruptured as a result of overloading in tension, the rivet hole being decisive for the location of the fracture. Plastic deformations preceding the fracture are clearly visible. The substantial weakening of the cross-section caused by severe corrosion undoubtedly contributed to the premature failure of the component.

Need to Establish Causes of Difficulties Immediately

If difficulties arise, irrespective of their nature, their causes should be investigated immediately. In this way, it is often possible to prevent extensive damage by discovering deficiencies in their early stages of development and rectifying them – usually at relatively little cost.
Thus, for example, with derailments of the travel carriages (especially if repeated), with increased flange friction of the travel wheels, with local differences in the ease of movement of the jib adjustment system or of the slewing gear, an extensive check of the relevant axles, joints, etc. should be carried out. Sometimes it will also be necessary to measure accurately their geometrical position within the system. As a result of even relatively weak impacts, such as may occur when running against buffers and to which no significance is often attached by operating personnel, it is possible, in some circumstances, for small, barely visible displacements or deformations to take place, and these can then manifest themselves by increased wear, derailments, etc.

Inspection of Crane Installations before Taking into Operation for the First Time and after Constructional Changes

Finally, in connection with the question of inspections, special attention is drawn to the fact that, according to the regulations for the prevention of accidents, it is necessary, apart from checks repeated from time to time, for an inspection to be carried out by an expert recognized by the employers' liability insurance company not only before taking a crane installation into operation for the first time, but also after every design or constructional modification.
Modifications subject to obligatory tests are specifically.

- changes in the load capacity of a crane installation,
- extension of booms beyond the hitherto permissible length,
- changes in transmissions,
- transfer of rail-bound crane installations to other crane tracks,
- lengthening of existing crane tracks.

In cases of doubt, the employers' liability insurance company must be consulted.

6.1.8. Repair of Crane Installations

Importance of Taking the Causes of Damage into Account

When repairing a crane installation, it is necessary to be aware of the fact that lifting equipment is a dynamically highly stressed structure. The safety requirement for personnel and equipment necessitate that the damage be rectified in the best possible way, i.e., not merely to restore the original condition, but also to prevent repetition of the damage. A repair is unsatisfactory if its effect on an installation cannot be predicted with certainty and, above all, if the causes of the previous damage have not been taken into account.
The high alternating stressing of the steel structure and the mechanical equipment, the age of the installation, and the relevant regulations are factors which set the standards for repairs. In principle, a distinction must be made between actions which must be performed without permission and repairs in the spirit of the regulations for accident prevention. In this regard, recourse should be had to the regulations mentioned at the end of Section 6.1.7 concerning inspections after design or constructional modifications to crane installations.
VDI guidelines 2382 for the repair of crane installations [10] deal in great detail with the influence of customary repair methods on crane installations. These guidelines classify repair work in terms of whether it is permissible or not.

Repair Work for which Permission must be Obtained

Permission must be obtained for drilling, welding and flame-cutting work of all kinds on load-bearing structures.
Among others considered to be load-bearing structures are crane-track girders, crane-track supports, crane girders, end girders, trolley frames, cab suspensions, gantries or portals, jibs, etc.
With regard to the obligation of obtaining permission for drilling work on load-bearing structures, it must be mentioned that every hole results in stress accumulations, i.e., a notch effect, in the adjacent material. A decision as to whether holes

are permissible must therefore be left to a specialist.

Welds or flame-cuts—however minor they may be—always cause a change in the structure of the material and invariably lead to additional stresses, as a result of which the fatigue strength of the material is reduced. In some circumstances, cracks are initiated which can eventually bring about fracture of the component concerned.

As a matter of principle, therefore, no tack welds must be performed on load-bearing structures, even if they are only for the temporary attachment of some kind of auxiliary equipment. Similarly, the welding-on of clamps and the like to load-bearing structures, e.g., for the attachment of pipelines or cables, is not permissible; for such purposes, clamps or adhesive-bonded fastenings are available.

The practice of using constructional components in the neighbourhood of a welding location to pre-strike arc-welding electrodes is strictly prohibited. It should be borne in mind that an arc—even of short duration—always causes a change in the structure of the material, and this is undesirable in any circumstances. An arc-striking plate should therefore always form part of a welder's equipment.

Proximity of Earth Connections to the Welding Location

Earthing cables should be brought as close as possible to the welding location. In all cases, care must be taken to prevent reverse current from flowing away via ball bearings, ball-bearing slewing rings, plain bearings, travel wheels, wire ropes, etc. Otherwise, current transfer within the bearings, from wire ropes to pulley sheaves, and so on will create burnt spots which, sooner or later, will inevitably lead to destruction of the components concerned.

Checking of Materials for Suitability for Welding

As in the above-mentioned VDI guidelines, it must also be stressed here that, prior to carrying out a repair by welding, the materials to be welded must undergo a metallurgical investigation, unless it is quite clear that they are suitable for welding and are not aged.

It is often necessary in the course of repairs to remove and replace large and heavy parts of an installation. Although, in general, suspension of the load from a hook or in a sling presents no difficulties, there are also cases where special transport lugs must be welded on. However, some bad experiences suggest the need for caution. Thus, not every material is suitable for transport lugs, and certainly use should never be made of random odd pieces of waste material for the purpose. Actually, the workshop concerned should be provided not only with the dimensions of the transport lugs required, but also with an indication of the material from which they should be made. A condition, of course, is that the component to be lifted is also made of a material suitable for welding.

The problems associated with welding cold-worked profiles have already been indicated in Section 6.1.7 in connection with the appearance of cracks on cold-deformed portal-type gantries.

Requirement for Expert Welding of Load-Bearing Structures

Numerous errors are also made in the execution of the weldments themselves. With incorrect electrode guidance, with damaged or damp electrode coatings, or with inadequate seam preparation, satisfactory results cannot be obtained. The execution of perfect weldments, the selection of filler metal, the treatment of welding electrodes, the preparation of welding seams, welding at low ambient temperatures and many other factors demand a wealth of experience which the user of crane installations does not generally possess.

The welding equipment available in the crane user's plant does not itself ensure the execution of high-quality welded connections. Even the best fitter occasionally entrusted with welding work does not possess the necessary expertise. Welding work on load-bearing structures should therefore always be entrusted to the manufacturer of the installation concerned, as he has at his disposal the appropriately qualified personnel.

Figure 47 shows a fractured component from the jib support of a slewing crane. The failure of the component was partly due to its material, which was unsuitable for welding because of its high sulphur content, and partly because

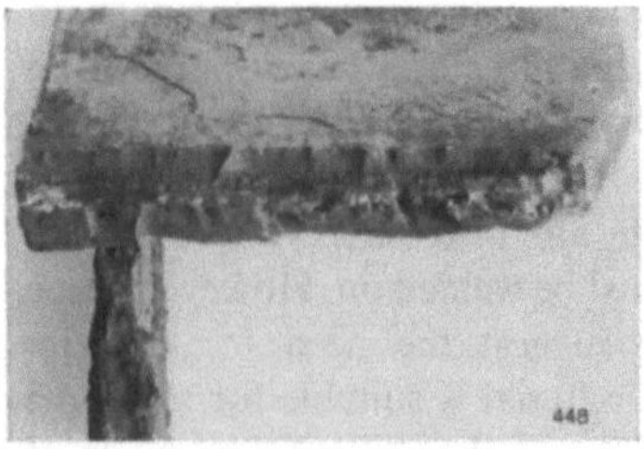

Figure 47. Fracture of this component from a slewing crane was due partly to its plate material, which was unsuitable for welding because of its high sulphur content, and partly to an unsatisfactory repair weldment in the region of strong segregation. As a result, the jib of the crane crashed down and was completely destroyed.

of the unfavourable position of the weld, which was badly executed during the course of a repair in the region of strong segregation. As a result of the fracture of this component, the jib crashed down and was completely destroyed.
The older installations are mainly riveted structures. Since expert riveaters are not always available (Figures 48 and 49), the obvious solution is to use high-strength bolts for repair work. In such a case, however, the greatest care must be taken. The surfaces of the overlapping components are often scarred as a result of corrosion, in which event the preliminary conditions for a high-strength screwed connection are no longer obtained. The use of fitted bolts, instead of high-strength bolts, is then more suitable. Here, of

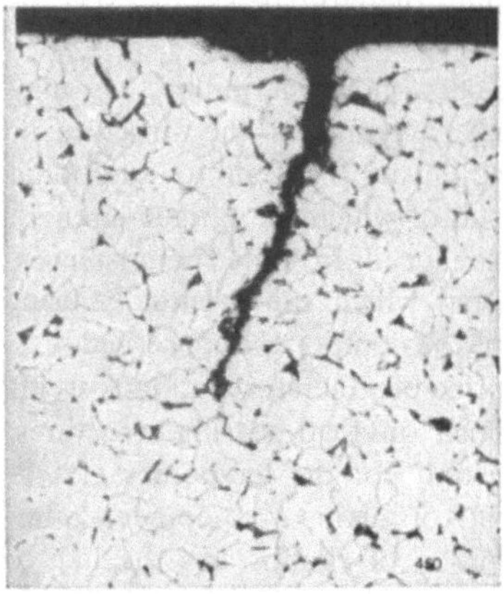

Figure 49. Photomicrograph of a polished section from the region of another rivet hole in the plate shown in Figure 48. Clearly visible, though relatively minor, is the surface damage from which the crack emanates. The crack is filled with corrosion products, indicating that it has been in existence for some time.

course, the hole must be correspondingly reamed out.

Expert Installation of High-Strength Screwed Connections

The necessary conditions for a high-strength screwed connection can be fulfilled only with completely new joints. Even so, it must be stressed that these connections require special measures to be taken and should be made only by firms that can satisfy the necessary conditions of expertise and personnel.

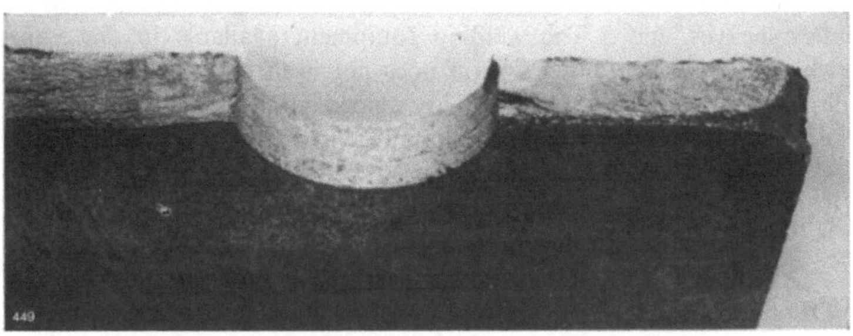

Figure 48. The fracture of this plate section from the jib of a slewing crane emanated from an old incipient crack at the edge of the rivet hole (right top), characterized by a strongly aged layer of rust. It appears that the crack later developed into a long-term or fatigue fracture as a result of a sudden overload. A closer examination revealed that the first crack originated from damage to the rivet-hole edge presumably sustained during riveting (see Figure 49).

Contact Surfaces of Components to be Joined

Before joining, the contact surfaces of components to be riveted together must be cleaned carefully to remove rust and must then be provided with a corrosion-inhibiting coating. This latter treatment does not, however, apply to high-strength screwed connections.
Thus, whilst the forces in riveted connections are transmitted directly by the rivets themselves, which are stressed in shear, high-strength bolts serve merely to press together the components to be joined, force transmission being effected at right-angles to the bolts solely by the friction of the components pressed against each other. The friction areas of high-strength screwed connections must therefore be cleaned meticulously before installation to ensure that they are free not only from rust, but also from grease, paint and dirt.

Sealing of Joints between Screwed or Riveted Components

For both screwed and riveted connections, a strong recommendation is proper sealing of the joints between components and also of all gaps between components and bolt or screw heads immediately after installation, using a suitable plastic material or a viscous primer as a protection against the ingress of moisture.

Corrosion Protection and Painting of Structural Steelwork

Protection of a repaired steel structure from corrosion requires great care. Attention should be paid to the complete removal of rust from old parts of the installation, since the corrosion-inhibiting coating does not adhere to surfaces that are incompletely derusted, as rusting below the coating will lift it up again after a short time. The best conditions for lasting corrosion protection are offered by a sand-blasted surface. However, if the surface to be derusted is within narrowly defined limits, manual removal of rust is usually preferred for economic reasons, i.e., by a hammer and wire brush, or by means of a mechanical rapper. The quality of the work is decisive for the quality of corrosion protection. A single primer coat is inadequate, as it does not cover the peaks of the cleaned material. For

Figure 50. Part of a U-section from the upper chord of a loading bridge. Traces of the simultaneous effects of crevice corrosion and mechanical stressing are clearly discernible (see also Figure 45).

good corrosion protection, a second primer coat must therefore always be provided.
Figure 50 shows a U-section, flame-cut lengthwise, from the upper chord of a loading bridge. A further component was riveted onto the web visible on top. The original paint coating on the upper chord had largely distintegrated and was discernible only in a few places. Moisture could therefore penetrate into the gap between adjacent components and, under the influence of this moisture, coupled with mechanical stressing in operation, cracks developed as a result of corrosion fatigue, which finally led to fracture of the upper chord of the loading bridge. With regard to the corrosion protection of steel structures, recourse should be made to the publications in references [11–13], as well as to DIN 55928: Protective Coating of Steel Structures.

Precautionary Replacement of Components Subject to Wear in the Case of Damage

In connection with the repair of damaged gear transmissions, it is necessary to consider whether the damaged component alone should be replaced or whether, at the same time, other components subject to wear, such as bearings and individual transmission stages, should also be replaced. In some cases, undoubtedly, further operating life can be predicted for one or the other component, but in some circumstances, precautionary replacement measures still offer economic advantages. Thus, the life of the transmission as a whole is increased, further damage by progressive wear need not be anticipated in the foreseeable future and the additional costs incurred in the event

Figure 51. The jib of this crane buckled during a gust of wind as a result of undetected previous damage to the luffing gear.

of damage (e.g., opening up the transmission, removal of wire ropes and of the overload protection device and similar equipment) do not, so far as it is possible to judge, have to be faced again in the near future.

Installation and Removal of Rolling Bearings

Considerable experience is required for proper handling of rolling bearings. Thus, when installing or removing them, errors are often made which, although hardly noticeable at first, result in damage to the bearing. The most favourable method of assembling or dismantling bearings entails the use of appropriate pressing and pulling equipment. If need arises, small bearings can also be fitted with the aid of a clean, squarely cut-off piece of tubing by careful hammering (always utilize a rubber hammer or an intermediate piece of hardwood). In all cases, the greatest care must be taken to ensure that the forces applied act directly only on the race to be pressed in or pulled out. For example, for fitting the outer race in its seat in the housing, the piece of tube employed should have a diameter corresponding to that of the outer race, whereas its diameter should be appropriately smaller for fitting the inner race on the shaft. In no circumstances must fitting of the outer race be effected by the action of forces on the inner race—and *vice versa;* otherwise, damage to the races and rolling elements and premature failure of the bearing will be the inevitable consequence.

Care should also be taken to ensure that the pressing force acts centrally, so that the rolling bearing is not tilted in its seating.

It is often possible to produce one's own simple and effective fitting aids at low cost. If, for example, the shaft is provided with a thread, the inner race of a rolling bearing can easily be pushed onto its seat on the shaft with the aid of the shaft nut and a suitably cut piece of tubing.

The drawing of a rolling bearing onto its seat on the shaft can sometimes be greatly facilitated by prior heating of the bearing in an oil bath. Excessive force must never be applied.

Deficient Press Fits

Press-fit seats on shafts and housings should be clean. Damage must be avoided in all cases. If necessary, the places concerned must be carefully smoothed.

Bearing races subjected to loading on their periphery sometimes tend to become displaced and hence require a reliable press fit. If the seating becomes loose and the ring thereby begins to wander, the rapid and complete destruction of the press fit usually coincides. In the initial stage, such a deficiency can usually be remedied by adhesive bonding or other means. With somewhat looser press fits, a repair is sometimes still possible by spray-coating. In no circumstances must loose press fits be "repaired" by roughening, e.g., by punch-marking the shaft or housing seat.

Shaft and Housing Seals

When assembling bearings and gear transmissions, care should be taken to ensure that shaft and housing seals are perfect. Damaged seals must be repaired or replaced.

Tab Washers

As a matter of principle, tab washers for screws should be used only once, since repeated bending of their lugs in one direction and another leads to fracture and hence makes the locking device ineffective.

Oil and Grease Fillings

Charging of bearings and gear units with the oil or grease prescribed by the manufacturer must not be forgotten. With grease-lubricated rolling bearings, the free space in the housing should be from a half to two-thirds filled with grease. With regard to the lubricants used, the manufacturer's specifications should be observed.

Repairs to Ball-Bearing Slewing Rings

In many cases it is possible to repair damaged ball-bearing slewing rings. If the need arises, the races can be re-turned, hardened and ground by the manufacturer concerned. Provided with new rolling elements, these ball-bearing slewing rings will be fully serviceable again.

Matching of Driving Travel-Wheels

In the event or repair, the drive wheels of the travel-mechanism should be matched. Differences in outside diameter result in differences in peripheral speeds, resulting in skewed running, heavy flange friction and increased risk of derailment of the crane installation. In addition, the drives become overloaded.

Deposition Welding on Wheel Tyres

Worn wheel tyres are often restored to their nominal dimensions by deposition welding. The question here arises as to whether the tyre should first be removed from or left on the body of the wheel.

In the former case, it should be noted that the shrunk-on tyre, after removal and weld deposition and before shrinking-on again, must be stress-annealed. However, on no account must the fit be re-machined.

After deposition welding, it will be found that the original diameter of the bore of the tyre has become smaller, due to shrinkage stresses generated during welding. If the bore is re-machined, loosening will later take place in the shrink fit because the shrinkage stresses induced by the welding operation will be removed again to a certain extent by stress-annealing or by the heating required for shrink-fitting. With operational stresses, a tyre with a re-machined bore can become completely loose from its seat.

With the alternative method of deposition welding with the tyre in the drawn-on condition, it would seem reasonable to conclude that, in these circumstances, a firm shrink fit will be obtained, as the weld-deposited material constitutes, as it were, a second and at the same time hot-applied tyre. In practice, however, matters are not quite so simple. According to tests by Hofes [14], deposition welding on tyres left on the body of the wheel does, in fact, impose at least some risks—a finding that has been repeatedly confirmed by cases of damage caused by loosening of wheel tyres repaired in this way.

Accordingly, it is recommended that deposition welding should be effected with the tyre removed from its shrink fit. However, as previously indicated, it is essential in this case that the tyre is stress-annealed after welding and that the bore remains unaltered, i.e., that it is not re-machined.

Storage of Spare Parts

A store for spare parts must also be planned The extent of the required stocks of spare parts cannot, however, be stated in general terms, as it depends on a number of factors. Among these are conditions of operation, the type of crane installation, its age, condition and susceptibility to damage of its equipment, and many other considerations.

With types of crane that are built in large numbers, e.g., tower cranes, the user will often be able to confine himself to a relatively small selection of spare parts, particularly in cases where the crane manufacturer concerned keeps a stock of spares within a reasonable distance. In all cases, however, even if the manufacturer resides in the same location, at minimum such important wearing parts as a complete set of brake linings, appropriate carbon brushes, etc., as well as a set of spare wire ropes, should always be kept in the user's stores, because there may be a shortage at the manufacturer's premises which would temporarily lead to delivery difficulties.

It is recommended that an agreement should be reached with the manufacturer regarding the user's own stock of spare parts. The basis for making repeat orders at the proper time should

be the records kept of deficiencies established and repairs carried out. Furthermore, after every issue of spare parts from stores, repeat orders of the parts concerned should be made immediately.
It should, however, be noted that it is not enough merely to keep an adequate stock of spare parts. Thus, the parts must also be stored in an orderly manner and be accessible at all times, as well as protected from damage, moisture and dirt. Above all, bright components must be protected from corrosion and, if necessary, must be greased.

Special care is required for rolling bearings. Generally, however, rolling bearings are supplied by the manufacturer in packings that protect them from corrosion and the ingress of moisture; for this reason, they should always be stored in their original packings. Even so, these bearings should not be subjected to storage conditions involving high humidity and temperature fluctuations and should be kept in a place which is free from dust, dampness and (particularly important) vibrations.

6.2. Tower Cranes

Despite its special external features, the tower crane, by virtue of its basic construction and principle of operation, belongs to the group of slewing cranes and, in the main, its equipment resembles that of other slewing cranes. Accordingly, the same possibilities present themselves for errors. Indeed, during manufacture and operation and during maintenance and repair, the same, usually elementary technological rules are frequently thoughtlessly ignored.

For this reason, reference should be made to Section 6.1, in which guidelines and recommendations given for loss prevention of slewing cranes are also basically valid for tower cranes. However, damage to tower cranes which must be regarded as typical for this type often occurs and therefore merits some special remarks.

The reason for these cases of damage ultimately lies in the fact that tower cranes are often moved from one building site to another. The necessary dismantling and re-erection of these cranes and constantly repeated adaptation to the local conditions of a building site, as well as different modes and methods of construction, require particular care and expertise on the part of the personnel concerned. Unfortunately, however, these qualities are often lacking.

Finally, attention must be drawn to the fact that tower cranes, which are made as lightweight structures because of the possibility of a quick change of site, are particularly vulnerable to rough operation.

Practical experience of damage cases demonstrates how absolutely essential are such guidelines; in fact, on an average, one in every seven tower cranes is subjected to damage each year.

6.2.1. Guidelines for the Planning of Tower Cranes

By no means inconsiderable is the number of damage cases which simply have their origin in the incorrect selection of a crane for a particular site, either when purchasing a crane or when selecting one out of an existing fleet of cranes from a contractor's yard.

The basis for selection of a tower crane must be the structure to be erected. Its dimensions in height above ground level and depth below ground level yield the basic measurements affecting the type of crane to be selected. Naturally, it is not possible in practice to purchase or hire a special type of crane for every structure. However, every building company should have an idea of its main fields of activities and their requirements.

Problems Concerning Tower Cranes without a Driver's Cabin

When erecting small buildings, such as one- or two-family houses, use is often made of small, low-cost tower cranes without a driver's cabin. When building a row of houses, however, it has been shown that such cranes, which are merely equipped with remote control, can be used "rationally" only if the crane driver somehow finds himself a makeshift place on the tower, so as to avoid having to follow the movement of the crane by running after it on the scaffolding, since observation of the building site from the lower control stand is no longer possible.

Figures 52 and 53 indicate clearly that the selection of this type of crane was incorrect from the point of view of the structures concerned. Such improvised "control platforms" as those shown do not satisfy the requirements that must be met for a control stand. Thus, apart from the accident risk for the driver, his freedom of movement is certainly restricted by such arrangements. As a matter of course, he should (or at least it is hoped that he should) be greatly concerned in such a case for his own immediate safety. As a result, his attention is divided and the safety of man, machine and building is endangered, particularly if the well-being of the crane driver is impaired by unfavourable climatic influences.

The later addition of a driver's cabin has often led to complications during erection and dismantling of the equipment and is sometimes not even permissible from the point of view of statics. Such reconstructions or additions should, if they are necessary, therefore be left to the crane manufac-

Figure 52. This crane without a driver's cabin is evidently inadequate for the size of the building. To get a better view of the building site, the crane driver is standing on a jammed-in piece of board and, if his attention wanders for a single moment and if he should accidentally trip, a serious accident and extensive damage could result.

Figure 53. This crane was also not matched to the size of the projected building. Improvised control platforms such as that shown do not conform to the regulations for accident prevention and certainly do not contribute to loss prevention.

turer. Actually, some manufacturers have recognized the problems associated with tower cranes without a driver's cabin and have stopped producing them.

Determination of Crane Size

When selecting a crane, the same attention should be paid to both the depth of a building below and its height above ground level, and consideration should be given to the load capacity of the crane necessary for the various areas of the building site during all stages of erection, taking into account stone packets, skips for concrete and prefabricated components. It is absolutely essential to have, prior to selection of the crane, a really clear idea of the method of building the structure to be erected. For example, if the decision is suddenly made at a later time to utilize heavy precast concrete components, instead of concrete slabs cast on site, it should not then be discovered that the capacity of the crane available is insufficient for the load involved. As a result, recourse may then be had to more or less questionable improvisations which, virtually without exception, will lead to very severe damage, sometimes even to toppling over of the crane, and, invariably, damage to the building.

Special Lifting Equipment for Specific Tasks

The current widespread use of large-area formwork or shuttering for concrete or of subassemblies has imposed the need for special lifting equipment for the crane. For instance, heavy subassemblies with eccentric centres of gravity must be supported by functionally correct lifting equipment (Figure 54), so that they can later be put down safely and without damage. In such cases, moreover, it is recommended that the crane rope be double-reeved to permit "creep" lowering, millimetre by millimetre, when locating shock-sensitive subassemblies.

Observance of Scaffolding Regulations

When selecting a tower crane in respect to its reach, account should also be taken of the necessary scaffolding and supporting structures of the building, as well as of the prescribed protective scaffolding. The radius of every tower crane which stands outside the building and travels alongside it must therefore be larger than the

Figure 54. Example of special lifting tackle for centre-of-gravity compensation, with an intermediately suspended tackle block for precision lowering.

longest transport path from the edge of the building by at least the depth or width of the scaffolding plus the prescribed safety distance of 50 cm between the scaffolding and the furthest protruding parts of the tower or its travel carriage (Figure 55). In this connection, care should also be taken that winch ropes cannot be caught in protruding parts of the scaffolding (Figure 56).

An example of how things should not be arranged is given in Figure 57; a collision between the crane and scaffolding, with all its unforeseeable consequences, is only a matter of time.

In this context, special attention is drawn to the scaffolding regulations in DIN 4420 and to UVV scaffolding.

Under restricted space conditions, above all with buildings along streets, where the erection of a large crane running on rails meets with difficulties, the use of "climbing" cranes can be advantageous.

A detailed discussion of the guidelines given above, as well as further points of view concerning the accident-prevention regulations which must be observed for various building projects, can be found in a publication by H. Göttling [15].

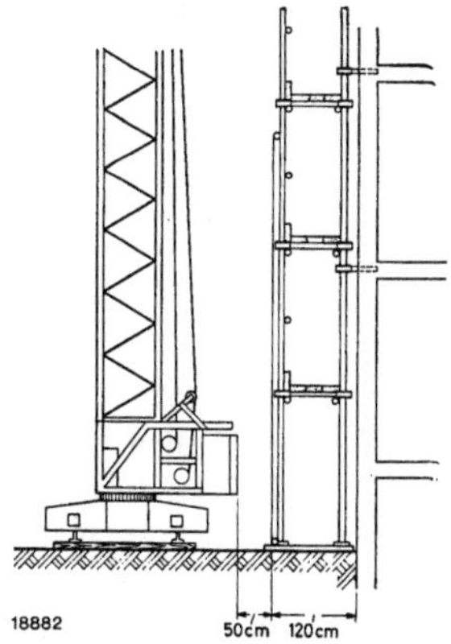

Figure 55. Prescribed minimum distances between the crane and scaffolding or between the crane and building.

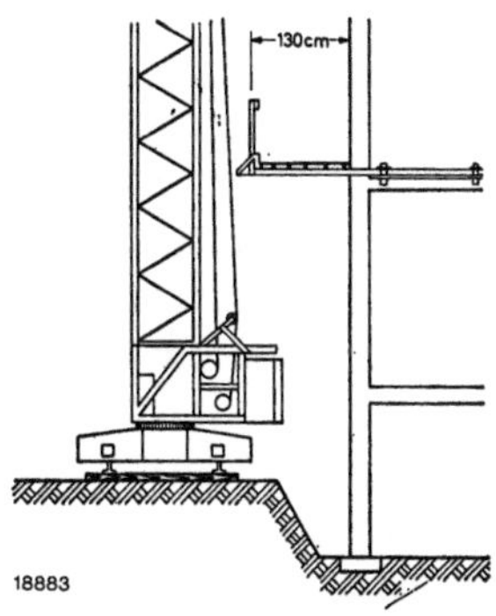

Figure 56. When using overhanging scaffolding, there is a possibility of bringing the crane closer to the building and achieving better utilization of its reach. However, care should be taken to ensure that there still remains a sufficient distance between the wire ropes of the crane and the scaffolding.

Figure 57. Here, evidently, planning mistakes were made. The arrangement of the scaffold poles within the slewing range of the crane is contrary to all regulations and spells danger.

6.2.2. Erection, Commissioning and Dismantling of Tower Cranes

The great majority of all tower cranes are employed on building sites, where they are used for limited periods and must be dismantled and re-erected. If a crane is subjected to this procedure consecutively during the erection of several small buildings, perhaps even several times a year, it should, however, not be assumed that it forms part of the everyday duties of the building-site personnel, from whom it is therefore unreasonable to expect extensive experience in the erection and dismantling of a tower crane. This lack of experience, coupled with disregard of assembly instructions and often carelessness are frequent causes of severe damage during the already inherently dangerous erection and dismantling procedures.

Unreliable, insufficiently trained or inexperienced personnel should never be entrusted with the erection and dismantling of tower cranes. A tower-crane driver, who perhaps has had a certain amount of experience in the operation of a crane, is for that reason unlikely to have the ability to erect or dismantle it expertly.

Erection and Dismantling of Cranes by the Manufacturer's Customer Service

Some firms with a large fleet of cranes employ their own specially trained and therefore experienced personnel for the erection and dismantling of their equipment. Unfortunately, this procedure is usually not possible for smaller building contractors, and it is recommended that they should avail themselves of the customer service of the manufacturer for carrying out erection and dismantling of their cranes. In the long run, this procedure is certainly more economical than the possibility of catastrophic damage and, with it, cessation of operations.

Discussion of Erection Instructions with Personnel

Because of the variety of equipment, it is not possible to give uniform guidelines for the erection of tower cranes. It is therefore stressed that the manufacturer's operating or erecting instructions must be referred to. If a crane is to be erected or dismantled by the user's own personnel, it is strongly recommended that the construction manager responsible should discuss these instructions thoroughly and repeatedly with the personnel entrusted with carrying out the work before every erection or dismantling operation. This procedure may appear somewhat exaggerated, but it should be appreciated that the inadvertent omission of a single bolt may result in serious injuries to or the death of a coworker.

Convenient Erection Instructions for Building-Site Personnel

Moreover, it is desirable that the erection instructions from the manufacturer should not only be reproduced in the operating instructions themselves, but should also be issued for use by personnel on the building site in the form of handy, durable and easily cleanable information sheets. On the other hand, it appears worth recommending the duplication of instructions in block form for use as check lists.

Two more recommendations should be mentioned. The first of these is concerned with the fact that it is stated as a final point in many erection instructions that dismantling of the tower crane should logically follow erection in reverse order. For this reason and for completeness it is recommended that special instructions should also be prepared for dismantling of the crane.

The second recommendation is to supplement the erection and dismantling instructions by simple and clear sketches.

Foundations and Securing of Track Installations

One of the prerequisites for trouble-free crane operation is a properly laid out track installation.

Track installations for tower cranes must be laid out in such a way that no load is imposed on the building excavation and that the slope of the building excavation cannot cave in. Building excavations must therefore be provided with a slope that depends on the type of soil.

As a starting point, a pressure-distribution angle of 30 deg. should be taken for loose or built-up soil and 45 deg. for undisturbed firm ground. The angles of slope themselves can be obtained from DIN 18300. The borders of the slope must

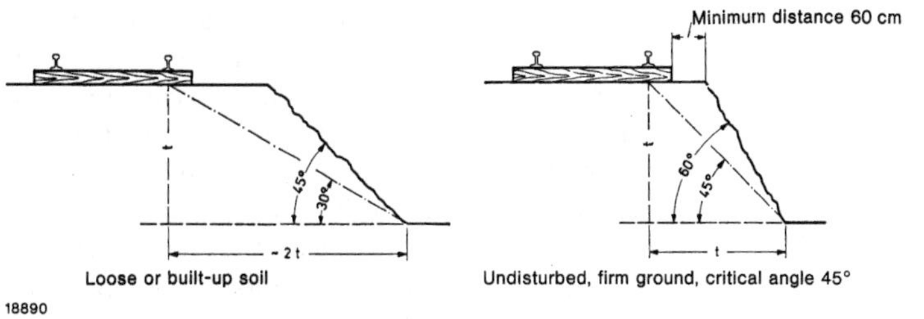

Figure 58. Angles of slope to be observed, according to DIN 18300, when erecting tower cranes at building excavations.

always be kept free from any kind of load for at least 60 cm, reckoned from the edge, as shown in Figure 58.

The stability of a tower crane can be guaranteed only if the track installation runs horizontally. Track settlements of only 1% (or even $1/2$% with cranes having the ballast high up) can make stability doubtful. Tracks sloping longitudinally (gradients) entail the risk of the crane running down, striking the end-stops and then toppling over.

The track installation is decisively dependent on the substructure. It is possible either to construct a provisional substructure, e.g., on wooden sleepers, and then have to put up with under-packing during operation or to select a safe measure, such as concrete foundations. This latter measure will initially involve higher expenditure, but offers a safe track installation during the whole of the building period. Concrete-strip foundations are particularly economical in cases where they can be left under the top soil, with which they must still be covered and do not have to be taken up again.

Sleepers can often be laid directly on load-resistant, undisturbed ground or on a firm street surface. Less load-resistant types of ground, e.g., banked-up soil, must first be compacted by stamping or similar means.

The requirement is therefore to push away the top soil, spread gravel or rubble, and compact. Also recommended are concrete foundations, either cased or flattened with a shovel. The sleepers are then exposed to the air, so that there is no risk of their rotting away. With gradients, strip foundations are usually necessary in any case. With longitudinal sleepers, tie-rods must be provided as a matter of principle. If these basic rules are disregarded, it is possible that the crane will topple over even during the erection stage (Figures 59 and 60).

Particularly critical is the situation when stationarily operated cranes, which attain their complete stability only after "spindling down", are pressed into the refilled area of the building excavation. Any giving way of the support for the spindle means immediate toppling over of the crane (Figure 61).

Figure 59. This crane toppled over because the slope of the building excavation was much too steep and the crane track was located too near the edge of the slope.

Figure 60. This crane toppled over because the underpacking of the track was deficient and not adequately filled in.

Figure 61. This tower crane toppled over because one of its support spindles sank into the refilled area of the building excavation.

If, for reasons of space, it is not possible to bank up the building excavation and, because of the reach of the crane, it must be brought as near as possible to the building, it is necessary to provide the periphery of the building excavation with steel-sheet piling or other reinforcement, which must also be able to carry the loads imposed by the crane.

In recent years, back-staying by so-called injection anchors has gained increasingly in importance, since this technique enables even very deep building excavations to be safeguarded by a series of anchors lying one on top of the other. In this way, the erection of even the heaviest tower cranes is possible at building excavations. In many cases, it is recommended that safeguarding the building excavation by means of this method should be carried out by a specialist firm.

If it is necessary to extend the crane track into the building excavation, the load-carrying structure must not only be calculated statically, but also be designed on a technological basis. Thus, the structure must be capable of accommodating, in addition to vertical forces, horizontal displacements arising from, for instance, braking and starting forces. Frequently, braking and starting forces are underestimated, and the consequences are noticed only when the whole structure is displaced and the crane lands in the building excavation. Figure 62 shows a representative example of a well-designed load-carrying structure.

Figure 62. Representative example of a wooden structure for the erection of a crane, set down on a strip foundation, in the region of the building-excavation edge.

Figure 63. Representative example of a reinforced-concrete substructure for the track of a tower crane used for building a sewage-treatment plant.

Special building projects also require special measures for the erection of a crane. Local conditions are so varied that it is impossible to lay down generally valid rules. A representative example of a well-designed substructure for the track of a tower crane for use in a special project is shown in Figure 63.

Construction of Tracks

The need to use suitable materials for rails and sleepers should be self-evident. Figures 64 and 65 give some examples of the sort of arrangements that should not be used.

Only identical rail profiles are an assurance of safe crane operation. If, because of a lack of co-ordination or for economic reasons, different profiles are used, damage to the travel wheels will certainly occur and possibly even toppling over of the crane. Here, then, economic considerations are unwarrented.

To avoid unwelcome surprises, the recommendations of the supplier should be followed in regard to the selection of crane rails. In general, only rails with a horizontal head are permissible. Worn-down rail heads are useless.

The length of sleepers depends on the track gauge and the special stipulations of the manufacturer in his operating instructions. As a rule of thumb, the sleepers should be 25% longer than the track gauge and, in addition, an excess length of at least 30 cm on both sides should be provided.

When laying tracks in a curve, the data of the crane manufacturer in regard to the reduction in track gauge must in all circumstances be observed.

The normal distance between sleepers is between 50 and 70 cm, measured from the centre of the sleepers, unless smaller distances are specified in accordance with operating instructions, particularly with large cranes. One sleeper each must be provided immediately before and after a rail joint.

The rails must be bolted to baseplates but, as a matter of principle, chamfered baseplates, such as those employed on the Federal railway system, should not be used; otherwise, the rail will be tilted and the travel wheels of the crane will make line contact only when rolling. The result will be severe wear or damage, as exemplified in Figure 66. Sleeper nails must not be used for fastening rails. Both travel rails must be earthed by means of a metal wire or strip having a cross-section of about 50 mm^2.

Figure 64. Safe crane operation on such weather-worn and broken sleepers cannot be assured. Reliable fastening of the rails to these sleepers has become quite impossible. Contrary to regulations, the rails have been nailed to the sleepers without baseplates and, at the sleeper in the foreground, the rail fastening is completely missing.

Figure 65. At this rail joint, two rails of different profiles abut and, instead of two sleepers, there is only one squared-up piece of timber. Moreover, the rails are not even bolted to this "sleeper", while fastening of the fishplates to the rails with only one bolt each is quite inadequate. Also remarkable is the wide gap in the rail joint. Under such conditions, crane operation should be suspended immediately or, better still, should never have been started.

Figure 66. This travel wheel was damaged by running on a crane track as deficient as that shown in Figure 65.

Safeguarding of Track Terminations

Robust end-stops or buffers should be attached firmly to both rail terminations ahead of the last sleeper (Figure 67), i.e., not on an overhanging piece of rail. These end-stops must not be offset with respect to each other, so as to ensure that the crane, if the need arises, will make contact with both stops at the same time. The necessary equipment for actuating the prescribed emergency travel-limit switches, such as ramp rails, etc., must be fixed a head of the rail ends so that the crane comes to a halt after slow-down before running against the end-stops or buffers. It is recommended that the length of the rails should be made several metres more on both sides than that corresponding to the required travel distance and also that, over a distance of about 1 to $1^1/_2$ metres from the end-stops, the top edges of the rails should be covered with gravel. The trip mechanism for the travel-limit switches should then be positioned so that, when they are actuated, the crane will come to a halt shortly before the layer of gravel.

Figure 68. This baulk of timber, intended to serve as an end-stop, in no way satisfies requirements. When overrunning the limit switch, the baulk offers no resistance to the crane. Also, the two wooden pegs rammed into the ground ahead of the last sleeper cannot in such a case prevent the crane from running off the rails. Furthermore, contrary to regulations, the rails have been nailed to the sleepers without baseplates.

Numerous cases of catastrophic damage have been caused by cranes running over unprotected rail ends and then toppling over because either no travel-limiting devices were installed at all or the last track section with the safety limiting system had not yet been installed or had already been removed, despite the fact that the crane had started or had not yet finished operating.

A completely inadequate track-termination arrangement is shown in Figure 68. Also, scaffolding chains are unsuitable for fastening end-stops based on baulks of timber or sleepers. Such "security devices" are all too easily pushed aside

Figure 67. Rail end-stops and limit switch installed in accordance with regulations.

by the crane if the emergency limit switches are inoperative or are missing—a by no means surprising possibility in view of such improper methods of safeguarding track terminations.

Erection and Dismantling of Tower Cranes

It is important to bear in mind that not only tower cranes from different manufacturers, but sometimes also different types of crane from the same manufacturer require basically different methods of erection and dismantling. It is therefore strongly recommended that recourse be made to the relevant instructions of the crane manufacturer.

Unfortunately, far too little attention is paid to the fact that tower cranes, in addition to being subjected to high loading during operation, must sustain extraordinarily high stresses during erection and dismantling.

The major proportion of all damage arising during the erection or dismantling of tower cranes is caused by mishandling when erecting or bringing down the supporting tower or jib. Incorrect rope guidance; latching devices that have not been inserted; omitted locking pins, irrespective of whether they have, contrary to assembly instructions, been removed too early or were never put in; unsupported jib ends; and various other omissions have repeatedly led to severe accidents and damage.

Even such a simple requirement as leading the jib-supporting rope over the jib mounting during raising or lowering the jib is all too often disregarded. The result in this case is that, when pulling up the jib, the rope tension, because of the missing leverage via the jib mounting, can increase to such an extent that the tower will finally buckle (Figure 69).

Figure 69. The corner struts of this tower crane buckled because the jib-supporting rope for pulling up the jib had not been led over the jib mounting.

Filling and Emptying of Ballast Containers

Filling of central and counterweight ballast containers is not always carried out at the right time and in the exactly prescribed amount. By the same token, removal of ballast is also sometimes not undertaken at the proper time.

In this connection, it must be pointed out that rusted-through or otherwise holed ballast containers are unsuitable for filling with gravel or sand as a ballast material. It has happened before that a perforated ballast container has emptied itself—an occurrence which, in one instance, was discovered only after the crane had overturned.

After Erection, Check First the Functioning of the Crane and Safety Equipment

After the crane has been erected, all crane movements and safety equipment must be checked with the greatest care for proper functioning before initiating operation proper of the crane. Attention must also be paid to ensuring that the directions of movement are identical when operating from the upper driver's cabin or the control cubicle.

Examples of Common Cases of Damage

To emphasize and supplement the recommendations given above, some random examples from the daily reports are given in the following of damage cases to cranes during erection and dismantling:

- After pulling up the upper section, the crane collapsed inwards, because locking arrangements between the upper and lower sections of the tower had not been carried out in accordance with regulations.

Figure 70. This tower crane overturned as a result of inexpert dismantling.

- When pulling up the jib, the tower buckled. The tip of the jib had not been disconnected from the tower.
- After erection of the tower, the intention was to install the jib. Because of insufficient lifting of the tip of the jib, the top of the tower was overloaded and buckled inwards.
- When connecting a crane to the electrical power supply, the terminals were reversed. As a result, when the driver attempted to lower the jib, it was instead pulled up over the top of the crane and was badly deformed.
- The middle section of a travel carriage distorted and cracked, and spacer struts deformed because of settling of the crane track as a result of the sandy soil.
- A tower crane collapsed as it was being dismantled, because the (excessive) ballast had not been removed.
- When letting down the upper section of a tower, the mast-withdrawal cable broke and the upper section slid down without restraint. (A later investigation revealed that the cable had evidently been placed incorrectly, so that a number of outer wires had been badly worn as a result of chafing against the steel structure.)
- Jamming occurred when drawing in a telescopic tower head. The erection cable was slackened, the upper section slid down and consequently the cable snapped. The tip of the jib, which had already been let down, penetrated into the ground, causing the crane to be pushed backwards and overturn (Figure 70).
- As a result of forgetfulness, the jib-supporting rope attached to the tip of a jib was not loosened and was not tied to the lower section of the mast. The jib-supporting rope therefore lacked the counter-pull necessary for tilting the mast and, when moving the crane, it suddenly crashed to the ground.
- During transport, the tip of a jib protruded above the driver's compartment of the transport vehicle and ran into the wall of a house when the vehicle turned into a side street.

6.2.3. Guidelines for the Operation of Tower Cranes

With tower cranes in particular, an extraordinarily large number of accidents are due to the fact that the cranes are often operated and maintained by unsuitable personnel.

Importance of Crane-Drivers' Courses for Tower-Crane Operation

The recommendation to employ as crane drivers those persons who have completed a crane-driver's course is particularly valid in the case of tower cranes. Such courses have been held successfully for years not only by tower-crane manufacturers, but also by various associations, the builders' guild and the building industry.
Some mishandling faults which, with heavy slewing cranes or loading bridges, have sometimes resulted in comparatively small damage, have often led to complete loss with tower cranes of relatively light construction.

Figure 71. Six men are attempting to pull a concrete container outwards, so as to be able to concrete a cellar wall. In this way, they risk overturning the crane. (The quality of the illustration is not very good because of the photographic conditions.)

Prohibition of Oblique Pulling

Oblique pulling of loads presents a particularly great danger to tower cranes. If the reach of the crane is, for instance, insufficient when concreting, it is better to utilize suitable equipment for bringing in and distributing the concrete or to transfer it to wheelbarrows (Figure 71).
The pulling out of loads from scaffolding stages must also be regarded as oblique pulling and can lead not only to crane damage, but also to collapse of the scaffolding, because the load can get caught in the scaffolding and, when it is freed, will inevitably have a pendulum movement. Before erecting scaffolding, consideration should therefore be given to the position at which and for what crane loads landing platforms will become necessary. Naturally, platforms on several floors must be offset from each other, since the crane can only put down its load vertically from above.

Prohibition of Pulling Scaffold Poles out of the Ground

Just as the oblique pulling and dragging along of loads is highly dangerous and therefore prohibited, so too is pulling out of jammed loads with the aid of a tower crane. The same prohibition applies to the pulling of scaffold poles and planks out of the ground. Apart from the risk of overloading, it is possible that the jib, when the load is suddenly freed, will spring back so sharply that it will recoil over the top of the tower and perhaps even overturn the whole crane.

Other Mishandling Faults

The following two examples of damage are representative of many, almost identical occurrences. They demonstrate that careless drivers often overestimate the load-carrying capacity of their equipment or, by interfering with the adjustment of safety equipment, e.g., the overload-protection device, almost provoke, so to say, the crane into overturning. In this regard, it is necessary to arrange for periodic repetition of instructions and explanations.
The first example concerns a tower crane having a maximum load capacity of 2.8 tons at a radius of 8.2 m. Using this crane, 5.3 m^3 of timber weighing more than 3 tons was handled in one go and, in addition, the radius was far greater than 8.2 m. The load could not be lifted successfully with the aid of the hoisting rope, but by adjustment of the jib. The crane toppled over and was wrecked, while the imprudent driver lost his life.
A further danger of overturning exists if, for instance, instead of a stone skip belonging to the crane, another larger skip supplied by the brickworks is suspended. Such was the case with the tower crane shown in Figure 72. An added complication was that it was required to deliver the load to a bricklayers' scaffold a considerable distance away and, when the overload safety device responded, an attempt was made to reach the scaffold by direct actuation of assembly switchgear.

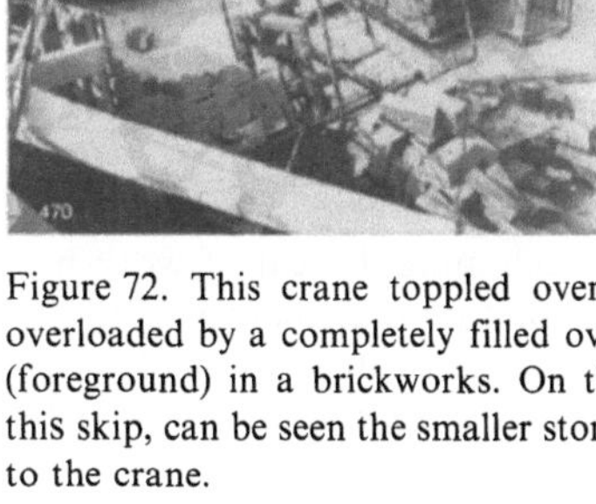

Figure 72. This crane toppled over because it was overloaded by a completely filled oversize stone skip (foreground) in a brickworks. On the right, next to this skip, can be seen the smaller stone skip belonging to the crane.

Excessive Reliance on Overload Safety Devices

Quite a number of cranes overturn simply because the crane driver, relying on proper functioning of the overload safety device, has lifted a load which, as he has found out later, was too heavy. Whatever may be the reasons for failure of the overload safety device in such cases, crane drivers should never place too much reliance on safety equipment, but should rather, if there is any doubt, handle the crane several times with smaller loads.

Prohibition of Braking by Reverse Current

The rule for all tower cranes is that they should be controlled in such a way as to ensure shock-free movements. On principle, it is prohibited to brake the run-out of slewing or travel movements by the application of reverse current.

Operation in the Vicinity of Electric Overhead Lines

If, during operation, it is not possible to keep a sufficient distance away from electric power lines, the appropriate power station should be contacted to agree on suitable protective measures. These measures can include the following:

- switching off the line;
- erecting a barrier of poles;
- insulating by means of rubber hoses (possible with lines up to 380 V);
- moving and re-erecting the line or laying cables, since new buildings often make such procedures necessary in any case;
- restricting the slewing range of the crane by additional built-in switches, in which case a gently acting slewing brake must also be provided. (After working hours, free slewing movement must be ensured or special measures adopted. See Section 6.1.5. Measures for the Prevention of Storm Damage to Crane Installations.)

In some circumstances, the provision of protective measures is much more difficult if work has to be carried out in the immediate vicinity of street or railway overhead lines. In such cases, the relevant transport authorities must always be informed.

The adoption of unauthorized measures is strictly forbidden.

If contact of any wire rope of a crane with an electric overhead line has occurred—even if of only 220 V—the rope must always be replaced because, although there may be no outward signs of damage, internal damage may have taken place.

6.2.4. Inspection of Tower Cranes

At this stage, it would appear appropriate to summarize officially prescribed inspections of tower cranes.

6.2.4.1. First Inspection

When purchasing a tower crane, no notice is sometimes taken, despite all regulations, of the fact that these cranes must be checked by an expert (first individual inspection) before the crane is put into operation for the first time. Proof of this first individual inspection can be dispensed with only if proof of a type inspection for the crane is available (see para. 31 of the regulations for the prevention of accidents to tower cranes).

Every crane buyer should therefore always be absolutely satisfied in good time that the crane concerned is, in fact, provided with proof of a type inspection. If this precaution is omitted when purchasing a crane, disagreeable surprises may sometimes be in store for the purchaser on the occasion of the first individual inspection by an expert. Not every firm is entitled to manufacture tower cranes.

A compilation of all tower cranes for which a type inspection is recognized in the German Federal Republic (according to the position on June 15, 1970) is given in reference [15].

Even if a type inspection is available, some crane buyers still arrange for a TÜV inspection of the crane before taking it into operation for the first time, so as to discover, from the expert point of view, any deficiencies or faults present when the crane is being delivered and to have them remedied by the supplier under guarantee.

6.2.4.2. Inspection after Structural Modifications

In accordance with para. 32 of the relevant accident-prevention regulations for tower cranes, operation of these cranes after structural modifications must not be resumed until they have been subjected to an inspection by an expert.
Modifications for which such inspections are obligatory include:

- an additional extension to the jib, not covered in the type inspection;
- changes in the load-carrying capacity of a crane;
- changes in drives;
- changes in brakes or brake systems;
- certain repairs after accidents to cranes, particularly after overturning, etc.

In connection with repairs after which an inspection is obligatory, see Section 6.1.8 dealing with repair work for which permission must be obtained.
In cases of doubt, information can be obtained from the employers' liability insurance companies or TÜV.

6.2.4.3. Regular Inspections by Skilled Persons

According to para. 33 of the accident-prevention regulations for tower cranes, these cranes must be inspected after each erection and as the need arises—but at least once a year—by a skilled person. The results of the inspection must be entered in the crane control book.

6.2.4.4. Regular Inspections by Experts

According to para. 34 of the accident-prevention regulations for tower cranes, these cranes must be inspected by an expert at least every four years. The results of these inspections must be entered in the crane check book and noted in the crane control book. Application for the inspection must be made by the user to the relevant authority, e.g., TÜV. (A standing order is recommended.) In this regard, the following distinctions are made:

Skilled persons:
machine engineers,
machine foremen and
crane foremen, particularly of the manufacturer.

Experts:
the representatives of
Technical Supervisory Associations (TÜV),
the Office for Work Protection, Hamburg, and
the Technical Supervisory Offices in Hessen.

It must be specially emphasized that ultimately the user is always responsible for rectification of faults that have been discovered by skilled persons or experts.

Bibliography to 6

[1] Töpfer, B.: Geradfahreinrichtungen an Verladebrücken. Fördern und Heben 15 (1965), No. 2, pp. 89/99.

[2] Hanisch, F.: Zahnradschäden und ihre Beeinflussung durch Schmierstoffe. Erfahrungsberichte, No. 2. München: Allianz Versicherungs-AG 1965.

[3] Bartel, A.: Betriebsschäden an Zahnradflanken mit besonderer Berücksichtigung der Entstehung grober Pittings. Lecture held at the 5th Works Manager Conference of Allianz Versicherungs-AG, München, 1960.

[4] Hofmann, F.: Krane und Ausrüstungen. Materialfluß im Betrieb, No. 17. Düsseldorf: VDI-Verlag GmbH 1969.

[5] Hofmann, F.: Auslegerkrane. Erläuterungen zur Unfallverhütungsvorschrift. Materialfluß im Betrieb, No. 15. Düsseldorf: VDI-Verlag GmbH 1965.

[6] Ritter, W.-D.: Sturmschäden an Kranen und Verladebrücken – Ursachen und Maßnahmen zu ihrer Verhütung. Der Maschinenschaden 40 (1967), No. 5, pp. 163/172.

[7] Poklekowski, G., Schulz, W., Hanke, J.: Eine nach dem Ultraschall-Echoverfahren arbeitende Kranabstands-Meßeinrichtung. Der Maschinenschaden 44 (1971), No. 4, pp. 129/133.

[8] Grothus, H.: Vorbeugendes Instandhalten industrieller Anlagen. Der Maschinenschaden 37 (1964), No. 3/4, pp. 23/39.

[9] Bartel, A.: Passungsrost bzw. Reiboxydation – besondere Verschleißprobleme. Erfahrungsberichte, No. 1. München: Allianz Versicherungs-AG 1965.

[10] VDI 2382: Reparaturen an Krananlagen. Schweißen – Heften – Brennschneiden – Bohren. Berlin and Köln: Beuth-Vertrieb 1965.

[11] Oeteren, K.A. van: Korrosionsschutz von Stahlbauwerken durch Anstriche. Der Maschinenschaden 38 (1965), No. 9/10, pp. 153/161 and No. 11/12, pp. 193/202.

[12] Seils, A.: Optimaler Korrosionsschutz der Stahlbauwerke – ein wesentlicher Faktor ihrer Wirtschaftlichkeit. Der Stahlbau 3 (1968), pp. 72/81.

[13] Oeteren, K.A. van: Besondere Probleme des Korrosionsschutzes von Stahlbauwerken. Der Maschinenschaden 43 (1970), No. 1, pp. 7/14.

[14] Hofe, H. von: Können auftraggeschweißte Radbandagen locker werden? Der Maschinenschaden 40 (1967), No. 5, pp. 173/175.

[15] Göttling, H.: Der Verantwortungsbereich des Bauunternehmers bei der Verhütung von Schäden im Turmdrehkranbetrieb. Der Maschinenschaden 44 (1971), No. 2, pp. 41/54.

7. Industrial Furnaces

7.1. General Concepts and Damage Survey

7.1.1. Terminology of Industrial Furnace Construction

Because of the multiplicity of designs and the wide variation in the whole field of process technology [1, 2], uniformity of the terminology of industrial furnaces is an essential condition for communication between manufacturers, users and insurers. Definitions of the most important basic parameters, such as dimensions, planning, design and operational data for industrial furnaces with fuel firing and electric resistance heating are given in DIN 24201 [3].
A VDMA study group (Equipment for Industrial Furnace Construction), taking account of earlier work (e.g., VDMA Instruction Sheet 24202), has formulated [4] the following definition of industrial furnaces:
The term "industrial furnace" comprises, in a collective sense, all equipment used in industry or trade having the essential characteristic that, in a compartment enclosed by walls (the working space), the material to be treated (the work) is subjected to the action of energy in the form of heat until such time as certain processes, whose simplest form is the accumulation of heat, have taken place in the work or on its surface. If a working space is lacking, the equipment, under otherwise equal conditions, can no longer be considered an industrial furnace, but rather an industrial heating installation.

Classification of Industrial Furnaces

The classification of industrial furnaces according to AIW (Study Group Industrial Furnace Construction and Heating Installations) proposes the grouping shown in Figure 1.
The sequence is not in order of importance. Manufacturers will mainly be guided by the columns in Tables 1 to 4 in Figure 1, and users more by those in Tables 5 and 6.
For brevity, all the individual kinds of classification are provided with code numbers and some also with symbolic representations. With the coding, all essential characteristics of an industrial furnace can be expressed in terms of numbers from the six different tables.

Examples:
Gas-heated cylindrical rotary kiln for cement
53.52.61.08.09.23
Oil-fired tube furnace for mineral oils
52.03.64.37.03.52

7.1.2. Survey of Damage

For the following survey of damage to industrial furnaces, 36 cases of damage were evaluated from industrial-furnace installations of the most varied types during the years 1966 to 1971. The total cost of the damage amounted to more than DM 5,000,000.

Discussion of Damage Locations

Approximately 75% of the costs of damage must be allocated to the two locations of damage "basic structure of furnace" and "brickwork/rammed lining", the proportion of the costs for the former location being somewhat higher than

1 Furnace type

Group	Furnace type	Symbol	Ref. Nr.
Stationary Furnaces	chamber, pit		01
	cubicle		02
	trough, tub		03
	shaft		04
	bogie hearth		05
	hood		06
	lifting hearth		07
	tilting hearth		08
	retort		31
Continuous Furnaces	shaft		51
	tubes		52
	rotary tube		53
	drum		54
	rotary hearth		55
	inclined hearth		56
	pushing, pulling		57
	trip latch rod	1060 1061 1062	58
	vibrator hearth	1071	59
	pulley, roller hearth		60
	tunnel		61
	walking beam		62
	step grate		63
	conveyor belt conveyor chain	730 735	64
	paternoster	734	65
	endl. charge pulling: straight passage		66
	endl. charge pulling: passage reversing		67
	dropping shaft		68

2 Charge bedding

Group	Subgroup	Charge bedding	Symbol	Ref. Nr.
Vessels	charge (env.) liquid	crucible, tub removable	626	01
		hearth, tub fixed		02
		nest of tubes	746	03
	charge lumpy	vessel rack		04
		muffle, pot gas-tight		05
Fixed bases	Stationary hearth	plain		11
		profiled		12
		beams, benches		13
		rails		14
		hurdles, noses slats		15
		thwarts grates		16
		muffle, tubes fixed		17
Movable bases	rolling hearth	pulleys, rollers without drive	1076	21
		pulleys, rollers with drive	1074	22
		rollers, balls running		23
	slide hearth	hearth movable	747 740	31
	carrying hearth	transport basket grate, plate	1057	41
		bogie	750	42
		belt	730	43
		chain	735	44
		suspension	1058	45
Others		without base in the furnace		51
		suspended		52
		others		61

3 Heating

Group	Subgroup	Heating	Symbol	Ref. Nr.
Electric	arc			11
	inductive	iron core closed		21
	inductive	iron core, part. closed		22
	inductive	without flux guide		23
	capacitive			31
	indirect resistance heating, metallic	helices, bands		41
	indirect resistance heating, metallic	cartridge		42
	indirect resistance heating, non-metallic	solid: rod, tube grit		43
	indirect resistance heating, non-metallic	melt: salt		44
	conductive			51
	special	infrared beam	858	52
	special	electron beam		53
Combustibles	gasiform, direct	burner	515 E	61
	gasiform, indirect	rad. tube rad. surface		62
	liquid, direct	burner	515 E	64
	liquid, indirect	rad. tube rad. surface		65
	comb. gas/liquid	burner	515 E	66
	solid	semi-gas firing		67
	solid	burner		68
Combined		gas/electr.		71
		liquid/electr.		72
Other means of heating		gasiform		81
		therm. oil		82
		hot water		83
		vapour		84
Others				91
none				99

4 Envelope

Group	Subgroup	Envelope		Ref. Nr.
Without forced circulation	gasiform	air		01
		vapours (explosive)		02
		vacuum		03
		inert. prot. atmosphere	A	04
		exo-therm. prot. atmosphere	B	05
		endo-therm. prot. atmosphere	C	06
		atmosphere rich in hydrog.	D	07
		combustion gas		08
	liquid	oil		11
		salt		12
		metal		13
		water and aqueous solution		14
	solid	powder		21
		ores		22
		others		23
With forced circulation	gasiform	air		31
		vapours (explosive)		32
		inert. prot. atmosphere	A	33
		exo-therm. prot. atmosphere	B	34
		endo-therm. prot. atmosphere	C	35
		atmosphere rich in hydrog.	D	36
		combustion gas		37
	liquid	oil		41
		salt		42
		metal		43
		water and aqueous solution		44

5 Thermal procedure

Thermal procedure	Ref. Nr.
melting	01
drying	02
heating, warming	03
holding, soaking	04
heat-treating	05
cooling	06
reducing	07
sintering	08
burning	09
melting-out	10
refining	11
evaporating	12
coating	13
bonding	14
oxidizing	15
roasting	16
burning burning-off	17
baking	18
calcinize	19
degasing	20

6 Production range

Production range	Ref. Nr.
ores	01
steel, iron	12
non-ferrous metals	13
light metals	14
precious metals	15
hard metals and rare metals	16
glass, enamel	21
ceramics	22
cement, lime and gypsum	23
further inorganic materials	29
rubber	31
plastic materials	32
fibrines	33
wood	34
victuals, luxury and provender	35
further organic materials	39
gas	41
liquids	51
mineral oils	52
combustible	61
waste, dust	71

Figure 1. Classification of industrial furnaces according to AIW (shortened version).

Table 1. Survey of the causes and locations of damage to industrial furnaces.

Cause of damage	Distribution of the number of damage cases [%]	Location of damage	Distribution of the number of damage cases [%]
Product faults	47	Basic structure of furnace	35
Operational faults (with explosions)	36	Furnace tubing	8
		Brickwork/rammed lining	41
Extraneous influences (e.g., thunderstorms, cable breakage)	9	Furnace accessories	16
Unclarified causes	8		

that for the latter location. Also, the proportion of the costs for furnace tubing is higher than that for furnace accessories.

The damage evaluation was based less on damage which occurred during continuous operation of industrial furnaces, but primarily on damage which took place during manufacture, assembly and commissioning. The following guidelines for loss prevention therefore mainly concern these activities.

7.2. Planning, Design, Manufacture and Assembly of Industrial Furnaces

7.2.1. Guidelines for Planning

Prevention of Heat and Welding-Seam Cracks in the Furnace Shell

A careful determination of the operating temperatures and also of the maximum possible temperatures during abnormal operating conditions is essential.

If, when introducing new processes, the operating temperature to be expected can be determined only by calculation, a check on this calculation should, if possible, also be carried out by pilot tests on a semi-technical scale. When doing so, parameters and other variable factors whose influence is not exactly known from earlier operational experience should be altered as widely as possible to enable the maximum temperatures to be expected during later operation to be assessed realistically.

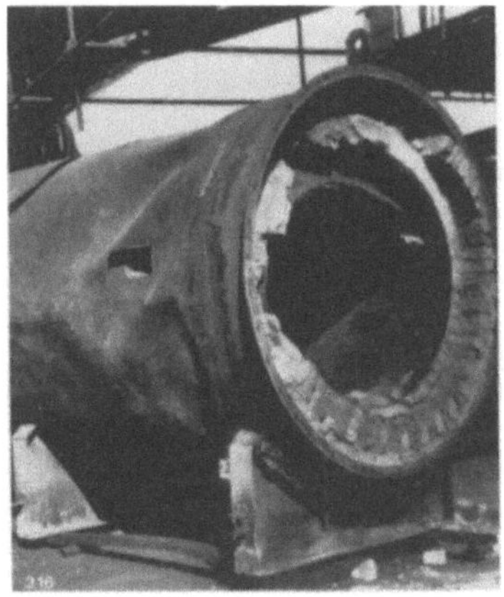

Figure 2. Combustion chamber for an air heater after a prolonged deficiency of cooling air. Since emergency cooling was not provided and the burners were not turned off, despite a high excess temperature in the cooling jacket, the equipment was completely destroyed.

Prevention of Overheating of Cooled Components of the Basic Furnace Structure, Such as Suspensions, Load-Carrying Constructions, Induction Coils and Tubing

The amount of heat to be carried away must be carefully determined [5] by precise thermodynamic estimates of the energy input, possibly of the heat of reaction, as well as of heat transfer. Such cooling media as water or air, or heat-removing process streams, e.g., liquid or gaseous products, must be in adequate quantities and with not too low a safety factor.

To cope with an always possible outage of the normally used cooling medium, emergency cooling must be provided, e.g., an independent gravity-fed supply of cooling water in the event of disruption of electricity supplies. Also, care must be taken that the dimensions of the cooling system, such as the cross-sectional areas of the pipes, are adequate. Moreover, starting up of the emergency cooling system in abnormal operating circumstances must take place automatically within a sufficiently short time.

Figure 3. Part of a tube (material G-X 25 CrNiSi 20 14) from a cracking furnace. Corrosive attack by fused sulphide set in from the gasoline-vapour side and led to break-through within 85 hours, the high chloride content (of the steam) exerting an accelerating effect.

Prevention of Damage to Brickwork/Rammed Linings

The careful determination of operating temperatures and possible maximum temperatures is also necessary for abnormal operating conditions.
Here too, a practical determination of the operating temperatures and of the maximum temperatures to be expected is recommended in all circumstances since, in particular, the selection of refractory materials presupposes an exact knowledge of maximum temperatures [6, 7, 26].
To safeguard the often very expensive brickwork and rammed linings during subsequent operation under conditions of unexpectedly high temperatures, facilities for emergency emptying of the charge should be provided.

Prevention of Corrosion of Built-in Components and Furnace Tubing

With furnace tubing that may also have to carry process steam, inadequate steam quality can lead to corrosion by fused salts. The presence of chloride ions may, in some cases, make it necessary to reckon with stress corrosion cracking of austenitic chrome-nickel steels during the condensing phases. Adequate water conditioning must therefore be taken into account in the planning stage. In most cases, the only way to ensure attainment of the required steam quality is to provide a complete desalination plant [8, 9].
Since even a brief inflow of excessively high concentrations of harmful impurities can lead to damage, constant monitoring of steam quality, e.g., by the provision of conductivity-measuring stations, is essential. In this regard, it is worth following the VGB guidelines for feed and boiler water [10]. If a limiting value, empirically derived, is exceeded, the operating personnel must be warned by an alarm installation. Should it prove impossible to re-establish the rated value within a short period, an emergency shut-down becomes necessary, to enable measures for flushing out harmful constituents from the furnace tubes to be instigated as quickly as possible.
If, during the planning stage, it becomes obvious that harmful constituents in the fuel (e.g., sodium, vanadium and sulphur) cannot be avoided, the design department must be informed accordingly, so that due account can be taken later in connection with materials selection and design.

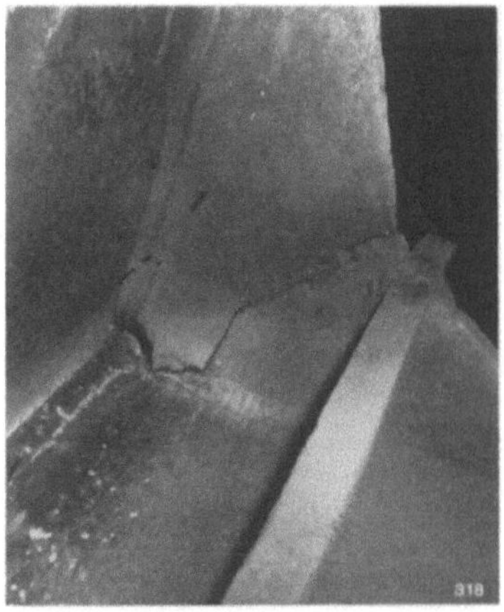

Figure 4. Fractured stiffening rib between the furnace head and shell of a cylindrical rotary kiln. At this constructionally weak location, mechanical and thermal overloading led to failure of the design.

7.2.2. Guidelines for Design

Avoidance of Thermal and Welding-Seam Cracks in the Furnace Shell and Internal Furnace Components

The correct selection of basic and supplementary materials must be founded on the maximum temperatures stipulated by the planning department [25]. At the same time, care must be taken to include an adequate safety factor to prevent materials from being stressed to the limit and to ensure that the life attained corresponds with the planned life of the installation [9, 11].
The selected specifications of the materials must be laid down in the design records clearly and without any possibility of misunderstanding. This requirement should be observed particularly if, for instance, on a building site abroad, the normal means of communication available at home must be dispensed with and the personnel carrying out assembly must rely entirely on information given in the design records. When considering the possibilities of thermal expansion of all constructional items in calculation and design, expansion of the charge under heat must not be overlooked. Exceptional operating conditions, such as emergency switch-off or emergency emptying of the charge, can lead to increased mechanical stresses. Since such deviations from normal operational

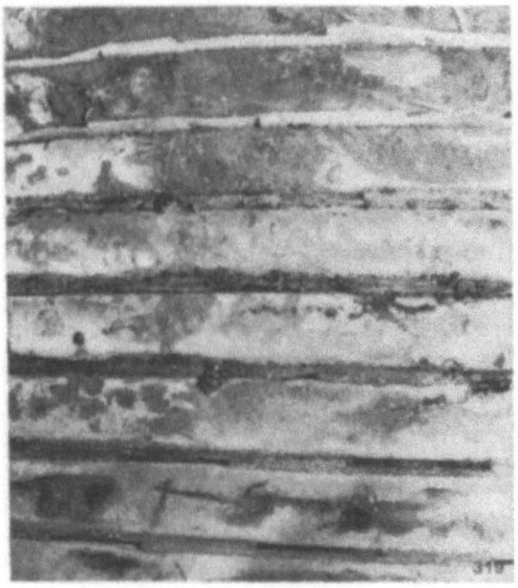

Figure 5. Overheated coil of an induction furnace following bursting of a cooling-water hose. The insulation between individual windings was also rendered useless by exposure to excessively high temperature.

Figure 6. Section of a copper conductor of an induction coil which was melted by excessively high temperature and began to leak. After failure of the normal cooling system, the emergency cooling proved inadequate and insufficiently safe in operation.

conditions are unavoidable in principle, they should also be taken into account in the design layout.

Prevention of Overheating of Suspensions, Load-Carrying Constructions, Induction Coils and Tubing

A careful calculation of heat transfer must also be carried out for abnormal operating conditions [5, 11]. Cooling media in sufficient quantity must be supplied to all structural components exposed to heat; also, special care must be taken with regard to pressure losses. With parallel circuits or branch lines of coolant, uniform distribution of coolant flow must be ensured by an adequate number of measuring stations.

In the event of failure of the normal cooling system, an emergency coolant feed, independent of energy supply, must be provided. Particularly important are adequate dimensioning and reliable automatic cutting-in of such an emergency cooling system.

The correct distribution of thermal energy supply is effected by determining the number of burners and the flame length to be expected, and by ensuring sufficient distance from one burner to another and from the burners to constructional parts of the furnace [12].

Avoidance of Damage to Brickwork/Rammed Linings

The difference in the coefficients of expansion of various lining materials, insulating layers, and the load-bearing structure must be borne in mind. Especial attention should be paid to providing the correct number and the correct design configuration of expansion joints [13, 26].

The careful selection of highly heat-resistant anchor bolts and lining reinforcements must be made by considering the maximum possible temperatures and allowing adequate safety factors. Moreover, insufficient coverage of the anchor bolts or reinforcements by the refractory material can cause them damage and hence can lead to collapse of the rammed lining.

The correct selection of the refractory materials proper must be undertaken with the following service criteria in mind [6, 7]:

- Compressive strength under heat
- Resistance to sudden temperature changes
- Tendency to shrinkage
- Expansion
- Resistance to decomposition
- Resistance to wear

The wall thickness will depend on the maximum possible furnace temperatures and the permissible temperatures for the insulating interlayers and furnace shell.

Apart from the indications given above regarding the correct thermal layout of the individual furnace components, it is also necessary to calculate the static strength of rammed linings and built-in parts, as well as of the load-bearing structure proper.

Avoidance of Destruction of the Basic Furnace Structure and/or the Brickwork or Rammed Linings through Detonations or Explosions

Of particular importance is exact design throughout all stages of the measuring and control equipment and the circuits for the safety interlocks. For example, lighting of a burner should be possible only if the functions interlocked with lighting up, such as furnace atmosphere, fuel pressure, calorific value of the fuel and, depending on the type of furnace, the throughput of the charge, are within the correct range [14, 15]. If measurement of these correct values is made a prerequisite for burner ignition, it is also necessary that failure of one of these functions during normal operation results in shutting off the fuel supply.

Figure 7. Bottom of a cylinder-type tube furnace broken through in a downward direction. Mishandling of the fuel-oil control caused pressure fluctuations in the fuel supply to the burners and extinction of the flame. An automatic burner monitor was not yet in operation for automatic switch-off of the fuel supply, so that it was possible to form an oil/air mixture which was explosively ignited by the hot furnace walls. Because of the cylindrical construction, substantial damage did not occur to the furnace shell, but the bottom and upper cover were completely destroyed.

All these necessary safety interlocks must be stipulated in a special circuit diagram (the "safety chain").

Fuel lines and the corresponding operating fittings must be so designed (e.g., by additional heating arrangements) that the fuel will reach the burners in the correct physical condition, i.e., with the correct viscosity. Special attention should be paid to facilities for water removal (undesirable condensate formation) [16]. Particularly with combined firing, e.g., gas and oil, the fuel lines and operating fittings must be clearly marked to avoid confusion [17].

With fuel-line stopcocks operated by detachable socket-spanners, the socket and the stem of the stopcock should be designed in such a way that, when these elements are engaged, there can be no possibility of mistaking the position of the stem. Incorrect positioning of a single stopcock in a fuel line can cause the inadvertent admission of gas into the furnace and severe damage by explosion.

The size of explosion flaps must be in accordance with DIN 2922 [28].

As a further safety measure, it is recommended that the design should include intentional points of fracture in the form of welding seams of somewhat lower strength at the most suitable locations. In this way, total destruction of the furnace can be prevented in the event of an explosion.

7.2.3. Guidelines for Manufacture and Assembly

All loss-prevention aspects, which must be taken note of during "planning" and "design", must also be made known to the responsible assembly personnel, so that due attention can be paid to them in the practical execution of an installation.

Prevention of Thermal and Welding-Seam Cracks in the Furnace Shell and Internal Furnace Components

The welding of alloyed components must be carried out only by qualified welding personnel in accordance with the appropriate welding instructions (e.g., preheating, stress-relief anneal-

Figure 8. Welding seam on a tube of material 18 Cr 37 Ni (centrifugally cast), later provisionally repaired by welding. Because of inadequate heat treatment of the weld-sensitive material, thermal stress cracking occurred next to the circular welding seam.

ing). If insufficient data are given in the design records, they must be obtained from the design department.

The quality of welding seams, at least of those exposed to pressure, must be checked by radiography (X-rays or isotopes). Welding seams on components of lesser importance should at least undergo ultrasonic spot checks. In the case of welding seams on materials that are sensitive to thermal stress cracking, freedom from cracks in the region of the heat-affected zone should be tested by the magnetic-powder or "Met-L-Check" methods [18, 19].

Pressure tests on components subject to pressure must be performed in accordance with the calculated data in the design records (often 1.3 to 1.5 times the operating pressure) or the individual TÜV rules.

After complete assembly and successful pressure testing, the tightness of all components exposed to pressure must be checked until the final operating temperature is reached. In this connection, particular attention should be paid to screwed fasteners or joints which have had to be opened up again after the pressure test (e.g., for the removal of inserted plates and similar items) [15].

Prevention of Cracks, Break-outs or Collapse of Brickwork/Rammed Linings

Particularly careful attention should be given to storage of firebricks and refractory rammed linings on the building site before they are to be used (above all, protection against moisture or humidity).

The preparation of mortar and refractory ramming material must be carried out carefully and in exact accordance with the manufacturers' instructions. In particular, additions of water should not be made on the basis of "feel" or empirical values, but precisely in conformity with the suppliers' instructions. Formwork or dimensional gauges must be sufficiently well secured to ensure that neither the dead weight of bricks or ramming materials nor shaking-in can bring about a change in their position.

When applying ramming materials, good compaction by shaking, vibrating or spraying-on must be ensured. Setting times for mortar or ramming materials should not be reduced, even when time is at a premium. The setting periods prescribed by the suppliers should be observed meticulously and only after such periods may formwork or supporting structures be removed.

The setting process, particularly for ramming materials, takes place with the release of moisture. If such evaporation is not sufficiently obtained through joints and open areas, special "evaporation apertures" must be provided. Dry-

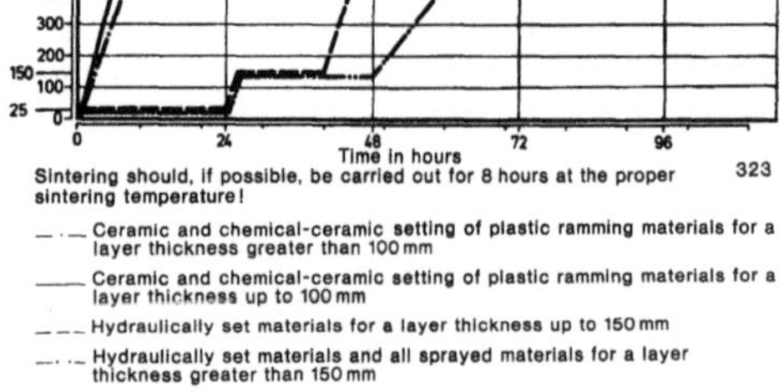

Figure 9. Examples of heating-up curves.

ing and heating up of a refractory lining must take place slowly and evenly. For this purpose, it is essential that a drying and heating-up curve is available from the design department or supplier, in which the gradients for the hourly temperature rise and dwell times are shown, as exemplified in Figure 9 [27]. With sintered ramming materials, the necessary sintering temperature often cannot be attained during the first drying operation. In such a case, special care should be taken to ensure that start-up proper of the furnace is effected with continued observation of the heating-up curve for the lining material, including adherence to the possibly still necessary dwell times.

7.3. Commissioning of Industrial Furnaces

When taking an industrial furnace into operation for the first time, a point is frequently reached where the responsibilities of the manufacturer's assembly personnel begin to overlap those of the user's operating personnel.
Clear instructions by the management, with precise delineations of responsibilities between commissioning and actual operating personnel, are absolutely essential for the avoidance of misunderstandings which may have disastrous results.
In case the operating personnel are still inexperienced and must be trained by the assembly personnel, instructions regarding responsibilities and duties must be recorded unambiguously in writing before the beginning of test operation.

Figure 10. Burst-open longitudinal joint between the furnace ceiling and side wall of a box-type reformer tube furnace after an explosion in the combustion space. The cause of the explosion was found to be mishandling of the burners without adequate automatic supervisory equipment. Misunderstandings between the operating personnel led to an attempt at lighting up at the wrong time.

7.3.1. Measures Prior to the First Heating up

Mechanical Operating Checks

All mechanically movable parts must be tested for proper operation in the "dry" condition, i.e., before the furnace is charged. Air flaps must move easily and must not be obstructed by foreign bodies; adjustment for "fully open" or "fully closed" must be correctly indicated on the outside or, in the case of remote control, must correspond with the respective positions of the controller.
Tipping, lifting and other materials-handling equipment must be tested for ease of movement on reaching the proper final position, and hydraulic systems must be checked for freedom from leaks and for correct oil filling. The pressures reached in hydraulic systems must lie within the correct range.
A check on the direction of rotation and the speed of fan drives must be carried out. Over- or under-pressures of blowers in the furnace interior must be measured and compared with the design data.

Heating System

All burners should be examined from the inside (combustion-space side) to see whether assembly corresponds with the drawings. For instance:
The central position of the burner lances and the correct distance of the burner mouth from the combustion-space walls and from internal furnace equipment must be observed.
Passages for combustion air must be checked for freedom from foreign bodies (e.g., fragments of brickwork); also, ease of operation and the possibility of locking swirl adjustments must be checked.
Sight-glasses and viewing ports must be usable.
External supply lines and fittings for fuel media should be colour-coded (e.g., gases = yellow, and air = blue) in accordance with DIN 2403 [17].
Prior to putting fuel lines into operation for the first time, compressed air or steam should be blown through them, up to at least the last fitting ahead of the burner, several times in succession,

until even the smallest foreign particles (dirt, welding beads, etc.) no longer emerge (see also Section 4.5. Steam-Carrying Pipelines).

Electrical Checks

All components which in operation carry electric current must be tested (e.g., with a megohmmeter) for correct resistance and insulation values with respect to non-current-carrying furnace components.

Functional Checking of Safety Equipment

Explosion flaps must be tested for ease of movement.
All mechanical, electrical and control interlocks must be checked in relation to the circuit diagram (the "safety chain") for correct functioning and operating sequence by short-time bypassing. For instance, the supply of fuel must be possible only when the correct fuel pressure (possibly also the correct fuel temperature), the correct combustion-air pressure, the correct throughput, the correct combustion-space atmosphere, and the functional readiness of the ignition equipment and flamemonitoring equipment have been established [20].

Final Inspection

Finally, an inspection should be carried out by both the manufacturer's and the operating personnel. During this inspection, the undamaged condition of internal components, brickwork and burners should once again be confirmed, as also should be the cleanliness of the furnace interior, including, perhaps, flue-gas passages difficult of access, as well as freedom from foreign objects, tools (and possibly even persons).
Immediately following this final inspection, all access openings, manholes and covers must be closed from the outside, and re-opening should be permitted only with the agreement of the responsible plant manager. An informal record of the final inspection by the joint personnel should then be prepared.

Functional Checks of Quick Shut-down Equipment

Analogously with the checking of all interlocks, the functioning of quick shut-down equipment and other emergency circuits should be tested (possibly by simulating operating conditions).
Only just before going into operation is it permitted to remove the last safety blanking-off plates before entry of the fuel to the furnace, after which a final check of the tightness of all joints in the fuel line is necessary [15].
The removal of electric safety earthing connections, the insertion of fuses, and the switching on of isolating or main switches must be carried out only on instruction from the responsible plant manager or his deputy.

7.3.2. Preliminary Measures during Heating up or Switching on

Before the start-up of heating, the *charge* must be correctly located in the furnace and in the correct quantity or it must pass through the appropriate furnace compartments at a minimum stipulated rate.
For *cooling equipment* which may be required for the removal of heat energy, a steady specified minimum supply of cooling medium must be available. In some cases, protective *envelopes* (e.g., shielding gases) must be available in steady supply [21]. Brickwork/rammed linings must reach at least a specified *predried state* before the temperature is increased [22, 27].
The combustion space must be *flushed out with fresh air* for at least 15 minutes with the blower at full output or giving the requisite minimum of scavenging air. At the same time, the attainable final pressure and the controllability of the combustion air should also be checked.
Prior to lighting up a pilot burner for the first time, the furnace atmosphere should be given a test for explosiveness. If it is found that there are still present gas mixtures which are capable of exploding, flushing must be repeated [29].
Pressure fluctuations or unstable operating conditions in the fuel circuit can be the cause of dangerous interruptions or irregularities in burner operation [30]. Before lighting up, an analysis should therefore be made to ascertain whether the calorific value of the fuel is sufficiently high for stable combustion. (By heating and perhaps draining the fuel lines, it is possible to ensure, for example, that heating gas is dry and free from

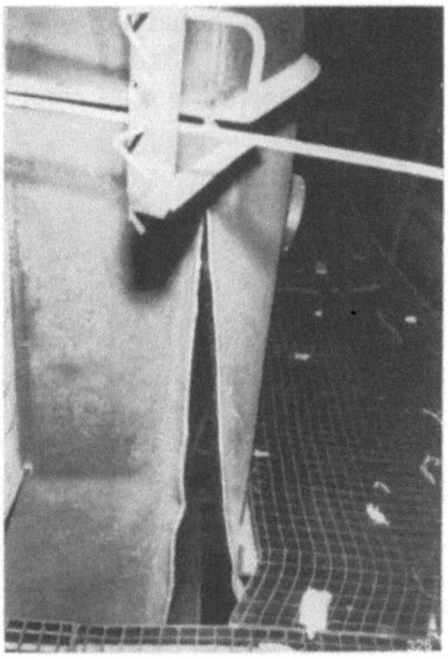

Figure 11. Burst-open flue duct after a gas explosion in the combustion space of an annealing furnace. The cause of the explosion was inadequate ventilation of the combustion space before re-ignition after the burner flame had been extinguished.

condensate or that heating oil is of the correct viscosity on reaching the burners) [16].

Regulations for lighting up must always be put in writing and the operating or starting personnel must be familiar with them; with safety burners, lighting-up regulations must be based on the circuits and interlocks of the lighting-up procedure [23, 24].

After lighting up for the first time and immediately after putting the burners into operation, the fuel pressure should be checked to see that it does not drop below the permissible minimum.

Before putting the first main burners into operation, all pilot and auxiliary burners must be lit and their stable operation—also with the maximum quantity of fresh air in the furnace chamber—must be checked.

With multiple burner installations, the sequence and heating power of the burners to be taken successively into operation must be determined by the responsible commissioning engineer or plant manager.

During the first operational phase of individual burners or of a number of burners, as well as of the total number of burners, their stable operation and the flame pattern must be supervised at short intervals (e.g., every half-hour or hour).

In many cases, the refractory lining does not reach the full setting or sintering temperature during the assembly period; in such an event, temperature increases up to the operating temperature must be adjusted to accord with the heating-up curves for the refractory lining (temperature gradients and dwell periods).

7.4. Operation of Industrial Furnaces

7.4.1. Continuous Supervision of Operational Data

For safe continuous operation, it is necessary to observe, either by means of local instruments or remote indication, a variety of operational data which must not show substantial deviations in their values and which must be continuously recorded. The limiting values for permissible deviations from these data must be marked on the indicating instruments and, if they are exceeded in either direction, the operating personnel must be warned by visual or acoustic alarms. If the rated value cannot be re-established within a definite period of time, manual or automatic emergency shut-down must follow. Under automatic operation, a time-delay relay with the appropriate adjustment (e.g., 15 seconds to several minutes) is interposed ahead of the emergency switchgear. To keep the cost of instrumentation and control, as well as the pressure of work on the operating personnel, within economical and meaningful limits, it is not possible to give general guidelines for the selection of those data which must be included in this range of strict supervision, but only to make, by way of examples, the following proposals:

Temperature pattern of the furnace charge. Measurements on the product should be made at the furnace exit.

Furnace-charge flow lines. If there are several flow lines of charge in the same furnace chamber, it is necessary to determine the total quantity, as well as the individual quantities of each flow line, so as to ensure even distribution. If only the total quantity is determined, it is possible that insufficient loading on one line or overloading on another line can lead to unpredictable damage.

Flue-gas temperatures. By measuring the temperature of the flue gases, it is possible to avoid exceeding the total heat capacity of a furnace and its internal components.

Flame monitors for gas-fired furnaces. Breakaway of the flame from the burner mouth or extinction of the flame without cutting off the fuel supply leads to a dangerous accumulation of unburnt explosive gas mixtures in the combustion space and to the risk of detonations or explosions, particularly in furnaces with only a few individual burners. By installing one or two parallel flame monitors, cutting off of the fuel supply can be ensured [20].

Supervision of process-steam quality. With tube furnaces fed by process steam, inadequate steam quality can lead to rapid corrosion of the furnace tubes by fused salts and hence to their destruction within a short time. The supervision of steam quality is best carried out by measurements of conductivity [8–10].

7.4.2. Operational Data to be Checked and Recorded at Approximately Hourly Intervals

The entire combustion space, the condition of brickwork and tubing, and also, as far as possible, the furnace charge should be observed by the operating personnel through viewing ports for irregularities.

The burner muffle and the adjustment of the burner or burners (flame length and position) must also be supervised through viewing ports by the operating personnel. With large single burners, the installation of television cameras (with air- or water-cooled lenses) to relieve the operating personnel by transmitting a picture on a monitor

Figure 12. Interior of a tube (168 mm o.d. × 11 mm wall thickness, material X 12 CrMo 91) from a reformer furnace which was locally overheated. Internal deposits led to inadequate heat dissipation of the externally heated tube and hence to rapidly progressing wearing away of the tube wall down to 50% of its original thickness.

Figure 13. Outer layer of scale ($\times 100$) on the tube section in Figure 12. These corrosion products of the material on the outside of the tube essentially consist of haematite (Fe_2O_3).

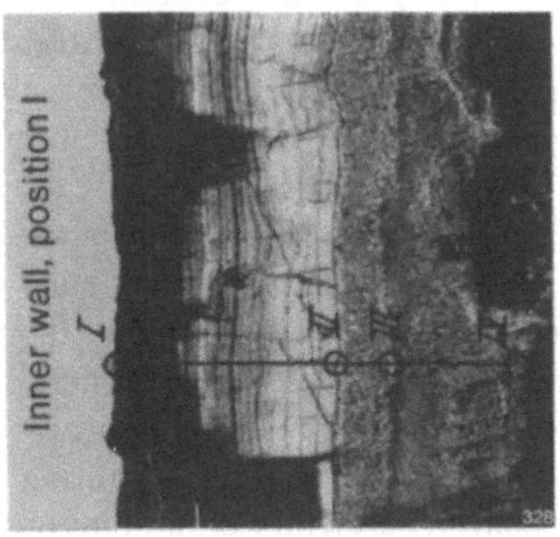

Figure 14. Internal layer of scale ($\times 20$) on the tube section in Figure 12. This layer essentially consists of iron sulphide.

screen in the control room has economic advantages.

Particularly highly temperature-stressed components in the interior of the furnace (e.g., the exterior wall temperature of tubes at critical locations in tube furnaces) must be supervised with respect to an even temperature pattern. The temperature-measuring points in the interior of the furnace should be so located that they can be observed during a visual check of the furnace interior [18, 19]. In this way, otherwise inexplicable deviations from the operating temperature (caused, for example, by burning away of a measuring point or the connecting leads) can be clarified.

7.4.3. Functional Checks to be Carried out at Longer Intervals, e.g., during a Scheduled Shut-down Period

Automatic protective or switching-off equipment must always be ready for operation. Since, in most cases, checks during continuous operation are not possible, functional checks of readiness should at least be made during scheduled shut-downs by appropriate actuation. To avoid dangerous stressing of the whole furnace system, such emergency shut-downs should not be effected from full-load operation but only after a reduction to part-load operation and then only when a complete shut-down has been planned in any case.

Apart from emergency shut-down proper, safety interlocks must be checked for functional readiness during actual operation. When carrying out

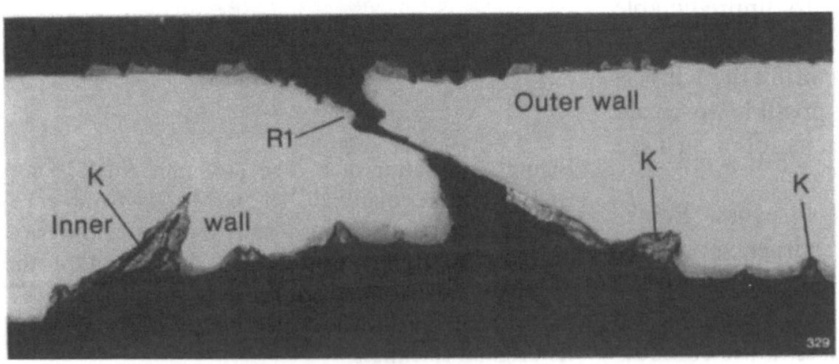

Figure 15. Magnified section ($\times 5$) through the tube wall in Figure 12. At this location, primarily internal corrosion had progressed so far that it resulted in breakthrough of the tube wall.

these checks, all alarm functions must be actuated by simulation of the various dangerous operating conditions, such as failure of blower air, cooling water, instrument air, etc.

Even if these procedures require a certain expenditure of time and repeated re-ignition of the furnace, a functional check of the exceedingly important safety equipment should be carried out on the occasion of a scheduled shut-down, especially with furnaces that operate on a continuous 24-hour basis. The failure of a single safety function in the event of an emergency during continuous operation can result in catastrophic damage to property and can endanger human lives [30].

7.5. Guidelines for Shut-downs and Revisions

7.5.1. Inspection of the Basic Furnace Structure

All parts of the furnace structure which are inaccessible during operation, particularly static load-carrying elements, mounting fixtures and built-in components, must be checked for reductions in wall thickness and for corrosion, scaling and burn-out. Mechanically moved parts must be tested for their ability to function properly.

All components which carry electric current during operation must be tested for proper resistance and for insulation values with respect to non-current-carrying furnace components.

7.5.2. Inspection of Brickwork/ Rammed Linings

Within the whole furnace interior, including the associated flue ducts, visual checks must be made for cracks, bulges, break-outs, reductions in wall thickness, wear or decomposition. The general strength and good adherence to the base can be checked by tapping. In doubtful cases, individual bricks or other samples from the brickwork should be removed for laboratory tests, e.g., with regard to compressive strength under heat [6, 7].

7.5.3. Inspection of the Tubing of Tube Furnaces

The wall thickness of all tubes must be measured ultrasonically. Suitable measuring points must be determined in a previously prepared measuring programme, and all these points must be checked entirely by ultrasonic tests. A comparison with the values of corresponding initial measurements or earlier revisions can be taken as a good starting point for deciding whether replacement of tubes has become necessary. A permissible minimum wall thickness, below which a replacement becomes mandatory in all circumstances, should be established unambiguously before the start of revision.

At particularly critical locations, such as tube bends, their internal condition, wall thickness, and possibly the formation of deposits must be checked radiographically. Here again, a comparison with an earlier radiograph from the same position can yield valuable information regarding operational influences (e.g., erosion, deposits, etc.) and the possible need for replacements.

Figure 16. Partially broken-away, cracked rammed lining detached from the base of a furnace cover. The cause of this destruction of the inner lining was found to be incorrect disposition of expansion joints.

Good documentation of measurement results from ultrasonic and radiographic tests is most helpful when deciding on reserve stocks and repeat orders.

Specially selected sites on the outside of tubes should be subjected to at least random crack tests by the dye-penetrant method.

If the exterior condition of the outside of tubes so permits, random measurements of the outside diameter of tubes should be made and compared with the results of earlier measurements to determine whether bulging has occurred.

If deposits are found on the outside or inside of tubes, they must not only be removed but must also be chemically analysed to gain an insight into their causes of origin. The removal of deposits on the inside of tubes must be effected by chemical processes or by burning off, to ensure efficient heat transmission and adequate cooling of the tubes in subsequent service.

Bibliography to 7

[1] Brunklaus, H.: Industrieofenbau. 3rd edition. Essen: Vulkan Verlag 1969.

[2] Pohle, H.-J.: Industrieöfen (Jahresübersicht) BWK 23 (1971), No. 4, pp. 177/180.

[3] DIN 24201: Industrieöfen mit Brennstoff- und elektrischer Widerstandsheizung (Begriffe). Aug. 1957.

[4] Stepanek, J.: Klassifikation der Industrieöfen. Techn. Mitt. 64 (1971), No. 3, pp. 79/82.

[5] Schwiedeßen, H.: Die Wärmetechnik des Arbeitsraumes brennstoffbeheizter Industrieöfen, ihre Theorie und deren praktische Anwendung. Special Publication. Essen: Vulkan Verlag.

[6] Schwiete, H.E.: Entwicklung und Stand der Prüfmethoden feuerfester Baustoffe. Mitteilungen aus dem Institut für Gesteinhüttenkunde der TH Aachen. Nov. 1961. Essen: Vulkan Verlag.

[7] Konopicky, K.: Die Entwicklungsrichtungen bei Verfahren und Öfen in der Stahlindustrie. Auswahl und Qualität der feuerfesten Baustoffe. Essen: Vulkan Verlag.

[8] Günther, T.: Schadensfälle an Apparaten und deren Berücksichtigung für neue Konstruktionen. Chemie-Ingenieur-Technik 42 (1970), No. 11, pp. 774/780.

[9] Herda, W., Swales, G.L.: Nickellegierte Werkstoffe im Chemieofenbau. Werkstoffe und Korrosion. 22 (1971), No. 9, pp. 759/767.

[10] Vereinigung der Großkesselbetreiber e.V., Essen: VGB-Richtwerte für das Speise- und Kesselwasser von Wasserrohrkesseln. Febr. 68. VGB-Dampftechnik, Essen.

[11] Gernhardt, P.: Über die Sicherheit heißer Konstruktionen. Verfahrenstechnik 4 (1970), No. 3, pp. 108/113.

[12] Narasimhan, K.S.: Berechnung der Flammenlänge für den Industrieofenbau. Erdöl und Kohle-Erdgas-Petrochemie 24 (1971), No. 7, pp. 471/472.

[13] Heß, K., Schweitzer, A., Stöckel, A.: Überlagerung von Gasdruck und Mauerwerkspannung im Sekundärreformer einer Ammoniak-Anlage. Chemie-Ingenieur-Technik 43 (1971), No. 6, pp. 397/402.

[14] Brinke, R.: Sicherheitstechnische Probleme an Gasfeuerungen. Mitteilungen der VGB 50 (1970), No. 1, pp. 44/50.

[15] DVGW Arbeitsblatt G 610: Technische Regel für den Einbau von Sicherheitseinrichtungen in gasbeheizte Industrieöfen. Frankfurt: ZfGW-Verlag.

[16] Wrobel, J.: Explosion eines gasbeheizten Raffinerieofens. Techn. Überwachung 12 (1971), No. 6, pp. 189/190.

[17] DIN 2403: Kennzeichnung von Rohrleitungen nach dem „Durchflußstoff". March 1965.

[18] Nisbet, D.F.: Solving problems in steam-methane reformer heater. The Oil and Gas Journal. June 1971, pp. 64/69.

[19] Nisbet, D.F.: Case history: Failures in a steam-methane reformer furnace. Hydrocarbon Processing, May 71, pp. 103/105.

[20] Dignas, M.: 'Neue Gasfeuerungsautomaten für Gebläsegasbrenner mit Flammenüberwachung. Gas + Wärme International 20 (1971), No. 9, pp. 310/317.

[21] VDI-Richtlinie 2046: Sicherheitstechnische Richtlinien für den Betrieb von Industrieöfen mit Schutz- und Reaktionsgas. Dec. 1969.

[22] Fuchs, W.E.: Siliciumcarbidmassen – ihr Einsatz in Dampferzeugern, in Müllverbrennungsanlagen und in Hochöfen. Der Maschinenschaden 44 (1971), No. 6, pp. 205/215.

[23] DIN 4787: Ölbrenner-Begriffe, Anforderungen, Bau, Prüfung. Oct. 1967.

[24] DIN 4788: Gasbrenner-Begriffe, Anforderungen, Bau, Prüfung. Febr. 1966.

[25] DIN 28070: Chemieöfen mit Ausmauerung. Richtlinien für die Konstruktion der Öfen. June 1966.

[26] DIN 28071: Chemieöfen mit Ausmauerung. Richtlinien für die Ausmauerung. June 1966.

[27] DIN 285: Feuerungsanlagen, Dampferzeuger, Industrieöfen und Schornsteine. (Richtlinien für Trocknen, Anheizen und Inbetriebnahme.) Jan. 1970.

[28] DIN 2922: Explosionstüre 450 × 450. April 1968.

[29] Freytag, H.: Handbuch der Raumexplosionen. Weinheim: Verlag Chemie 1965.

[30] Berufsgenossenschaft der chem. Industrie: Sicherheit im Chemiebetrieb. Düsseldorf: Econ-Verlag 1954.

8. Selected Components and Machines

8.1. Threaded Fasteners

During the period 1964 to 1970, the Allianz Centre for Technology (AZT) investigated 132 cases where the failure of threaded fasteners had caused damage to machines. The distributions of the causes and nature of failure are shown in Table 1.

Discussion of the Causes and Nature of Failure

As can be seen in Table 1, a high proportion of damage to threaded fasteners is caused by product faults. These faults are concerned with design and manufacture, particularly with regard to the wrong choice of material, inadvertent use of the wrong material, and faulty surface finish or heat treatment.

It can also be seen that, with respect to the *nature of damage*, fatigue and creep failure predominate, indicating that the method of production of some bolts and nuts is inadequate for their use (Figure 1) or that they are subjected to excessive dynamic loads due to inadequate dimensions, relaxation of preloads or misalignment. However, the detrimental effect of corrosion accounts for one of the lowest proportions of failure, leading to the conclusion that the choice of suitable materials or combinations of materials to resist corro-

Figure 1. Fatigue failure at the first load-carrying thread, caused by the severely scored and sharp-edged nature of the root of the thread.

Table 1. Causes and nature of damage to threaded fasteners.

Cause of damage	Number of damage cases [%]	Nature of damage	Number of damage cases [%]
Product faults	50	Fatigue failure	40
Operational faults	40	Creep failure	20
Assembly faults	10	Sudden failure	
		brittle	10
		plastic	20
		corrosion	10

sion presents less difficulty. *Sudden plastic failures* consist of primary failures and of secondary consequential failures which can, for instance, occur when remaining and previously undamaged bolts fail through overload after fatigue failure of other bolts.

Failures of bolts and screws caused by failure of locking devices are of significance, owing to the possibility of extensive resulting damage. The cases investigated by ATZ of such damage originating from failure of locking devices were found to be due predominantly to repeated use of the same device, as well as to the use of unsuitable materials for their construction.

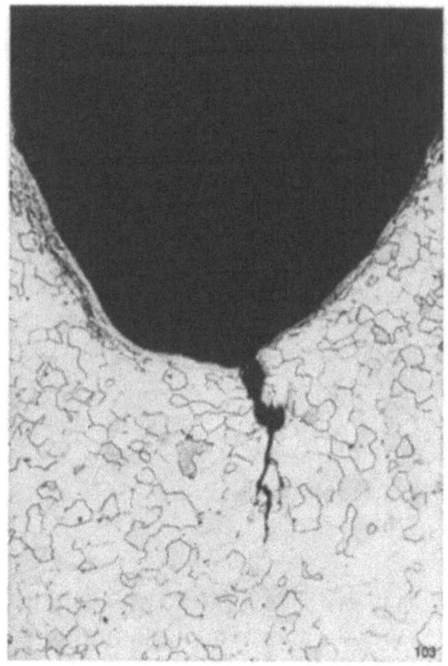

Figure 2. Fatigue crack at the root of bolt thread, caused by too small a radius and by score marks from machining (× 50).

8.1.1. Design, Manufacture and Assembly of Threaded Fasteners

From the investigations carried out by ATZ, the following main points for loss prevention have emerged:

The primary considerations in the design of a machine or installation naturally include the nature and magnitude of the forces that have to be transmitted [1, 2]. In this regard, it has become evident that estimated or purely theoretically calculated stresses frequently deviate appreciably from those in actual practice. Safety is therefore enhanced if loadings are determined on the basis of measurements taken during operation under known conditions.

The nature of the forces to be transmitted—either static or fluctuating—is of decisive importance in the design of threaded fasteners [3]. Over prolonged periods fluctuating loads can be transmitted only by resilient bolts or nuts. In this respect, it is essential to specify the precise preload to be applied, taking into account the reduction that will occur as a result of subsequent settling.

The choice of materials, apart from the forces applied, also requires temperature and corrosion effects to be taken into consideration. The appropriate heat treatment must be specified for heat-treatable materials and must be carried out to achieve the expected strength.

The method of locking a threaded fastener to prevent loosening or release assumes great significance in respect to durability [4, 5]. For this reason, the designer has to make a suitable choice from the multitude of locking methods available. However, a correct choice is facilitated by new knowledge gained about the mechanics of loosening of threaded fasteners and the development of appropriate systems [6].

The ever-recurring cases of inadvertent use of the wrong material during screw manufacture indicate that the material should be checked for agreement with the one specified on the drawing before starting production. In addition, the best design specifications are of little use if they are counteracted in production by, for example, scoring of the shank of a waisted bolt or too small a radius at the root of a thread (Figures 1 and 2).

That the provision of identification marks can also be the cause of failure is confirmed by a case in which titanium-alloy studs were inscribed on the shank with an electric engraver. This area should have been the least stressed part of the stud because of the special method of thread production used, but the engraving process resulted in local hardening of the shank, giving rise to stress concentrations and a fatigue fracture at the site of the inscription.

Particular attention must be paid during assembly to the fact that locking devices should not be

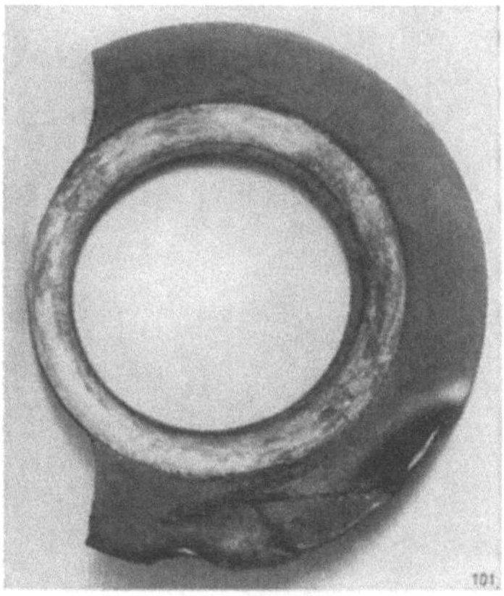

Figure 3. Fatigue failure of a lock-washer as a result of repeated re-use.

Figure 4. Longitudinal cracks (arrowed) in the region of local fusion and hardening, caused by repeated heating of the bore with a welding torch. The resulting stresses even overlapped the notch effect due to the badly scored bore.

used more than a few times; otherwise, there is a risk of their failure (Figure 3). The omission of locking devices is tantamount to gross negligence, and the omission should, at least, be noticed during the final inspection.

The special responsibilities of fitters include application of the correct preload: Tightening of threaded fasteners purely by feel entails the risk of overstrain in the case of small screws while, with larger screws, there is the danger that the requisite preload is not attained.

It is often not realized that, even with the use of a calibrated and properly maintained torque wrench, there can be appreciable scatter of the value of preload obtained, due to friction at the threads and under the head of the screw. Procedures governed by tightening angle give less scatter if the necessary angle is determined by trial measurements of elongation. Similar considerations apply to the use of powered screw-drivers or nut-runners [7]. The least scatter of preload is obtained with hydraulic tightening procedures, which do not depend on torque.

If the preload on alloy bolts is produced by heating the shank and subsequent tightening of the nut, the heating operation should not be carried out by means of a welding torch because of the consequent risk of local hardening and crack formation (Figure 4). As an example of a suitable means of applying heat, the insertion of a heating element into the bore of casing joint bolts on steam turbines has proved successful.

8.1.2. Threaded Fasteners in Service

The checking of at least the most essential threaded fasteners for possible loosening is necessary during routine inspections in the works. However, retightening of slack screws and bolts should not be carried out until the cause of loosening has been established beyond doubt and has been remedied. Checking of threaded fasteners for cracks can be effected by non-destructive ultrasonic testing, provided that at least one end of the bolt is accessible. For instance, the casing joint bolts of steam turbines are investigated in this manner during routine inspection [8].

The expert execution of repairs is also included in the responsibilities of the works or their contractors. In this regard, many sins are committed, particularly when time is at a premium, and among these are again the wrong choice of material, inadequate or excessive preloading of bolts tightened only by feel, and missing or repeatedly used locking devices.

Examples of Damage Caused by Failure of Threaded Fasteners

The intention was to retain a gearbox shaft in the fixed bearing by a retaining cover secured by a screw provided with a spring washer between the cover and screw head. It is well known, however, that spring washers are unreliable locking devices, at least with fluctuating loads. Furthermore, the cover was not secured against turning on the shaft end. Loosening of this inadequate locking device therefore occurred, but could have been prevented, for example, by using two screws positively locked against each other.

Repeated use of a lock-washer led to severe mechanical damage to a single-stage reciprocating gas compressor. Fatigue failure of a lock-washer for the retaining screw of a crosshead pin occurred, causing the screw to loosen itself. As a result, the end of the connecting rod became released and, after striking the piston-rod retaining nut with great force, was severely deformed.

On a 730-kW motor of a ventilating fan, the bolts securing a fibreboard grid to a cover were locked by toothed washers and by cement. This method of locking did not prove adequate in service because

- a satisfactory preload could not be obtained, owing to the inevitable settling down of the fibreboard;
- the outside diameter (18 mm) of the toothed lock-washers was 2 mm larger than that of the bolt heads. A locking action by engagement of the tooth was therefore not possible, as the major portion of the teeth projected beyond the bolt head;
- the cylindrical counterbores in the grid were so smooth that there was no adequate key for the cement.

As a result, the bolts became loose, and some unscrewed themselves to such an extent that they became entirely detached and caused a short-circuit. This damage prompted an examination of the remaining motors of the same type operating in the plant, and it was found that a number of retaining bolts had also become loose in these motors and that the cement had partly dropped out. The locking arrangements were therefore modified as follows:

A locking fluid was applied to the threads of the bolts.

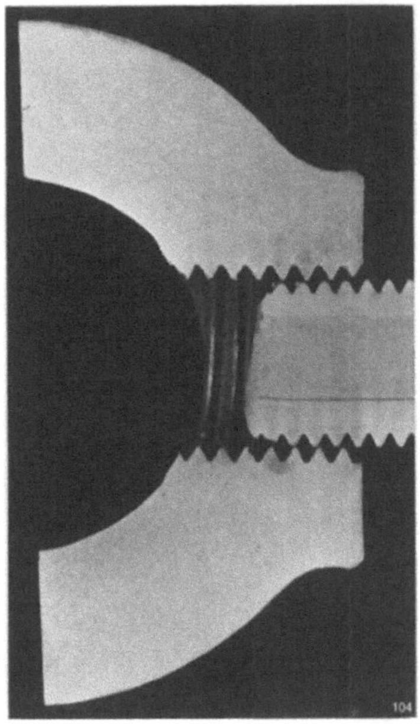

Figure 5. A squeezed and unsuitable tie-bolt could be screwed only to the extent of five threads into a lifting eye for slinging a transformer. The tie-bolt was pulled out of the thread under load, and the transformer was severely damaged in the ensuing crash.

The surfaces of the counterbores and bolt heads were roughened to provide a key for the cement between them.

As a result of failure of studs on a steam-turbine stop-valve, the valve body tilted sideways and the emergent jet of steam badly injured three of the operators. Investigations showed that studs made of three different materials had been used, i.e.,

open-hearth steel, specification approximating St 70;

open-hearth heat-treatable steel, specification C 35;

Bessemer steel, specification St 00 or St 34.

All studs made of the Bessemer steel had broken, as this steel is not suitable for operating conditions of this type. The design using ordinary studs without a waisted shank also had adverse effects. It must be assumed that the studs were replaced several times in the course of the long period of operation of the steam-turbine set. Apparently,

attention was not paid to the correct choice of material. Also, before the occurrence of damage, some of the studs had cracks, which were undoubtedly incurred during previous starting-up operations.
The damage indicates that, in cases of doubt, the constructional materials of bolts, screws and studs should be checked before the fasteners are installed.

Bibliography to 8.1

[1] Wiegand, H., Illgner, K.H.: Berechnung und Gestaltung von Schraubenverbindungen. Berlin, Heidelberg, New York: Springer-Verlag 1962.
[2] Weber, H.: Untersuchungen über die Schraubenbeanspruchungen bei exzentrischer Belastung. Konstruktion 23 (1971), No. 4, pp. 134/140.
[3] Wiegand, H., Strigens, P.: Betrachtungen zur Gestaltung und Haltbarkeit von Schraubenverbindungen. Der Maschinenschaden 42 (1969), No. 4, pp. 103/112.
[4] Bauer, C.O.: Sicherung von Schraubenverbindungen aus nichtrostenden Stählen. Werkstoffe und Korrosion 21 (1970), No. 6, pp. 463/473.
[5] Paland, E.G.: Die Sicherheit der Schrauben-Muttern-Verbindung bei dynamischer Axialbeanspruchung. Konstruktion 19 (1967), No. 12, pp. 453/464.
[6] Junker, G., Meyer, G.: Sicherung von Schraubenverbindungen, Möglichkeiten der Beurteilung von Schraubensicherungen. Maschinenmarkt 75 (1969), No. 39, pp. 818/823.
[7] Junker, G.: Reihenuntersuchungen über das Anziehen von Schraubenverbindungen. Draht-Welt 56 (1970). Suppl. Schrauben, Muttern, Formteile, pp. 122/141.
[8] Schmitt-Thomas, Kh.G.: Zerstörungsfreie Prüfung in der Schadenverhütung. München and Berlin: Allianz Versicherungs-AG 1968.

8.2. Shafts and Axles

Shafts are used to transmit torque and are subjected to torsion and bending. Axles are employed to support stationary, oscillating or rotating machine parts, do not transmit any torque, and are predominantly subjected to bending.
Table 1 lists the causes and location of damage to these machine components in the period 1968 to 1970.
The most frequent failures of shafts and axles are due to fatigue arising from rotating or unidirectional bending moments and from torsional loads. In some cases, fatigue failures occur where bending and torsion have been superimposed.

Figure 1. Fracture originating from a sharp corner inside a feather keyway.

8.2.1. Design and Manufacture of Shafts and Axles

Depending on the mechanical loads (static or dynamic) on a shaft or axle during operation, the stresses (torsional and/or bending), deflection (through bending, bearing misalignment and twisting) and vibration (torsional and/or flexural) must be taken into account [1 to 4]. In this connection, the following standards are applicable:

DIN 114	Shaft diameters for line shafting.
DIN 537	Shaft diameters for cranes and hoists.
DIN 667, DIN 668	Round steel, drawn or turned.
DIN 669	Steel shafts, drawn or turned.
DIN 748	Cylindrical shaft ends.
DIN 1448, DIN 1449	Conical shaft ends.
DIN 5491-65	Spline shafts
DIN 42946	Cylindrical shafts for electrical machines.

Surface Pressure

The permissible surface pressure on journals or shrink fits must not be exceeded. As a guide, indications of permissible surface pressures are given in Table 2.
In general, permissible surface pressures depend on the hardness and surface finish of the materials. In addition, a number of operating conditions have to be taken into consideration, i.e., operating temperature; thermal stresses; corrosive effects such as pitting, contact between dissimilar metals, fretting corrosion and corrosion fatigue; erosion and wear.

Incorrect Assumptions of Loadings

Incorrect assumptions about actual loading and operating conditions are especially dangerous in that they can give rise to failure of a whole batch of components. Designers must therefore be able to monitor their designs in operation and service

Table 1. Percentage distribution of the causes and locations of damage to shafts and axles in the period 1968 to 1970.

Cause of damage	Number of damage cases [%]	Location of damage	Number of damage cases [%]
Product faults	60	Design recesses	80
Operational faults	30	Corrosion sites	15
External influences	10	Other locations	5

Table 2. Permissible surface pressure [kg/mm²] for non-sliding surfaces according to [5]; permissible surface pressure for sliding surfaces according to [2].

Material	Permissible surface pressure for various types of loading		
	static [kg/mm²]	fluctuating [kg/mm²]	shock [kg/mm²]
Steel	800 to 1,500	600 to 1,000	300 to 500
Steel, hardened	1,500 to 1,800	800 to 1,200	400 to 600
Cast steel	800 to 1,000	500 to 900	250 to 350
Cast iron	700 to 800	450 to 550	200 to 300
Malleable iron	500 to 800	300 to 550	200 to 300
Chilled iron	1,000 to 1,500	700 to 1,000	350 to 500
Tin bronzes	300 to 400	200 to 300	100 to 150
Red brass	250 to 350	150 to 250	80 to 120
Rolling- and forging-quality brass	300 to 450	250 to 300	100 to 150

Figure 2. Double bending-fatigue fracture of the shaft of a cooling-water pump, caused by stress concentrations and hardening as a result of building up by welding.

for a prolonged period in order to determine whether the calculated or assumed loading and operating conditions actually apply.

Incorrect assumptions about stress concentrations at shoulders, recesses, threads, keyways and holes particularly endanger parts subjected to dynamic loads, such as shafts and axles. The analysis of failures in Table 1 indicates that the majority of failures of shafts and axles have at least involved design-induced stress concentrations as a contributory cause. The bibliography includes references to extensive data, particularly those based on experimental work, which provide reliable information about the stress concentrations to be expected [1, 6, 7].

Choice of Materials

The working medium or environment is of considerable importance for the choice of materials. Material charts (e.g., [21]) give comprehensive information about which materials are suitable for use under certain corrosive conditions. Further information about the reduction of fatigue strength under corrosive conditions (corrosion fatigue) can be found in [8, 9, 10].

The resistance to wear can by improved by appropriate surface treatments (hardening, nitriding) or by hard chromium plating [2, 11]. (The chromium deposit does not provide protection against corrosion as it usually includes microscopic cracks.)

Fits

Incorrect dimensioning of the seats of shrunk-on wheels, pulleys or couplings leads to loosening, fretting corrosion and frequently to consequent fracture of the shaft. The frictional grip must be maintained under the maximum temperatures and centrifugal forces encountered in operation (Table 3).

Inadvertent Use of Wrong or Faulty Materials

The effort expended on identification and avoidance of wrong material is in no relation to the possible damage by materials faults.

Table 3. Guide to the smallest relative interference for ISO interference fits according to [2] (see also DIN 7154).

Fit	H7·s6	H7·t6	H7·u6	H7·x6	H7·z6	H7·za6	H7·zb6	H7·zc6
1,000 × interference	0,40	0,63	1,0	1,6	2,5	3,15	4,0	5,0

Figure 3. Bending-fatigue failure of a boiler feed-pump, originating from a recess with an inadequate root radius.

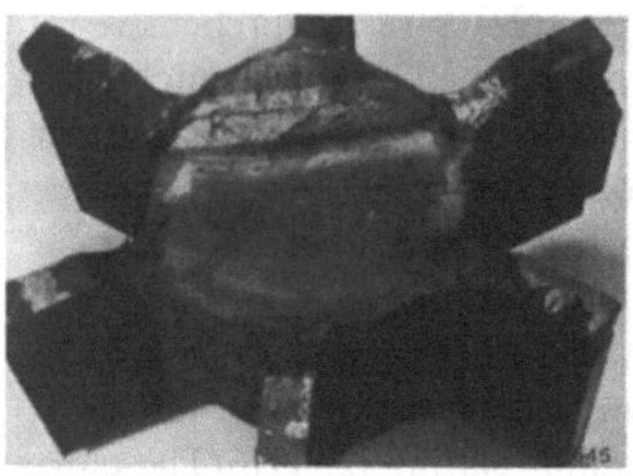

Figure 4. Bending fatigue fracture of a fabricated electric-motor shaft, initiated by stress concentrations at the end of the fillet welds.

Adequate identification of blanks prevents these mistakes. Inclusions and impurities, which occur only rarely in modern steel production, can be detected easily by ultrasonic or, if necessary, by X-ray tests. Faulty heat treatment can be detected metallographically or, better still, by carrying out notch-impact or tensile tests.

Manufacture

Deviations from the dimensions and tolerances specified on the workshop drawing and from the specified surface finish are not permissible. Appropriate inspection procedures must ensure strict discipline in the workshop, always presupposing that the data given in the drawings correspond in detail to actual requirements, so that there can be no latitude for discussion between the design office and workshop. For production-engineering metrology, see [15].

Damaged, badly mounted or resilient tools must not be used. Chatter marks and scoring can lead to cracks and fractures due to high stress concentrations.

Hardening cracks and distortion can be avoided by correct quenching conditions.

Excessive grinding-wheel pressure, inadequate cooling, and burning by grinding wheels of unsuitable grade lead to the generation of excessive heat and the development of grinding cracks.

Hardening and grinding cracks can be detected by non-destructive testing, in particular by the magnetic-flux test.

Welding repairs are permissible only to a limited extent (at low-stressed cross-sections). Preheating before welding and heat treatment after welding will generally be required to ensure the correct condition of the material in the heat-affected zone. Building up by welding is best performed under an inert gas with a filler metal of the same kind or by metal spraying.

Broken shafts made of non-alloy or low-alloy steel can be repaired by thermite welding. In the case of electric arc welding with a filler metal of the same type (high-alloy steels), the root weld should be made under an inert gas (through the bore of the shaft), if possible.

Static and dynamic balancing of shafts, complete with pulleys, wheels and couplings, is essential for avoiding vibrations in operation.

In individual cases, an overspeed test of the shaft, complete with pulleys, wheels and couplings, is necessary, using an overspeed equal to about 1.25 times the operating speed. A non-destructive test of the rotor must be carried out after the overspeed test.

Large forgings for shafts exposed to high operating temperatures should be subjected to a hot

Figure 5. Torsion-fatigue failure of a shaft, originating from a fillet at the shoulder.

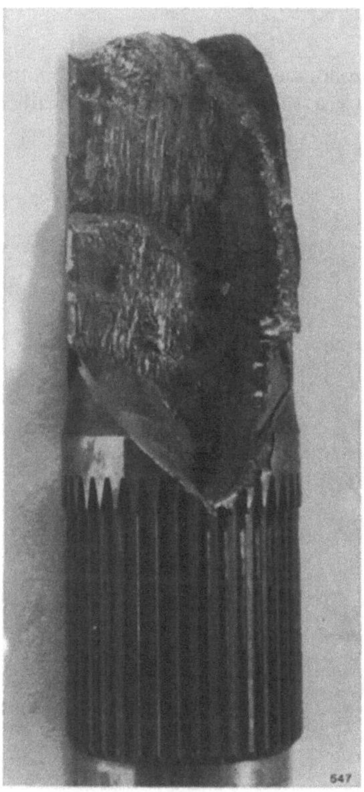

Figure 6. Torsion failure of a spline shaft caused by overload.

test run. This test releases internal stresses which are caused by the heat treatment and would lead to distortion of the shaft in operation without the prior hot test run.

Installation

Care must be taken to avoid mechanical damage. The journals must be supported on rollers, wood or lead sheeting to ensure that the sliding surfaces remain unharmed. Accurate alignment of the shaft in its bearings is a necessary condition for smooth running and low loading of bearings and foundations. Alignment of shafts at the operating temperature of the particular machine concerned has proved especially advisable for rigidly coupled sets of machines.

Radial and axial clearances are specified by the design office. The clearances set by the fitters should be checked by inspection with the aid of a prepared check list. Such records give valuable comparisons in the event of damage or for subsequent inspections.

Distorted shafts can be straightened mechanically by chasing shallow grooves or, better still, by local heating on the convex side. When using heat for straightening, it is essential to avoid concentrated local heating; otherwise, detrimental changes in the structure of the material can occur. After straightening, shafts operating at high temperatures should be subjected to a hot test run at a temperature about 50 °C higher than the operating temperature. Any renewed distortion during this test run can be eliminated by further straightening after completion of the run.

Transport

Adequate protection against mechanical damage must be provided during transport, e.g., soft supports for the journals, wooden boarding, and means of safeguarding against slipping out of place.

Corrosion, especially in the case of transport by sea, can be avoided by corrosion-inhibiting oils, oiled paper and plastic coverings.

There is a risk that lifting equipment may be operated by unqualified personnel. It is therefore advisable to specify the places where ropes or stackers are permitted to handle the load. Strong

wooden boarding has also proved useful for this purpose.
In special cases, wooden blocks for setting down the axles or shafts should be added to the consignment.
Large parts must be clearly marked with their weight.

8.2.2. Shafts and Axles in Operation

The operating instructions should include the operational limits for which a shaft is designed, i.e., the maximum rotational speed, torque, temperature, load, and thrust. Similarly, the maximum permissible load must be specified for axles. Personnel must be familiar with the operating instructions, which must be accessible to them at any time.
Inadmissible movements of the shaft due to prolonged running at the critical speed, shaft distortion or unstable operation should be recognizable by the operating personnel; only in this way can dangerous operating conditions be avoided. Caution must be exercised when passing through the critical speed in the case of distorted shafts, because of resonant vibrations.
Shafts are generally designed for use with a particular working medium. Inadmissible corrosion may take place if the medium is changed. If necessary, the manufacturer should be consulted or an experienced specialist in materials if the shaft material is known.
It is not permissible, for instance, to use for pumping river water with a high salt content or sea-water a pump designed for well-water and having a shaft made from unalloyed Ck 45 steel. On the other hand, it is permissible to use a pump with an austenitic-steel shaft, e.g., X 10 CrNiMo 18 12 for this purpose.
Damage from corrosion can also occur in boiler feed pumps with shafts made from an unalloyed steel when changing the pH-value at 23 °C from 9 to 8 through dosage with ammonia and hydrazine [16].

Maintenance

Poor lubrication endangers not only the bearings but also the journals.

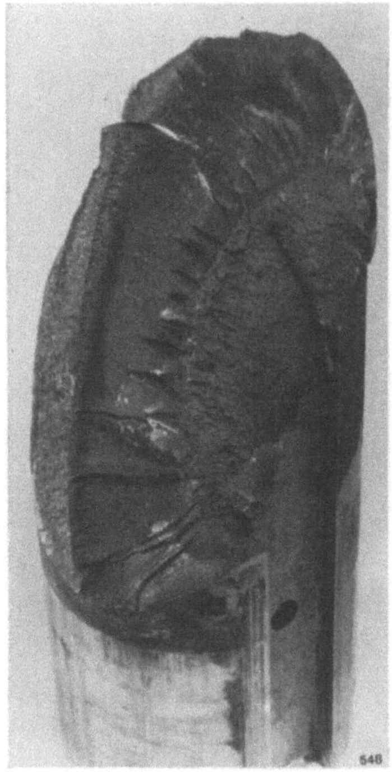

Figure 7. Torsion-fatigue failure of an electric-motor shaft, initiated by a sharp corner inside a feather keyway.

Figure 8. Fractures around a feather keyway in a shaft, caused by stress concentrations.

Contaminants and foreign bodies in the working medium reduce clearances, increase wear and lead to fouling. For this reason, filters and strainers installed by the manufacturer should not be modified or removed.
Inadequate precautions during prolonged shutdowns can lead to corrosion, particularly in narrow crevices. Adequate drying and, where applicable, cleaning are essential. If this cannot be carried out on the assembled units, they must be opened up or dismantled as far as necessary. The use of corrosion-inhibiting oils and plastic covers is imperative under corrosive conditions (high humidity or air pollution).
Checking of mechanical condition must be carried out as far as possible by non-destructive testing. In some circumstances, complete exposure of the shaft or axle will be necessary, possibly also with removal of pulleys, wheels and couplings. The shaft or axle must be examined for wear, corrosion, fretting corrosion and formation of cracks. No general rules can be given about the permissible limits of these phenomena as they depend substantially on the severity of the duty.
Ultrasonic testing frequently enables crack detection to be effected with only partial dismantling.

Supervisory Equipment

Measurements of vibration are still the best way of monitoring a shaft in operation.
It is easy to carry out vibration measurements at bearings, although such measurements are less sensitive than those carried out on the shaft because of the damping effect of the oil film in the bearings.
Vibration measurements on bearing pedestals can be performed by means of mechanically operating contact-type vibration recorders or electrodynamic vibration detectors. Contact vibration recorders are primarily suitable for on-site use, whereas electrodynamic detectors are suitable for both mobile and stationary use.
Shaft vibrations are measured either relative to the bearings or to the machine casing (using contactless detectors operating on an inductive or eddy-current principle) or absolutely (using electrodynamic detectors with a contact slipper). As a rule, shaft vibrations are measured with stationary equipment. On-site measurements of shaft vibrations should, if possible, be carried out by specially trained personnel.
Bearing-pedestal vibrations, and particularly shaft vibrations, should, if possible, be measured by two detectors at 90 deg. to each other (preferably vertical and horizontal), to enable the maximum deflection of the shaft to be determined.
Recording instruments should be employed for permanent supervision. Where vibration measurements are linked with certain existing safety devices, care must be taken in the case of shafts running above their critical speed that, for instance, rapid tripping of the safety device initiated by severe shaft vibrations at the resonant frequency does not lead to damage through clearance bridging.
Special measurements give very useful information about vibration characteristics (frequency analysis and determination of the out-of-balance force vector). These measurements should be performed only by specially trained personnel or by a team of vibration-measurement experts. Limits for bearing-pedestal vibrations and guidelines for shaft vibrations are given in references [17 to 19].

Bibliography to 8.2

[1] Hänchen, R.: Neue Festigkeitsberechnung für den Maschinenbau. München: Hanser Verlag 1970.
[2] Niemann, G.: Maschinenelemente. Vol. 1. Berlin, Heidelberg, New York: Springer-Verlag 1963.
[3] Biezeno, C.B., Grammel, R.: Technische Dynamik. Vols. 1 and 2. Berlin, Göttingen, Heidelberg: Springer-Verlag 1953.
[4] Den Hartog, J.P.: Mechanische Schwingungen. Berlin, Göttingen, Heidelberg: Springer-Verlag 1952.
[5] Dubbel. Taschenbuch für den Maschinenbau. Vols. 1 and 2. Berlin, Heidelberg, New York: Springer-Verlag 1970.
[6] Peterson, R.E.: Stress concentration design factors. New York, London, Sidney: Wiley & Sons 1953.
[7] Neuber, H.: Kerbspannungslehre. Berlin: Springer-Verlag 1958.
[8] McAdam, D.J.: Influence of cyclic stress on corrosion. American Institute of Mining and Metallurgical Engineering. Technical Publication No. 329.

[9] Lochen, R.E., Miller, E.R.: Stress corrosion of 12% chromium stainless steel. Industrial and Engineering Chemistry Vol. 51, No. 6 (June 1959).
[10] Spähn, H.: Maßnahmen zur Verzögerung oder Verhinderung der Schwingungsrißkorrosion. Metalloberfläche 16 (1962) No. 12, pp. 369/373.
[11] Wiegand, H., Heinke, G.: Beitrag zum Verschleißverhalten galvanisch abgeschiedener Nikkel- und Chromschichten sowie chemisch abgeschiedener Nickelschichten im Vergleich zu einigen Stählen. Metalloberfläche 24 (1970), No. 5, pp. 163/170.
[12] Richtlinien für die Prüfung von großen Schmiede- und Gußstücken für Dampfturbosätze. Frankfurt/Main: Verlags- und Wirtschaftsgesellschaft der Elektrizitätswerke 1965.
[13] Kreitz, K., Ivens, G.: Routinemäßige Prüfung von Schmiedestücken mit Ultraschall. Materialprüfung 2 (1960) No. 7, pp. 240/253.
[14] Martin, E.: Prüfung schwerer Schmiedestücke. Stahl und Eisen 85 (1965), No. 19, pp. 1173/1182.
[15] Hütte. Taschenbuch für Betriebsingenieure (Betriebshütte). Vol. 1. Berlin and München: Ernst & Sohn 1964.
[16] Honold, E.: Korrosionserscheinungen an Kesselspeisepumpen. KSB Technische Berichte No. 4, pp. 16/21.
[17] VDI-Richtlinien 1056. Beurteilungsmaßstäbe für mechanische Schwingungen.
[18] VDI-Richtlinien 2059. Beurteilung der Wellenschwingungen von Turbomaschinen.
[19] Wachter, J., Zischka, E.: Lager- und Wellenschwingungen von Großturbosätzen. Energie 20 (1968), No. 12, pp. 401/406.
[20] Peters, G.: Tastloses Messen der thermischen Verkrümmung von Turbinenrotoren. Konstruktion 20 (1968), No. 10, pp. 411/413.
[21] DECHEMA Werkstofftabellen. Deutsche Gesellschaft für chemisches Apparatewesen, Frankfurt/Main.

8.3. Plain and Rolling Bearings

From the variety of machines insured, an investigation was made of 1400 cases of primary damage to rolling bearings and 530 cases of primary damage to plain bearings, where the causes of damage could be established (Table 1) [1]. The most important specific causes of damage to bearings are indicated in Table 2.

8.3.1. Choice of the Most Suitable Bearing

The optimum type of bearing for a given application should be provided, i.e., the most reliable in the light of technical knowledge, not the cheapest. Although the cost of damage to bear-

Table 1. Causes and distribution of cases of damage to rolling and plain bearings.

Cause of damage	Distribution of the number of damage cases [%]	
	Rolling bearings	Plain bearings
Product faults	*30.1*	*23.4*
Faulty workmanship	14.4	10.7
Errors in planning, design and calculation, technological shortcomings	13.8	9.1
Wrong bearing-component materials	1.9	3.6
Operational faults	*65.9*	*69.6*
Mishandling errors, faulty maintenance, failure of supervisory equipment	37.4	39.1
Wear in continuous operation	28.5	30.5
External influences	*4.0*	*7.0*
(such as damage ensuing from irregular running)	4.0	7.0

Table 2. Specific causes of damage to bearings.

Cause of damage	Rolling bearings	Plain bearings
Product faults	1. Wrong assumptions about loading conditions 2. Faulty assembly	Loosening of the bearing metal as a result of defective bonding; inadequate provision for adjustment of the bearing in its housing
Operational faults	Faulty maintenance, particularly failure to replace bearings promptly during inspections and overhauls	Disregard of genuine indications of irregularities during operation; failure or wrong interpretation of signals from supervisory equipment; failure to inspect bearings
External influences	Foreign bodies, particularly those introduced by contaminated lubricants	Overloading of bearing by occasional irregular running caused by other factors; and damage arising from deficiencies in the oil supply, such as a temporary lack of lubrication when changing over to an auxiliary pump during start-up and shut-down, reverse running of compressors and pumps (see 8.6. Oil Supply), and contamination of lubricating oil

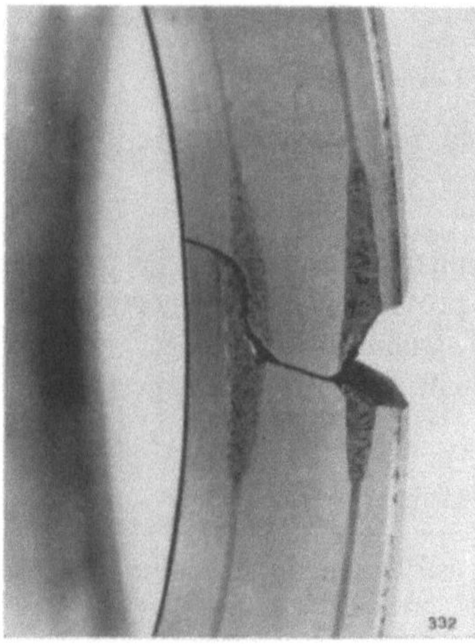

Figure 1. Damaged track of a rolling-bearing outer race (see also Figure 2).

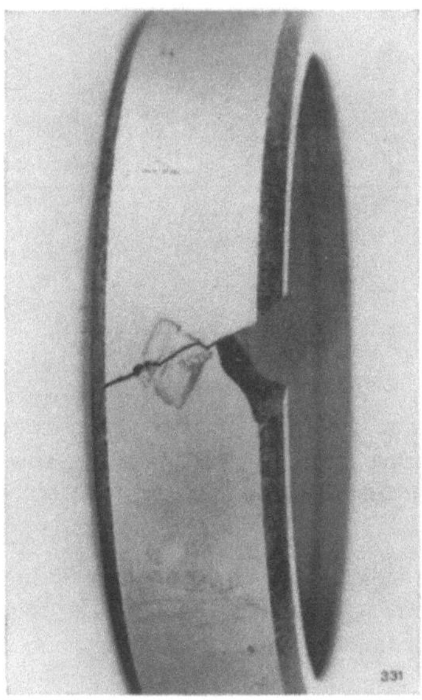

Figure 2. A crack in the outer race of a rolling bearing which had been incorrectly wedged in the bearing housing by a shim. As a result, local deformation of the outer race was caused, with consequent increased surface pressure and fatigue effects. Fragments broke away from the outer race during the ensuing fatigue failure.

ings is itself not significant, severe damage can result to machinery when clearances are no longer maintained; in fact, the destruction of entire rotary elements and housings, with correspondingly long periods out of operation, can be caused.

8.3.1.1. Guidelines for the Choice of Rolling Bearings

Recommendations for installation and choice of the type of bearings should come from the rolling-bearing manufacturer. For this purpose, it is essential that the manufacturer of the machine gives details of the most unfavourable loading conditions occurring during transient adjustment procedures, as the rolling-bearing manufacturer usually does not have the relevant detailed knowledge about the mode of operation of the machine. The majority of cases investigated, where damage to rolling bearings took place after a short time in operation relative to the calculated nominal life of the bearings, resulted from over-optimistic and simplified assumptions about loads and temperatures by the machine manufacturer and hence also by the bearing manufacturer, and not from defective bearings. This was demonstrated by the successful remedial measures applied after damage. Such short periods of overloading of the components of rolling bearings produce local plastic deformations which lead to irregular running of the rolling elements or to constraining forces between the cage and rolling elements. In turn, these additional imposed loads lead to dynamic overloading of the tracks of the races in the region of the points of deformation or to excessive elongation of the cage. After a short period in operation, the well-known flaking phenomenon, caused by exceeding the fatigue strength of the material, occurs. An example of failure due to overload caused by axial distortion is shown in Figure 3.

8.3.1.2. Guidelines for the Choice of Plain Bearings

Plain bearings of a type capable of running under emergency conditions are to be preferred. Experience gained from cases of damage indicates that the three-layer bearing shell is superior to all conventional types of bearings with a simple white-metal lining. The bearing housing, in which the bearing shell must seat perfectly, can be made of steel or cast steel. A steel bearing shell is preferable because of its good bonding properties with bronze. The emergency back-up layer is usually made of special G-SnBz 14 bronze. The surface layer consists of LgSn 80 white-metal, previously WM 80, with thicknesses of only 0.1 to 1.5 mm, as compared with the previous thicknesses of 5 to 10 mm. This reduction in thickness has had an appreciably detrimental effect on the emergency running properties of these white-metal bearings, as quickly sustained damage to the bearings by oil starvation or irregular running immediately removes the thin white-metal layer in the damaged region completely and the self-healing action associated with earlier types of white-metal bearings is not possible. Even during shut-down of the machine, i.e., despite proper functioning of safety devices, severe damage can be suffered by the steel bearing shell if there is no bronze back-up layer, with consequent damage also caused to the journal and possible formation of cracks due to thermal stresses or plastic deformation in the region of the bearing.

As rigid shafts in large machines are becoming more and more rare, owing to the rapid increase in dimensions and unit powers, spherically mounted radial and axial bearings, which adapt to shaft deflections during transit through the critical speed, are generally preferable to rigidly mounted bearings. The movement of the spherical bearing body during operation under load should be facilitated by pressurized lubrication between the bearing body and the bearing seat. These requirements apply especially to thrust bearings because, with these bearings, local overloading due to misalignment is second only to design-induced overloading as the cause of failure (Figure 4).

Thrust collars must either be turned integrally with the rotor or must be shrunk onto a taper in a similar manner to couplings. Assembly by means of a key with a sliding fit is not acceptable as a loose fit can lead to fretting corrosion of the key, thrust collar and shaft and finally to cracks in the shaft. After bearing metal has been cast into spherically seated bearing bodies, the spherical surface and the joint face of the bearing body must be checked for distortion by using engineers' marking ink to assess their fit in the bearing seat. Distortion at the seating and joint surface frequently leads to seizure of radial bearings in their seats.

Figure 3. Top: inner race; bottom left: outer race; bottom right: inner race. The fatigue fractures are evidence of axial overloading of this angular contact bearing during operation. The cage was stretched and torn by distortion.

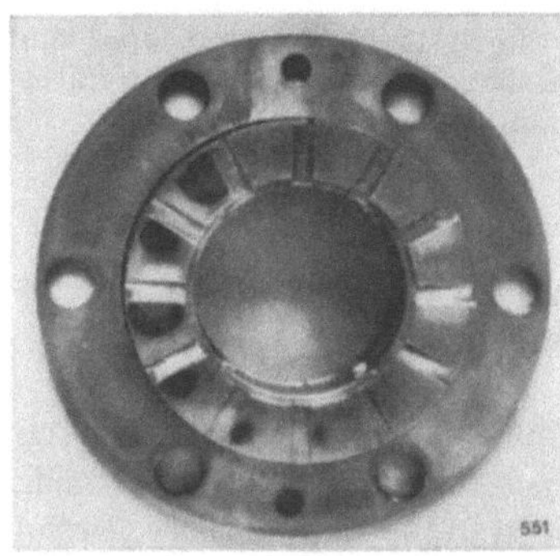

Figure 4. Bolted rigidly to the bearing housing, this thrust pad ring was subjected to excessive loads along approximately half its circumference as a result of having been seated at an angle relative to the thrust collar.

Figure 5. A white-metal bearing with typical signs of dynamic overload.

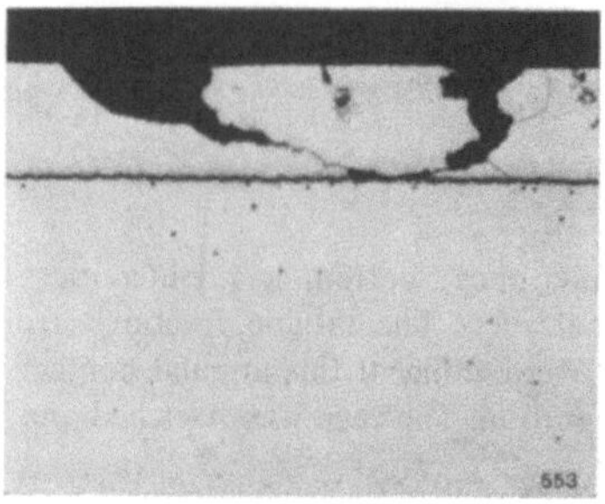

Figure 6. This section through the white-metal layer and bearing-shell material shows incipient fatigue cracks extending along the grain boundaries and break-away of the bearing metal without overheating—an indication of dynamic overload.

8.3.2. Guidelines for the Production of Plain Bearings

Manufacturing faults in plain bearings play a significant part only in respect to defective bonding of the bearing metal to the bearing shell. Careful degreasing in boiling solvent; etching of the surface in an acid bath with subsequent neutralizing; flawless tinning before casting the white-metal lining; careful preheating of the bearing shell, particularly in the case of cast bronze linings; and application of suitable bonding agents between steel, bronze and white-metal to overcome any lack of adhesion between these metals. As all these operations demand reliable execution with the highest standards of quality, they are best left to specialist firms. In the case of large white-metal bearings, the bearing shell must be annealed in a hydrogen-free atmosphere; otherwise, blistering of the white metal can occur in operation.

Ultrasonic and electrical resistance tests for detecting bonding flaws in cast bearing-metal linings have proved successful. It is essential, however, to repeat these tests after several hours of operation of the bearing under normal load, as separation of the cast bearing-metal lining usually develops just far enough under operating conditions, i.e., after subjection to oil pressure, temperature changes and vibration, to make detection by visual inspection, tapping or pressing of the bearing shell or by ultrasonic testing more reliable than on new bearings. Separation frequently manifests itself by local lifting of the bearing metal by pressure oil between the cast metal and the bearing shell. This separation between the bearing metal and shell particularly reduces the dynamic load-carrying capacity of plain bearings and therefore, in the event of irregular running due to other causes, endangers safe operation or even shut-down of the machine.

Figure 7. Damage to rolling bearings while stationary by so-called brinelling occurs quite often. The plastic deformations in the track of this ball-bearing race are clearly visible in the illustration. The vibrations and shock responsible can emanate from the machine itself or from adjacent installations. To prevent such damage to stationary rolling bearings, load-relieving devices or means of protection during transport are essential.

Figure 8. Damage to a roller bearing as a result of incorrect assembly. The break-aways at the rim were caused when the bearing was not fitted by direct pressure on the inner race, but improperly by hammer blows on the outer race, the impacts then being transmitted by the rollers to the inner race. Rolling bearings must be fitted only with suitable tools. Transmission of force by the rolling elements and their tracks leads to damage.

8.3.3. Guidelines for the Installation of Bearings

8.3.3.1. Rolling Bearings

The installation of rolling bearings must be carried out strictly in accordance with the instructions of the manufacturer and without subjecting the bearing components to shock and impact because of the sensitivity of the through-hardened bearing races and rolling elements to brittle fracture (Figure 8). When rolling bearings are used in a machine, the design must make provision for some form of protective device which prevents movement of the rotating elements relative to the housing during stoppage and transport of the machine and which must be fitted during assembly at the manufacturer's works and removed before commissioning. Only in this way can destruction of the bearings by external shock during stoppage or transport of the machine be prevented (Figure 7).

8.3.3.2. Plain Bearings

The installation of plain bearings has changed radically in recent years. Thus, modern plain radial and thrust bearings are, like rolling bearings, supplied as finished parts and do not require hand-finishing during installation. To facilitate subsequent inspections, it is recommended that records for each bearing should be compiled during installation and revisions.

The correct fitting of the tilting pads in thrust bearings of the Michell or Kingsbury type, to suit the direction of rotation of the machine, must always be checked in conjunction with the manufacturer's drawings. The tilting pads and any spare pads should have stamped on them identification numbers corresponding to their position in the thrust-bearing housing. The tilting pads must be located in the bearing housing in such a way that they cannot twist when the bearing contacts the thrust collar. During assembly in the course of revision and repairs, an engineer of the user's company should countersign the inspection record to confirm correct fitting of the tilting pads and thrust bearing, and also correct adjustment and functioning of supervisory and safety devices.

8.3.4. Commissioning of Bearings for Large Machine Installations

Before commissioning the installation, the lubrication system must be checked for faultless operation (see 8.6. Oil Supply). The oil pressure ahead of the bearings and the bearing temperatures are recorded while the rotational speed is slowly increased up to the maximum overspeed for the machine with the full-capacity auxiliary oil pump in operation. When it has been established that this pump provides an adequate oil supply over the entire speed range, adjustments for trouble-free take-over of the oil supply by the main oil pump are made. During this procedure, care must be taken that the full-capacity auxiliary oil pump cuts out or can be cut out only after the non-return valve on its delivery side has closed and also that this pump cuts in automatically in response to a signal from the oil-pressure monitor. Testing of the bearing auxiliary oil pump starts with the machine stationary and continues by slowly increasing the rotational speed of the machine to establish that the pump is capable of delivering an adequate supply of oil right up to the rated speed.

Figure 9. Severe scoring of a white-metal bearing as a result of dirt introduced by the jacking oil during commissioning (see Figure 10).

If planetary gearboxes are fitted, the delivery pressure for the entire bearing-oil supply and adjustments of the bearing-oil pressure monitor must be set to comply with the minimum oil pressure specified by the manufacturer of the gearbox. Even brief starting and shut-down procedures during initial commissioning with an inadequate oil-delivery pressure for a planetary gearbox can destroy the bearings of the planetary gears and hence, in most cases, the entire gear train. It is particularly important that the full-capacity and bearing auxiliary oil pumps also deliver oil for the bearings of the planetary gearbox at the pressures specified for a given speed range (see also 8.5. Gearing).

8.3.5. Guidelines for the Operation of Large Machine Installations

Technical information is available to enable the user to assess the reliability of the types of bearings offered. Additional expenditure on bearings and oil supply (see 8.6. Oil Supply), particularly on the layout of the auxiliary oil supply and the associated protective devices and interlocks, will improve the availability and reliability of the entire installation. Supervision and confirmation of satisfactory assembly by the countersignatures of an experienced engineer on inspection records during important stages in production and assembly at the manufacturer's works and on site are earnestly recommended. The assembly and inspection records shown in Tables 3 and 4 can be used for this purpose.

8.3.5.1. Supervision of the Installation before Commissioning

The manufacturer's recommendations for the limiting values of monitoring and safety equipment must be submitted to the user prior to commissioning the installation. A prerequisite is that the user must provide trouble-free operating conditions during commissioning, notably in respect of the functioning of monitoring and safety devices. Particularly during commissioning (the stage in the operation of the machine with the highest risk), the use of well-intentioned temporary expedients dictated by a shortage of time must be rejected, especially in the case of instrumentation and monitoring and safety equipment.

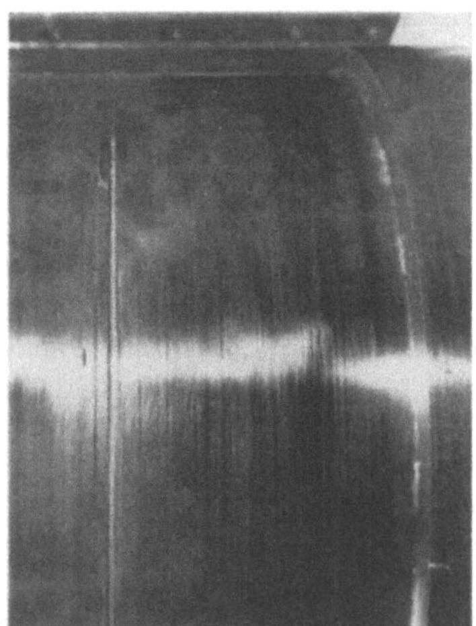

Figure 10. The sharp grooves in the journal of a 300-MW turbine were caused by foreign bodies and were discovered in the course of a bearing inspection during commissioning. The bearing monitoring equipment fitted did not respond. Examination showed that the foreign bodies consisted of particles of dirt introduced into the bearing by the high-pressure jacking-oil system.

Table 3. Assembly and inspection record for radial bearings.

1. *Design data*

Type:	Cylindrical, hydrostatic, twin-wedge, multi-wedge, fixed-pad or tilting-pad radial bearings	
Manufacturer:		
Journal diameter:		mm
Length/diameter ratio:		
Mean and maximum permissible bearing loads:		kg/cm^2
Top clearance ψ:		%
Joint-line clearance φ:		%

2. *Oil supply*

Orifice diameter: mm

Setting of throttle in oil supply in terms of turns from the closed position:

		Stationary	Barring	Normal speed
Bearing-oil supply pressure between throttle and bearing	$[kg/cm^2]$			
Shaft-jacking oil pressure between non-return valve and bearing with jacking pump in operation	$[kg/cm^2]$			
Oil-film pressure between non-return valve and bearing with jacking pump out of operation	$[kg/cm^2]$			
Steady-state bearing-oil supply temperature			[°C]	
Steady-state bearing-oil exit temperature			[°C]	
Steady-state bearing-metal temperature			[°C]	

3. *Functional checks when stationary*

Lift of shaft with jacking pump in operation: mm

4. *Monitoring and safety*

		Alarm	Shut-down
Oil exit temperature	[°C]		
Bearing-metal temperature	[°C]		
Bearing-oil pressure ahead of bearing	$[kg/cm^2]$		
Flow monitors			
in inlet	$[min^{-1}]$		
in exit	$[min^{-1}]$		
Bearing-housing vibrations	[µm or mm/sec]		
Shaft vibrations	[µm or mm/sec]		

8.3.5.2. Monitoring and Safety Facilities

Optimum loss prevention is achieved by a complete set of monitoring and safety devices. The optimum monitoring and safety facilities for bearings listed in Table 5 can be derived from equipment operating on various principles and produced by different manufacturers. Alternatives are also given in the chart. The recommended levels at which alarm and shut-down signals come into operation have been based on past experience of failures. (Approved limits for various types of bearings and machines can be given on request.)

Table 4. Assembly and inspection record for axial bearings.

1. *Design data*

Type:	Taper-land, Michell or Kingsbury axial bearings	
Manufacturer:		
Number of pads:	Total area:	cm^2
Mean sliding velocity:		m/sec
Maximum thrust:		kg Direction:
Normal thrust:		kg Direction:
Axial clearance, permissible axial movement of rotor in thrust bearing:		mm

2. *Oil supply*

Orifice diameter: mm

Setting of throttle in oil supply or exit in terms of turns from the closed position:

		Stationary	Barring	Normal speed
Bearing-oil supply pressure between throttle and bearing	[kg/cm^2]			
Steady-state bearing-oil supply temperature			[°C]	
Bearing-oil exit temperature at loaded side			[°C]	
Bearing-oil exit temperature at unloaded side			[°C]	
Bearing-metal temperature at loaded side			[°C]	
Bearing-metal temperature at unloaded side			[°C]	

3. *Functional checks when stationary*

Axial movement of shaft in bearing: mm

4. *Monitoring and safety*

		Alarm	Shut-down
Oil exit temperature at loaded side	[°C]		
Oil exit temperature at unloaded side	[°C]		
Bearing-metal temperature at loaded side	[°C]		
Bearing-metal temperature at unloaded side	[°C]		
Bearing-oil pressure ahead of bearing	[kg/cm^2]		
Flow monitors			
in inlet	[min^{-1}]		
in outlet	[min^{-1}]		
Displacement	[mm]		
Thrust	[kg]		

8.3.5.2.1. Monitoring and Safety Equipment for Rolling Bearings

Monitoring of Bearing-Race Temperatures (1.2 in Table 5)

The outer race of rolling bearings is usually stationary in the bearing housing, though in some applications the inner race can be stationary on a stationary shaft. In both cases, technical considerations restrict continuous monitoring of the temperature of rolling bearings in practice to *measurements at the stationary race,* i.e., either the outer or the inner race.

It is true that a marked increase in the temperature of the bearing races occurs only when incipient damage at the tracks or to the balls or rollers has taken place. Even so, damage to rolling bearings can be detected by monitoring the tempera-

Table 5. Chart of monitoring and safety devices for bearings with approved limiting values for alarm signals and shut-down pulses.

Monitoring and safety facilities		Rolling bearings				Plain bearings			
Variable measured parameter	Point of measurement	Radial		Axial		Radial		Axial	
		Alarm	Shut-down	Alarm	Shut-down	Alarm	Shut-down	Alarm	Shut-down
1. Temperature [°C]	1.1. Bearing housing	70**	90**	70**	90**				
	1.2. Bearing races	90*	120*	90*	120*				
	1.3. Bearing metal at minimum lubricant-film gap					90*	120*	90*	120*
	1.4. Lubricant ahead of bearing								
	1.5. Lubricant in bearing	80**	100**	80**	100**	80**	100**	80**	100**
	1.6. Lubricant after bearing	80**	100**	80**	100**				
2. Pressure [kg/cm²]	2.1. Lubricant ahead of bearing	0.5* (−30%)	0.3* (−50%)	0.5* (−30%)	0.3* (−50%)	0.5* (−30%)	0.3* (−50%)	0.5* (−30%)	0.3* (−50%)
	2.2. Max. lubricant-film pressure					160**	200**	160**	200**
3. Displacement [mm]	3.1. Shaft relative to bearing							0.1**	0.3**
	3.2. Shaft relative to housing			0.05**	0.1**			0.2**	0.5**
4. Vibrations [mm/sec]	4.1. Air-borne sound								
	4.2. Sound conducted by solid bodies, shock pulses								
	4.3. Bearing-housing vibrations					10	20		
	4.4. Absolute shaft vibrations								
	4.5. Relative shaft vibrations					20*	60*		
5. Loading [kg/cm²]	5.1. Bearing segments/bearing housing								
	5.2. Bearing housing/bearing pedestal								

* ALLIANZ recommendations ** alternatives

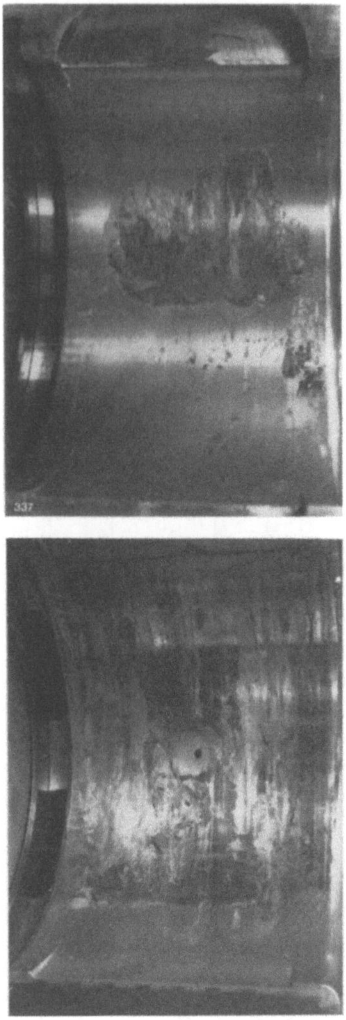

Figure 11. Damage to a radial bearing of a 300-MW turbine caused by inadequate vertical clearance during commissioning. The upper illustration shows the top bearing shell and the lower illustration the bottom bearing shell. In normal radial bearings for steam turbines, vertical clearances of the order of 0.08 to 0.2% of the shaft diameter have proved satisfactory in operation.

ture of the inner or outer race in time to avoid consequent damage to the machine. Equipment for monitoring bearing-race temperature has become available in recent times, and manufacturers of rolling bearings can take into consideration the optimum arrangement of such temperature-sensing devices in advance [2]. As a rule, the same temperature limits for alarm and shut-down signals as for plain bearings can be used for rolling bearings. If the steady-state operating temperature of a rolling bearing lies between 50 and 80 °C, it is recommended that the alarm signal should be set for about 90 °C and the shut-down signal between 120 and 150 °C. The manufacturers of the machine and rolling bearings will specify the most suitable temperature limits for every special rolling bearing on a machine. Monitoring of the temperature of the outer or inner race is considered the safest way of checking the operation of a rolling bearing today.

If the bearing-race temperature cannot be monitored, as is sometimes the case when retrofitting the equipment involved, *monitoring of the bearing-housing temperature* (1.1 in Table 5) is recommended as the second choice. The temperature sensor should be located as close as possible to the seat of the stationary race, which, as already mentioned, is usually the outer race. Because of the greater opportunity for heat dissipation, the limit for the alarm signal should preferably be set to 70 °C and that for the shut-down signal to 90 °C.

If both these methods of monitoring bearing temperature have to be eliminated, the remaining procedure is *measurement of the lubricant temperature* in or after passing through the bearing (1.5

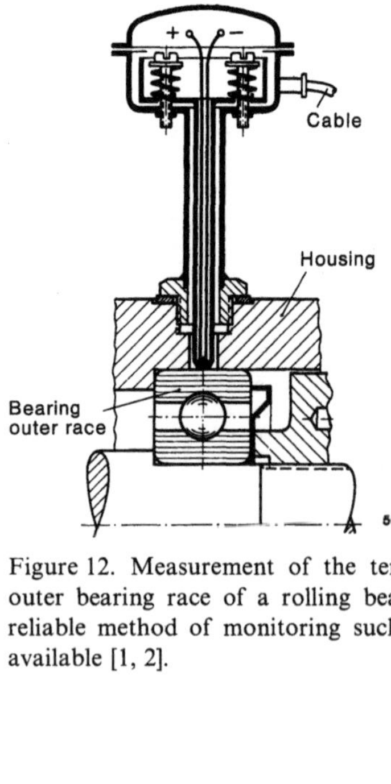

Figure 12. Measurement of the temperature of the outer bearing race of a rolling bearing is the most reliable method of monitoring such bearings so far available [1, 2].

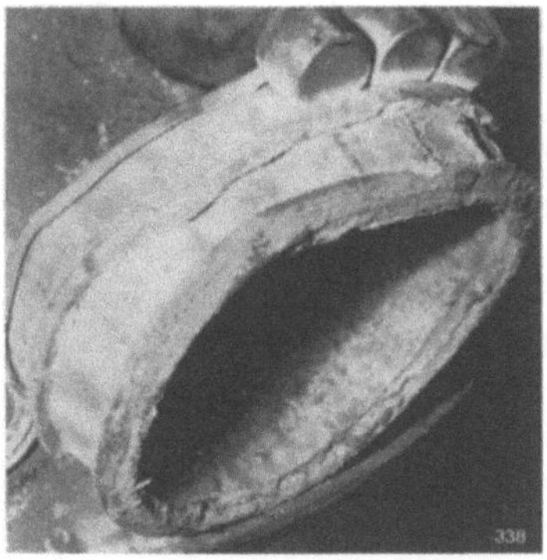

Figure 13. Owing to inadequate preload on the adaptor sleeve, the bearing inner race was able to rotate on a blower shaft. The resulting heat generated by friction caused the inner race to expand, and the rollers were pressed into the track. The shaft, supported only by the adaptor sleeve, then ran until the sleeve was destroyed.

and 1.6 in Table 5). For rolling bearings, the recommended limits are at least 10 °C above the maximum steady-state operating temperature, i.e., about 80 °C for the alarm signal and 100 °C for the shut-down signal. Measurement of the lubricant temperature is generally the least suitable method of monitoring a rolling bearing because the heat generated by friction is only slightly increased as bearing failure develops and a large amount of heat is transmitted by the lubricant to the bearing housing.

Another requirement is *monitoring the lubricant pressure ahead of the bearing* (2.1 in Table 5). It is recommended that the alarm signal should be set to at least 0.5 kg/cm^2 and the shut-down signal to 0.3 kg/cm^2, if there are no other sources of oil pressure in the bearing system. In general, the alarm signal should be set about 30% below the normal pressure and the shut-down signal 50% below the normal pressure.

Monitoring of the displacement of the shaft relative to the housing (3.2 in Table 5) has become accepted only for axial bearings. Because of the very slight play in axial rolling bearings, the alarm signal should be set for a displacement of 0.05 mm and the shut-down signal for an axial displacement of 0.1 mm. Measurements of displacement can be carried out by contact-type instruments or, better still, by contactless inductive or eddy-current instruments.

The *monitoring of shock pulses from rolling bearings* has recently been advocated strongly by one manufacturer [3]. This procedure is concerned with monitoring the sound transmitted through solid bodies, using the aperiodic components of the transmitted sound as indications of the occurrence of local irregularities in the tracks or on the balls or rollers. In this connection, numerical limiting values based on past experience cannot at present be specified, as equipment for this purpose is not yet in regular operation but has been used only for intermittent tests on rolling bearings.

8.3.5.2.2. Monitoring and Safety Equipment for Plain Bearings

Bearing-Metal Temperature (1.3 in Table 5)

Measurement of the bearing-metal temperature in the region of the narrowest lubricant-film gap is the most suitable method of monitoring hydrodynamic plain bearings. Where sliding surfaces of white-metal and steel are in contact, as

Figure 14. Heating up of the journal by rotation in the adaptor sleeve took place to such an extent that the shaft at the journal became plastic and was sheared by the torque still being transmitted by the drive.

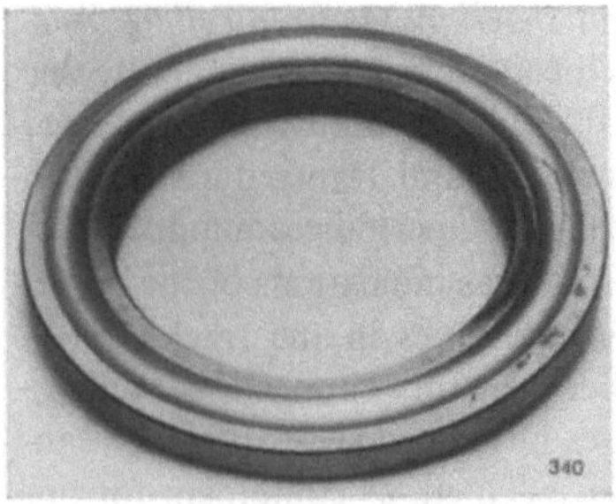

Figure 15. Because this race was loose on the shaft and could rotate with it, severe local overheating resulted and led to thermal stress cracking.

Figure 16. Cavitation occurs when there is a sudden pressure reduction at locations of unsteady oil flow. This purely mechanical, locally limited destruction of the bearing material can now be distinguished from electrically induced erosion with assurance by submitting the material to suitable examination. This thrust-bearing element exhibits damage by cavitation at the outer edge A and behind the step at B and C [4].

Figure 17. Section from the thrust-bearing segment in Figure 16. Cavitation at the step is here clearly visible. At this location, pressure oil at about 100 kg/ cm^2 issuing from the hydrodynamic lubricant film is decompressed to the bearing-oil pressure of about 0.5 kg/cm^2. Such damage can be prevented by increasing the bearing-oil pressure in the thrust-bearing housing and by providing a more ample oil supply.

is customary on large machines, a limit of 90 °C is recommended for the alarm signal and 120 °C for the shut-down signal. Both these limits are based on experience and on location of the temperature sensor about 1.5 to 2 mm below the sliding surface of the white-metal in the region of the narrowest lubricant-film gap. These limits do not apply if the temperature sensor is located elsewhere, as is sometimes the case where installation of the sensor is difficult or where retrofitting is involved. Comparative measurements of bearing-metal temperatures are available for axial bearings in particular and indicate that, in the centre of a tilting pad, a bearing-metal temperature already about 30 °C lower than that at the exit is recorded, so that, with the low permissible range of approximately 100 to 120 °C, a temperature limit for measurements at the centre of the pad of 70 to 90 °C is obtained. As this bearing-metal temperature is usually considered entirely harmless, monitoring of bearing-metal temperature at such an unsuitable location is not recommended, since a clear indication of danger to the bearing does not become obvious. With axial bearings, only by monitoring bearing-metal temperatures at minimally three or four tilting pads distributed around the circumference can unambiguous indications be obtained. Inclined seating of the bearing causes overloading at only a few pads along the circumference and failure of these pads, which may not happen to be monitored by a temperature sensor, increases the load on the remaining pads so rapidly that monitoring of these pads frequently gives warning of bearing damage too late.

There is no point in monitoring bearing-metal temperature solely by means of recording instruments without any provision for a continuous alarm signal when the limiting value is reached and for a shut-down signal when the alarm limit is exceeded. A number of cases of severe damage have indicated that recordings of temperature do not receive the attention they merit from operating personnel. Because increases in bearing-metal temperature and hence bearing damage occur rapidly, they require the very quick automatic shut-down of the machine when the already high limit of 120 °C is reached.

Lubricant Temperature
(1.5 and 1.6 in Table 5)

Measurement of the lubricant temperature in or after passing through the bearing is the second-best method of monitoring plain bearings. The temperature sensor should be located so that, as far as possible, it measures the temperature of the lubricating oil which has participated actively in lubrication of the bearing and which issues laterally from the bearing approximately in the region of the narrowest lubricant-film gap. This very hot oil can be led to the temperature sensor by means of collecting channels in the bearing itself or adjacent to it. A quick-response monitoring point for plain bearings is obtained in this manner. The limits of 80 °C for the alarm signal and 100 °C for the shut-down signal are selected lower than those for the bearing-metal temperature because, in the event of damage, temperature changes are reflected by the lubricant with some delay or not to the full extent, owing to mixing with oil that has not passed through the locations of damage.

Lubricant Pressure ahead of the Bearing
(2.1 in Table 5)

For the lubricant pressure ahead of the bearing, the alarm signal is set to 0.5 kg/cm^2 and the shut-down signal to 0.3 kg/cm^2. If the pressures in the lubricating system are appreciably different, an alarm signal at about 50% of the rated pressure and a shut-off signal at about 30% of the rated pressure applies. If the installation includes

Figure 18. A thrust-bearing disc was attached to this shaft merely by a sliding fit. The feather key worked itself into the shaft material in a counter-rotational direction and material broke away from the shaft.

Figure 19. View of the broken shaft journal of a steam turbine, showing severe fretting corrosion in the region of the shrink fit of the thrust-bearing disc.

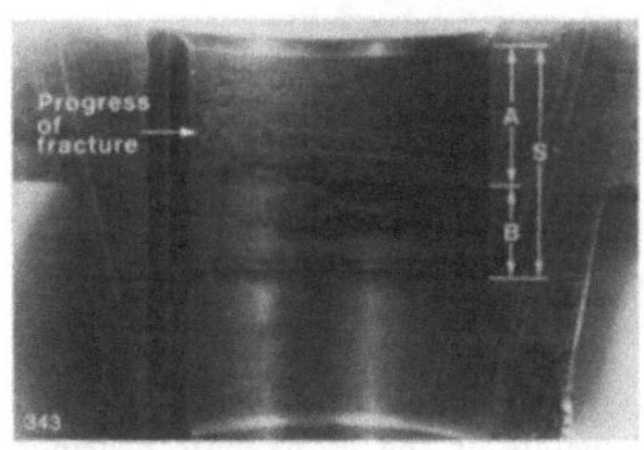

Figure 20. This section through the thrust-bearing disc (see also Figure 19) clearly shows severe fretting corrosion and a pattern of the progress of fracture in the shaft.

The shrink fit of the thrust-bearing disc referred to in Figures 19 and 20 was insufficient with 0.08%, instead of 0.15%, relative to the shaft diameter. The surface pressure in the shrink fit should exceed 9 kg/cm^2 to avoid fretting corrosion. The total of waviness and roughness depth at the surfaces participating in the shrink fit should not exceed 6 to 7 μm per surface. Since, with greater roughness depth, plastic deformations occur at the peaks and hence give rise to a reduction in shrink-fit tension, these requirements for shrink fits are fulfilled by grinding the surfaces at the shaft journal and the seating surfaces of the disc.

planetary gearboxes with rotating planet-wheel carriers, the minimum lubricant pressure specified by the gearbox manufacturer must be used for the shut-down signal. Even a brief period of operation below the minimum lubricant pressure can cause the planet-wheel lubrication to fail through inability to overcome the effect of centrifugal force and can therefore lead very quickly to severe damage at the planet-wheel bearings, usually to the planet wheel axles. It must be borne in mind that bearing-oil safety devices for the entire machine must be adjusted to comply with the minimum requirements specified by the manufacturer of the planetary gearbox (see also 8.5. Gearing).

Maximum Lubricant-Film Pressure
(2.2 in Table 5)

Measurement of pressure of the lubricant film actually provides the most meaningful indication of conditions in hydrodynamic and hydrostatic plain bearings. For technical reasons, monitoring of this pressure has been adopted only in isolated cases, particularly with hydrostatic bearings. With all types of bearings having a hydrostatic starting or low-speed load-balancing oil system, such as radial bearings for large turbo-machinery and thrust bearings for large water turbines, it is possible merely to measure the shaft-jacking oil pressure during continuous operation by means of an adequately damped pressure gauge to monitor the lubricant-film pressure with the desired degree of approximation to the maximum lubricant-film pressure.

Experience indicates that *sudden changes* in the lubricant-film pressure are a sign of bearing damage. *Gradual changes* suggest changes in the alignment of the machine.

The maximum permissible lubricant-film pressure can be derived from the known mechanical properties of the white-metal. For safety reasons, it is recommended that the absolute limit of about 330 kg/cm^2 should be halved to about 160 kg/cm^2 for the alarm signal and should be set to 200 kg/cm^2 for the shut-down signal. The choice of these limits gives adequate safety margins for the possibility that the very localized gauging point is not situated exactly at the position of the maximum lubricant-film pressure of the bearing.

Measurement of Displacement
(3.1 and 3.2 in Table 5)

Measurement of displacement is used particularly on axial bearings and can be carried out by me-

Figure 21. Bearing damage caused by electrolytic corrosion of bearing metals at the crankshaft bearings of a diesel engine. From the driving end (top) to the opposite end, damage to the lower shells of the crankshaft bearings becomes less. The engine drives a generator via a magnetic coupling.

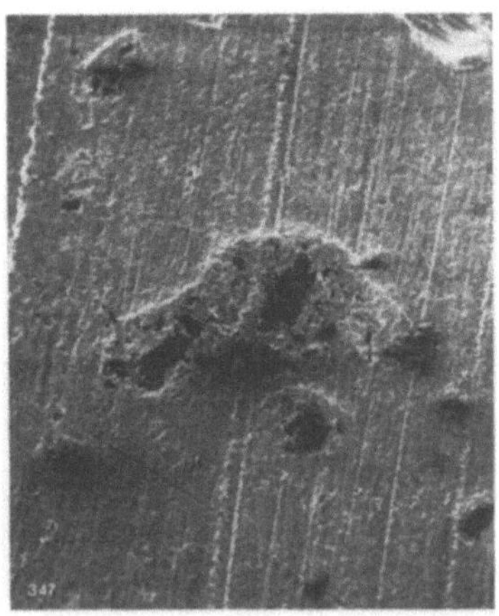

Figure 22. This scanning electron photomicrograph of the sliding surface of the bearing indicates that destruction was not caused mechanically but by removal of the upper lead-tin bearing-metal layer by an action akin to corrosion.

chanical contact-type or contactless inductive measuring systems. An axial displacement of 0.1 to 0.2 mm should initiate an alarm signal, as an axial displacement of 0.3 to 0.5 mm is usually sufficient to result in very severe damage to the sliding surfaces of the tilting pads; however, experience indicates that immediate shut-down prevents consequent damage by instantaneously relieving the load on the axial bearing during run-down of the installation. The measurement of displacement necessitates careful and accurate adjustment of the monitoring instruments during assembly or commissioning.

On the basis of the numerous cases of damage investigated, it is recommended that the zero position of the rotor should be set as follows:

The direction of loading, i.e., the direction in which the maximum axial thrust occurs, is established. Using suitable aids, the stationary rotor is moved against the tilting pads in the direction of loading until the pads lie flat against the thrust collar. This position of the thrust collar corresponds to the condition where failure is initiated in a hydrodynamic axial bearing. This position of the rotor is now designated the zero position. A negative displacement past the zero position then indicates wear of the sliding surfaces of the tilting pads. No difficulties in the interpretation of the measurement of displacement during operation are encountered if the monitoring equipment is adjusted in this manner.

In the case of axial bearings which require the provision of monitoring and safety equipment for both directions of loading, it is strongly recommended that the limiting position for the opposite direction of loading should also be established in the same manner. The axial play in a thrust bearing (see also 3. in Table 4) is included in the assembly and inspection record, and any increase found in the axial play during subsequent inspections is concrete evidence of a damaged bearing.

Measurements of displacement can be carried out between the shaft and bearing housing or between the shaft and bearing pedestal or machine casing. The monitoring of displacement between the shaft and bearing housing is generally recommended, as measurement here entails the least amount of play in the machine, i.e., measurement without unnecessary disturbing influences.

Bearing-Housing and Shaft Vibrations [5]
(4.3, 4.4, 4.5 in Table 5)

Because of the different rotational speeds of machines, the limits for measured vibrations are here given in terms of vibration velocity [mm/sec]. For bearing-housing vibrations, it is recommended that the alarm signal should be set for 10 mm/sec and the shut-down signal for 20 mm/sec.

These limits are so low that a time relay is required to allow a certain period of operation at these limits, so that no unnecessary shut-down signals are given when passing through critical speeds.

The two possible measurements of shaft vibration, i.e., absolute and relative (4.4 and 4.5 in Table 5), are also given in terms of vibration velocity. The alarm signal should preferably be set for 20 mm/sec and the shut-down signal for 60 mm/sec.

If these limits make it impossible to achieve satisfactory setting of the monitoring equipment for any given running conditions of the machine, the following procedure is recommended:

- Determination of the maximum bearing and shaft vibrations occurring throughout the entire speed range.

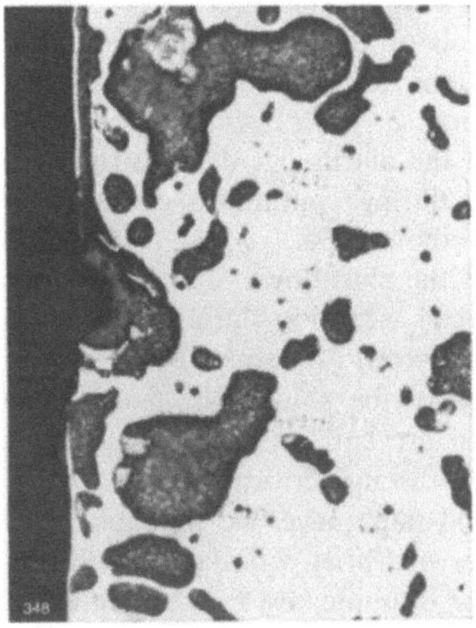

Figure 23. The bronze layer bared by erosion exhibits no signs whatever of melting or heat-affected zones, thereby excluding the possibility of the passage of electric current or a lack of lubricant. The section shown confirms the finding of an examination to the effect that it is a matter of erosion while stationary, without any temperature influence.

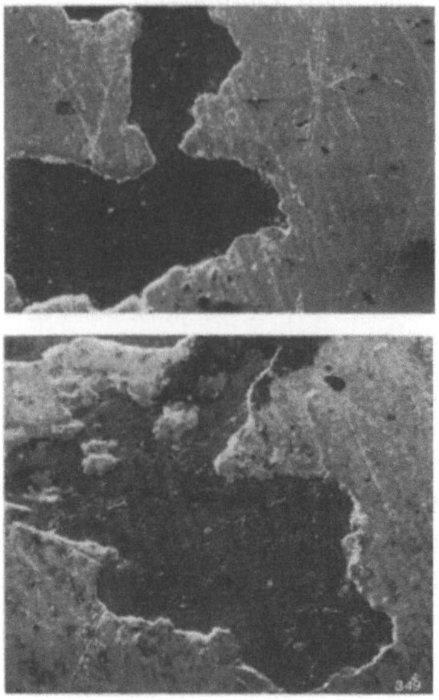

Figure 24. In a laboratory experiment (upper illustration), damage to the bearing was produced electrolytically, using a weak current flow, without heating. This damage was similar to that of the damaged bearing (lower illustration), indicating that the cause of damage to this bearing was due to weak stray currents while stationary, with oil having a small content of water. As a remedial measure, the passage of current through the crankshaft must be prevented.

- Determination of the maximum bearing or shaft vibrations during trouble-free continuous operation.
- Initiation of the alarm signal when the level of vibration during continuous operation is exceeded by about 50%.
- Initiation of the shut-down signal when the normal level of vibration during continuous operation is exceeded by about 100%.
- Transmission of the shut-down signal via a time relay to prevent unintended shut-down when passing through critical speeds.

The alarm signal must never be inhibited by a time relay; otherwise, brief periods of rough running caused, for example, by broken blades or foreign bodies, will not be detected.

The recommended limiting values prevent damage to the bearings of machines. However, upward or downward changes in the normal vibration level always point to changes in the state of balance of the machine, assuming trouble-free operation of the instrumentation. For this reason, in every case of a permanent change or a series of repetitive changes in the vibrations of the bearings or shafts of a machine, supplementary investigations should be carried out to establish whether damage to the machine has occurred.

Load Measurements on Axial Bearings
(5.1 and 5.2 in Table 5)

As measurements of bearing-metal temperature and displacement in axial bearings practically always give warnings only when damage to the bearings has already occurred and can merely prevent consequent damage to the installation, load cells are frequently employed to measure the load on axial bearings, especially in the case of new machines or prototypes. No limiting values for such measurements can, of course, be given. This method of monitoring is also unsuitable for detecting damage to bearings caused by faulty assembly, such as misalignment of the bearing housing or incorrect fitting of the pressure pads, or by an inadequate oil supply to the bearings, as such damage also occurs without overloading the bearing. These load measurements are therefore more of a check on the regular thrust load in the machine than a means of monitoring the bearing itself.

8.3.5.2.3. Inspection of Bearings

During Operation

Daily

Taking readings of bearing temperature, bearing-oil pressures and possibly oil flow, and comparing them with previous measurements.

Checking the rotation and lubricating action of oil rings in oil-ring bearings by visual inspection.

Checking oil levels in bearing housings and oil tanks, checking for leakage from bearing housings, and checking the operation of breathers or extractor fans.

Half-Yearly

Cleaning the insulation of electric-motor bearings and generators and testing their resistance, More frequent cleaning in the case of high rates of contamination.

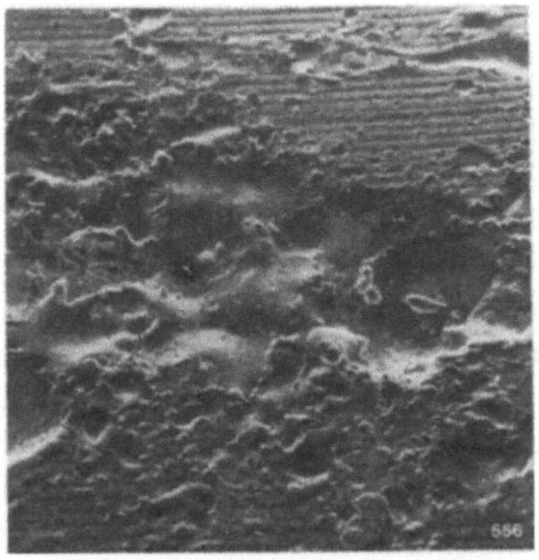

Figure 25. Scanning electron photomicrograph of a fusion crater in white-metal caused by the passage of electric current.

Yearly

Examination of oil and, depending on the result, cleaning, reclaiming or changing the oil.
Checking the radial and axial play of radial and axial bearings by lifting or pushing the rotor and entering the results in the inspection record.

Special Checks prior to and during Shut-down

For checking *rolling bearings* during operation, two techniques, both possibly carried out at low speed, have proved of value. The first of these comprises listening to the sounds made by the bearing by means of a stethoscope or other suitable acoustic device, while the other (more recent) technique is based on shock-pulse measurements. After shut-down, rolling bearings should be dismantled, at least to the extent of removing shields or other items to permit an inspection to be made of the oil or grease filling in the bearing housing. This inspection enables the correct seating of the outer and inner races of the bearing to be checked and also reveals damage to the cage and to raceways in the outer and inner rings whilst the machine is turned over slowly. Such simple inspections of rolling bearings increase operational reliability without loss of availability, such as would result from extensive dismantling. It is recommended that, during overhauls, rolling bearings should be replaced when they approach their expected life, which should, in principle, be stated by the manufacturer. The replacement costs of a still undamaged bearing bear no relation to the possible damage to the machine and the losses arising from down-time that may be caused by failure of a bearing.
With *plain bearings*, an inspection during commissioning after several hours of operation still remains the best measure for loss prevention. An omission of the customary bearing inspection for reasons of availability of the machine is most inadvisable. Damage to the journal of a 300-MW turbine by foreign particles is shown in Figure 10,

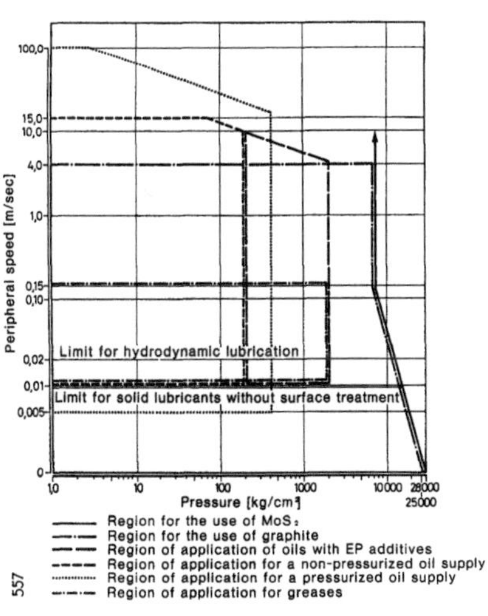

Figure 26. The diagram provides a guide to the selection of lubricants for plain bearings.

Figure 27. Linear marking or scoring of the inner race of a cylindrical roller bearing caused by the passage of electric current.

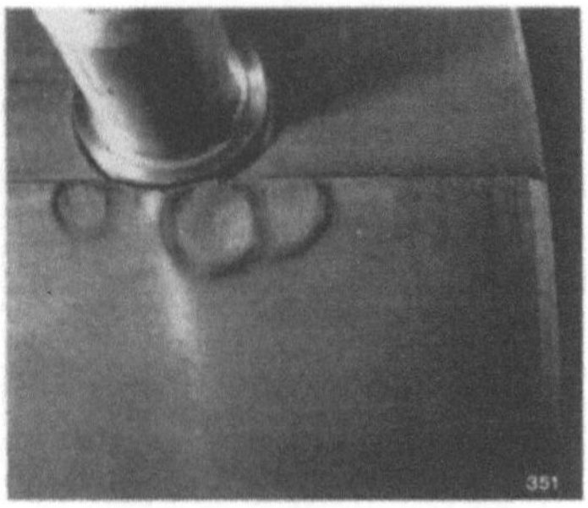

Figure 28. Marks produced by the passage of electric current on a thrust pad of an axial bearing of a 150-MW steam turbine. The white-metal of this thrust bearing was melted away by high currents through the shaft as a result of damage to the generator. The annular marks indicate the passage of high currents.

and it should be noted that damage of this nature to shafts and bearings can be detected early enough by bearing inspections to prevent further damage. On the other hand, the monitoring equipment does not give a warning of such dangerous conditions.

The wear pattern in radial and axial bearings indicates whether they have been fitted correctly and are adequate to cope with the loads encountered. The wear pattern is also the best indication of whether the oil supply to a plain bearing is adequate for lubrication and cooling.

Bonding of the bearing metal is best checked by tapping and by ultrasonic testing.

The operational reliability of the bearings of the machine can be investigated in the course of short shut-down periods during normal operation by taking samples of oil from the oil-tank drain, by inspecting the oil sumps in the bearing housings, by checking play at the journals without opening up the bearing, and finally by removing the top bearing shell and possibly also the bottom bearing shell without dismantling the machine.

Bibliography to 8.3

[1] Huppmann, H.: Schäden an Gleit- und Wälzlagern. Möglichkeiten der Schadenverhütung durch Überwachung und Konstruktion. VDI-Berichte No. 141 (1970), pp. 97/105.

[2] Frenzel, J.: Achslager-Temperaturüberwachung für schnellfahrende Schienenfahrzeuge. BBC-Nachrichten 53 (1971), No. 1/2, pp. 52/55.

[3] Botö, P.A.: Feststellen von Lagerschäden durch Stoßimpuls-Messungen (nur für Wälzlager). From: Kugellagerzeitschrift 167.

[4] Möhle, H.: Einige Gedanken zur Kavitation in Gleitlagern. Der Maschinenschaden 40 (1967), pp. 125/33.

[5] VDI-Richtlinie 2056 and VDI-Richtlinien 2059 (draft).

8.4. Couplings

The causes and patterns of damage to couplings are surveyed in Tables 1 and 2. Cases of damage to mechanical and hydraulic couplings are examined separately.

Discussion of the Causes of Damage

From analysis of the cases of damage examined in Tables 1 and 2, indications are given in the

Table 1. Survey of the causes and patterns of damage to mechanical couplings.

Cause of damage	Distribution of the number of damage cases [%]	Damage pattern	Distribution of the number of damage cases [%]
Product faults	27	Forced fractures	60
Planning and design faults, technological shortcomings	13	Scuffing	18
Manufacturing faults	7	Mechanical and corrosive surface damage	15
Materials faults	3	Incipient cracks	5
Assembly faults	3	Bending, deformations	2
Repair faults	1		
Operational faults	*62*		
Mishandling faults	38		
Maintenance faults	24		
Extraneous influences	*11*		
Foreign bodies	4		
Overloading	7		

Table 2. Survey of the causes and patterns of damage to hydraulic couplings.

Cause of damage	Distribution of the number of damage cases [%]	Damage pattern	Distribution of the number of damage cases [%]
Product faults	*34*	Bending	47
Assembly faults	19	Fatigue failures	20
Planning and design faults, technological shortcomings	11	Clearance bridging	20
Materials faults	4	Forced fractures	13
Operational faults	*30*		
Mishandling faults	19		
Maintenance faults	11		
Extraneous influences	*36*		
Foreign bodies	24		
Climatic influences, particularly cold	12		

following of the specific causes of damage which occurs repeatedly to different types of couplings, but which could be avoided.

Rigid Flanged Couplings with Forged-on or Drawn-on Flanges

Insufficient parallelism of the flange faces is a manufacturing fault arising from turning work on the coupling flanges and the shaft. Particularly prone to this cause of damage are couplings with drawn-on flanges, because the tolerances of the coupling seat on the shaft and the flange face can in some circumstances be added together.

Another cause of damage is loose seating of drawn-on coupling flanges on the shaft ends. This design or manufacturing fault is due to inadequate shrink-fit dimensions or to reworking on site without a knowledge of the required shrink-fit quality.

Axially Movable Gear Couplings with Coupling Sleeves Having External or Internal Teeth

With oil- or grease-lubricated couplings, damage can be caused by unfavourable sealing as a result of faulty design, so that lubrication of the coupling in operation is no longer sufficient.

Other causes of damage are clogging of the coupling teeth by dirt from a contaminated supply of pressure oil; and insufficient operational smoothness as a result of inadequate balancing of the individual coupling elements, especially if they are displaced when their teeth mesh under load at full speed [1].

Flexible Couplings with Deformable Transmission Elements

If the design lacks limitations for unloading the deformable transmission elements, these elements will be destroyed when the limits of loading are reached.

Hydraulic Couplings

Damage can arise from overloading by a hold-up in oil filling when the operating temperatures are too low, i.e., during start-up of hydraulic couplings.

Assembly faults primarily involve *deficient alignment* of all types of couplings. For couplings on electric motors and generators, the insulation via flexible intermediate elements assumes particular importance. Local welding in the region of the coupling, caused by the passage of electric current, is extremely dangerous, since the axial mobility of the coupling becomes more difficult, thereby resulting in damage to axial bearings.

Of *mishandling faults,* the most important is deficient lubrication of gear-type couplings.

With regard to *maintenance faults,* mention must be made of faulty or entirely omitted checks on alignment.

Damage from *extraneous influences* often arises from foreign bodies in the lubricant and from overloading by the driving or driven machines, particularly in connection with damage to these machines which leads to sudden jamming of their rotors in the housings.

8.4.1. Guidelines for Loss Prevention during the Manufacture of Couplings

Every coupling should satisfy the following minimum requirements:

- Complete disconnection of the coupling after removing the axial bearing or bearings of rotors with permissible axial displacement of the rotors in their housings or of housings in their guides, without disassembly of housings or bearing pedestals and without changes in alignment.
- Removal, when opening up a coupling, of all loosened components, such as the coupling bolts, centring disc, coupling sleeve and spring elements, without having to remove the rotors and without disassembly of unaffected components.
- Checking of concentricity and radial and axial alignment with testing equipment which is suitable for the coupling and is supplied by the manufacturer, and which makes use of dial gauges exclusively.
- Stationary components belonging to the coupling must, if the need arises, permit running of the driving and driven machines independently of each other for commissioning; balancing in operation; and emergency or part-load operation.
- For all types of couplings, the installation of the machines on *one* foundation made from

one material is to be preferred. The installation of driving and driven machines and gearboxes on baseplates made of different materials, such as reinforced concrete and structural steel, is unacceptable; so, too, is installation on the top of an oil tank or on connecting platforms of steel beams or concrete because, with such an arrangement, differences in the height of the various shafts are certain to be brought about by temperature effects, and these differences in height exert stresses on all types of couplings.

- Depending on the type of driving and driven machine, the coupling should be designed with a safety factor of 1.5 against distortion and crack formation, based on the *maximum* starting, short-circuit or driving torque. The effects of excessive torque should be neutralized by the provision of intentional fracture locations with shear elements so designed that there should not be any ensuing damage by imbalance to the machines, particularly to the rotors.

8.4.1.1. Couplings with Drawn-on Flanges

Couplings with drawn-on flanges must be secured in their conical seat by at least two keys displaced by 180 deg. In this way, difficulties in balancing of the shaft with the coupling flange can be avoided. Taper pins, used in large numbers for attaching the coupling flange to the shaft, should be spaced evenly on the circumference. Taper pins have the disadvantage that, in some circumstances, the coupling flange becomes inadmissibly distorted on the coupling.

Drawn-on couplings must be provided with a shrink fit that stresses the material to its permissible limit while taking centrifugal force into account. The shrink fit must not permit the coupling to become loose on the shaft, even at the maximum practicable overspeed of the set, since the coupling usually does not come back to its proper original position. Rough running after such temporary loosening of drawn-on couplings will then enforce shut-down of the set and extensive corrective measures, always involving removal of the rotors.

The safest flanged coupling is the type with forged-on flanges. The coupling bolts and nuts must be weighed, balanced with one coupling half and clearly allocated by numbering. Both coupling halves and the centring ring must be clearly marked in their relative positions.

8.4.1.2. Gear-Type Couplings

Lubrication by oil circulation is preferable to lubrication by oil or grease filling. Oil- or grease-filled couplings lose their content of lubricant if the O-ring seals leak—a condition that comes to light only during a visual inspection because of soiling of the coupling guard. Destruction of the teeth of the gear coupling, caused by the tooth flanks working into each other in the normal operating position, sometimes combined with welding by friction, generates inadmissible axial forces that act on the coupled machines during start-up and run-down and can lead to damage to axial bearings through overloading and severe damage by rubbing to the coupled machines.

Gear couplings with forced oil lubrication must be so designed that the pressure oil does not deposit particles of dirt in the gear teeth by centrifugal force, but continuously cleans the gearing (using an oil filter with elements rated at 25 to 50 μm).

The design of gear couplings must be such that the coupling sleeve can be removed without disassembly of the axial bearings of both rotors and

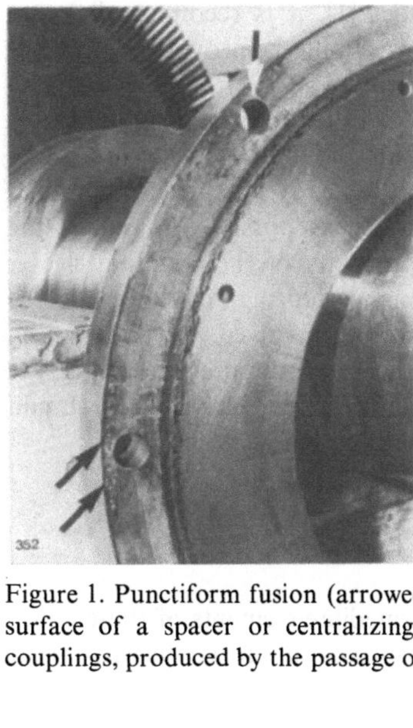

Figure 1. Punctiform fusion (arrowed) at the bearing surface of a spacer or centralizing disc of flanged couplings, produced by the passage of electric current.

Figure 2. A gear coupling with clear signs (arrowed) of local heating in the region of high surface pressure on the flange of the divided coupling sleeve. This heating resulted from the passage of direct current, and on the left can be seen the torn-off shaft after a bearing had melted away from the same cause.

without displacement of the housings. In special cases, it is sufficient to ensure the ability of the coupling sleeve to move over the coupling hub of the driven machine and over that of the driving machine.

The coupling hubs should be removable individually without having to disassemble the rotors, so that they, as well as seals and one-piece bearings, can easily be replaced.

All coupling components and the coupling as a whole, with the coupling sleeve in the fully assembled condition, should be dynamically balanced by the manufacturer to balance quality Q1, in accordance with [3]. Since this requirement still presents difficulties [1], it is recommended that the coupling sleeve should be provided with a circumferential groove or bores to permit rebalancing in operation at full speed and maximum torsional loading under conditions of perfect alignment of the set at a steady operational temperature. In this way, it is possible to reduce the effect of the eccentricity of the coupling sleeve, caused by the manufacturing tolerances of the four sets of gear teeth involved, on the running smoothness of the set and on the load of the radial bearings concerned. Gear quality should be at least 6 according to DIN 3962.

8.4.1.3. Flexible Couplings

The flexible intermediate elements of these couplings must be inspected at short intervals. Wear weakens these elements and leads to rough running and torsional vibrations of sections of the installation. In addition, the requirements set out under 8.3.1.1 and 8.3.1.2 are valid.

8.4.2. Guidelines for Loss Prevention during Assembly

8.4.2.1. Alignment of Couplings

Bending moments must be avoided by ensuring parallelism of the coupling flanges. There must be no error in the angularity of the flanges. Shear forces must be prevented by aligning the axes of the shafts. There must be no vertical or lateral offset. These *ideal* stipulations can be fulfilled *approximately* by careful alignment. Fundamentally, alignment instructions apply to all rigid, movable and flexible couplings and to all types of machines to be coupled. The manufacturers of the couplings and of the individual machines must state the amount by which it is necessary to deviate from the generally desirable zero misalignment in the cold condition, so that, when shafts are displaced under load and temperature, the desired zero misalignment will be achieved in continuous operation. If no data are available from the manufacturer for the intentional deviation from zero misalignment in the cold condition, zero misalignment should, in principle, be obtained.

From experience of damage to couplings and bearings, particularly gear couplings, it is strongly recommended, during commissioning of a new machine, that a coupling check should, wherever possible, be carried out on the set when it is at operational temperature.

If the running smoothness of a set is unsatisfactory after taking it into operation for the first time, or after starting it up again, this examination is the first step in the search for the cause of the trouble. With large sets, measurements of displacement at the bearing housings should be made with the aid of precision levelling equipment or a hose level balance if rough running occurs when increasing speed and applying the load, and if a frequency analysis of the bearing (or, better still, of shaft vibrations) has shown that rough running occurs at the frequency of rotational speed or twice the rotational speed, i.e., that it is the result of imbalance.

The Following Preparations Should Be Made for Alignment:

The rotors should be supported in the lower bearing shells, which should be amply lubricated by hand with the oil used in operation, and firmly secured against rotation in the bearing cradle.
The housings of the machines must be loaded as in operation by placing the top halves in position. For heavy rotors, the auxiliary bearing-oil pump and the shaft-jacking pump should be held in readiness to facilitate turning over the rotor when aligning.
Every alignment check must be preceded by a concentric-running check at the coupling flanges. Also, the rotors must be axially secured to prevent larger axial displacements with consequent rubbing of the rotors on fixed housing components. During measurements, axial displacements of, for example, half the measuring range of dial gauges do not affect the results of the alignment procedure proposed.

8.4.2.2. Concentric-Running Check

Each coupling half is tested for concentric running with respect to the bearing housing by means of a dial gauge. In this way, it is possible to discover rotor distortions or assembly faults of the coupling halves. The limiting value is 0.01 to 0.02 mm. To exclude the possibility of the shaft being displaced in the bearing, it is desirable to carry out a second check measurement with a dial gauge in the region of the bearing.
When the results of the concentric-running check have established the satisfactory condition of the coupling flanges, the aligning equipment from the machine or coupling manufacturer is installed on the coupling flanges. In his machine specification, the customer must include sturdy measuring fixtures for the use of dial gauges and definite measuring positions on the coupling flanges. A dial gauge is adjusted radially on a marked reference track on the other coupling side.
It is convenient to adjust the dial gauge in such a way that about half of the measuring range is available in either direction.

Figure 3. The pronounced one-sided deformation of the coupling teeth points to an assembly fault.

Two dial gauges are applied axially, also with about half the measuring range, to the marked reference track on the other coupling side. The measuring tracks must be provided by the coupling manufacturer; if they are missing on existing installations, they must, once and for all, be unambiguously established.
The permissible values of the axial gap must be stated for the given diameters of the axial measuring tracks. Both rotors are turned at the same time, resting freely in their bearings for measurement. Lowering by extrusion of a film of oil, with the rotor at rest, is controlled by means of a radial dial gauge in the region of each bearing; in this case, the shaft-jacking oil and the bearing-oil pumps are switched off after turning the rotors.
For rotors with only one bearing between two flanged couplings, the rules laid down by the manufacturers apply. In one case, the rotors are held in a centralizing fixture without an auxiliary bearing and are aligned axially. In another case, the rotors can be aligned radially and axially, with an auxiliary bearing, as with rotors having two bearings between the couplings. The auxiliary bearing should always be located in the region of the coupling flange to approximate the deflection line of the rotor during operation.
The numerical values of all dial-gauge measurements for radial and axial alignment are recorded in Tables 3 and 4 at the top, bottom, right and left measuring positions, i.e., always after rota-

Table 3. Alignment formulae for radial displacement—4 turns, each through an angle of 90 deg.

Measuring position	1. Measurement	2. Measurement	Example	
Vertical displacement				
Top	a_1	a_2	2.5	2.8
Bottom	b_1	b_2	3.1	2.8
Difference	a_1-b_1	a_2-b_2	−0.6	0
Half of difference ≙ vertical displacement	$\frac{a_1-b_1}{2}$	$\frac{a_2-b_2}{2}$	−0.3	0
+ Left-hand coupling lies higher than right-hand coupling − Left-hand coupling lies lower than right-hand coupling			Left raised by 0.3 mm	
Lateral displacement				
Right	c_1	c_2	3.8	2.8
Left	d_1	d_2	1.8	2.8
Difference	c_1-d_1	c_2-d_2	2.0	0
Half of difference ≙ lateral displacement	$\frac{c_1-d_1}{2}$	$\frac{c_2-d_2}{2}$	1.0	0
+ Left-hand coupling lies right from right-hand coupling − Left-hand coupling lies left from right-hand coupling			Left-hand coupling by 1 mm towards the left	
Check against measuring errors	$\frac{a_1+b_1}{c_1+d_1}=1=\frac{a_2+b_2}{c_2+d_2}$		$\frac{2.5+3.1}{3.8+1.8}=1=\frac{2.8+2.8}{2.8+2.8}$	

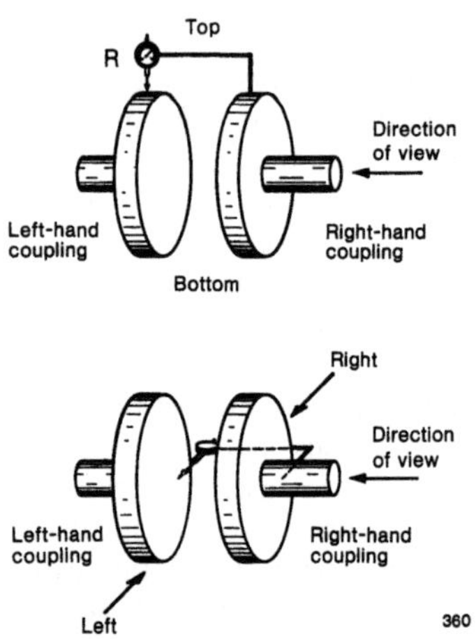

Figure 4. Explanation of Table 3.

tion through 90 deg. during *simultaneous* turning over of the two shafts to be coupled together.

8.4.2.3. Radial Misalignment

In Table 3, a_1, b_1, c_1, and d_1 are the indicated numerical values of the dial gauge R (Figure 4) at the positions a = top, b = bottom, c = right, and d = left. The measured values are inserted into the equations, and vertical and lateral displacements are obtained.

Axial Gap

The axial gap is determined by readings obtained from two dial gauges AI and AII (Figure 5), the first reading with gauge AI on top being designated e_1 and that of the gauge AII at the bottom h_1. The vertical and lateral gaps of the coupling are obtained as indicated in Table 4, axial displacement during measurement having no disturbing effects (in the example, axial displacement = 0.2 mm).

For adjusting alignment, the radial displacements and axial gaps can be reduced by changing the shims at the bearings or housings concerned until the alignment limits are reached.

In the case of couplings for spur gearing, displacement of the pinion under load must be taken into account. If the pinion, subject to direction of rotation and transmission of torque, runs under

Table 4. Alignment formulae for axial gap—4 turns, each through an angle of 90 deg. In the example, the assumed axial displacement of the rotors during measurement is 0.2 mm.

Measuring position	Gauge A I	Gauge A II	Example	
Vertical gap				
Top	e_1	g_1	5.0	4.2
Bottom	f_1	h_1	6.2	5.0
Vertical gap	$\frac{(f_1-e_1)-(g_1-h_1)}{2}$		$\frac{(6.2-5.0)-(4.2-5.0)}{2}=1\text{ mm}$	
+ Coupling gap larger on top – Coupling gap larger at the bottom			Coupling gap opens on top by 1 mm	
Lateral gap				
Left	i_1	l_1	4.0	3.6
Right	k_1	m_1	4.8	4.0
Lateral gap	$\frac{(k_1-i_1)-(l_1-m_1)}{2}$		$\frac{(4.8-4.0)-(3.6-4.0)}{2}=0.6\text{ mm}$	
+ Coupling gap opens left – Coupling gap opens right			Coupling gap opens left by 0.6 mm	

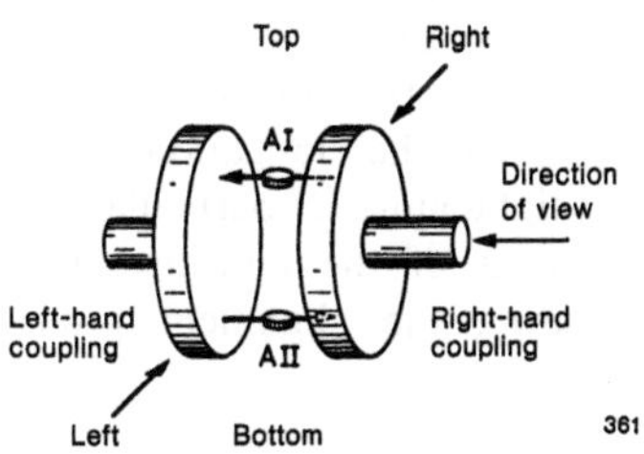

Figure 5. Explanation of Table 4.

load in the upper bearing shell and it is assumed that the pinion can lift itself upwards by two-thirds of the vertical bearing play, then the gear shaft must, during alignment, lie lower by this amount with respect to the shaft to which it is coupled, so that zero misalignment is established during operation. The lateral displacement of the pinion under load can, as a rule, be neglected. If the bearing of the pinion is raised for the purpose of alignment, by a shim having a thickness corresponding to twothirds of the vertical bearing play, zero misalignment can be carried out in the usual way and, after alignment, this shim is then removed. For planetary gearing, the alignment instructions of the manufacturer concerned apply. Particular care is necessary for the alignment of power-dividing gears, which must, in all cases, be aligned strictly in accordance with the manufacturer's instructions (see 8.5.4.1).

8.4.2.4. Tolerances for the Alignment Check

- Maximum permissible gap for axial alignment: 0.02 to 0.05 mm.
- Maximum permissible vertical and lateral radial displacement: 0.03 to 0.06 mm.
- Maximum permissible error in parallelism of coupling flanges: 0.03 mm.
  For couplings with very small dimensions and plain bearings on both sides, these tolerances are reduced to between 0.01 and 0.02 mm.

8.4.2.5. Guidelines for Final Assembly

After the last alignment check, which should establish that the limiting values have not been exceeded, the two cleaned metallic coupling halves are brought together without forcing and are bolted together. The coupling bolts are inserted without grease or other lubricating medium, so that no solid material can enter into the gap between the coupling halves. The threads and bearing faces of nuts, i.e., all parts subject

Figure 6. The screwed fasteners of the divided coupling sleeve loosened during operation. The fact that the threads of the coupling bolts are intact demonstrates that these screwed connections could become loose (see also Figures 3, 7, and 9).

Figure 7. After forced breakage of the transmission shaft at its weakest point, the other side of the gear coupling flew off with the coupling hub and part of the sleeve.

to sliding friction, are carefully treated with solid lubricants, whereas the counter-surfaces must not be lubricated to ensure that the coupling bolts will not turn on their supporting surfaces. Coupling fasteners are tightened crosswise, if possible in three steps. With gear couplings, the coupling sleeves are slid on or the components of the coupling sleeve are bolted together and secured axially.

If, because of changes in the axial play of rotors in their housings, remachining of a spacer disc or manufacture of a new disc becomes necessary, the old disc must first be measured carefully and the new disc produced accordingly or the old disc turned down in strict parallelism.

In every case, after even tightening of the fasteners of a rigid flanged coupling, it is recommended that a concentric-running test should be carried out with the bearing of the lighter shaft removed. This test will show whether distortion of the shaft has been caused by bolting together the coupling halves. For safety reasons, use of the earlier test involving oscillation of the coupled shafts, with the bearing journals freely suspended by a crane, should not be countenanced.

With gear couplings, the movability of the sleeve on the gearing during one revolution of the rotors should be tested. This procedure for verifying the free motion of the gear sleeve is one of the best means of checking machines with gear couplings during inspections or supervisions.

The assembly record must include the final alignment values, the vertical and lateral displacements of the corresponding bearings and bearing housings that had to be carried out, the final radial clearance between rotors and housings at shaft break-throughs, and the elongation of the coupling bolts after tightening. Of particular importance is realignment of the housings after bearing changes during the alignment procedure.

Figure 8. That damage originated in the gear coupling is proved by the forced fracture of the torn-off support lugs of the planetary gearbox, whose gearing and bearings were still in perfect condition, even after the occurrence of the damage.

Apart from the proposed alignment procedure with three dial gauges, as described, there are other alignment procedures which make use of feeler gauges or micrometers. However, these measuring devices are not free from individual disturbing influences engendered by the specialists carrying out the measurements, and it can easily happen that their results are not amenable to being checked by the supervisory personnel of the manufacturer or user. It is in the interest of all concerned that, for checking alignment, all the reliable measuring equipment stipulated here and the markings on the flanges are available. If requested by the user, every manufacturer will inform him of the most suitable alignment procedure for his type of machine. With large machine sets, alignment, or possible realignment, is so costly that it is well worthwhile for manufacturers and users to devote more attention to this important assembly procedure.

8.4.3. Guidelines for the Commissioning of Couplings

Commissioning of a machine set after installation or after repair and revision will show whether the coupling components have been assembled correctly and whether their alignment has been successful. If the running smoothness of the set is good or within the expectations of the manufacturer and user, and if the subsequent checks of bearings and the teeth of gear couplings show no signs of abnormal wear after several runs under load and several days of continuous operation, there are no grounds for suspecting an error in alignment. If there are running disturbances at the rotational frequency and distortions of the coupling, as can be established by a concentric-running test or impeded mobility of a coupling sleeve at operating temperature, the shafts must be uncoupled again and realigned. It is recommended that alignment be carried out by a different specialist, so as to eliminate individual errors or negligence, as can occur with simplified alignment procedures.

After the first runs under load, gear couplings must be checked in respect of axial limits, to determine whether the position of the coupling hubs in the coupling sleeve is correct in view of the relative elongations of the rotors and their axial displacement in the axial bearings as a result of operating conditions. The maximum relative movements of the rotors must be measured during commissioning and compared with the expected design data, and must be entered in the measuring record.

Figure 9. Designed to fail in shear and located between the transmission shaft and the shaft of the driven machine, this flexible pin-type coupling on the output side prevented further damage.

Unwanted transmission of axial thrust by gear couplings, caused by incorrect axial adjustment of the couplings hubs, can lead to substantial damage to turbo-sets and gear units. Unwanted transmission of axial thrust can also arise from inadequate lubrication during operation or from soiling of the teeth of the coupling after a long period in operation. This danger is not announced early enough by devices for monitoring the axial position of the rotors, whereas it can be shown in good time by measurements of axial thrust or of bearing-metal temperature. Inadequate axial mobility of gear couplings can give rise to the most serious damage to driving and driven machines because even after response of the axial-bearing monitor, i.e., during run-down of the installation with the damaged axial bearing, the distortion condition responsible remains, so that the usual unloading of the axial bearing after switching-off does not take place.

After inspection and realignment, it is possible, with gear couplings which have not been checked, that a wear step on the teeth will immediately cause damage to the axial bearing because of its new axial position.

8.4.4. Supervisory and Protective Equipment for Couplings

Measurement of bearing-housing vibration: Limiting values according to VDI-Guidelines 2056.
Measurement of shaft vibration: Danger-limit values according to VDI-Guidelines 2059.
Metal temperature of axial bearings at several tilting pads: With correct disposition of the temperature sensor in the region of the trailing edge, alarm at 90 °C and switch-off at 120 °C.
Axial-thrust measuring equipment: Alarm when the nominal thrust is exceeded.
Equipment for measuring axial position: Alarm or switch-off when exceeding axial play, as determined by displacements in the stationary condition, by 0.1 to 0.5 mm.

8.4.5. Guidelines for the Inspection and Revision of Couplings

Recommended Special Checks during Short Outages

Inspection of bearings, with examination of the load patterns of radial bearings in the upper and lower bearing shells.
Inspection of axial bearings, with examination of the load patterns of the pads in the normal direction of thrust and in the opposite direction.
Visual and sonic checking of shrunk-on coupling flanges and hubs.
Measurement of coupling-bolt lengths after load removal to determine their suitability for further use.
Assessment of the condition of the gearing of coupling hubs, coupling sleeves, and axial limits.
Examination of coupling surfaces to check for signs of the passage of electric current and fretting corrosion.
Alignment and concentric-running check, and comparison with the alignment and concentric-running values recorded during the last check.

Recommended Maintenance Work

Maintenance is particularly important with oil- and grease-filled gear couplings, whose sealing elements should be renewed routinely and whose oil or grease filling topped up or renewed. During inspections at monthly or half-yearly intervals, the coupling guard should be examined for traces of extruded grease, because this is the only way of indicating failure of the sealing elements, which in most cases are not visible.
With forced oil lubrication, the teeth of gear couplings must be cleansed of solids thrown out of the oil.
Flexible couplings must be checked for wear of rubber, plastic, or metallic elements, for cracks on flexible discs and spring grids, and for fretting corrosion on spring packets.
The insulation of couplings must also be examined regularly when checking the insulation of the bearings.

Bibliography to 8.4

[1] Möhle, H.: Beurteilungsmaßstäbe für hochtourige Zahnkupplungen. Konstruktion 17 (1965), No. 2, pp. 61/66.
[2] VDI-Richtlinie 2240: Wellenkupplungen, June 1971.
[3] VDI-Richtlinie 2060: Wuchtung.

Figure 10. The loose seat of a coupling flange of a flexible pin-type coupling resulted in severe fretting corrosion.

8.5. Stationary Gearing

From stationary gear transmissions of all types covered by insurance, 134 cases of damage have been examined. From the results, Table 1 has been compiled, which shows the distributions of the causes and locations of damage. A survey of the origins and manifestations of damage is given in Table 2.

These surveys show that 50% of all damage to gears could be discovered at an early stage by better supervision, particularly by inspection of the gears and bearings, without the occurrence of subsequent damage, which arises to a large extent when pieces break away.

Since, as indicated in Table 1, damage to gears amounted to 60% of the number of cases of damage to gear transmissions, it has been examined separately.

Table 1. Distributions of the causes and locations of damage to stationary gears.

Cause of Damage	Distribution of the number of damage cases [%]	Location of Damage	Distribution of the number of damage cases [%]
Product faults	*40*	Gears	60
Faults in planning, calculation and design	12	Axial and radial bearings	19
		Shafts	10
Assembly faults	9	Casings	7
Manufacturing faults	8	Fasteners	3
Materials faults	7	Seals	1
Repair faults	4		
Operational faults	*43*		
Maintenance faults	24		
Mishandling faults	19		
Extraneous influences	*17*		
Foreign bodies	8		
Disturbances from the driving or driven side	7		
Disturbances from electrical supplies	2		

Table 2. Conditions governing the origins and manifestations of damage to stationary gears. The remaining cases cannot be allocated unambiguously and were usually discovered during revisions.

Origin of damage	Distribution of the number of damage cases [%]	Manifestation of damage	Distribution of the number of damage cases [%]
Mechanical overload	33	Forced fractures	50
Jamming by foreign bodies or broken-off gear teeth	16	Changes in the load pattern of gears and other surface damage	22
Lack of lubricant	10		
Loosening	8	Fatigue and long-term failures	16
Running roughness	6	Incipient cracks	6
Wear	4	Distortions and bent shafts	6
Distortion	4		

Table 3. Survey of the causes and manifestations of damage to gears. The high percentage of forced fractures indicates that supervision of gearing is still inadequate. Inspection of gears is the best means of loss prevention, particularly if it includes non-destructive testing for cracks.

Cause of damage (gears)	Distribution of the number of damage cases [%]	Manifestation of damage (gears)	Distribution of the number of damage cases [%]
Product faults	*41*	Forced fractures	56
Faults in planning and design	11	Fatigue failures	17
Materials faults	8	Changes in load pattern	16
Repair faults	4	Incipient cracks	16
Assembly faults	10	Distortions	4
Manufacturing faults	8		
Operational faults	*40*		
Maintenance faults	21		
Mishandling	19		
Extraneous influences	*19*		
Foreign bodies	6		
Driving and driven machines	10		
Overload from electrical switching	3		

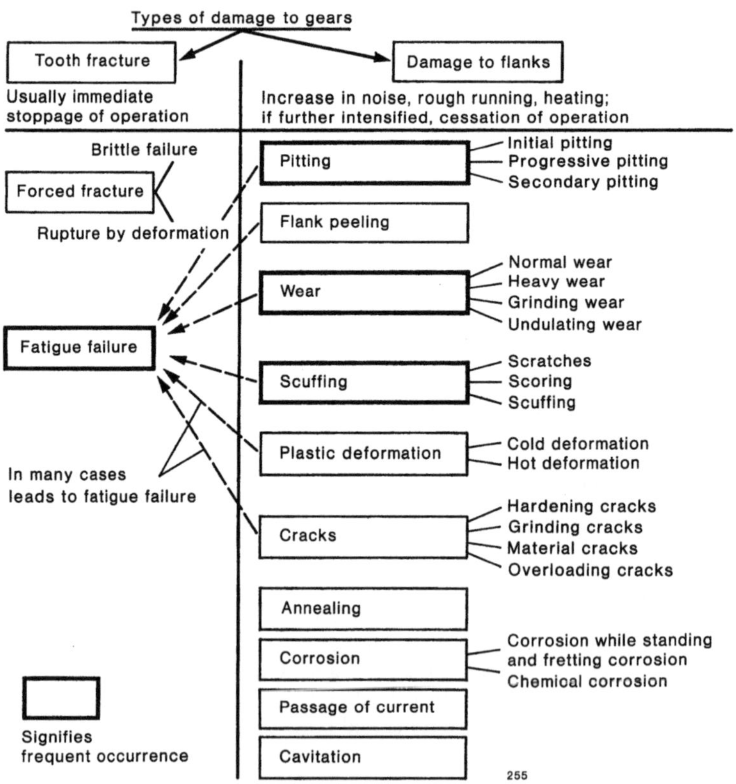

Figure 1. Survey of types of damage to gearwheels.

8.5.1. Types of Damage to Gearing

To complement the above evaluation, it is desirable to discuss the most important components of gear transmissions, particularly gearwheels and bearings, together with indications of the typical types of damage to which they are subject. A survey of the types of damage to gearwheels is given in Figure 1.

8.5.1.1. Damage by Broken Teeth

Tooth breakage usually results in immediate cessation of operation. Measures to prevent breakage are therefore especially important, and safety factors used in calculations must be put higher than those required to prevent damage to the tooth flanks.

Forced Fracture

Appearance (Figure 2)

The fracture surface is usually rough (crystalline) with brittle failure, or smooth and bulging with rupture by deformation. Can also occur in the form of tooth-edge breakage when teeth bear on one side. Splintering at the tooth crest of hardened gears is generally without further consequences.

Causes

Exceeding the tensile strength of the material by sudden overloading through jamming by foreign bodies, with extreme torque shocks, by power switching or incorrect synchronization, and often by mechanical damage to the driving or driven machine.

Countermeasures

Torque limitation by a slipping or shear-pin coupling. Gearing of a "short-circuit-proof" design.

Fatigue Failure

Appearance (Figures 2 and 3)

The fracture surface is fine-grained, frequently with annular shell-like markings around the

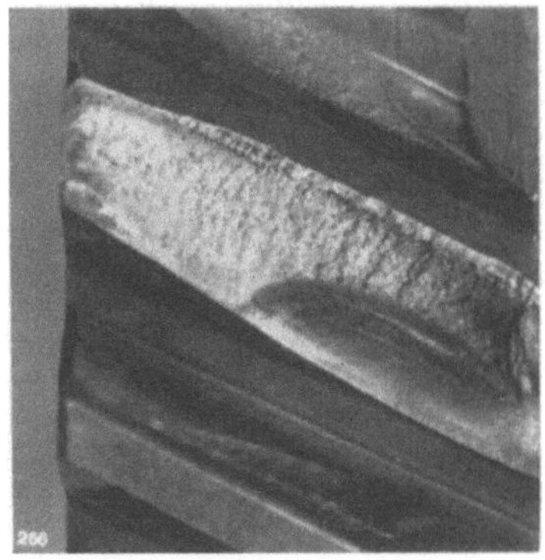

Figure 2. Fatigue failure and forced fracture of a case-hardened helical gearwheel. The proportion of fatigue failure is 30%, the remaining 70% being forced brittle fracture. Fatigue failure can be recognized by shell-like markings and forced fracture by the crystalline structure of the material.

Figure 3. Fatigue failure of an intermediate spur gear, originating from front and rear flanks, since both are loaded (alternating loading of the root of the tooth).

origin of failure, and often with brown areas of fretting corrosion. Final fracture takes place after sufficient weakening of the cross-section of the tooth and is in the form of a forced brittle failure. The origin of failure is on the tooth flank, which is stressed in tension, so that, with planetary and intermediate gears, fatigue failure is possible on both sides.

Causes

Materials fatigue by overloading of gearing and also by vibrations, by an inadequate load-bearing pattern, such as load-bearing on edges, by insufficient dimensioning or by an unsuitable gear material. The point of origin is often from cracks due to earlier torque shocks, cracks arising from machining and hardening faults, slag inclusions and forging defects in the material. If, under nominal load, fatigue failures have not occurred before about 10^7 load cycles, the gearing can usually be considered fatigue-proof. Any failures which occur later can generally be traced back to other causes, such as shocks, overloading, foreign bodies or rough running.

Countermeasures

Torque limitation (see under *Forced Fracture*); careful attention to design, machining, hardening, checking of materials, and assembly. Heat-treatable materials can carry a greater load by surface-hardening. Large, smooth fillets at the roots of teeth reduce stress-concentration peaks.
Fatigue failures also occur as tooth-edge fractures, especially with hardened gears, if there are faults in the load-bearing pattern. Helical gears with a large helix angle are liable to sustain this type of damage. The remedial measure is to use crowned teeth.

8.5.1.2. Damage to Tooth Flanks

Pitting

Appearance (Figure 4)

Hole-like excavations are produced in the flank, often just below the pitch circle. *Initial or running-in pitting* after first operation of the equipment develops at "hard bearing positions" and does not extend further if more even load-bearing of the flanks is achieved by running-in wear. *Progressive pitting* finally results in a pronounced reduction in the load-bearing surface of the flanks, leading to plastic deformation of the flanks and, in some circumstances, to fatigue failure. With gas-nitrided gearwheels, pitting of the flanks does not occur in practice, but peeling does.

Causes

Materials fatigue when exceeding the permissible Hertzian contact pressure, usually with unhardened gears, often arises from external overload, vibrations, an unsatisfactory load pattern (e.g., load-bearing on edges), inadequate dimensioning or an unsuitable material. If, under nominal load, no pitting occurs before about 3×10^8 load cycles, the gearing is proof against fatigue, and any pitting that develops later can be ascribed to external causes, not present previously, such as overload, vibrations or faults in alignment.

Figure 4. Pitting on a heat-treated helical gearwheel, mainly under or in the vicinity of the pitch circle, conditional on the amount of pressure and the sliding characteristics in this region.

Countermeasures

Torque limitation; careful attention to design, machining, selection of materials, and checking of load-bearing pattern during commissioning. Initial pitting with slowly running unhardened gears will be prevented by adding MoS_2 to the lubricants. With heat-treated gears, surface-hardening increases the load limit by a factor of from 4 to 8. Smooth flanks and fewer errors in the involute tooth profile have a positive influence. The use of heavier oils and some synthetic lubricants can be of advantage.
It should be noted that, unless they are removed, pitting particles are often subsequently pressed into sound flanks, giving rise to *secondary pitting*.

Flank Peeling, Flaking

Appearance (Figures 5 and 6)

Peeling of the flank surfaces of case-hardened or induction-hardened gears takes the form of extensive scale-like flaking (Figure 5). With gas-nitrided gears, sharper-edged, fissured splintering of the hardened layer takes place (Figure 6).

Causes

Materials fatigue (see under *pitting*). Also brought about sometimes by faults in the material or in the heat treatment, or during machining. The stresses which are then present cause the early appearance of flank peeling.

Countermeasures

Better control over material, heat treatment and machining. In other respects, as for pitting.

Wear

Appearance and Causes (Figure 7)

With *normal wear,* smoothing of the tooth flanks takes place. Roughness from machining disappears as a result of mechanical contact. *Heavy wear* occurs with the lack of or with an unsuitable lubricant. Hardening of the gears and a lubricant of higher viscosity have the effect of reducing wear.
With *grinding wear,* scoring in the direction of the involute or of sliding and substantial metal removal are caused by hard contaminants in the lubricant.
With *incorrect meshing,* e.g., by too small a centre distance, squeezing or scraping occurs at the root or the tip.
With *gas-nitrided gearing,* the outer nitrided layer can wear as a result of overloading or vibrations, causing steps of from 2 to 10 μm to appear in the flanks.
Wavy flank wear or undulation, perpendicular to the direction of sliding velocity, e.g., on bevel gears of heavy road-transport vehicles and construction machinery, is caused by vibrations.

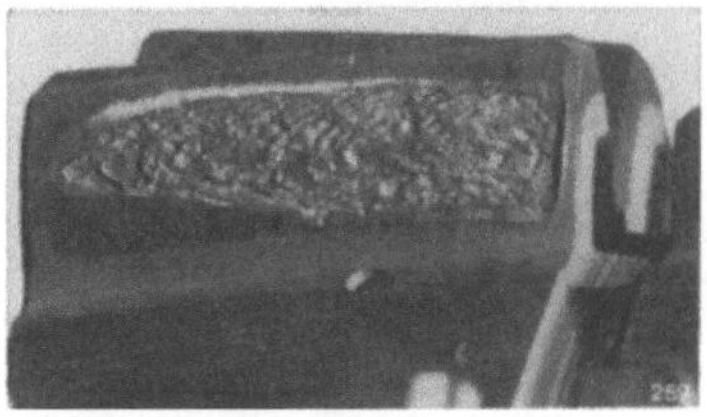

Figure 5. Severe pitting on a case-hardened, ground helical gearwheel led to almost complete peeling of the flank.

Figure 6. Flank peeling of a gas-nitrided 31 CrMoV 9 sunwheel of a planetary gear. Excessive forces of a worn gear coupling caused overloading of that part of the helical teeth which had to sustain the axial thrust brought about by the thermal elongation of the shaft of a turbo-compressor.

Figure 7. Tooth-flank wear on a gas-nitrided helical gearwheel. The difference in height between the dark and light zones is about 5 μm.

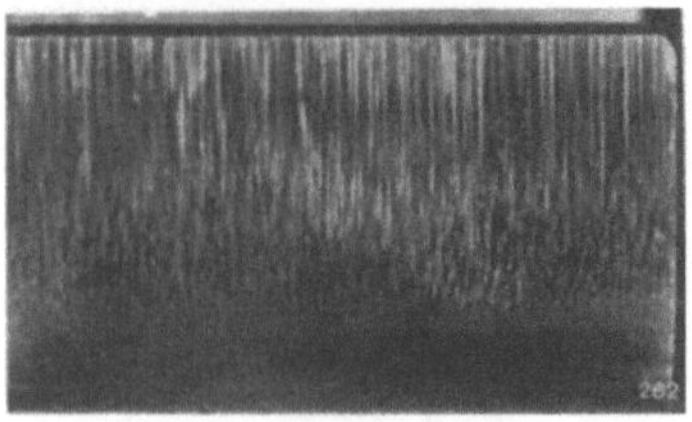

Figure 8. Scoring and signs of scuffing on a case-hardened spur gear.

Scuffing

Appearance (Figure 8)

In its preliminary form, scuffing starts with a few linear indentations or *scratches*. Subsequently, these scratches develop, in the direction of the sliding motion, into *scores* with a depth of from 3 to 5 μm and finally, in the heaviest form of damage, into streaked bands of roughness (*scuff marks*), originating at the root or tip of the tooth in the region of the highest sliding velocity. Scuffing of lesser roughness can also occur during running-in and, like running-in pitting, running-in scuffing does not normally extend further.

Causes

Fusion of the metallic surface following local collapse of the lubricating film at positions of high Hertzian contact pressure and sliding velocity as a result of overloading, an incorrect choice of material or lubricant, or excessive roughness. Scuffing usually shows up soon after the start of operation.

Countermeasures

The use of oils with scuff-inhibiting extreme-pressure additives (high load rate in the FZG test according to DIN 51 354 [8], also Ryder and IAE tests) and of oils with a higher viscosity; attention paid to a favourable load pattern, avoidance of overloads, and tooth-tip relief; the use of hardened materials, particularly gas-nitrided materials, which are substantially more resistant than those that have merely been heat-treated; low flank roughness (an electro-deposited copper layer, 3 to 5 μm in thickness, promotes smoothing out of asperities and running-in); small module.

With turbine gears, scuffing occurs only rarely, despite the high velocities involved, because of the thick hydrodynamic lubricating film.

Plastic Deformation

Appearance (Figure 9)

Plastic deformation gives rise to flattening or wavy deformation of the flank, often with burring at the tooth tip or faces.

Causes

Continuous overload, usually with unhardened materials or faulty hardening treatment; shock-loading by pulsation of lightly loaded auxiliary drives. Plastic flow of the tooth form results from excessive heating arising from inadequate lubrication.

Countermeasures

Torque limitation; use of hardened or surface-hardened materials.

Figure 9. Plastic deformation as a result of dynamic tooth forces.

Surface Cracks

Appearance and Causes (Figure 10)

Hardness cracks are usually long drawn-out cracks on smooth surfaces of high-alloy case-hardening steels and heat-treatable steels with a high carbon content, and are the result of hardening faults.

Grinding cracks are fine, often reticular cracks which sometimes can be detected only by etching. These cracks are caused by overheating during grinding, particularly with case-hardened gears.

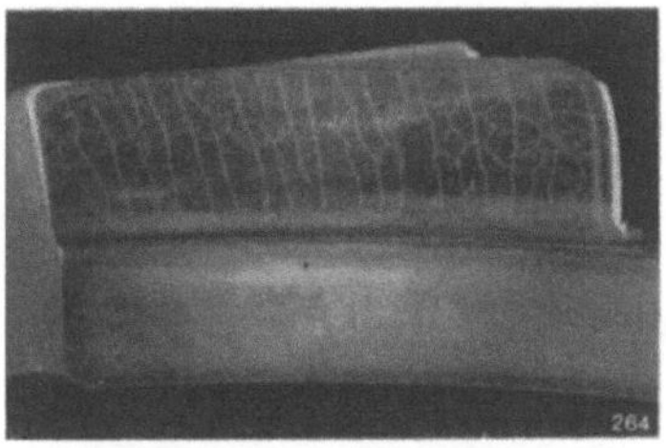

Figure 10. Reticular grinding cracks on a case-hardened helical gear; etched.

Figure 11. An overheated case-hardened helical gear. Loss of oil cooling and breakdown of the lubricating film resulted in wear and scuffing, causing heating up and plastic deformation of the material.

Materials cracks result from fine lines of slag and from forging folds.
Overloading cracks are caused by excessive torque shocks and are especially likely to occur with hardened gears. These cracks can then become the origin of fatigue failures.

Annealing

Appearance and Causes (Figure 11)

Annealing (or, rather, soft annealing by overheating) causes the gearing to change colour, usually to blue (temper colours), to lose hardness, to deform plastically or to scuff as a result of overheating. On the whole, the tooth form is preserved, in contrast to hot plastic deformation.
This annealing is caused by excessive friction during overloading, a lack of lubricant, breakdown of lubricant-cooling equipment, or excessive speed. There is no relationship to the frequently occurring discolouration of gas-nitrided gears, which originates from a chemical reaction of the steel with oxygen when opening the nitriding furnace and is of no importance.

Corrosion

Appearance and Causes

When equipment is subjected to vibrations during non-operative periods, i.e., when there are relative vibrations of machine components in static contact, *fretting corrosion* takes place in the presence of water or oxygen and manifests itself as brownish-red spots. The same conditions apply to fretting corrosion of gears and gear couplings. Moreover, corrosion of the gearing can occur if a gearbox is stored for long periods without corrosion protection or if operation is discontinuous, because moist air is sucked into the interior of the box as it cools down. Also, water vapour from the lubricant condenses on the gearbox walls. If the chemical additives in the oil are unsuitable, the flank surfaces can become porous and roughened or pitted by *chemical corrosion.*

Passage of Current through Gearing

Appearance (Figure 12)

The passage of electric currents through gearing during operation generally produces on the active flanks wide areas which give the tooth a sandblasted appearance, but which are, in fact, composed of an immense number of small craters. With higher currents, larger burn marks, with

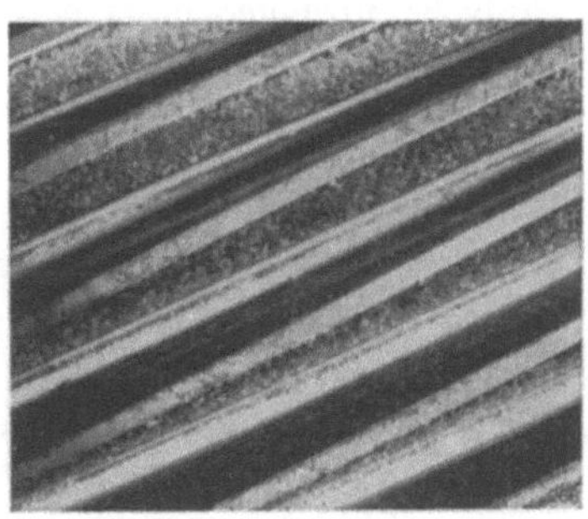

Figure 12. Cratering caused by the passage of current at the leading and trailing flanks of a gas-nitrided herringbone planetary wheel in a gear transmission between a generator and exciter.

temper colours at their rims, will be noticed. If current passes during stationary periods, only a few burnt spots can be observed. The formation of fusion craters and the heating zones can be substantiated by materials inspection.

Causes

With electric motors and generators, the shaft usually has a potential with respect to earth, especially in the case of machines running in plain bearings. Discharge of current takes place through a plain bearing having the thinnest oil film, sometimes only after first passing through the teeth of gears. The teeth of planetary or epicyclic gears often offer the path of least resistance because the annuli with internal teeth are usually in metallic contact with the casing. When electric welding is being carried out on the set while stationary, the passage of current is caused by unsuitable earthing connections.

Countermeasures

Provision of earthing slip-rings; in special cases, insulation of gearboxes from foundations; attention to ensuring that the leads of measuring instruments do not bridge the insulation; use of insulated shaft couplings, especially if there is gearing between two electrical machines.

Cavitation

Appearance (Figure 13) [11, 19]

The surface of the tooth (even, in some circumstances, the rear flank) has a sand-blasted appearance which can easily be attributed mistakenly to the passage of electric current. However, as shown in Figure 14, the small pores produced by cavitation are relatively smooth at the bottom. Materials inspection does not show any thermally influenced rim zones at the craters. In the long run, the flank surface is destroyed by cavitational wear.

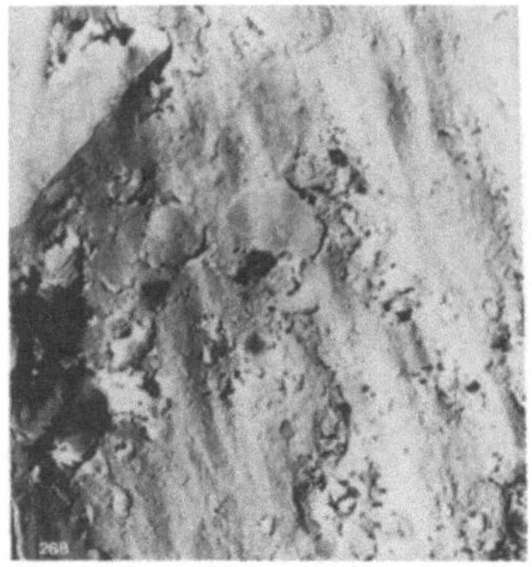

Figure 14. Electron micrograph of the bottom of a crater from Figure 13, showing the relatively smooth surface characteristic of cavitational attack ($\times$ 10,000).

Causes

With fast-running gears—also with the hardened gears—cavitation can occur only if several circumstances cooperate, i.e., vibration from the set as a whole or from the gearbox (vibration cavitation); additives in the oil (e.g., against oxidation, corrosion or foaming); gas dissolved in the oil (e.g., air or gaseous hydrocarbons). The collapse of gas bubbles in the lubricating film between the gear teeth under subatmospheric pressure destroys the surface of the material.

Figure 13. Cavitation on a gas-nitrided herringbone sunwheel of a planetary gear.

Countermeasures

Use of pure mineral oils without additives; precautions to keep gases away from mineral oils; reduction of vibrations.

8.5.1.3. Damage to Plain Bearings of Gears

(see also 8.3. Bearings,
8.6. Oil-Supply Installations)

The following discussion is limited to bearing damage which is specific to gears, particularly planetary gears.

Wear in the form of scoring by particles of dirt is the most frequent cause of failure of plain bearings. Such contaminants are deposited locally by centrifugal force, resulting in a lack of lubricant (Figure 15). Manifestations of fatigue of the white-metal layer in fast-running gears often have their origin in shaft vibrations or faulty gearing. The fatigue strength of the bearing metal is exceeded by the vibrational loading, causing lattice-shaped break-outs from the cast lining (p. 365 Figures 5 and 6).

Radial annular thermal cracks on bolt-type planet spindles of epicyclic gears (Figure 16) are particularly noticeable at changes in cross-section of the steel bolts and are caused by a faulty procedure for casting bearing metal around the bolts, with insufficient preheating of the bolts prior to casting.

Figure 15. This oil-inlet hole in the planet spindle of a planetary gear has been partly blocked radially towards the outside (towards the top of the illustration) by oil contaminants as a result of the effect of centrifugal force.

Passage of Current

Appearance (Figures 17 to 19)

With planetary gears running in plain bearings and a rotating planet carrier, electric currents emanating from the fast-running side, i.e., from the sunwheel, are led away directly via the gearing and the internally toothed annulus connected to the casing, while currents emanating from the slow-running side, i.e., from the planet carrier, are conducted to the casing via the planet bearings and the gearing. A passage of current through the planet spindles therefore indicates that the source of current must lie on the slow side of the transmission [9], while a passage of

Figure 16. Radial thermal cracks on the circumference of a bolt-type planet spindle, caused by insufficient preheating of the steel bolt prior to casting the bearing metal around it.

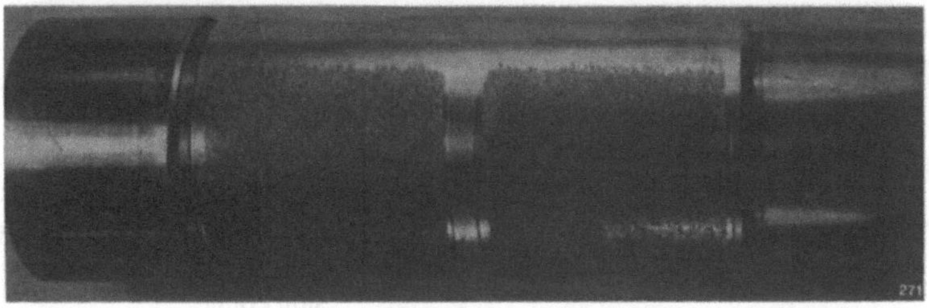

Figure 17. Cratering, caused by the passage of electric current, of the load-bearing surface of a planet bolt surrounded by a white-metal layer.

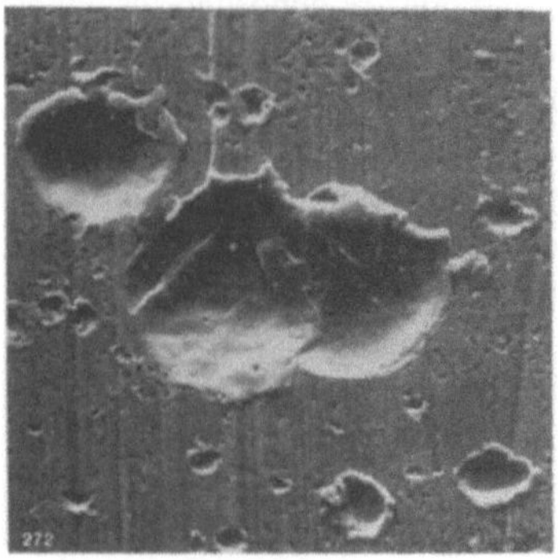

Figure 18. Fusion craters in white-metal caused by the passage of current.

current through the gearing indicates the fast-running side as the source of the current.

For *countermeasures*, see *passage of current through gearing* (p. 368).

Figure 20. Destruction by cavitation of a steel/bronze/white-metal planet bearing spindle after 500 h in operation. Note: The arrow shows the direction of sliding.

Cavitation

Appearance (Figure 20)

With gear transmissions having offset axes, the appearance of bearings attacked by cavitation is no different from that associated with other machines (see Section 8.3). With planetary gears, cavitation occurs almost exclusively on the planet spindles and not on the slow-running bearings of the planet carrier. With oil flowing radially towards the outside, cavitation always proceeds, seen from this viewpoint, in the direction of rotation of the planet wheel, starting with a roughened appearance of the bearing metal, followed by break-outs of the metal. The often strip-shaped areas of damage frequently run in the direction of oil flow.

Causes

Vibrations, either from the coupled machines or the transmission itself, cause relative movements between planet wheels and planet spindles, so that, in the low-pressure area of the lubricating film immediately behind the oil inlet, these strong pulsations cause real underpressures and hence the formation of gas or vapour bubbles whose implosive collapse, on rising lubricating-film pressure, attacks the bearing metal. This phenomenon is often attributable to the use of mineral oils containing additives, e.g., for inhibiting ageing, corrosion or foaming.

Countermeasures

Use of pure mineral oil, particularly with good air-separation capabilities (test according to DIN 51 381); provision of oil tanks of larger size

Figure 19. This polished section shows the heat-affected zone in the vicinity of a fusion crater.

and with baffle plates to improve air or gas separation; reduction, if possible, of vibration excitation [10, 11] (see Section 8.6).

8.5.1.4. Other Damage to Gear Transmissions

Damage to Shafts and Shaft-Hub Connections
(Figures 21 and 22)

In practice, damage to shafts occurs only as a result of shock overloading, i.e., jamming or "short-circuit" stressing. Unhardened shafts are usually plastically twisted (Figure 21), while hardened shafts shear or exhibit spiral-shaped cracks, since the maximum shear stress lies at an angle of 45 deg. relative to the axis (Figure 22). With surface-hardened shafts, these cracks are generally only as deep as the hardened layer, so that the remaining shaft cross-section still permits operation under emergency conditions. Damage specific to shaft-hub connections in gearing takes place only if toothed rims, i.e., relatively thin-walled rings, are shrunk onto the wheel bodies. These rings can move axially and radially. The frequently employed method of testing adhesive strength by exerting axial pressure can often yield values which are too high because, when running under load, elastic deformation of the ring zone momentarily in mesh creates conditions more unfavourable than those obtaining during the test.

Figure 21. Forced fracture of an unhardened gear-transmission shaft after the occurrence of damage by jamming.

Figure 22. Spiral-shaped cracks on the shaft of a gas-nitrided sunwheel of a planetary gear transmission, caused by a "short-circuit" shock with decaying torsional vibrations in the shaft system.

Damage to Auxiliary Drives

Turbine transmissions in particular often incorporate auxiliary drives for lubricating- and control-oil pumps, for turbine governors and tachometers. Usually, only small powers are required to be transmitted, so that, from the point of view of strength, the gearing is greatly overdimensioned. Even so, such unhardened gears can be subjected to damage in the form of fatigue failures or plastic deformations, e.g., from impacts. Damage of this kind can almost always be traced back to vibrations or torsional pulsations, particularly with governor and tachometer drives. In such cases, a reduction in tooth backlash helps and perhaps also hardening of the gears or the use of plastic gears. Moreover, the installation of an oil nozzle with a retarding action or, if there are signs of resonance, the use of a different number of teeth can have a remedial effect. With oil-pump drives, the causes of vibration can be oil-pressure pulsations generated by sharp bends in the pipeline, sudden changes in cross-section in the pipeline system, pulsating relief or pressure-reducing valves, or rhythmic migration of air bubbles through the oil pump. Here, the causes of vibration must be removed; an improvement can be achieved by hardening of the gears and the choice of a different number of teeth.

8.5.2. Planning and Design of Gearing

The information necessary for the selection or construction of gearing or gear sets which satisfy the conditions of the application concerned should be available in the planning stage.
In this connection, the following information should be given by the enquirer or potential customer:

a) Nominal and maximum power according to data from the manufacturer of the prime mover.
b) Rotational speeds or one speed and the reduction or step-up ratio.
c) With variable speed, details of maximum torque.
d) Overspeed (e.g., 10 to 15% with steam and gas turbines, 200 to 400% with water turbines); in the case of planetary gears, for example, this information is important because of loading of the planet bearings by centrifugal force and the supply of lubricant.
e) The types of machines the gear transmission is required to operate; its location (in the open or in a hall) and the range of ambient temperatures; the foundation to be provided.
f) The type of gear to be considered (planetary, power-dividing, spur gearing; horizontally or vertically offset axes; bevel gears, etc.).
g) Subjection to shocks and overloads (see shock factors in VDI guideline 2151, AGMA standard 421.06) and how frequently.
h) Daily operating period; total service life.
i) Maximum noise level (see VDI guideline 2159) and vibration level (see VDI guideline 2056) required.
j) Type of couplings provided on both sides; expected shaft displacements in the axial direction.
k) External paintwork; corrosion protection until commissioning.
l) Oil pumps on the gearing or with external auxiliary drives.

These data are important to enable account to be taken of overloads or torsional shocks which are known to the user, at least qualitatively. Noise and vibration levels must be decided upon right from the start. Couplings which are too large or too heavy can present alignment difficulties by tilting of a transmission shaft mounted in plain bearings by an amount equal to the bearing play. If this condition is maintained during operation, vibrations by imbalance arise. The customer should regard it as important not to receive under-dimensioned gear transmissions. If extreme shocks or short-time overloads are expected during operation, it is expedient from the beginning to incorporate torque- or power-limiting devices in the installation, such as motor-protection cut-outs and shear-bolt, slipping or fluid couplings with overload protection.
After designing the transmission, the gear manufacturer should provide the customer with the following data:

a) Dimensions of the transmission and of the shaft ends.
b) Quantity and type of oil; normal and minimum oil pressure; permissible oil temperatures at inlet and outlet.
c) Permissible bearing temperatures for the selected method of bearing-temperature measurement.
d) Alignment conditions, alignment reference surfaces, alignment tolerances.
e) Measures and equipment for the maintenance, supervision and inspection of the gearing.
f) Noise generation, e.g., according to VDI guideline 2159.
g) Generation of vibrations, e.g., according to VDI guidelines 2056 and 2059.

8.5.3. Manufacture of Gear Transmissions

In special cases, customers can obtain information regarding the workmanlike manufacture of gear units from the makers themselves; alternatively, it can be obtained with the aid of quality-assurance companies, which take as a basis their own acceptance guidelines. Large firms and trade associations have also formulated their own guidelines [12].
For checking *materials quality,* DIN 50 049 and the following guidelines can be used:

- General quality specifications and chemical composition, e.g., DIN 17 100, DIN 17 200, DIN 17 210 [13].
- Strength values according to DIN 1602

(concepts), and DIN 50 145 (tension), DIN 50 106 (compression), and DIN 1605 (mechanical metal-testing).
- Hardness according to DIN 50 351 (Brinell), DIN 50 103 (Rockwell), and DIN 50 133 (Vickers).

In addition, the material can be examined, before and after machining, for freedom from *slag inclusions and cracks* by X-ray and ultrasonic test methods or (only on its surface) by magnetic or dye-penetrant crack-detection methods (e.g., "Met-L-Check"). These last two methods are commonly used on finished gears, whereas X-ray and ultrasonic methods are employed on workpieces before cutting the teeth because gear teeth no longer give a clear echo pattern.

Case-hardened ground teeth can be inspected for *grinding burns and cracks* by etching.

Tooth-flank roughness, which is important especially in the vicinity of the root of the tooth, can be determined, but binding specifications regarding its value are not yet in existence (DIN 4761, 4762).

Gear errors (involute, pitch, accumulated pitch, radial run-out, helix angle) should lie within the tolerance appropriate to the field of application (DIN 3960, 3961, 3962, 3963, 3964, 3967, 3971, 3975, 3976) [14]. For marine gears, British Standard 1807, with three quality grades A1, A2 and B, is preferred.

Moreover, with hardened gears, *hardness* is of significance. When the gears are case-hardened, the hardness value should generally be HV 550 to 850 kg/mm^2 or HRc 50 to 65, depending on the carbon content of the case-hardened layer. With gas-nitrided gears, the hardness value should be HV 700 to 800 kg/mm^2 for aluminium-free steels; otherwise, it should be HV 900 to 1,100 kg/mm^2. Also, the *hardness depth* should be from about 0.8 to 1.5 mm for case-hardening, and from about 0.3 to 0.7 mm for gas-nitriding, though the definition of hardness depth is not yet uniform. With surface-hardened gears, too high a test load results in penetration of the hardened layer, depending on hardness depth, and hence yields dubious results.

With induction- and flame-hardening, the important point is that hardening should embrace the whole of the tooth, including the root fillet; otherwise, fatigue failure can occur at the hardness boundaries.

Imbalance of gear-transmission components can be controlled in accordance with VDI guideline 2060, when the degree of quality Q of the machine coupled to the transmission must be matched. With turbo-transmissions, for example, the degree of quality of the turbo-machine is $Q=2.5$ ($e\,\omega=2.5$ mm/sec).

8.5.4. Acceptance, Installation and Commissioning of Gear Transmissions

8.5.4.1. Alignment

Alignment is important for the life of a gear transmission since faulty alignment can result in the generation of vibrations that damage couplings and bearings. Moreover, one-sided tooth contact, with overloading of the gear, can occur. Particularly with turbine installations or high-speed machinery, alignment problems are in the foreground. The object of alignment is to bring the shaft ends to be coupled into a concentric position during operation and hence to take into consideration bearing play, oil-film thickness and thermal expansion (see Section 8.4.2).

With epicyclic gears running in plain bearings, attention must be paid to ensuring that both shafts run approximately central with respect to the casing or the bearings. By contrast, with spur gearing in plain bearings with offset axes, both shafts are displaced in the opposite direction as a result of the forces exerted by the teeth. Data regarding the position of the shafts during operation must be provided by the gear manufacturer. An inclination of the sleeve by 5 minutes of arc, i.e., an offset of about 1.5 mm over a length of 1,000 mm, is given as a guide for the maximum permissible error in alignment for high-speed crowned gear couplings. The usual radial misalignment tolerances are smaller, i.e., a shaft misalignment of 0.02 mm.

8.5.4.2. Load-Bearing Pattern

The greatest attention should be paid to the *load-bearing pattern* or load-marking of the gearing, which should first be determined during idling by thinly covering the tooth flanks of the gearwheel with engineers' marking ink and then rotating the wheel several times. If, for this purpose, use is made of special lacquers which leave a layer of a few microns in thickness on the tooth flanks, the load-bearing pattern can also be examined under partial or full load, since these lacquers are oil-proof and wear away only after a short run under load. While the load-bearing pattern during idling usually comprises only 60 to 80% of the theoretical load-carrying width of the tooth and 20 to 40% of its theoretical load-carrying height, more or less the whole of the active flank should be marked, for instance, under nominal load [20] (Figure 23). The surprisingly high elastic deformation of the teeth, even with hardened gears, leads to the disappearance of errors in the load-bearing pattern in view of the fact that gaps can usually be measured only in microns. Epicyclic gears designed for load equalization have, in addition, elastic transmission components which can deform as a whole, so that, in this case, larger differences in the load-bearing pattern between idling and running under load are permissible. The load-bearing pattern of spur gearing with offset axes or of bevel gears should

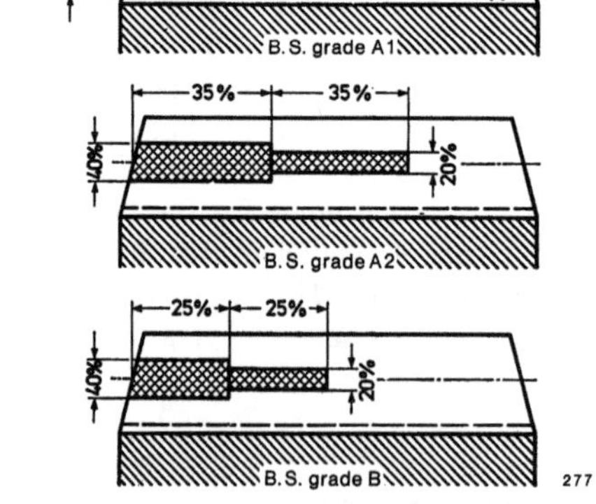

Figure 23. Specification of a load-bearing pattern according to B.S. 1807 for the unloaded flank. The distribution of the pattern can also differ from that shown. Only the load-bearing area is specified. B.S. grades A1 and A2 apply to transmissions with a pitch-circle velocity of over 50 m/sec.

be checked not only in acceptance tests, but also after taking them into operation, because large deformations of a gearbox casing can take place during the installation, for example, of a transmission with offset axes, thereby creating faults in the load-bearing pattern. Any unevenness in the base-plate must then be compensated by shims. Conversely, the method of distorting the casing by means of shims is often utilized for correcting a load-bearing pattern. Transmissions whose pinions are supported in the upper bearing shell during operation should also be examined in this condition with regard to their load-bearing pattern by raising the pinion shaft with lifting tackle. This procedure applies particularly to transmissions whose bearings have been rescraped for correcting the load-bearing pattern. Large oil pipelines should be connected up without imposing any stresses, possibly with a securely held pipe compensator. The deformation of oil-pump casings should be checked with dial gauges when connecting the pipelines.

Faults in the Load-Bearing Pattern

Faults in the load-bearing pattern can arise if spur gears or gear transmissions exhibit the following frequent defects (see Figure 24):

a) Edge bearing through faults in flank direction or helix angle.
b) Edge bearing through faults in parallelism of the shafts or incorrect crossing angle.
c) Axial run-out or wobble, whereby, after a rotation of 180 deg., the flank bears on the other end of the face width (alternating edge bearing); other angles are also possible.
d) Tip bearing or root bearing, caused by faults in the pressure angle of the involute (base-circle fault).
e) Waviness in the direction of width, undulations, also extending obliquely across the flank, caused by faults in the machine or defects in or faulty clamping of the milling cutter or gear hob.
f) Undulations in the direction of the involute, caused by inaccuracies in or faulty clamping of the gear hob or cutter.
g) As the load-bearing pattern is normally generated by the contact of *two* faulty gears, only a check with a master gear can clarify the

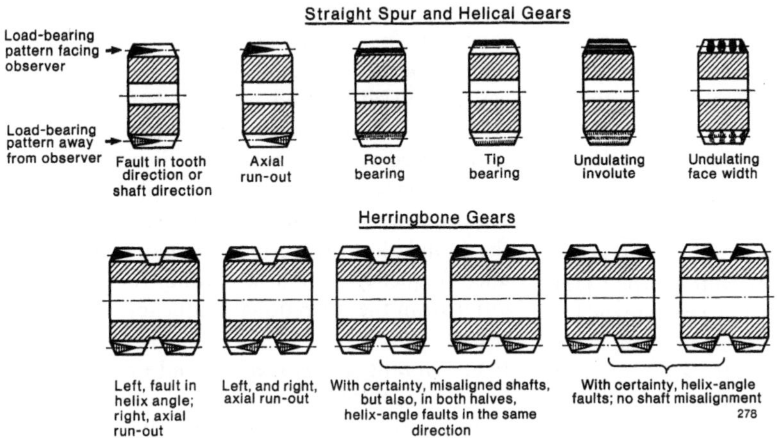

Figure 24. Typical faults in the load-bearing pattern from idling tests with (top) straight spur and helical gears and (bottom) herringbone gears.

type of fault. With a very high gear ratio and hence slender pinions, faults in the load-bearing pattern are caused by twisting and bending of the pinion, i.e., heavy load bearing on the sides of the teeth facing the coupling of the pinion.

8.5.4.3. Selection of Lubricants

Experienced manufacturers of lubricating oils can nominate reference lubricants for similar transmissions. Difficulties can arise if only one oil is to be used for the whole installation. The question of oil viscosity, measured in degrees Engler (E) or centistokes (cSt) at 50 °C, must be settled. Also, a decision must be made as to whether an absolutely pure mineral oil should be used or an oil with additives against ageing, corrosion or foaming, or an oil with additives against scuffing (chlorine, sulphur, and phosphorus compounds), the latter being available as a mild-action or strong-action gear oil (EP oil) (examination with FZG test [8], DIN 51 354, see also gear lubricants according to DIN 51 509). In most cases, slow- and medium-speed transmissions have gear oils with additives against scuffing; high-speed transmissions (such as turbine gears) are generally lubricated with turbine oils, i.e., mineral oils with additives only against ageing, corrosion and sometimes foaming. (Anti-foam additives should, if possible, be avoided because of deterioration of air-separation capability.) The higher is the peripheral speed of a transmission, the lower is the viscosity that can be selected, thereby reducing transmission losses. Thus, slow-running transmissions, such as worm gearing and marine diesel transmissions, operate with oil viscosities from 6 to 10 °E/50 °C (45 to 76 cSt); turbine transmission operate with oil viscosities from 2.5 to 4.5 °E/50 °C (17 to 33 cSt), while some *worm gears operate with oil viscosities from 8 to 50 °E/50 °C (60 to 380 cSt)*. Running-in oils for improvement of the load-bearing pattern and tooth surface are generally not used with stationary transmissions. Transmissions with a peripheral speed at the pitch circle of over 15 m/sec are provided with pressure-oil lubrication.

8.5.4.4. Oil-Supply Requirements Specific to Gearing (see 8.6. Oil Supply)

The *oil tightness* of the transmission must be checked, but a final check can be undertaken only when the transmission operates under full load, since it is often only the hot (i.e., thin) oil that starts to exude. With oil-filled transmissions, a regular check must be made of *oil level* which, incidentally, can drop considerably during operation because a considerable proportion is flung around centrifugally. A check must be carried out in the interior of the transmission to see whether the *oil-injection lines* spray into the proper places; the smallest bore diameter should generally be not less than 2 mm because of the risk of clogging by contaminants. If an internal *coat of paint* has been applied to the transmission, the paint must be guaranteed oil-proof, as a peeling paint layer can clog the filter.

The provision of a high tank for emergency lubrication is usually pointless with epicyclic gearing, which requires a certain minimum oil pressure to be lubricated at all. Thus, with normal tank heights up to 8 m, a circulating pressure of less than 0.7 atg is generated, so that, to be effective, such a tank, although inherently advantageous, must be placed sufficiently high up.

As distinct from the conditions prevailing with spur gearing having offset axes, planetary gears with a rotating planet carrier require a *minimum oil pressure* for lubrication. For lubricating the planet wheels, the oil must penetrate radially towards the rotating shaft against centrifugal force and must at least be capable of overcoming centrifugal pressure.

In addition, it is necessary to check the *temperature difference* of the lubricating oil between the inlet and outlet; if the difference exceeds 40 °C, it indicates that insufficient oil is being put through the transmission. In such a case, the oil pressure should, for example, be increased. Also, in most cases the *temperature in the oil sump* of oil-filled transmissions or the *oil-outlet temperature* of transmissions with pressure-oil lubrication should not exceed 100 °C; otherwise, the oil will age too quickly and hence will require frequent changing; alternatively, scuffing of the gears must be expected.

8.5.4.5. Vibrations

It is useful to check the transmission for *vibrations* during the first run. The magnitude of the amplitude and frequency and the vibration rate [mm/sec] are of interest (critical values are given in VDI guideline 2056). From measurements of frequency, it is possible to draw conclusions regarding the cause of excitation. Thus, shaft vibrations at the frequency of rotation are generally produced by imbalance or faults in alignment. By the same token, vibrations at double the rotational frequency can arise from alignment faults with double gear couplings. With high-speed transmissions, usually if they are lightly loaded, vibration frequencies below the rotational frequency, particularly at about half this frequency, can be caused by "oil whip"—shaft vibrations which are oil-film-dependent. When running up to full speed or shutting down turbine sets, critical shaft speeds in bending of the coupled shaft line, critical torsional frequencies and resonance with the foundation are passed through. This resonance frequency should, according to DIN 4024, differ from the operational frequency by at least ±20%.

If the vibrations observed depend on the power being transmitted, imbalance can usually be excluded as the cause of vibration. Vibrations from the coupled machines are often transmitted to the gear unit and, in such a case, it is necessary to take measurements and recalculate the set as a whole, treating it as an assembly capable of vibrating. Faulty gearing more rarely generates disturbing vibrations, but rather noise, since pitch or tooth-form faults generally emit high-frequency vibrations in the audible range. The low-frequency radial run-out or cumulative pitch errors cannot generate large vibration amplitudes perpendicular to the axis of the machine.

8.5.4.6. Gear Noise

Gear noise increases with the nominal power of the transmission [15] and with increasing speed. Small sound-reflecting rooms increase the impression of noise. Noise measurements should be carried out in accordance with DIN guideline 2159. In contrast, noise is often indicated as acoustic pressure level, instead of acoustic power level, in which case the distance of measurement must also be known. Noise measurement in frequency bands permits localization of the cause of noise. With gear transmission, it is usually the tooth frequency (with spur gears, the number of teeth multiplied by the rotational speed [rev/sec] of the shaft) which is prominent. A remedial measure is suitable reworking of the gears or the provision of a sound-absorbent hood, which reduces the noise level by from 10 to 20 dB, or by from about one-half to two-thirds.

8.5.5. Loss Prevention during Operation

Optimum Supervision

To prevent damage, it is expedient to monitor, in the order of their importance, the following parameters:

- Oil inlet pressure with transmissions lubricated by oil circulation.
- Casing or shaft vibrations.
- Oil temperature and level with oil-filled transmissions.
- Temperature difference between oil inlet and outlet and bearing-metal temperature, oil inlet temperature.
- Pressure difference at the oil filter.

Oil Inlet Pressure with Transmissions Lubricated by Oil Circulation

The supervision of *oil pressure* has been described in Section 8.5.4.4. A drop in pressure to 0.4 kg/cm^2 above the minimum permissible oil pressure should start the auxiliary pump. At 0.2 kg/cm^2 above the minimum permissible oil pressure, the set should be switched off. With all other gears, the minimum oil pressure can be taken as 0.8 kg/cm^2. With both switching procedures, a visual and audible alarm must be given. A drop in oil inlet pressure with clean filters is always a sign of dangerous bearing-clearance enlargement or broken oil-spray nozzles at the gears.

Casing or Shaft Vibrations

Monitoring of vibrations of the transmission casing or, better still, of the transmission shafts in the vicinity of the bearings is now the most effective method of detecting at an early stage incipient damage to the gears, as well as to bearings and couplings. Vibrations are measured by contactless inductive techniques or in accordance with eddy-current methods. The sensors are attached to the casing and measure either the absolute casing vibrations or the relative shaft vibrations [17]. Usually, the vibration amplitude or the vibration rate is determined. Measuring instruments which measure only in definite adjustable frequency bands have found wide usage, particularly in the U.S.A. Frequently, it is even possible to dispense with recording or indication of measured values and be satisfied with an alarm or switch-off pulse on exceeding the set limiting value. The usual casing vibrations normally lie in the region designated "good" according to VDI guideline 2056. Since the permissible amplitude s reduces with increasing frequency, it is more to the point to select for alarm initiation (visual and audible) the vibration rate v_{eff} as the limiting value, which can lie in the region designated "still permissible".

Monitoring of radial shaft vibrations, relative between the shaft and bearing housing, is more delicate, since, as a result of the elasticity and damping of the oil film of the bearings and the relation between the stationary and rotating mass, shaft vibrations are several times higher than those of the casing. Turbine gears with the favourable vibration-rate value of 2 mm/sec on the casing can exhibit shaft vibrations relative to the casing of 20 mm/sec. The limiting values in VDI guideline 2059 can be used as danger criteria for plain bearings [17].

It is useful to register the vibration on a recorder with a chart speed of about 60 mm/hr. An additional recording of loading, on the same chart, is advantageous, as it makes it easy to recognize changes in vibration which are load-dependent.

Oil Temperature and Level with Oil-Filled Transmissions

With oil-filled transmissions, *monitoring of vibrations* is also the most effective procedure. All *oil-temperature* monitors have the disadvantage of responding too slowly, so that, in most cases, gearing and bearings have already suffered severe damage before a rise in temperature has been detected. Partial tooth breakage or heavy pitting often do not change temperature conditions at all. If oil-sump temperature is monitored, use should be made of a thermometer whose limiting-value contact, set at about 10 °C above the normal operating value, triggers an alarm.

Temperature Difference between Oil Inlet and Outlet and Bearing-Metal Temperature, Oil Inlet Temperature

More effective than the supervision of oil inlet or oil outlet temperature is recording the difference between these two temperatures, as, with constant oil quantity, it corresponds to the transmission losses. Extreme bearing or gear

wear can manifest itself by *oil heating*. However, it must not be overlooked that the temperature difference is also influenced by the power throughput or the speed. The supervision of plain and rolling bearings is described in detail in Chapter 8.3 on bearings.
The supervision of planet spindles is difficult. For this purpose, it is possible to provide thermally released pins which, during rotation, strike a key that triggers an alarm or preferably a shut-down signal.

Pressure Difference at the Oil Filter

The pressure difference at the oil filter should be monitored by an alarm release. It the pressure is too high, the filter (with a mesh size usually from 0.06 to 0.08 mm) must be cleaned or a switch-over made to the clean reserve filter.

To summarize, it can be said that, with gear transmissions, the operating oil pressure or the oil level and vibrations offer the most suitable and important possibilities for supervision. All other measures are not very effective because of their delayed response.

8.5.6. Inspection of Gears

When out of operation—at least yearly—an inspection of the gear transmission should take place, i.e., an examination without dismantling.

Checks

Functioning of the oil supply, particularly switching-in of the auxiliary pumps at the specified nominal pressures; oil-level control; functioning of all alarm and switch-off signals.
Oil ageing and contamination by determining the neutralization value (DIN 51 558), the saponification value (DIN 51 559), the content of solid foreign matter (DIN 51 592) and the water content (DIN 51 582); possible topping-up or change of oil; analysis of oil contaminants, particularly of metallic constituents.
Oil filter for iron-containing residues from the gearing or white-metal or bronze from the bearings. It is useful to provide a "swarf catcher" at the oil outlet from the gear unit in the form of a pocket or a magnetic rod.
Alignment of both transmission couplings and, if necessary, re-alignment.
Of great advantage for checking gears, particularly epicyclic gear transmissions, is a hole, about 20 mm in diameter and normally plugged, at a suitable position in the casing, through which the gearing can be inspected accurately by means of a glass-fibre endoscope [21].

8.5.7. Revision of Gears

Revision of gears, i.e., disassembly with checking of components, should be carried out at least after about 15,000 h in operation or roughly every two years. The results must be recorded on measurement charts, photographs, imprints on transparent adhesive film, replicas from cast plastics [18]. The revision can be performed in accordance with the following *check list:*

8.5.7.1. Spur, Bevel and Worm Gearing

- *Cracks* in the casing: Optical checking, perhaps with the dye-penetrant method.
- After opening up the casing: Marking the direction of rotation with chalk on bearings and gears, and marking the active flanks on pinions and gearwheels. Checking whether bearings and gears are correctly installed and, particularly with axial bearings, whether the thrust collar is in the correct position.
- Measurement of *backlash* by inserting feeler gauges while the rear flanks are touching. Winding through lead wire which is subsequently measured. Normal backlash S_{perp} (perpendicular to the flank) should be approximately

  $S_{perp}=0.035\ m_n+(0.05\ \text{to}\ 0.1)\ \text{mm},$

  where m_n is the normal module [mm]. A somewhat larger backlash is permissible, provided that the direction of the externally applied load does not alternate.
- Checking of the *load-bearing pattern* (see Section 8.5.4.2 and Figure 24). Documentation with transparent adhesive tape. Subsequently, lifting the set of gears out of the casing.
- Measurement of bearing clearances and checking with new values. A clearance of from

0.1 to 0.2% of the shaft diameter is customary with plain bearings, and usually 0.5 mm with axial bearings. With plain bearings, determination of the shaft and internal bearing diameters. With rolling bearings, determination of the radial and axial bearing clearances with a dial gauge. For records of bearing measurements, see Section 8.3 on bearings.

- Checking of *gear teeth* for *cracks* by the dye-penetrant and magnetic methods. *Pitting* of moderate severity can also be recorded with transparent adhesive tape after painting with engineers' marking ink, so as to enable a comparison to be made during the next revision in regard to severity and position; the teeth should therefore be numbered. The same consideration applies to *scoring* and small scuff marks, which can also be determined in the direction of the face width with a roughness-recording instrument.
- Checking of gear teeth for *pitch* with a manual pitch-measuring instrument, two-flank and single-flank rolling-test equipment, or a Höfler tester. Checking of *involute profile* and *helix angle* with suitable measuring equipment when it is clear that there are good grounds for suspecting heavy *flank wear* or if the transmission runs very noisily. When measuring the involute, it is easy to observe flank wear (Figure 25) because a step, on which the wear can be read off, is usually formed between the uncontacted root portion of the involute and the active portion of the flank through engagement of the tip of the meshing gear.
- Checking of *radial run-out* or eccentricity of the gearing with a dial gauge via balls or rollers or with a Höfler instrument.
- Checking of the hardness of gearwheels only if there is a colour change after an increase in temperature.
- Checking of the *load-bearing pattern* of plain bearings. Detection of *cracks* by the dye-penetrant method (e.g., "Met-L-Check") and of *bonding faults* by tapping or ultrasonic testing.
- Checking of shaft/hub connections and of gear couplings for freedom from fretting corrosion. Fretting corrosion usually manifests itself as rust-red streaks radiating from the position of the seat towards the outside.
- Checking of oil pumps built onto the gear transmission and of other auxiliary drives.

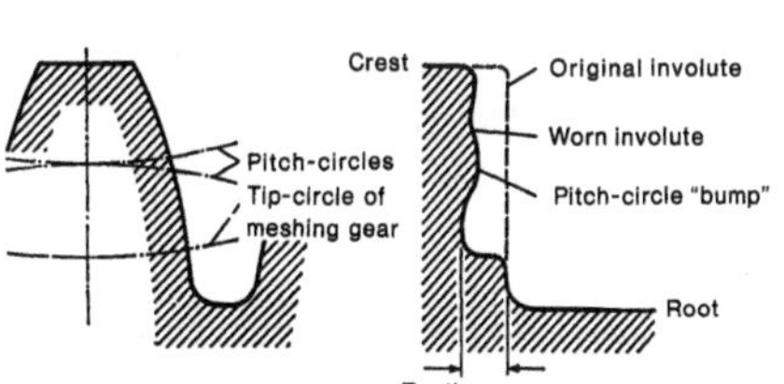

Figure 25. Diagram of an involute with wear of the tooth flank.

8.5.7.2. Planetary Gears

- Planet wheels have active flanks on both the front and rear flank. Distinguishing marks should be applied to the flank loaded from the outer annulus and to the flank loaded from the inner central wheel. The load-equalizing devices generally present, such as gear couplings or flexible elements, must be checked for fretting corrosion and wear, as also must the seats of the planet spindles of the plant carrier.
- The axial-movement limiter in gear couplings must be examined for pressure spots or deformations; if necessary, rotors must be re-aligned axially.

8.5.8. Emergency Repairs

To prevent production losses, it is often necessary to carry out temporary emergency repairs on gearing. The following possibilities exist:

- When the gears show *scuffing damage,* additions should be made to the oil, i.e., an admixture of additives in accordance with the data of the oil manufacturer, and operation should initially be under reduced load.
- When slow-running heat-treated gears show signs of incipient *pitting,* use should be made of an oil containing MoS_2 [7].
- When there are minor *tooth-edge breakages* or cracks and break-outs at the tip, the site should be ground back manually. With larger tooth breakages, there should also be a reduction in the power transmitted in accordance with the ratio of sound tooth width to original tooth width. With herringbone gearing, the undamaged helical half, for reasons of thrust

equalization, must not be taken into account in this ratio.

- With a *"hard" load-bearing pattern,* such as is the case with edge bearing, as a result of which damage can be expected, the involute should be ground back by hand with a corundum stone.
- When there is *damage to the flanks,* such as pitting, scuffing or denting, symmetrical gearwheels—even those of the herringbone type—can be turned round, so that the former rear flank becomes the active flank. Herringbone gears need not necessarily run in the direction of the "arrow".
- When there is *damage to the flanks* (e.g., pitting, scuffing or denting), it is often possible to regrind or rehob a gearwheel and, if it has been heat-treated, to gas- or bath-nitride it. The increased backlash is usually of no importance. With good original stress-relieving of the workpiece, there is a surface growth of only 0.02 mm with gas nitriding and a few microns with bath nitriding.
- *Bonding faults or cracks* in plain bearings can be repaired by soldering.
- *Incipient cavitation* on planet spindles can be inhibited by changing over to a pure mineral oil and improved oil-flow arrangements in the tank, so that the sucked-in oil is free from air or gas. An increase in inlet pressure has proved successful.
- *Cracks* in gear casings can be repaired and made proof against oil leaks by the locking-bolt ("Metalock") system without detriment to subsequent operation.

Bibliography to 8.5

[1] Niemann, G.: Maschinenelemente, Berlin, Göttingen, Heidelberg: Springer-Verlag 1960.
[2] Dudley, D.W., Winter, H.: Zahnräder. Berlin, Göttingen, Heidelberg: Springer-Verlag 1961.
[3] Winter, H., Brugger, H., Dick, F.: ZF-Norm 201 – Zahnradschäden. Zahnradfabrik Friedrichshafen 1963.
[4] Hanisch, F.: Zahnradschäden und ihre Beeinflussung durch Schmierstoffe. Allianz-Berichte für Betriebstechnik und Schadenverhütung, No. 2 (1965).
[5] Mauri, H.: Getriebeschäden an Hauptantriebsanlagen von Seeschiffen. Lecture held at Verein der Schiffsingenieure zu Hamburg on 12. Oct. 1967.
[6] AGMA Standard 110.02 (Dec. 1961). Nomenclature of Gear Tooth Wear Failure.
[7] Barte, A.: Betriebsschäden an Zahnflanken mit besonderer Berücksichtigung grober Pittings. Special issue 5. Works Managers' Conference of Allianz Versicherungs-AG 1960.
[8] Niemann, G., Rettig, H., Lechner, G.: Zur Prüfung von Getriebeölen im Zahnrad-Verspannungsprüfstand – Stand der Erfahrungen. Erdöl & Kohle 12 (1959), pp. 472/480.
[9] Salomon, T.: Elektrische Erscheinungen in der Schmierung. Mineralöltechnik 9 (1964), No. 6, 7, 8.
[10] Frössel, W.: Schwingungskavitation in Gleitlagern. Maschinenmarkt 66 (1960), No. 13.
[11] Möhle, H.: Einige Gedanken zur Kavitation in Gleitlagern. Der Maschinenschaden 40 (1967), No. 4, pp. 125/133.
[12] American Petroleum Institute API Standard 613 High Speed Special Purpose Gear Units for Refinery Services. August 1968.
[13] Stahlschlüssel, 9. edition Marbach/Neckar 1971.
[14] Apitz, Budnick, Keck, Krumme: Die DIN-Verzahnungstoleranzen und ihre Anwendung. Braunschweig: Vieweg & Sohn 1957.
[15] Opitz, H., Zumbroich, H., Timmers, J.: Serienuntersuchungen des Geräuschverhaltens moderner Hochlastgetriebe. Industrie-Anzeiger No. 96, 30. Nov. 1965.
[16] Bartel, A.: Passungsrost bzw. Reiboxidation—besondere Verschleißprobleme. Allianz-Berichte für Betriebstechnik und Schadenverhütung, No. 1 (1965).
[17] Peters, G.: Wellenschwingungsmessungen an Turbomaschinen, Erfahrungen und Entwicklungstendenzen. Der Maschinenschaden 44 (1971), No. 5, pp. 157/192.
[18] Hanisch, F.: Abdruckverfahren für die Oberflächenuntersuchung. Der Maschinenschaden 40 (1967), No. 1, pp. 11/22.
[19] Piltz, H.-H.: Werkstoffzerstörung durch Kavitation. Düsseldorf: VDI-Verlag 1966.
[20] British Standard 1807.
[21] Huppmann, H., Löbert, P.: Erfolgreiche Schadenverhütung in Kraftwerksbetrieben durch den Einsatz von Endoskopen. Der Maschinenschaden 44 (1971), No. 1, pp. 11/13.

Acknowledgement
Our thanks are due to Zahnradfabrik Friedrichshafen AG for placing at our disposal Figures 2, 5 and 10, and to the Research Establishment for Gears and Transmissions, TU Munich, for Figure 8.

8.6. Oil-Supply Systems for Machine Installations

With many machines, damage to bearings through lack of lubricant supply plays an important role. Apart from the low-pressure supply of lubricant to the bearings, the oil system has important tasks to perform in the control and supervisory system, particularly with fluid-flow machines. These tasks are today taken over by high-pressure oil-supply systems with operating pressures from 5 to 100 kg/cm^2. The functioning of control and quick-closing valves, as well as of other hydraulic equipment of a machine, can be endangered by leakages of the high-pressure oil supply; for this reason, hydraulic systems are generally laid out in accordance with the "rigid-pipeline" principle. On such machine installations, leakages in the high-pressure pipeline system or high-pressure fittings are frequently the cause of extensive oil fires which can also spread through the building and to neighbouring non-involved locations and can lead to long outages of the installation. Insofar as oil-supply systems are concerned, loss prevention therefore means prevention of damage to bearings by correct planning, construction and supervision of the low-pressure oil system and the prevention of oil fires by correct planning, construction and supervision of the high-pressure oil system in particular [1–4].

8.6.1. Requirements for Components of Oil-Supply Systems [5–8]

8.6.1.1. Requirements for Oil Tanks [5]

The *dimensions* of the oil tank should be such that, according to the oil requirements of the bearings and of the control and supervisory equipment, as well as the desired amount of oil in circulation, the volume of the tank will be sufficient to accommodate safely the total oil volume, including flow-back of the contents of any high-position tanks that may be present, during periods when the installation is non-operative.

Cuboidal tanks of welded construction and *cylindrical* tanks are currently offered as oil containers. Cuboidal oil tanks facilitate the incorporation of internal equipment and hence make it possible to provide an extensively integrated oil-supply system. Also, these tanks have the smallest dimensions for a given volume and are simple to co-ordinate into the installation as a whole. Cylindrical tanks are cheaper and are employed, particularly in foreign countries, with non-integrated oil-supply systems.

A most important requirement for an oil tank is *oil return* onto a bubble-breaker baffle above the operational oil level, especially if it is possible for the oil to absorb steam or gas, as can happen with steam and gas turbines and turbo-compressor installations with a sealing-oil system.

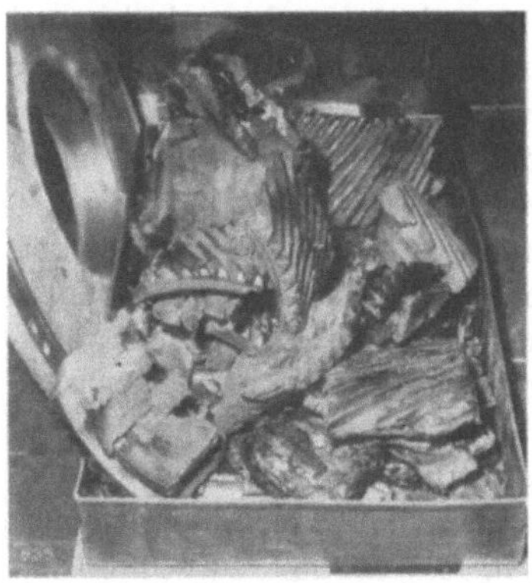

Figure 1. Damage to a planetary gear transmission of a turbine set after failure of the entire main and auxiliary oil supply. A period of 10 minutes elapsed before manual switching off of the generator from the grid and, during this period, the turbine set was operated without oil.

Figure 2. Damage to an axial bearing, with fusion of the white-metal, caused by failure of the entire oil supply.

If, however, steam or gas separation is not important or if the oil-supply system provides for carrying out degassing of sealing oil in an independent gas separator, oil return below the oil level is now preferred to prevent foaming and air absorption. A partition wall with a strainer should be located in the oil tank in such a way that the oil returning from the system is held back for as long a period as possible, so as to keep contaminants away from the suction branches of the oil pumps.

The *bottom of the oil tank* must have a single- or double-funnel configuration to permit separation of water and dirt in the region of the lowest flow velocities.

The *suction branches* of all pumps must be positioned so that air bubbles and sediment from the bottom of the tank cannot be sucked in and also so that a sufficient amount of oil is available for running down the set safely in the event of its being switched off by a signal indicating an inadmissibly low oil level.

The *drainage fittings* at the outlets from the bottom of the oil tank must be secured against unauthorized opening during operation. Mobile or stationary oil-cleaning equipment must be connected to these outlets in such a way that unintentional drainage of the oil tank to below the level of the suction branches of the pumps is avoided. For this reason, connecting lines must be arranged in the form of loops.

The *internal walls* of the oil tank can be left bare or covered with a plastic liner to protect them from corrosion. The lining must permit cleaning of the whole oil system by a pickling treatment [17].

For *outdoor installations* or for certain climatic conditions, heating equipment inside the tank may become necessary. Electric heating, with heating elements that can be replaced during operation, is preferable. If heating by steam or hot water is required, the heating coil must be introduced from above the oil tank, so that, for checking leak-tightness during operation, the tube bundle can be lifted upwards without having to drain the oil tank. When planning the installation as a whole, the necessary space requirements must be stated to avoid difficulties during disassembly. Installation of the heating coil horizontally below the suction branches of the main and auxiliary oil pumps is totally unacceptable. External heating of the oil tank with a tubular coil and appropriate insulation has proved to be quite suitable for outdoor installations.

The *cover* of the oil tank must be provided with a manhole to permit access to the interior of the tank for inspections and cleaning operations. All outlets from the tank, including those above the oil level, must be securely closed to prevent the ingress of water, dust and dirt, particularly when work is being carried out above the tank.

8.6.1.2. Integrated Oil-Supply Systems

The following components of an oil-supply system can be integrated in an oil tank or in its immediate vicinity:

- The separately or remotely driven main oil pump with its foot valve.
- The various auxiliary oil pumps.
- The oil coolers with their de-aeration lines and stop- and change-over valves on the oil and water sides.
- The oil filters for the high-pressure and bearing-oil systems with their stop- and change-over valves.
- The majority of fittings, such as pressure-reducing valves between high-pressure and bearing-oil circuits, stop-valves for isolating oil circuits or pumps, to enable repairs to be made to these components when the installation is not running or during operation.
- The oil-vapour exhaust.
- The fire-protection valve or the fire-protection slide valve for the remotely controlled isolation of the high-pressure oil system from the bearing-oil system.

This integrated arrangement of a high-pressure and a low-pressure oil-supply system within or adjacent to the oil tank is particularly preferred for reasons of fire protection. The internal installation of pumps and fittings ensures the smallest possible number of pipelines and connections outside the oil tank, prevents points of leakage at shaft seals, and leads to a very compact arrangement of the oil-supply system (hydraulic power pack). These packaged units, in terms of location, can best be separated from the hot parts of the installation by building a wall.

8.6.1.3. Main Oil Pumps

The main oil pump must be de-aerated either on the suction or the pressure side. The air bleeds must be kept open permanently. Bleed lines on the suction side must be led to below the minimum permissible oil level in the tank, while bleeds on the pressure side can be free, e.g., led to a bearing housing or to the oil tank.
The main oil pump must be provided with a foot valve in the oil tank, unless it is a separately or remotely driven submerged pump.
For use as main oil pumps, centrifugal and positive-displacement types, directly driven from the set, are available.

Centrifugal Pumps

Advantages

Safe design of pump without narrow clearances and with a very low probability of outage of the pump itself. No inadmissible pressure rise if the flow resistance in the oil system to be supplied increases. Substantial increase in delivery if the resistance in the oil-supply system drops.

Disadvantages

Difficulties with suction. Delivery drops rapidly with decreasing speed and hence an inadequate emergency oil supply to the bearings if the speed of the machine set is reduced. Correction of oil pressures entails a change in the geometry of the impeller or diffuser of the main oil pump, in that the leading and/or trailing edges of the impeller and/or diffuser blades must be set back.

Positive-Displacement Pumps
(Screw or Gear Pumps)

Advantages

Linear relationship between delivery and speed, i.e., a safe emergency oil supply to the bearings with decreasing speed down to zero. No increase in delivery if the resistance of the pipeline drops because of leakages.

Disadvantages

More liable to give trouble because of the narrow clearances between the displacement elements and the casings; damage by clearance bridging and overheating; heavy pressure drop in the high-pressure oil system in the case of sudden oil demands by the control elements, with no possibility of adjusting delivery rate and pressure. With machine sets that are liable to run in the reverse direction, a positive-displacement pump will, in such an event, operate against the non-return valve on the suction side, generally leading to bursting of the pump casing. For this reason, a safety valve must be provided between the non-return valve in the suction branch and the pump.

8.6.1.4. Requirements for Auxiliary Oil Pumps

It is customary to install at least one full-load auxiliary pump and one auxiliary bearing-oil pump as an emergency unit. This auxiliary bearing-oil pump must be so designed that the supply of bearing oil can be maintained for about half a day at normal speed, during speed reduction and, if necessary, also during operation via barring gear. An auxiliary bearing-oil pump that can take over the supply of bearing oil only during barring operation does not satisfy safety requirements, as postulated on the basis of experience of damage. These auxiliary oil pumps are preferably submerged types without a suction foot valve and hence with only a non-return valve in the pressure line. These pumps must be attached to the cover of the oil tank in such a way that they can be removed or exchanged without trouble during operation.

Choice of Drive Media

- A reliable three-phase a.c. supply for the full-load auxiliary oil pump.
- Steam for the full-load auxiliary oil pump or the auxiliary bearing-oil pump.
- Direct current for the auxiliary bearing-oil pump.
- Compressed air for the auxiliary bearing-oil pump.

Based on experience of damage, it is recommended that the full-load auxiliary oil pump should be driven by a three-phase a.c. motor and a second full-load auxiliary oil pump or the emergency oil pump by a steam turbine.

If the full-load auxiliary oil pump is driven by a steam turbine, the emergency oil pump should preferably be driven by an air motor. This latter recommendation applies particularly to installations without a reliable supply of direct current and to outdoor installations.

Generally, direct-current drive motors have proved their value for auxiliary bearing-oil pumps. However, in the event of trouble, the capacity of the batteries has tended to be insufficient to safeguard the supply of oil to the bearings during run-down of the set.

Both auxiliary oil pumps should be interlocked with the main oil pump via pressure controllers and with the aid of non-return valves on the pressure side, so as to avoid with certainty the mishandling faults which frequently occur, especially when starting up and running down and during disturbances.

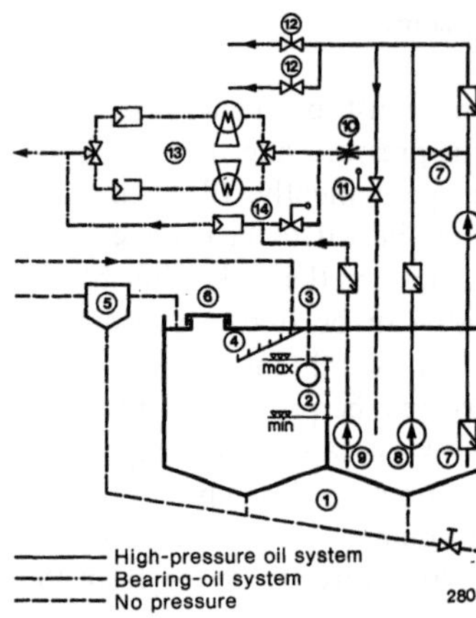

Figure 3. 1 Oil tank with rim for leakage oil, a bottom of double-funnel configuration for dirt and water separation, and secured sludge-removal valves. 2 Partition wall, with inserted strainer, in the oil tank. 3 Oil-level indicator with contacts for alarm and switch-off signals in the event of an inadmissible rise or fall in oil level. 4 Bubble-breaker baffle for the oil return, to improve separation of gases, air and steam from the oil return and to prevent foaming. 5 Oil-vapour exhauster with oil and steam separator; the separated oil/water mixture flows to the sediment in the oil tank. 6 Projecting manhole to prevent the ingress of dirt and water. 7 Main oil pump with foot valve on the suction side, non-return valve on the pressure side, and closable filling line on the pressure side. 8 Full-load auxiliary oil pump, as a submerged unit with non-return valve on the pressure side. 9 Auxiliary bearing-oil pump or emergency oil pump with non-return valve in its pressure line, which underpasses the oil cooler and feeds the bearing-oil system via a single filter. 10 Adjustable pressure-reducing valve, which reduces the pressure of the oil in the high-pressure line to the required bearing-oil pressure. 11 Relief valve for preventing an inadmissible pressure rise in the high-pressure oil system, e.g. after extensive control operations. 12 Fire-protection slide valve or fire-protection valve for several separate high-pressure oil circuits, e.g., separately for live-steam emergency shut-down and control valves, for reheater emergency stop-valves and for cross-over valves. 13 Oil cooler and oil filter for the bearing-oil circuit, with common uninterrupted switch-over of these units, each designed for 100% duty. 14 Safety valve which, when difficulties arise with the oil coolers, oil filters and change-over valves, opens a bearing-oil by-pass via a single filter. During operation with the emergency oil pump, this safety valve is closed.

Requirements for the Interlocking of Auxiliary Oil Pumps

Manual switching-in of the full-load auxiliary pump and of the auxiliary bearing-oil pump must always be possible.

Provision should be made for manual switching-in of the shaft-jacking oil pumps for hydrostatic unloading, if bearing-oil pressure is present from the full-load auxiliary pump or the auxiliary bearing-oil pump.

Manual switching-off of the shaft-jacking oil pumps should be possible at normal speed and when the shaft is stationary or comes to rest after switching off the barring gear, when the time between switching off the barring gear and complete stoppage of the rotor can be taken into account by a time relay.

Manual switching-off of the full-load auxiliary pump and the auxiliary bearing-oil pump should be possible only if the set is not in operation.

Manual switching-off of the full-load auxiliary pump and the auxiliary bearing-oil pump during operation of the machine should be possible only if the supply of oil has definitely been taken over by the main oil pump or upon a signal for automatic switching-off of the full-load auxiliary oil pump when the non-return valve on the pressure side of the auxiliary oil pumps has been closed by the overriding delivery pressure of the main oil pump. Circuits which permit manual switch-

ing-off of the full-load auxiliary oil pump without guaranteed take-over of delivery by the main oil pump must in all circumstances be regarded as inadmissible. This proposed interlocking arrangement utilizes the open position of the non-return valve of the main oil pump and the closed position of the non-return valves of the auxiliary oil pumps as a precondition for releasing the manual switching-off facility or for the automatic switching-off of the auxiliary oil pumps.
Provision should be made for automatic switching-in of the auxiliary oil pumps by pressure controllers when the machine is operating and/or for closure of the weight-loaded non-return valve in the pressure line of the main oil pump. Figure 3 shows an oil-supply system with fire-protection valves and the arrangement of the main oil pump, the full-load auxiliary oil pump and the auxiliary bearing-oil pump, as proposed here.

8.6.1.5. Requirements for Oil Coolers

The oil coolers are generally in a parallel circuit with uninterrupted switch-over by a double three-way valve and with continuous de-aeration of the oil spaces into the oil tank. Installation of the oil coolers inside the oil tank has proved to be advantageous, but provision must be made for the removal of the nest of cooling tubes without having to interrupt operation of the installation. In the majority of cases, only the bearing oil is cooled.
The oil-cooler tubes must be in accordance with the guidelines given in Section 5.2 (Condensers).
Selection of the pressure on the oil and water side is governed by the following criteria: Whilst at one time penetration of cooling water into the oil system had to be prevented at all cost, emphasis is now laid, for reasons of environmental protection, on providing a higher cooling-water pressure to prevent the ingress of oil into the cooling water.

8.6.1.6. Requirements for Oil Filters

It is recommended that the bearing-oil filters should be arranged in series behind the oil cooler and that a common emergency by-pass for the oil cooler and oil filter should be provided. In an emergency, this by-pass is opened and an alarm is triggered by an automatic safety valve which is adjustable during operation, so that the supply of oil to the bearings is not interrupted by difficulties with the oil cooler, the cross-over valve and the filter. After the by-pass has been in operation, and hence after an unfiltered supply of bearing oil, it is absolutely essential to carry out a bearing check. Bearing-oil filters with a mesh size of 0.2 mm and, because of the requirements of the control and protective equipment of the machine, high-pressure filters with a mesh size of 0.04 mm, should be designed as double units. After cleaning edge-type filters, there is a tendency towards an increased amount of small dirt particles which previously were held back by larger foreign bodies, whereas filters with inserts or filter elements which are removed for cleaning do not cause this sudden increase in dirt. In any event, magnetic inserts should be provided; these inserts can be taken out during short stoppages and retain even fine iron abrasive particles.

8.6.1.7. Requirements for Fittings

Axial non-return valves or non-return flaps with closure aids—weight- or spring-loading—should be provided with internal or external damping, so as to prevent pressure shocks in the oil-supply system, with the inherent risk of damage to valve seats and possibly deformations.
Non-return flaps with external indication of position and facilities for attachment of position sensors for automatic plant and normal interlocking are preferable to closed seal-free valves, since the latter lack the possibility for position indication and control.
Spring-loaded or weight-loaded non-return flaps should be installed horizontally, so that closure starts as soon as the pressure is reduced and hence avoids slamming the flap.
Vertically installed non-return flaps, particularly foot valves in suction lines, are closed by the weight of the oil column above the foot valve. These valves must never have a de-aeration vent; otherwise, they can empty themselves, and will then not deliver when switched on and will perhaps run dry. Despite the pressure loss

incurred, it is recommended that closure of a foot valve should be aided by the application of a light spring pressure.

Non-return flaps or axial valves in the pressure lines must be provided either with a small bleed or filling hole in the valve disc or, better still, with their own filling or bleed line with stop valve, as this provision makes it possible to remove pumps during operation for the purpose of overhaul.

Constant-pressure, pressure-reducing, cross-over and relief valves must be easily accessible during operation for maintenance work and checking and must be located away from hot components. Preferably they should be designs which permit replacement of seals and readjustment without leakages in the normal operating position. Since all these fittings are partly installed in suspended pipelines, their casings must not be made of cast materials susceptible to brittle fracture. Valve bodies made of grey cast iron are not acceptable.

8.6.1.8. Requirements for Oil-Vapour Exhausters

In the simplest case, vents protected from dirt and without an exhauster will suffice. If there is risk of admission of gas to the oil with steam-turbine, gas-turbine and turbo-compressor installations, an exhauster is used to deliver the oil vapour from the oil tank into the open air via an oil separator. The power of the exhauster must be sufficient to ensure that the oil-filled spaces of the machine installation, particularly the bearing housings, are under subatmospheric pressure. Also, the exhauster must be powerful enough, during non-operational periods of the installation, to generate subatmospheric pressure in the remaining bearing housings if one bearing is opened up. This requirement is especially important with hydrogen-cooled generators, since accumulations of hydrogen in bearing housings constitute an explosion hazard.

8.6.1.9. Requirements for Oil Accumulators [16]

Recently, apart from spring-loaded piston-type accumulators, hydro-pneumatic oil accumulators have often been used in the vicinity of control elements to balance short-time peaks in oil de-

Figure 4. The damage to this plastic bag of an oil accumulator was caused during installation.

mand. These hydro-accumulators consist of a steel bottle, in which a quantity of fluid is stored under pressure, and a plastic bag which is filled with nitrogen and displaces the stored oil quantity on pressure reduction in the oil system. These accumulators must be easily accessible, must be remote from hot components and, after installation (horizontally or vertically), must always have the oil valve facing downwards. Also, they must be mounted in robust fixtures, so that, in the event of rupture of the pipeline at the oil connection or fracture of the gas valve, the accumulator will not be pulled out of its mountings by the recoil and sent hurtling unrestrained through the plant.

The checking of hydro-pneumatic accumulators must be carried out with suitable filling and test equipment strictly in accordance with the operating instructions of the manufacturer of the accumulator. For loss prevention, the following recommendations should be observed:

- Only nitrogen must be used for filling the plastic bag.
- Welding, brazing or mechanical repairs of any kind must not be carried out on the accumulator casing.
- There must be no permanent connection of the nitrogen bottle with the gas bag of the accumulator.
- Spare accumulator bags should be stored in a cool, dry and dark place, and should be unstressed and unfolded, but suspended at the gas valve and blown up to their normal length with air.

If a check of a hydro-pneumatic accumulator, which should take place about once a month, shows it to be defective, the plastic bag must as a rule be renewed. Particularly important is draining of the accumulator on the oil side according to instructions, to exlude the possibility of oil spraying out. Lubrication of the accumulator casing and of the exterior of the bag with the medium that will later enter the accumulator from the hydraulics side has been shown to be advantageous as, in this way, damage to the accumulator bag is prevented.

8.6.1.10. Requirements for Pressure-Carrying Oil Pipelines

Whilst it is customary in the U.S.A. to locate pressure-carrying pipelines inside the pressure-free oil-return lines, this costly constructional method has not found favour in Europe. The special separation of pressure-carrying oil pipelines from hot components is recommended, e.g., by laying pipes in channels in the foundation.

Pipeline systems under pressure are generally made with welded connections, sometimes with welded socket joints because of reduced soiling of the oil system during assembly. Simple prewelded flanges with open soft seals between them are especially unsuitable, since experience has shown that, particularly under vibrational stressing of the pipeline or through assembly faults, such as inexpertly tightened flange fasteners, the seals can burst radially by oil pressure and consequently can produce substantial leakage in the form of spray.

To make spray leakage impossible, the use of flanges with recesses, in which are located soft seals, is recommended or, if simple prewelded flanges are employed, armoured or metal-covered gaskets. In general, screwed or flanged connections should show up assembly faults by pronounced leakage without extensive radial spraying of leakage oil.

Vibrational stresses should be kept away from compression or cutting-ring screwed joints by clamping the pipes.

Pipelines must be checked during the commissioning stage in respect of their movements caused by thermal expansion and by vibrations. Thermal expansion leads to sticking of pipelines at break-throughs of ceilings, walls or gratings. Vibrations lead to abrasion of pipelines at these break-throughs and give rise to inadmissible and unknown overstressing at welded or screwed positions. Both these modes of stressing endanger the operational efficiency of high-pressure pipelines and can result in the occurrence of sudden leaks which can cause oil fires.

Flexible connections between rigidly laid pressure-oil lines and the machine should be avoided. However, should their use become imperative for operational reasons, armoured pressure hose made of an oil-resistant plastic material is preferable to corrugated-pipe connections.

Unpressurized oil-return lines must be provided with fire- and impact-resistant sight-glasses at the bearing housings for the control of oil discharge. The use of plastic viewing flanges in these pipelines is to be avoided because, under thermal influence, they can be responsible for extensive leakages from the bearing-oil system, which carries large amounts of oil. The possibility of complete inspection of the oil pipeline system during operation, particularly of the high-pressure oil system with all flanges, screwed connections and fittings, is absolutely essential to discover with certainty leakage sites during operation. Such leakage sites often do not show themselves in the cold state—when expansion conditions are different. The remedying of leaks at flanges or screwed connections by retightening the flange fasteners or coupling nuts can be carried out *only once*. If leakage occurs again, the installation must in all cases be shut down to ascertain the cause of leakage, as experience has shown that there must be some reason other than loosening of screwed connections. Usually, with flanged connections, the cause of leakage stems from incipient tearing of seals, followed shortly by their rupture, or, with pipe couplings, from cracks in the pipe in the invisible area under the coupling nut.

It is recommended that the high-pressure oil system should be capable of being isolated from the low-pressure system, so as to enable, during operation with barring gear and during short non-operative periods, repair and maintenance work to be carried out on the high-pressure oil system and the connected control and safety equipment.

The best oil-pipeline system is the shortest possible system with few bends. For this reason, obliquely laid pipelines are to be preferred. The oil system as a whole should so designed that, with only a few valves and by-pass lines, critical components can be isolated for repairs in an emergency, without necessitating an interruption of operations.

For the design and assembly of such pipeline systems, models have proved especially useful. During production, particular attention must be paid to soiling of pipes and to their cleaning.

As a matter of principle, pipes should not be bent with a sand filling because, if certain temperatures are exceeded, the sand will become so firmly attached to the interior of the pipe that its removal throughout cannot be reliably effected by chemical or mechanical cleaning. When cleaning pipes with steel shot, care should be taken to ensure that the internal surface of the pipe is completely dry, so as to exclude possible adherence of the steel shot.

All pressure controllers must be installed in permanently de-aerated pipeline branches in such a way that their functioning can be checked during operation without having to dismantle them and also that their reponse level can be re-adjusted.

Measuring probes must be so installed in the oil-supply system that they dip into the flowing medium. They must be of solid or welded (not brazed) construction.

8.6.2. Supervision and Protection of Oil-Supply Systems

A distinction must be made between two danger criteria:

- Dependability of the supply of lubricating and control oil to the bearings and control and protective equipment of the set [9].
- Security against fire of the oil-supply system [10, 11].

8.6.2.1. Supervisory and Protective Equipment for the Oil-Supply System

Supervision of Oil Level

Integrated float-type devices in the cover of the oil tank, capable of easy replacement and with facilities for reading the oil level during topping up of the tank and with manual control of adjustable alarm and switch-off pulse transmitters, are preferable to the sight-glass type of oil-level indicator.

If sight-glasses break, they can be the cause of oil fires and, if there is a fire, they will feed the fire further with oil. Connecting branches for external magnetically operated oil-level gauges must never be located below the level of the suction branches of oil pumps.

The following recommendations are made for supervision of oil level:

Alarm pulses:
Oil level higher than the operational and stationary level marks.
Switch-off pulse:
Oil level lower than the permissible operational level mark.

Supervision of Oil Pressure

Each pressure, power and control-pulse circuit and each oil inlet to a bearing or sealing position must be provided with at least one pressure-gauge connection for checking oil pressure at this location. These control connections permit functional checking of the entire oil-supply system during commissioning, during disturbances, and after repair work or revisions. In particular, dosage of oil quantity when effecting adjustments by fixed or adjustable throttles can then be carried out with assurance if the pressure-gauge connections are placed between the throttle and consumer locations.

Safety Equipment for High-Pressure Circuits

Pressure gauge with limit-contact	as stage 1
Pressure controller	as stage 2

Stage 1: Alarm and switching in the full-load auxiliary oil pumps.
Stage 2: Alarm and switching in the full-load auxiliary oil pumps and the auxiliary bearing-oil pumps, switching off the set.

Safety Equipment for the Bearing-Oil Circuits

Immediately before entry into the bearing furthest removed from the oil pumps:

Pressure gauge with limit-contact	as stage 1
Pressure controller	as stage 2

Stage 1: Alarm and switching in the full-load auxiliary oil pumps.
Stage 2: Alarm and switching in the full-load auxiliary oil pumps and the auxiliary bearing-oil pumps, switching off the set.

Safety Equipment for the Sealing-Oil Circuits

Differential-pressure measurement with limit-contact between the sealing oil and medium to be sealed	as stage 1
Pressure controller or measurement of oil level in high-position tanks with minimum-level contact	as stage 2

Stage 1: Alarm and switching in the full-load auxiliary oil pump or sealing-oil pump.
Stage 2: Alarm and switching in the full-load, sealing-oil and bearing-oil auxiliary pumps, switching off the set.

Supervision of Oil Temperature

Temperature sensors must be provided before and after oil coolers and after each bearing and sealing location. The measuring position for bearing-oil temperature after the cooler is recommended as a protective and safety monitor, with alarm initiation when the temperature falls below 35 °C or exceeds 50 °C. With automatic or remotely controlled installations, these limiting values can be utilized for starting interlocks or triggering switch-off pulses. The oil temperatures past the bearings serve as elements for bearing protection (see Chapter 8.3).

Supervision of the Current Consumption of Electrically Driven Oil Pumps

The ammeter of electrically driven oil pumps should be provided with marks corresponding to the normal current consumption of the drive motors of the pumps while the main oil pump is running and when it is shut down. In this way, the satisfactory functioning of the oil pumps can be checked during operation.

8.6.2.2. Prevention and Extinction of Oil Fires

Disposition of Pipelines

It is advantageous to arrange the supply and discharge of oil from below fittings, so that leakage oil cannot drip onto hot pipelines and fittings.
All pipelines, even if they are laid in channels in the foundation, should be arranged in such a way that fire extinction can be carried out easily and from several places.

Fire-Protection Valves

The fire-protection valve or fire-protection slide valve with remote control is installed in the oil-supply system in such a way that, in the event of a fire, the main oil pump and all auxiliary oil pumps will feed only into the bearing-oil system and not into the high-pressure oil system (see circuit diagram in Figure 3)[12].
The fire-protection valve should be able to close only when the emergency shut-down and control valves of the set are already closed, in case the high-pressure oil is required to back up the closure movements of the stop-valves. With nuclear power stations, subdivision of the high-pressure oil system to the live-steam, reheat and cross-over valves is recommended, so that, after closure of the live-steam valves, reactor steam can continue to be led to the condenser. The return of high-pressure oil to the oil tank is permissible only as long as the minimum bearing-oil pressure is maintained, since only then can the main oil pump supply sufficient oil to the bearings during run-down of the set if the auxiliary oil pumps are out of operation as a result of a fire.

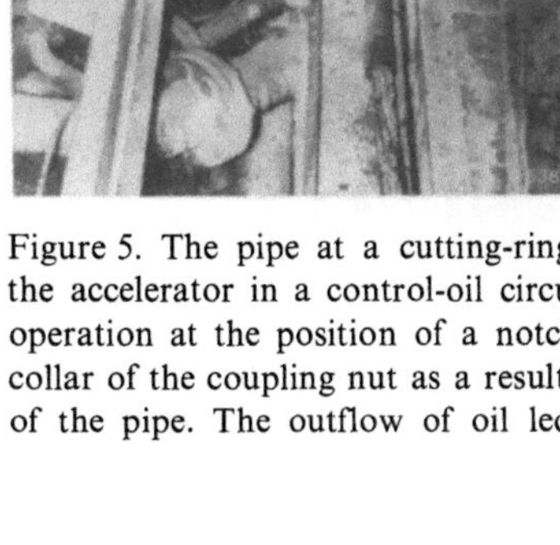

Figure 5. The pipe at a cutting-ring screw joint of the accelerator in a control-oil circuit broke during operation at the position of a notch caused by the collar of the coupling nut as a result of faulty laying of the pipe. The outflow of oil led to an oil fire.

Fire-Protection Emergency Circuit

The fire-protection emergency circuit closes the fire-protection valve and switches off the set. If there is no fire-protection valve, it switches off the full-load auxiliary oil pump. This circuit is conditional on correct dimensioning of the auxiliary bearing-oil pump for oil supply at full speed.

Early Detection of Oil Fires [10]

The entire oil-supply system can be supervised by ionization-type fire alarms or by photometrically operated smoke alarms. An ionization alarm responds only to combustion gases after ignition, whereas a photometric alarm can announce the presence of oil vapour and smoke before ignition takes place. Since surfaces at temperatures from about 80 °C upwards favour the evaporation of oil constituents and, at the same time, the generation of combustible vapours and gases which can be ignited by contact with very small surfaces at higher temperatures, and particularly by sparks or open fires, a smoke alarm has advantages. If leakage oil is absorbed by porous insulation layers, such as glass wool or sprayed asbestos, surface temperatures of about 600 °C can be reached in the course of several hours [2]. Air circulation promotes smouldering, which can, after a further supply of leakage oil, result in an open fire. Since the time interval from the inception of oil vapour and smoke until ignition can amount to several hours, smoke alarms should preferably be located in the region of the oil tank, the auxiliary oil pumps, oil-pipe crossings or

Figure 6. Damage caused by an oil fire to a turbine control station near the front bearing pedestal.

Figure 7. Cable runs and distribution boxes ignited by an oil fire.

neighbouring pipes, and hot components. The installation of a remotely controlled extinguishing, foam or flooding plant for the oil-supply system, including the oil-distribution network, or a fully automatic sprinkler plant is urgently recommended, particularly for oil-supply systems which, during operation or for fire extinction, are difficult of access or can be approached only for short periods.

Extinction of Oil Fires [10]

Smouldering or glowing fires in the region of insulations can be controlled with CO_2 or powder extinguishers. The removal of oil-contaminated insulation is essential in all cases; otherwise, there will always be a potential fire hazard. Experience has shown that oil fires involving a large supply of oil from a leakage point can best be extinguished with water. Often there is too long a period of delay before an oil fire is fought with water, because of fears of the effect of thermal shock on hot components of the installation. However, pipelines can be protected adequately from thermal shock by sheet-metal-clad insulation, and fittings and casings by sprayed-on insulation, provided that extinction is not attempted with a concentrated jet of water but with a water curtain. For this reason, hoses should be provided with adjustable nozzles. Well proven in service

Figure 8. The coupling nut of a screwed pipe connection became loose during operation. The resulting outflow of oil led to an extensive oil fire.

are wall hydrants of the foam type. Extinguishers and respirators should be stored, ready for use, in a room, not itself endangered by the effects of fire and smoke, in the vicinity of the oil-supply system.

Protection by Fire-Resistant Control Fluids [13–15]

The most comprehensive protection from oil fires is obtained by filling the high-pressure circuit with a fire-resistant or non-flammable control fluid. The same technical recommendations, without restrictions, apply to the lay-out of a supply system employing one of these special fluids instead of lubricating oil.
The quality of the paint coatings, the nature of the seals and other non-metallic materials used in the system must be agreed upon with the manufacturer of the control fluid. Because of the narrow temperature range in which these fluids can be utilized, it is strongly recommended that the control-fluid tank should be provided with heating arrangements, regulated in accordance with the temperature of the fluid, and should be thermally insulated. Nowadays, control-fluid systems are normally equipped with electrically driven main oil pumps, designed for 100% continuous duty. Since, at present, these systems are still rarely encountered, we recommend that, where there is an interest, we should be consulted, as our insurance covers systems with all control fluids currently available, and users and manufacturers of these systems have experience going back up to ten years at their disposal.

8.6.3. Commissioning of Oil-Supply Systems [18]

After assembly of an oil-supply system, all oil-carrying components must be subjected to intensive cleaning, so that the supply of oil to bearings and to the control and protective systems should be free from dirt. Cleaning is best effected by flushing with the lubricating oil to be provided for operation. For this purpose, a minimum amount of 60% of that required for operation should be employed. The following tasks are important for flushing with oil:
All *bearing-oil throttles* must be fully open. In special cases, the immediate supply to the bearings is stopped by means of blanking-off plates, and flushing is carried out via a by-pass line, so that dirt cannot enter the bearings themselves.
The *upper shells of the load-carrying bearings* are removed, so that the oil inlet bores of the load-carrying bearings can be directly connected with the bearings pedestals. By leading the flushing oil directly into the bearing pedestals, the entry of dirt between the shaft and bearing shell should be prevented.
The *oil lines* for the *shaft-jacking oil* are disconnected from the bearings and are freely led into the bearing pedestals, so that dirt cannot get into the lower bearing shell.
All *regulating slide valves* in the control and safety systems are removed to permit the maximum through-flow of oil.
Oil heating can be effected by the existing heating equipment in the oil tank or by provisional hot-water heating in an oil cooler.
Flushing is carried out with the full-load auxiliary oil pump.
The oil is heated to between 70 and 90 °C and is cooled several times to 30 °C. By this alternating-temperature stressing of the whole oil-supply system, the intention is to break away *residues of sand in pipe bends* and *scale particles from welded areas*. Mechanical shaking of the oil-supply system has the same effect. A recommended

procedure is banging the oil-supply pipeline system with a hammer, either manually or, better still, with a pneumatic hammer, especially at pipe bends and welding sites.

After a flushing operation lasting 24 hours, it is possible to install *oil filters, oil strainers* and *bearing-oil strainers,* so that dirt particles in the oil circuit can be separated. It is desirable to draw off a sample of oil from the outlet of the oil tank for checking the degree of contamination.

Centrifuging and filtering of oil during flushing provide continuous cleaning of the whole oil system. All components of the oil-supply system, including the second oil cooler and all auxiliary oil pumps, should be cleaned by flushing with oil for a prolonged period.

As the *high-pressure pumps for the shaft-jacking oil system* are very sensitive to dirt, this system should be flushed only after the oil system has already been extensively cleaned. Filtering of the inlet oil to the shaft-jacking oil pump is strongly recommended if small piston pumps or gear pumps are used.

Flushing with oil is terminated by uncovering all bearing pedestals and carrying out careful inspection and cleaning of all accessible oil spaces, especially the dead spaces between the ribbing of the bearing housings. Also, it is highly desirable to remove the lower shells of the load-carrying bearings after flushing with oil and to clean them carefully.

Finally, the entire content of flushing oil is drained from the oil tank and the oil coolers. These components, too, must then be cleaned thoroughly with non-fraying materials before refilling with the oil to be used in operation. After regeneration and filtering, the flushing oil can safely be used again as fresh oil.

8.6.4. Inspection and Revision of Oil-Supply Systems

Dirt should be removed from the oil tank and a check of water content made monthly to yearly, depending on operational conditions and the type of installation.

Indications from Oil Analyses of Possible Damage:

White-metal or bronze abrasions in the oil:
Damage to bearings in the set.

Steel abrasions in the oil:
Damage to gears in gear pumps or transmissions.

Raw water in the oil:
Leakage in the oil cooler.

Condensate in the oil:
Blowing of leakage steam from the shafts seals of the steam turbine onto bearing housings and sucking it in by too a high a vacuum of the oil-vapour exhauster.

An *examination of the oil or special control fluid* should be made by the manufacturer at prescribed intervals, at least once a year.

Precise records must be kept of *oil consumption,* of first filling and subsequent topping-up operations, of admixtures of additives and regeneration.

An *inspection* should be made at least once a day, according to a check list which takes into account the peculiarities of the installation. During this inspection, particular note should be taken of movements of the oil-supply system caused by thermal expansion and vibrations, of position indications of pipeline suspensions and of possible leakages.

Comparisons should be made of the *oil pressures, oil temperatures* and *oil levels* in the system.

Decrease in bearing-oil pressure:
Increased play in the load-carrying bearings.

Decrease in the pressure of the high-pressure oil:
Increased clearance in displacement pumps, particularly axially with respect to the casing.

Increase in bearing-oil temperature past the cooler:
Dirt on the water side or insufficient cooling water.

Increase in bearing-oil temperature past the bearings:
Bearing damage.

Rise or fall in oil:
Fracture of a pipe in the coolers or in the heating coils in the oil tank.

Drop in oil level:
Leakage in the oil system.

Functional checks, involving switching on the full-load, bearing and sealing-oil auxiliary pumps, should be made once a day.

The *set values of the contacts* of pressure and temperature monitors, as well as of level monitors and position indicators on fittings, should be checked about once every three months.

Cleaning of the oil-cooler tube nest and of the oil filters should be carried out according to the degree of contamination; a check should be made once a year.

Checking the running noise, once a year, of pumps and drive motors normally used as stand-by because of possible damage by vibrations while standing.

Revision of the *auxiliary oil and main oil pumps* should be made during revision of the main set or during non-operative periods.

Bibliography to 8.6

[1] Feldmann, J.: Schäden in Kernkraftwerken – eine erste Analyse des Versicherers. Allianz-Berichte für Betriebstechnik und Schadenverhütung, No. 13 (1969).

[2] Lutz, H.: Der Ölbrand im Maschinenhaus des Atomkraftwerks Mühleberg. VGB-Mitt. (1972). Der Maschinenschaden 45 (1972).

[3] Huppmann, H.: Das hydrodynamische Gleitlager im Großmaschinenbau. Der Maschinenschaden 42 (1969), No. 1, pp. 1/9.

[4] Huppmann, H.: Schäden an Gleit- und Wälzlagern. VDI-Berichte No. 141, pp. 97/105.

[5] VDEW-Technische Richtlinien für Ölversorgungsanlagen von Dampfturbinen. 3rd edition. Frankfurt/Main: Verlags- und Wirtschaftsgesellschaft der Elektrizitätswerke 1969.

[6] Seelmeier, H.: Ölversorgungsanlagen für Turbomaschinen. GHH-Techn. Berichte 69 (1969), No. 3, pp. 2/13.

[7] Arnold, H.P.: Ölversorgungsanlagen für Turbomaschinen. Ölhydraulik und Pneumatik 14 (1970), No. 9, pp. 418/422.

[8] Schmierung von Dampfturbinen. Energie 15 (1963), No. 8, pp. 331/335.

[9] VDEW-Überwachungs-, Sicherheits- und Schutzeinrichtungen von Dampfturbinenanlagen. 1st edition. Frankfurt/Main: Verlags- und Wirtschaftsgesellschaft der Elektrizitätswerke 1967.

[10] Brandschutz im Kraftwerk. VGB-Tätigkeitsbericht für 1968/69, pp. 28/31.

[11] Brandschutz an Dampfturbinenanlagen. Empfehlung zum Vermeiden von Schäden an Dampfturbosätzen No. 12, VIK 1970.

[12] KWU-(Siemens)-Information No. 50.

[13] Whirl, S.F.: Fire-resistant turbine lubricating fluids. Lubrication Engineering 26 (1970), No. 5, pp. 174/170.

[14] Brandenstein, H.: Betriebserfahrungen mit einer Steuerflüssigkeit im Turbinenbetrieb. Energie 19 (1967), No. 5, pp. 163/165.

[15] Schober, J.: Schwerbrennbare Hydraulikflüssigkeiten. BBC-Mitt. 53 (1966), No. 12, pp. 142/148.

[16] KWU-(Siemens)-Information No. 93/65.

[17] Die Entrostung eines Turbinenölsystems mit Ammoniumzitrat. Energietechnik 19 (1969), No. 12, pp. 550/553.

[18] Siemens Montage- u. Betriebsanleitung Turbinen-Ölspülungen. KWU, Mülheim/Ruhr.

8.7. Diesel Engines

The following guidelines are based on experience of over 800 cases of damage which occurred to internal combustion engines during a period of four years.

Discussion of the Causes and Locations of Damage

In Table 1, *product faults* include faults in planning, design, manufacture and assembly. *Operational faults* embrace maintenance and mishandling faults. *Extraneous influences* comprise all those cases of damage which have occurred to installed engines in connection with damage to the driven machine.

The fact that pistons, crankshafts, and cylinders and cylinder liners are the components most frequently subjected to damage indicates a high proportion of maintenance faults (deficiencies in coolants and lubricants).

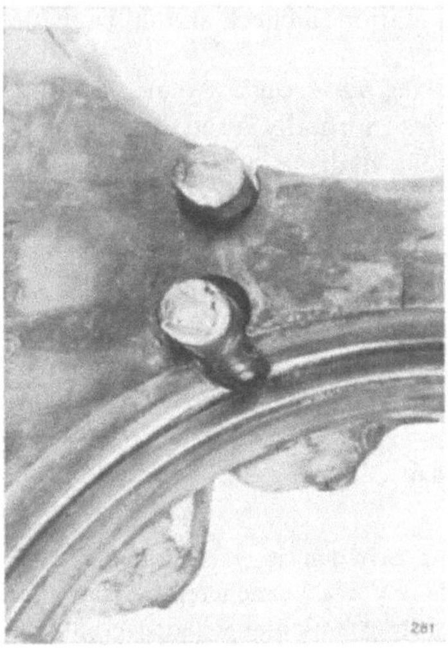

Figure 1. Crack in the fillet of the supporting area for the collar of the cylinder liner in the frame of a 1,200-hp diesel engine, caused by insufficient wall thickness.

8.7.1. Planning, Design, Manufacture and Assembly of Diesel Engines

Planning

During planning and design, manufacturers favour the use of components from a proven series and apply them, with small dimensional modifications, to a new, more powerful design. With large increases in power, or several small increases, components which have previously been satisfactory can fail.

Example:

When raising the power by increasing piston diameter, the cylinder liner must also become larger. If, now, the crankshaft and frame of a proven design are applied without change, the sealing areas of the cylinder-head joint become smaller, resulting in sealing difficulties. The reduced wall thickness can cause cracks emanating from the blind holes of the cylinder-head screws and at the bases of the supporting areas for the collar of the cylinder liner (Figure 1).

Table 1. Distribution of the causes and locations of damage cases.

Cause of damage	Distribution of the number of damage cases [%]	Location of damage	Distribution of the number of damage cases [%]
Product faults	20	Piston	24
Operational faults	43	Crankshaft	24
Extraneous influences	37	Cylinder and liner	22
		Bearings	12
		Frame	12

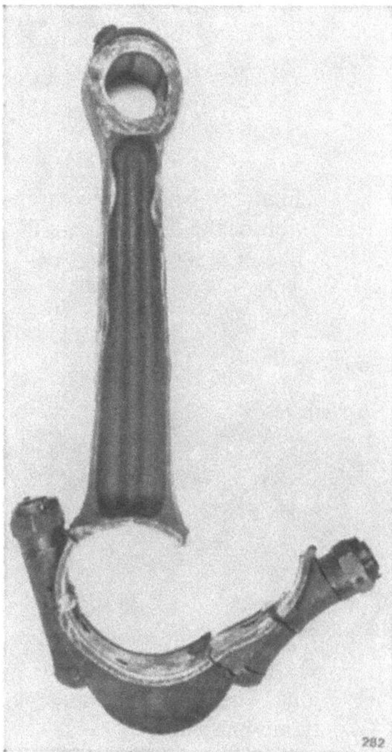

Figure 2. Fracture of a connecting rod of a 120-hp diesel engine at a point of insufficient strength.

Moreover, connecting rods (Figure 2) and crankshafts (Figure 3), as well as bearings, can be mechanically overstressed. Also, the increased thermal loading can result in damage.
The environmental conditions of the application must be taken into account. Dust-laden air requires special filters and sealed bearings, while humid locations require the provision of corrosion protection, with care taken to supply dry combustion air. If ambient temperatures are high, cooling should be augmented; also, combustion air should never be aspirated from the vicinity of the exhaust pipe.
If pipelines are different from those supplied as standard equipment (e.g., for combustion air, fuel, cooling water, and exhaust), pipe resistances must be recalculated and taken into account.

Design

As confirmed by experience, sharp changes in cross-section lead time and again to damage to moving and to stationary components (fillet radii in accordance with DIN 250). Damage from inadequate cooling can often be avoided by changes in design. Thus, cracks in the webs of cylinder heads (Figure 4), cracks in injector bodies and galling of pistons can often be traced back to deficient cooling inherent in the design.
Attention should be paid during the design stage to the selection of locking devices for screwed

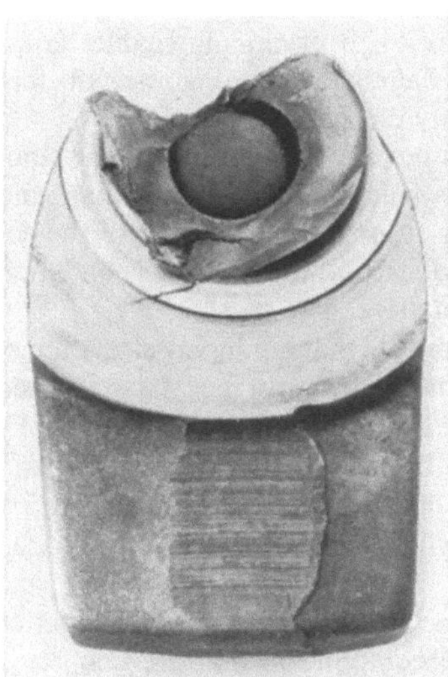

Figure 3. Fatigue failure of the crankshaft of a 23-hp diesel engine as a result of insufficient material thickness.

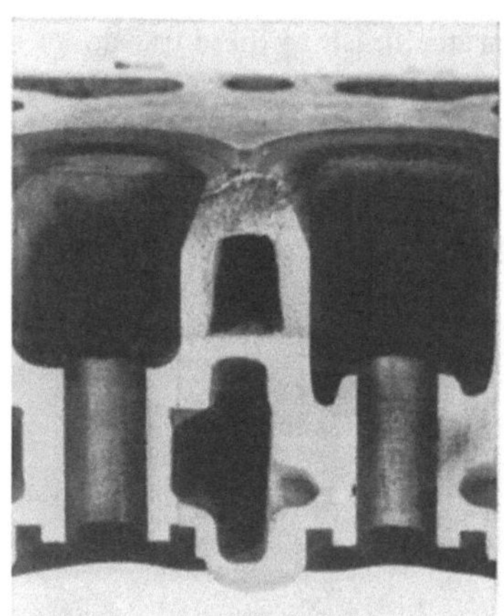

Figure 4. This web of the cylinder head of a 500-hp diesel engine was cracked as a result of deposits at the site of a large accumulation of material.

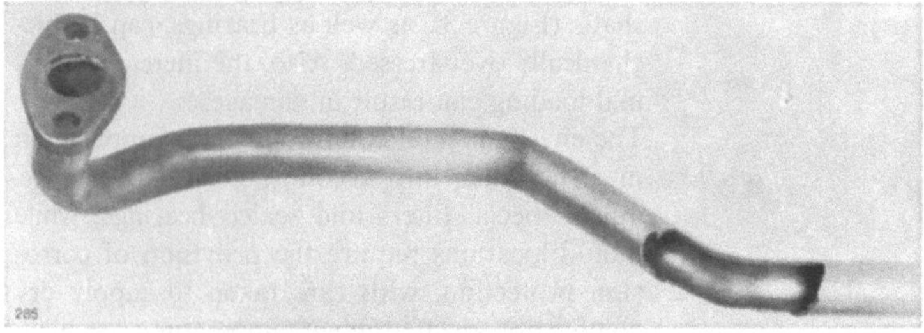

Figure 5. Fracture by vibration of the lubricating-oil line of a 500-hp diesel engine.

Figure 6. Failure of the camshaft pinion of a diesel engine. The pinion was not flame-hardened.

fasteners (see also Section 8.1. Screwed Fasteners), to avoid damage caused by loosening or unscrewing of screwed connections (Figure 5).

Manufacture

The fillet radii at cross-sectional transitions must in no circumstances be less than those stipulated in the drawing. Special care should be taken to obtain a good surface finish at these transitions. Scores and corners constitute notches which lead to stress peaks and hence to incipient cracks and fatigue failure.

Figure 7. Fracture of the crankshaft of a 15-hp diesel engine as a consequence of faulty heat treatment.

Faulty heat treatment or hardening permits development in the workpiece of stresses that can also lead to cracks or fractures. In this respect, crankshafts are particularly susceptible to damage (Figures 6 and 7).

Cast components must be free from defects. An ultrasonic test prior to machining enables faults to be detected, thereby eliminating rejects before the usually costly machining operations.

Of particular importance is cleaning of cast components to remove any residues of moulding sand because, during subsequent operation of the engine, such residues promote heat accumulation with crack formation.

Machine components that are highly stressed mechanically or thermally must not be weakened by holes (e.g., for attaching pipe clamps), by punched-in symbols for type identification, and by welding or other processes. Even with good mass balance, vibrations will arise, during operation of the engine, which experience has shown are especially likely to lead to fracture of those engine components that are connected to the foundation. These connections, e.g., at cooling-water lines, fuel lines, control levers, etc., must be particularly carefully made and tested (Figure 8).

Figure 8. Wear of cooling tubes of a 300-hp diesel engine as a result of poor brazing into the tube plate.

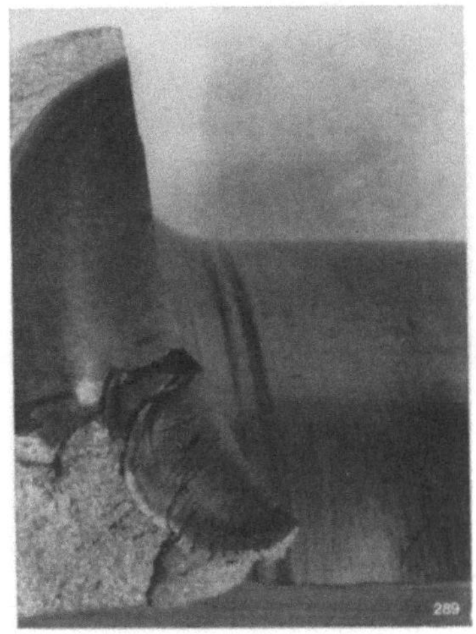

Figure 9. Fracture, caused by slag inclusions, of the crankshaft of a 450-hp diesel engine.

Materials faults in the form of inclusions and contaminants or arising from forging faults must be detected by non-destructive testing, e.g., by ultrasonic, radiographic, dye-penetrant or magnetic-powder techniques before machining the workpieces, or at least before their installation. To be able to assess whether a defect so found is still permissible from the point of view of the operational safety of the engine, it is necessary to draw up guidelines based on laboratory investigations or on experience gained from actual cases of damage. It is known that materials faults are mainly found in components which are prefinished by forging or pressing, particularly crankshafts (Figure 9), connecting rods, rocker arms and valves.

Assembly

During assembly, it often happens that individual components are forgotten or are installed in an incorrect position. Intensive training, especially when changing to a different type, as well as detailed assembly instructions (with check list), will help to avoid such mistakes (Figure 10).
Pipelines must be so marked (e.g., by colour-coding according to DIN 2403) that confusion cannot arise when connecting up.
Waisted screws must be tensioned with a torque wrench [15]. The correct prestressing of screws prevents loosening (Figure 11). Locking devices for screws and nuts must be used only once.
Control levers, handwheels, multi-way cocks, ad-

Figure 10. Fracture of the volume-adjustment lever of the fuel-injection pump of a 145-hp diesel engine, caused by too large a difference in diameters.

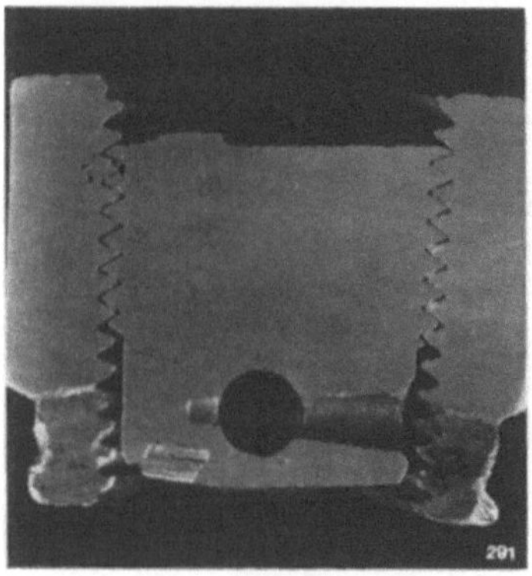

Figure 11. Fracture of a connecting-rod screw of a diesel engine as a result of excessive tensioning.

justing screws, etc. must be marked. Their function, particularly in the case of multi-way cocks, must be checked before and after assembly, mainly with regard to indications of their individual positions or operating instructions.
Foreign bodies in an engine, such as welding sputter, pieces of cotton waste, nuts, locking elements and washers, have repeatedly led to severe damage. Before closing up hollow spaces in the engine (sumps, cylinder heads, suction lines, etc.), as well as before the assembly of pipelines, a visual check for the presence of foreign bodies must be carried out. If there is any reason to suspect that foreign bodies are present, an internal examination with, for example, an endoscope should be made. A further aid to the removal of foreign bodies is to clean out the interior of the engine with flushing oil.

8.7.2. Operation of Diesel Engines

With diesel engines from about 100 hp upwards, operating personnel must be instructed by expert fitters. Also, in this connection, unusual operating conditions should be simulated.
During test operation and up to take-over of the engine by the user, responsibility for operation rests with the fitter of the manufacturer.
Of critical importance for the supervision of operation are instrument indications, whose movements must be confined within the permissible limits established by the manufacturer. Moreover, any changes in the normal running noise of the engine, as well as in the noise in the machine room, can give indications of incipient damage (see also Allianz Memorandum No. 7: Guidelines for the Damage-Free Operation of Diesel Engines of from 5 to 500 hp, 1971 edition, Allianz Order No. TVmb7).

Maintenance and Inspection

The basis of all maintenance work is the maintenance instructions of the manufacturer. If, during an inspection, irregularities or defects are discovered, they must be remedied immediately, irrespective of the normal maintenance schedule.

Engine Cooling

Irregularities in the cooling system can bring about extensive damage. Damage arising from defective cooling can be ascribed to the following main causes:
- Inadequately treated cooling water.
- Insufficient quantity of cooling water.

Cooling-Water Treatment

Untreated cooling water promotes corrosion damage to cylinder liners, particularly with larger engines (over 100 hp). Moreover, corrosion-inhibiting oil, which should be present in all cooling water, separates out of hard water (over 10° dH) and becomes ineffective; also, the separated oil forms a heat-insulating deposit on the cooling surfaces and impairs heat dissipation.

Data for a good cooling water are as follows:

Hardness:	5 to 9° dH.
pH-value:	7 to 8 (a pH-value below 7 indicates an acid, corrosion-promoting water).
Chloride content:	not more than 150 mg/l.

Simple means are available commercially for measuring hardness and pH-value:

Hardness:	Tablets.
pH-value:	Indicator papers.

Unsuitable water must receive appropriate treatment (Figure 12). For reducing hardness by 1° dH, 4 g of calcined trisodium phosphate should be added per 100 litres of water. For increasing the pH-value by one unit, from 40 to 50 g of calcined trisodium phosphate should be

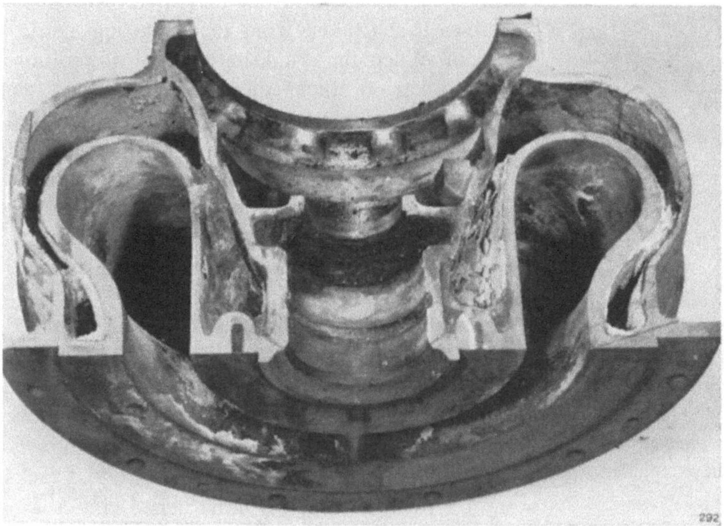

Figure 12. Ruptured casing of an exhaust-gas turbo-blower of a 1,300-hp diesel engine. The damage was the result of the formation of thick and hard deposits.

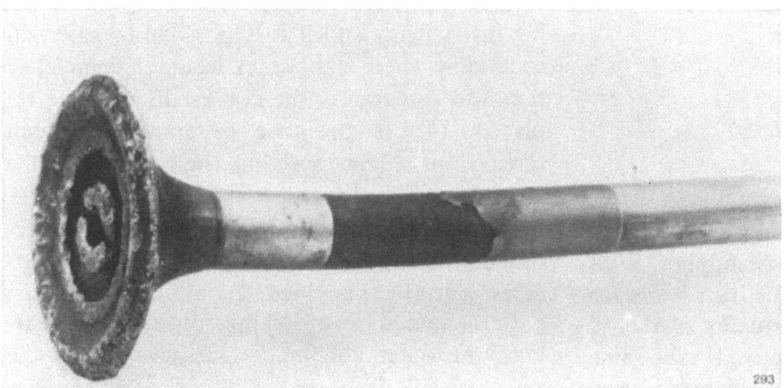

Figure 13. Damage to the head of this cooled exhaust valve of a 5,000-hp diesel engine was caused by insufficient cooling.

added to 100 litres of an emulsion of water and corrosion-inhibiting oil.

Insufficiency of Cooling Water

From past experience, it is clear that damage arising from insufficient cooling water can be traced not only to negligence on the part of operating personnel (e.g., failure to observe the cooling-water level), but also to leakages in the cooling system. Such leakages often occur at hose connections, owing to breaks in the material and rupture of hose clips, and also on honeycomb radiators as a result of mechanical damage or corrosion. Figure 13 shows an example of a component damaged as a result of insufficient cooling.

Engine Lubricating Oil

For every engine, the lubricating oil used should be of the type and properties recommended by the manufacturer. The quantity, as well as the condition, of the lubricating oil must be monitored and checked constantly (Figure 14). Samples of lubricating oils from internal combustion engines should be taken in accordance with DIN 51574. Data on lubrication requirements and lubricant consumption can be found in DIN 51500. Some indications regarding the marking of lubricant containers, lubricating equipment and lubrication points are given in DIN 51502. For used engine lubricating oils, the most important parameters and methods of checking them are as follows:

Figure 14. Insufficient lubrication was responsible for galling of this piston of a 150-hp diesel engine.

Viscosity at 50 °C	DIN 51550
Density at 15 °C	DIN 51757
Ash content	DIN 51575
Neutralization number	DIN 51558
Water content	DIN 51582

The minimum lubricating-oil requirements under which operation is still permissible are stipulated by the engine manufacturer or oil supplier. With engines that have been run under high loads for long periods, the quantity and quality of the oil must be controlled more frequently [4]. The same consideration applies to engines which are operated in dust-laden environments and, in this case, particular care should be taken regarding scheduled changes or cleaning of lubricating-oil filters [2].

Fuel

For the operation of diesel engines, only those fuels which satisfy the requirements of DIN 51601 should be used.

8.7.3. Revision of Diesel Engines

For the purpose of revision, the diesel engine must be dismantled to an extent sufficient to make all components (particularly drive components and bearings) accessible for inspection.

Crankshafts: Play must be measured (permissible values to be taken from the operating instructions). When the crankshaft is still in position, freedom from cracks should be checked by the dye-penetrant method but, when it is removed from the engine, the magnetic-particle method can also be employed, with attention paid to careful demagnetization. The largest and smallest diameters of the crankpins and journals (out-of-roundness) should be determined (permissible values from the operating instructions). The surfaces of the bearing journals should be examined for freedom from cracks (Figure 15). A check should be made for temper discoloration and scoring. In particular, fillet radii should be inspected for freedom from cracks, circumferential cracks being especially dangerous [9]. Cracks and surface damage on unhardened crankshafts may be removed only if the depth of the cracks is small (as determined by a crack-depth measuring instrument) and if it is possible to ascertain accurately whether the crack has been completely removed; otherwise, the crankshaft must be replaced [14]. In the case of crankshafts with screwed-on balance weights, the tightness of the screws must be checked.

Main and connecting-rod bearings (see also Section 8.3): With plain bearings, the running surfaces must be examined for abnormal loading patterns and signs of galling. With rolling bearings, the outer and inner races, as well as the

Figure 15. Thermal stress crack in the journal of a crankshaft of a 600-hp diesel engine.

rolling elements, must be free from temper discoloration and cracks. Should there be any change in the appearance of the running surfaces, replacement of the bearing is recommended. Visual inspection of the running surfaces should be followed by a check of the proper functioning of all points of lubrication.

Connecting rods: The smallest cross-section and regions of pronounced changes in cross-section should be inspected for cracks.

Gudgeon pins: The condition of the running surface of gudgeon-pin bushes should be checked.

Pistons: The piston skirt must exhibit an even bearing pattern. Pistons showing signs of galling should be replaced. Piston rings should be inspected for evidence of wear and for freedom of movement. Jammed piston rings should be freed, damaged piston-ring grooves should be turned out and wider piston rings should be fitted. The appearance of the piston crown gives an indication of the condition of the injection nozzles [10] (Figure 16).

Cylinder liners: Particular attention must be paid to deposits, to sites of galling on the running face of the piston, to cracks in the collar of the liner [6] (Figure 17), and also to corrosion on the water side [7]. Deposits on the water side must be removed as they impair heat dissipation and hence cooling. If there are heavy deposits on the water side, the water treatment must be investigated and changed. Signs of galling on the piston running face or cracks in the collar [8] make it necessary to replace the liner, as also does excessive wear. In this connection, the bores of cylinder liners should be measured at the top, in the middle, and at the bottom, both longitudinally and crosswise (permissible values from the operating instructions). Cylinder liners with extensive pitting on the water side by cavitation or corrosion [7] must be renewed if they cannot be turned (Figure 18).

Figure 16. Crack, caused by thermal overloading, in the piston crown of a 310-hp diesel engine.

Figure 17. Incipient crack (arrowed) in the collar of a cylinder liner of a 500-hp diesel engine, as a result of loosening.

Frame: The oil sump must be cleaned, and oil holes and pipes in the frame must be checked for cleanliness and unobstructed passage. The seating surfaces for the bearing shells in the bearing pedestals must be inspected for compression marks and fretting corrosion. Loose bearing shells lead to destruction of the white-metal layer; they manifest themselves by knocking. Seating surfaces must then be remachined and oversize shells fitted. Cooling-water spaces must be checked for deposits, corrosion and cracks. Any deposits found must be removed to ensure that the cooling effect cannot be impaired. Corrosion sets in at the height of the gudgeon pin when it is at top dead centre and in the region of the upper and lower cylinder-liner seats and seals. If the lower cylinder-liner seat leaks, cooling water will get into the engine oil, and the contaminated oil will then cause corrosion pitting of the bearings. Only in the rarest cases can corrosion scars in cylinder-liner seats be repaired (possibly

Figure 18. Signs of cavitation/corrosion on the cylinder liner of a 440-hp diesel engine.

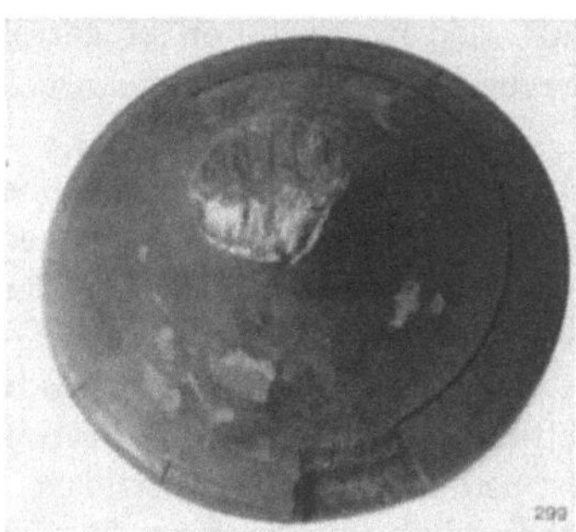

Figure 19. Fracture of a valve of a 5,000-hp diesel engine as a result of exceeding the hot strength of the unalloyed base material.

with iron cement or a similar material); if not, the frame must be renewed.

Cylinder heads must be inspected for deposits on the cooling-water side and for cracks in the webs. Cracks in the webs are usually caused by inadequate cooling as a result of insufficient cooling water, deposits or design faults. The reason for the shortage of cooling water must be discovered and removed [12]. The formation of deposits should be prevented by improved cooling-water treatment. Valve seats in the cylinder head and valve heads must be checked for damage, and valve stems and the associated guides for wear. Untight valves lead to poor compression and hence to bad combustion and reduced power. Valve heads with drag marks must be reground, and burnt and/or worn valves and guides must be renewed (Figure 19).

Injection nozzles: Defective injection nozzles can cause overheating as a result of poor atomization, with a consequent reduction in power (Figure 20). Such nozzles must be reconditioned or renewed by the manufacturer. The injection pump should be overhauled by the manufacturer as a matter of principle.

Exhaust-gas turbo-chargers: Deposits in the cooling-water spaces must be removed by pickling. Ruptured casings must be renewed. Soiled rotors must be cleaned, and damaged bearings or rotors must be reconditioned or replaced (Figure 21).

Control mechanisms: A check should be made of all gear- and chain-wheels on their shafts, as well as of the condition of such torque-transmitting elements as keys and the associated keyways (Figure 22). Loosening (giving rise to fretting corrosion) should be remedied by remachining or replacing the affected parts.

Gearwheels: Gearing should be inspected for drag marks, edge-bearing, pitting and incipient cracks. Drag marks and pitting indicate overstressing of the material. Pitting, although it cannot be prevented entirely, can be reduced in severity by using efficient lubricating media (addition of solid

Figure 20. Hole in the piston crown of a 250-hp diesel engine, caused by bad combustion.

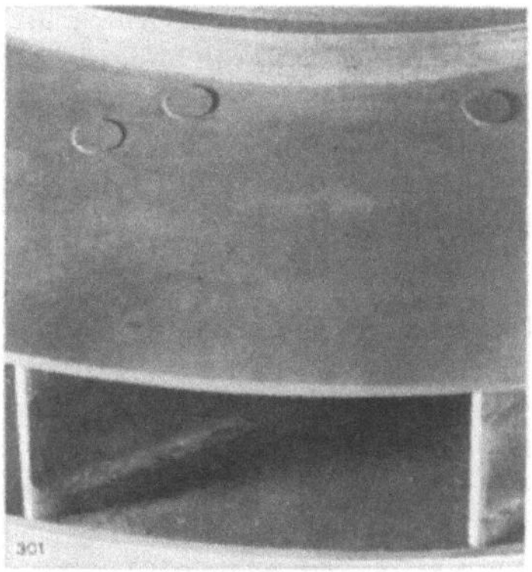

Figure 21. Loosened rivets on the impeller of the exhaust-gas turbo-charger of a diesel engine.

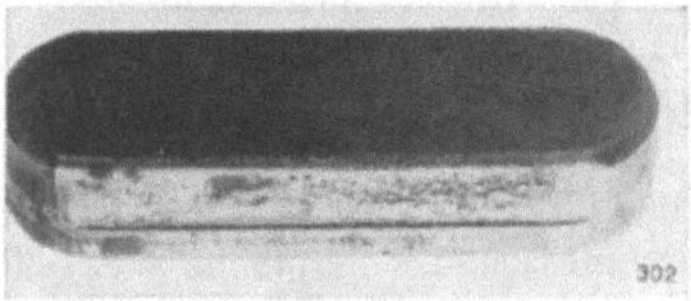

Figure 22. Deformed feather key from the camshaft drive of a 360-hp diesel engine.

Figure 23. Cam with break-out from the camshaft of a 225-hp diesel engine.

lubricants). Edge-bearing is the result of misaligned axes. It is necessary to determine whether there are manufacturing or assembly faults and to take remedial measures. Wheels with cracked teeth should be replaced.

Roller chains for the camshaft and oil-pump drives, etc. should be checked to see whether all rollers are still present and free from cracks. An examination should be made to determine whether and to what extent chain links have lengthened and possibly are liable to "climb". Damaged or worn roller chains should be replaced.

Camshafts: The condition of cam surfaces, which is a contributory factor to the satisfactory operation of engines, must be checked. Camshafts with worn cam surfaces or break-outs from the surfaces (Figure 23) should be renewed. The bearing locations of the camshafts must be examined for corrosion, and transitions of cross-sections for cracks.

Cam-followers with rollers should be inspected for wear, and rollers for break-outs from the running surfaces and other cracks.

Push-rods must be checked for bending and the condition of the spherical heads.

Rocker-arms: Should be inspected for distortion, satisfactory bearings, freedom of curved tracks from wear and break-outs, as well as perfection of setting arrangement.

Control elements must be checked for wear, distortion and breakage. Worn adjustment sleeves can jam and can allow overspeeding, as also can broken governor springs. Faulty parts must be replaced.

Couplings in the drives between the engine and the fuel injection pump and governor must be checked for wear of the torque-transmitting surfaces. Excessive play affects running of the engine. Worn parts must be replaced.

Cooling-water pumps: The casings and impellers of the pumps must be inspected for break-outs and erosion (cavitation), and the stuffing boxes for leak-tightness. Damaged impellers have an adverse effect on water circulation, while stuffing boxes which have not been retightened at the proper time and are worn occasion a loss of cooling water. Faults in pumps cause reduced cooling of the engine and its consequent risk of damage.

Pipelines must be checked for leak-tightness and, above all, for internal cleanliness. Leakage causes loss of operating media, with consequent exposure of the engine to dangerous running conditions. Thus, a loss of cooling medium impairs cooling, while a loss of engine oil impairs cooling and lubrication (Figure 24). Leaking oil not only

Figure 24. Split (arrowed) in a tube of the oil cooler of a 290-hp diesel engine, caused by dezincification.

fouls the engine and the environment, but also presents a fire hazard. Leakage must be stopped, and the missing quantity replaced with the prescribed engine oil. Dirty or clogged oil lines prevent the requisite amount of lubricating oil from being supplied to the engine and hence subject it to dangerous running conditions. Pipelines should therefore be kept clean. Fuel lines must also be checked. Loose lines cause loss of fuel and fouling, with a consequent fire hazard, as well as aspiration of air into the lines and disturbed running of the engine. With clogged lines, the engine receives less fuel and its output drops. Leakage and fouling must therefore be prevented. An loose exhaust line presents a risk of poisoning, while a dirty exhaust line leads to increased pipe resistance and to a reduction in engine output. Exhaust lines must therefore be kept tight and clean.

Cocks and valves must be checked for free through-flow and tight closure. Leakages entail losses. Stop elements must be disassembled, cleaned and reassembled with new packing.

Sealing surfaces must be checked for drag marks, coarse scratches and distortion. If necessary, sealing surfaces must be levelled or remachined.

Instruments: Oil-pressure gauges, thermometers, deficiency-monitoring safety devices, remote-indicating instruments, etc. should be checked for proper functioning. Any instruments which are damaged or no longer reliable should be replaced.

Bibliography to 8.7

[1] Reiche, R.H.: Bemerkenswerte Schäden an Dieselmotoren kleiner und mittlerer Leistung (5–500 PS). Der Maschinenschaden 44 (1971), No. 2, pp. 66/69.
[2] Allianz-Merkblatt No. 7. Hinweise für den schadenfreien Betrieb von Dieselmotoren mit Leistungen von etwa 5–500 PS, 2nd. edition 1971.
[3] Schemmrich, O.: Schäden und Schadenverhütung an modernen Dieselmotoren. Der Maschinenschaden 41 (1968), No. 5, pp. 177/185.
[4] Reiche, R.H.: Hinweise für den schadenfreien Betrieb von Dieselmotoren. Die Schadenauslese 38 (1968), pp. 2/6.
[5] Schmitt-Thomas, Kh.G., Schmidt, E.: Schäden an Leichtmetallkolben durch Schlagmarkierungen. Der Maschinenschaden 39 (1966), No. 1/2, pp. 23/24.
[6] Richter, G.: Gußfehler als Ursache von Rissen im Bund der Zylinderlaufbüchse eines Schiffsdieselmotors. Der Maschinenschaden 37 (1964), No. 7/8, pp. 152/154.
[7] Bullnheimer, H.: Kühlwasserseitige Anfressungen an Zylinderbüchsen und Gestellen schnelllaufender Dieselmotoren. MAN-Forschungsheft (1960), No. 9, pp. 20/32.
[8] Reiche, R.H.: Schäden an Zylinderlaufbüchsen von Verbrennungsmotoren. Der Maschinenschaden 32 (1959), No. 11/12, pp. 172/175.
[9] Richter, G.: Die Dauerhaltbarkeit von Kurbelwellen. Der Maschinenschaden 32 (1959), No. 11/12, pp. 167/171.
[10] Lusar, R.: Kurbelwellenbruch an einem Dieselmotor als Folge eines Kolbenbruches. Der Maschinenschaden 32 (1959), No. 9/10, pp. 143/145.
[11] Horton, G.: Die meisten Schäden an Dieselmotoren sind vermeidbar. Locomotive (1957), No. 7. Der Maschinenschaden 30 (1957), No. 11/12, pp. 181/182.
[12] Börsig, F.: Kühlwasserraum eines Diesel-Zylinderkopfes als Wirbos-Reaktor. Der Maschinenschaden 29 (1956), No. 5/6, p. 90.
[13] Kluth, P.: Vermeidbare Schäden an Verbrennungsmotoren. Der Maschinenschaden 29 (1956), No. 1/2, pp. 27/28.
[14] Kraemer, O.: Schäden an Dieselmotoren. Der Maschinenschaden 25 (1952), No. 5/8, pp. 89/95.
[15] Bark, R.: Beanspruchung und Behandlung von Pleuellagerbolzen. Der Maschinenschaden 24 (1951), No. 1/2, pp. 1/6 and No. 3/4, pp. 38/45.

8.8. Printing Machines

The number of cases of damage to the highly developed machines and automatic equipment used in the printing industry has risen threefold during the past ten years. Table 1 indicates the causes of damage to printing machines and their distribution. The following guidelines for loss prevention are primarily directed to works managers and printers, with examples of damage in the foreground.

Table 1. Distribution of the causes of damage to printing machines.

Cause of damage	Distribution of the number of damage cases [%]
Product faults	4
Operational faults	75
Extraneous influences (unclarified causes)	21

Discussion of the Causes of Damage

Of the high proportion (75%) of cases of damage described to operational faults in Table 1, 60% originate when preparing and adjusting machines during operation. *Loosening* is responsible for 12% of these cases of damage, while the remaining low proportion of 3% is caused by *lack of attention to protective equipment.*

Under the heading of extraneous influences, about half the cases of damage arise from wear and unclarified causes. Insofar as product faults are concerned, faults in machining are the major cause of damage.

Incidentally, mention should be made of the results of an investigation which indicated that most damage due to mishandling and faulty maintenance was caused by personnel in the age group between 25 and 35. The statistical evaluation of the damage shows that it occurred most frequently on Tuesdays and Fridays between 1 and 5 p.m.

Figure 1. The provision of large mirrors for checking lay-outs has proved particularly useful under cramped space conditions.

8.8.1. Loss-Prevention Guidelines for Users of Printing Machines

8.8.1.1. General Guidelines for Loss Prevention

Before installing machines, their space requirements must be established. In this connection, sufficient room must be provided for storing materials. Experience has shown that machines in cramped space conditions are more frequently affected by damage than those which are conveniently arranged (Figures 1 and 2).

Lighting fixtures (fluorescent tubes, bulbs) should not be positioned directly above the machine. Splinters of broken glass can get between roller beds, platens and cylinders.

8.8.1.2. Operation of Printing Machines [1]

Every printer must familiarize himself with the operating and maintenance instructions issued by the machine manufacturer and must observe them carefully. The works manager must corroborate that a repeated study has been made after long intervals. Instruction leaflets and hints issued by the machine manufacturer should be given to the operator.

Figure 2. Stools do not belong in a machine room and certainly never next to a machine. They impede the freedom of movement of the operator and the machine.

The operating instructions must be kept by the printing machine, so that they are handy at all times, particularly when changing personnel. Machine manufacturers regularly hold courses lasting a week for machine operators.

When preparing formes, locking-up must be against iron—never against lead or aluminium reglets.

After formes are prepared, all tools must be put in their prescribed place and checked for their full number. Any tools in excess of that number must be removed. A bright-yellow coating on tools has proved to be most useful.

Before starting operation, a check should be made to ascertain whether the safety devices provided are properly adjusted and fully functional, whether tools or loose components have been left in or on the machine (the handles of frequently used tools should be painted *yellow* for better recognition), whether the roller locks are firmly closed and whether anyone is in the vicinity of moving parts of the machine. The operator who switches on must first give a loud verbal warning of his intention and await an acknowledgment from all workmen busy on the machine.

Colour rollers must not be withdrawn while the machine is in operation.

Should irregularities (e.g., unusual noises) be noted while the machine is running, it must be stopped immediately and carefully examined.

When changing shifts, instructions must be passed on to the next shift—orally when one man hands over to another, and in writing to the next morning's shift for single- or two-shift operation.

Before starting work, the operator who takes over must check the machine and forme and retighten where necessary.

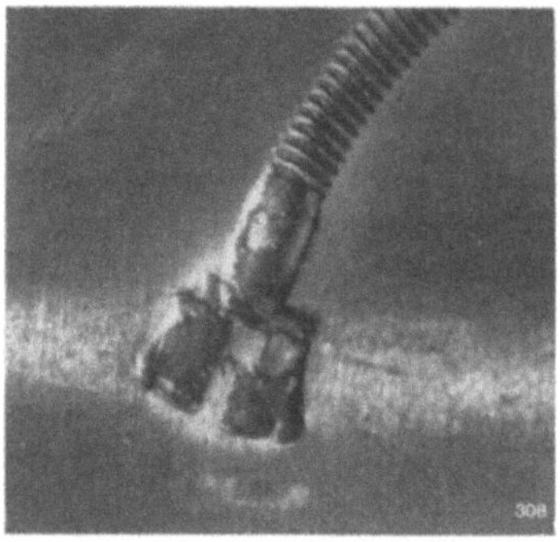

Figure 4. Impression in the cylinder of an offset printing machine of an M8 hexagon socket-head screw. This screw, from a spring-loaded swinging gripper, became loose, dropped into the machine and caused damage about 20,000 DM.

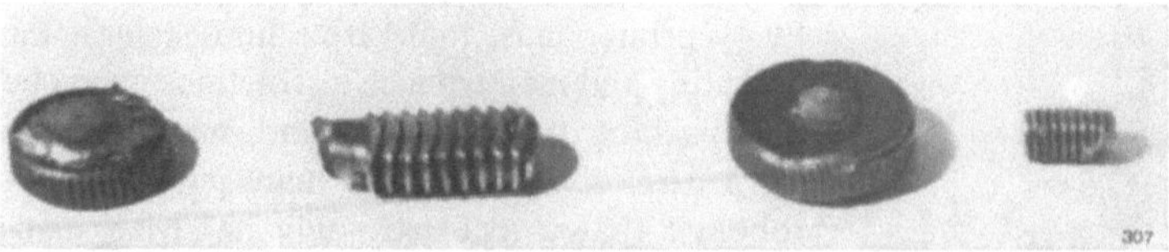

Figure 3. Unsecured setscrews and adjustment screws often drop into a running machine and cause severe damage. The illustration shows a crushed knurled screw from a single-colour offset printing press. The screw got into the machine and caused damage about 15,000 DM.

Figure 5. Yellow-painted tools cannot easily be overlooked. All tools should therefore be given a bright-yellow paint coating.

8.8.1.3. Lubrication of Printing Machines

Printing machines must be lubricated regularly in accordance with the instructions of the manufacturer.
Lubricating points marked with yellow paint must contain a sufficient but not excessive quantitiy of oil. Oil drips can soil the machine and, above all, the printed product; also, they can lead to bearing damage as a result of absorption of paper or dust particles.
In small and medium-size plants, it is advantageous if every printer himself lubricates the machine he operates.
In the case of multi-shift operation, the printer of the first shift must take care of all lubricating points. In the following shifts, only the main lubricating points will need attention and, if necessary, renewed lubrication. In large plants, the provision of a lubrication service is recommended.
Before starting a shift or before every printing run, it is necessary to check the correct level of oil in the containers of central lubricating installations. Oil must always be topped up through a strainer. If the oil content of a container inadvertently falls below the prescribed minimum level, work must not be resumed until all lubricating-oil lines have been filled and properly bled.
If several types of lubricating oil have to be used, the corresponding containers and cans must be clearly and unambiguously marked. The mistaken use of containers in which, for instance, roller-cleaning media are stored must be made impossible.
Great care should be taken to ensure that lubricating points are not clogged by pieces of paper, ink or dirt.

8.8.1.4. Maintenance and Cleaning of Printing Machines

A *machine card* should be prepared for every printing machine and should be kept up to date by the foreman or printing-works manager. An example of a suitable card for this purpose is the *machine-maintenance card AWF 3094* (obtain-

Figure 6. The deplorable design of an open lubricating point which induced the printer to carry out a "do-it-yourself" repair job with self-adhesive tape.

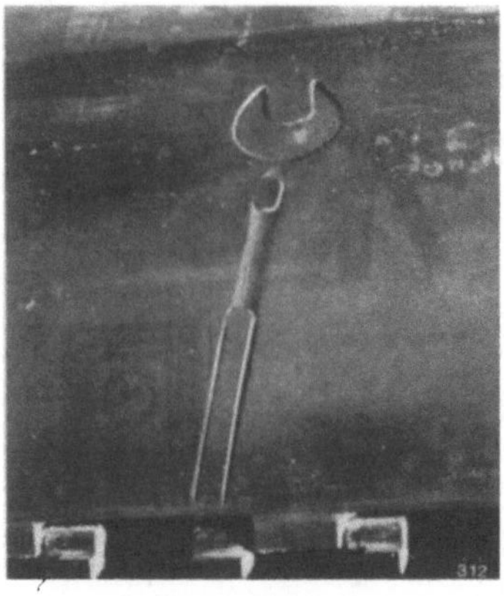

Figure 7. After finishing preparatory and adjustment operations, all tools must be collected at the prescribed location and checked for completeness.

able from Reuth-Vertrieb, Berlin and Cologne), which provides for the most important technical data, including information on the electrical drive, overhauls, inspections, repair work and costs, as well as the availability of spare parts.

8.8.1.5. Maintenance of Formes

All locking-up equipment and forme sticks must be disassembled, cleaned and checked after each printing run but at least at intervals of four weeks.

After cleaning, the screws of the locking equipment must be re-oiled lightly, unless a special cleaning medium with lubricating properties has been used.

Locking equipment showing compression marks or signs of heavy wear, or with jamming parts, constitutes a danger to the printing machine and must be replaced at once.

The printer or compositor has the right to replace poor materials.

8.8.1.6. Overhaul of Printing Machines

As already mentioned, it is necessary to prepare a card for each printing machine on which the most important technical data, including overhaul and repair work, as well as replacement parts fitted, must be entered.

Printing machines which are in continuous use, particularly in multi-shift operation, must undergo a general overhaul at least once a year.

General overhauls must be carried out by the manufacturer's fitters, or at least in their presence.

When their machines are being overhauled, the operators should be present, so that they can acquaint themselves with the different wear rates

Figure 8. Wooden wedges, which are inserted into the paper stack to compensate for unsuitable paper materials, must in all cases be tied on; otherwise, they can, and often do, get into the printing machinery with the sheets and can cause extensive damage. Fortunately, the number of foreign-body arresters installed as standard by some feeder manufacturers is nowadays increasing. The percentage of severe damage cases caused by wedges and tools finding their way into a printing machine is unusually high. The fact that it is possible to use one's own initiative in a printing plant is exemplified in this illustration of an older installation, to which a "home-made" deflecting rail had later been added.

of the various components and adjust future maintenance work accordingly.

8.8.1.7. Protection of Printing Machines against Fire

Burning solvents on printing machines must be extinguished with carbon dioxide appliances. These appliances must be kept in the immediate vicinity of the printing machines, since such fires spread very quickly.

Dry-powder extinguishers are more effective than carbon dioxide extinguishers.

They protect the fire-fighter from the heat radiated by the fire by their cloud of extinguishing powder. However, a disadvantage is that the powder tends to spread over a wide area and requires extensive cleaning operations after a fire to remove it.

Bibliography to 8.8

[1] Allianz-Merkblatt No. 5: Bedienung, Pflege und Wartung von Druckmaschinen. 3rd edition München and Berlin: Allianz Versicherungs-AG 1971.

[2] Reiche, R.: Schäden an Druckmaschinen und Maßnahmen zu ihrer Verhütung. Der Maschinenschaden 35 (1962), No. 9/10, pp. 133/144.

[3] Schäden an Druckmaschinen durch Bedienungsfehler. Der Maschinenschaden 37 (1964), No. 9/10, pp. 207/109.

[4] Beschädigung einer Schnellpresse durch einen liegengebliebenen Steckschlüssel. Der Maschinenschaden 38 (1965), No. 9/10, pp. 164.

[5] Schäden an Rotationsmaschinen. Der Maschinenschaden 39 (1966), No. 1/2, pp. 26/27.

[6] Wellenbrüche an Elektromotoren von Druckmaschinen. Der Maschinenschaden 40 (1967), No. 3, pp. 104.

[7] Schäden an Offsetdruckmaschinen. Der Maschinenschaden 41 (1968), No. 3, pp. 119/120.

[8] Schäden an Buchdruckmaschinen durch Fremdkörper. Der Maschinenschaden 42 (1969), No. 5, p. 158.

8.9. Frame Saws

Experience of damage to frame saws has shown that, compared with other types of machines, the risk of total loss is high. The proportion of cases of catastrophic damage, relative to all cases of damage considered, amounts to about 4%.
An analysis of a large number of frame saws insured over a long period has demonstrated that a yearly average of about 16% of these saws suffered damage. Excluded are small cases of damage up to 300 DM. Table 1 indicates the distribution of the causes of damage.

Table 1. Distribution of the causes of cases of damage to frame saws.

Cause of damage	Distribution of the number of damage cases [%]
Product faults	21
Operational faults, mishandling	50
Maintenance faults	27
Extraneous influences	2

The following guidelines for loss prevention will include indications for the maintenance of operational serviceability and will also draw attention to the planning and design of machines.

8.9.1. Guidelines for Loss Prevention during Planning and Design

To obtain high efficiency, when planning frame saws for partially or fully automatic sawmills, it is necessary, to stipulate, in the following order, their practical usefulness and safety in operation at relatively high productivity and, if possible, their ease of maintenance and ease of operation. The fulfilment of this concept yields the highest efficiency.
Depending on application requirements, i.e., number of cuts, cutting programmes, types of timber and their lengths, frame-saw manufacturers have developed, in the course of decades, their own basic types and have improved technical equipment through experience gained in practice. In the case of new orders, they usually offer individual advice and initial planning free of charge, based on types of machines in the final stage of development, and utilizing the latest experience. In this regard, it can also be advantageous to obtain tenders from competitors, so as to be able to make comparisons. Furthermore, information on similar installations recently delivered should be requested to acquire knowledge of experience gained in other plants. When the order is finalized, the stipulated characteristics, as well as guarantee agreements, must be confirmed by the manufacturer in writing.
Clarification of the question of the drive and the place of erection should be done in the planning stage. For preparation of the foundation, a plan with details of loading conditions and locations of holes for anchor bolts must be available. In some circumstances, depending on the condition of the ground, a soil test may be necessary to establish whether the subsoil is suitable as a place of erection.
Like other machines, frame saws have undergone a long period of development, during which progress has been made from slow-running machines operating at about 200 rpm and with feed rates up to 4 m/min, to modern high-efficiency machines with speeds up to 360 rpm and feed rates up to 15 m/min. However, there have also been setbacks. Thus, depending on the state of knowledge at the time, shortcomings in design frequently led to damage to bearings and not infrequently to fracture of drive shafts, flywheels, connecting rods or frames, with consequent severe and often dangerous damage to saw stands and other components of the installation. The causes are to be found not so much in the well-proven design of machine frame made of cast iron and of bolted construction, which ensures quiet running, but far more in the details.
For the prevention of damage and premature wear during operation by deposition of dirt and sawdust on parts of the saw stand, guides, bearings, racks and chains, or by jamming with short pieces of timber of offcuts, the problem has been to ensure elimination or interference with

Figure 1. This example of total loss was caused by an inexpert welding repair on a connecting rod.

throughput. With modern frame saws, a change has therefore been made in that the ribbing of the side components of the stand, and also guides and bearings, have been transferred to the outside and housed together with the mechanical lifting or feed devices.

A wedge-shaped design of the lower traverse favours disposal of waste material towards the vibratory conveyor, while additional offcut removers provide for unimpeded carriage of the cut pieces.

For increasing operational safety, it has proved useful to replace the plain bearings of the connecting-rod heads by adjustable rolling bearings. For the main drive, however, the change-over to rolling bearings has not been universally adopted because, in the event of damage or during an overhaul, removal of rolling bearings involves withdrawal of flywheels and, in some circumstances, complete disassembly of saw stands, with a high expenditure of time. If, despite this consideration, it is decided to use rolling bearings for safety reasons in heavy or high-speed installations, they are overdimensioned as a precautionary measure to achieve longer lives.

Flywheels conventionally made of grey cast iron often fractured as a result of internal casting stresses caused by their greatly varying wall thicknesses, with consequent severe damage to connecting rods, saw frames and stands. As a remedy, apart from a more favourable configuration from the casting point of view, different methods have been devised, such as dividing the wheel rim or screwing in balance weights and, for reasons of safety with large fast-running flywheels, the use of steel hubs and screwed wheel rims.

Next to the saw frame, the connecting rods are the most highly stressed components of frame saws. Since fracture of the connecting rod on one side generally leads to jamming of the saw frame, coupled with deformation of the other connecting rod, resulting in fracture of the saw stand, special care must be taken with their configuration from the point of view of design and stressing. Most cases of damage have been caused by fatigue failures of connecting-rod shafts of rectangular cross-section, originally left unmachined on the outside. Now, however, H-section connecting-rod shafts, machined all over, have found general acceptance as a means of preventing damage.

Figure 2. Because the load-carrying section was so reduced by the holes for fastening the connecting-rod shaft to the head on the crank side, it could not withstand the stresses imposed and eventually fractured.

Figure 3. With this screw-fastening design, stress conditions are more favourable than those with the method shown in Figure 2.

Furthermore, apart from completely forged connecting rods, there will today also be found types in which the shafts and heads are separately forged and butt-welded together, and also those in which the shafts and heads are made of rolled material and are screwed together. All designs can be operationally safe if they are shaped to take account of the stresses imposed and if they are carefully machined and faultless in regard to strength and material. Without doubt, connecting rods of wholly forged construction are the safest in operation.

Figure 2 exemplifies a typical design fault in the screwed connection of a connecting-rod shaft with the head at the crank side. Because of the location of the screw holes in the longitudinal cross-section of the shaft, the load-carrying section was so underdimensioned that a fatigue failure occurred as a result of overloading, with total loss of the frame saw. The screwfastening design was then changed in accordance with Figure 3 and has since proved its effectiveness.

The saw frames, consisting of an upper and lower cross-piece and the lateral frame pillars, are most frequently subjected to damage, problems and complaints, particularly in respect of the most highly loaded upper cross-piece. Depending on the characteristics of a riveted, screwed or welded frame design, loosening or fractures at the joints will occur eventually, as well as fractures of the drive shafts with accompanying damage to bearings. To reduce this susceptibility to damage to a minimum, saw frames are nowadays manufactured and assembled in a much more rigid or torsion-resistant form from profiled, sometimes preforged, individual components.

Further design improvements for loss prevention, such as fully automatic overhang adjustment, suitable design, disposition and driving of the rolls with overload slipping clutches, and also electro-hydraulic control of feed, roll lift and sawblade tensioning or additional central lubrication, will not be discussed here in detail.

8.9.2. Guidelines for Loss Prevention during Manufacture, Assembly and Commissioning

Manufacture

In the manufacture of frame saws, great care should be taken, above all else, to ensure that cast components, such as the baseplate, lateral members of the saw stand and cross-beams, as well as flywheels, are free from casting stresses or are stress-annealed and that they do not contain blowholes. In this regard, flywheels, with their great differences in material thickness between the counterweight, crank-arm and rim, present casting difficulties because the different shrinkage stresses during cooling can lead to cracks and fractures. A particularly careful stress-annealing treatment is here essential.

To avoid later difficulties during assembly or operation, machining of these components must be strictly in accordance with fitting tolerances. This requirement applies particularly to the seating of the drive disc and flywheels on the main shaft, as well as to the crank-pins in the flywheels themselves. Moreover, the shaft and pin must be ground and, to obviate fatigue failures, must not have any sharp transitions. In all cases, an ultrasonic and magnetic crack-detection test ahould be carried out after machining.

In principle, the steelworks concerned should be required to provide a test certificate in respect of the sound quality of the raw materials supplied for the manufacture of highly stressed frame-saw components, irrespective of whether extruded, rolled or preforged materials are involved. This requirement is particularly applicable to the manufacture of connecting rods and the individual components of the saw frame. Moreover, all these items should be ordered from 5 to 10 mm oversize, to allow for the absolutely essential machining of entire surfaces as a precautionary measure to eliminate the possibility of fatigue failures arising from surface defects.

Also, attention should be paid to providing clean transitions in the profiles of connecting-rod shafts and saw-frame components. Special care is needed when machining the upper cross-piece of the frame, which is the most highly stressed component of frame saws. After manufacture of the connecting rods and saw frames, they should be subjected, for reasons of safety, to another ultrasonic test for internal defects and to a magnetic test for surface cracks, as well as, depending on the design, to an examination of welding seams or screwed fastenings. After testing, the components should immediately be given a surface-protection treatment against corrosive attack.

Assembly and Commissioning

To avoid the possibility of mishaps during assembly at the final erection site, an important requirement is that the frame saw should first be assembled, with the co-operation of the fitter to be in charge at the erection site, at the manufacturer's works, ready for operation and tested for proper functioning, including the lubrication system and mechanical and electro-hydraulic equipment. This requirement is especially advantageous if the frame saw, because of its size and weight, must be transported in sections and reassembled.

When receiving frame-saw parts, particularly connecting rods, they must be examined carefully for damage suffered during transport and any complaints must then be made immediately in writing. If installations are not already available for unloading, transporting and erecting on the prepared foundation, it will be necessary to provide such equipment as mobile cranes, fork-lift trucks and lifting gear (e.g., block and tackle, hydraulic and mechanical jacks), as well as wire ropes with a sufficient load capacity.

During all stages of assembly until commissioning of the frame saw, the intended operating personnel should be present and should play an active part, so that they can familiarize themselves with the installation and receive strict training and instructions from the fitter in charge.

8.9.3. Guidelines for Loss Prevention by the User of Frame Saws

Loss Prevention until Take-over

Commissioning and adjustment or running-in of the installation are generally followed by an agreed period of test operation to prove serviceability and the stipulated characteristics in practice. During this period, the operating personnel, under the responsibility of the fitter in charge, should become progressively more acquainted with the operation, supervision and maintenance of the installation until they can take over responsibility themselves. As from take-over of the installation, operational risk, with the exception of damage under guarantee, is transferred to the user.

Loss Prevention during Operation

Handling

According to statistics, 77% of all cases of damage to frame saws can be attributed to a certain degree of carelessness, negligence, and also to incompetence, 50% of these cases being due to mishandling alone and 27% to faulty maintenance. These facts indicate that operating personnel, on the average, are not fully capable of coping with operational requirements or that too high demands are made on them. In view of the continuously increasing demands imposed by the technological progress of modern partially or fully automatic installations, it is becoming more and more necessary for operating personnel to possess the corresponding aptitude and reliability and to have undergone a proper training in good craftsmanship.

Well-trained specialist personnel will find it easier to absorb operating instructions in practice and hence will have a clearer picture of the course of operation, with the ability to recognize more quickly difficult situations and to act accordingly for the prevention of damage and breakdowns. Then cases of damage are only partly a direct consequence of mishandling, while many are due indirectly to neglect or to insufficient caution and attention during the course of sawing operations.

For example, failure of the hydraulic equipment for exerting roll pressure when cutting a log caused the log and pressure roll to be thrown upwards, resulting in forced fracture of the stand at the height of the upper cross-beam (Figure 5). A somewhat similar type of damage was occasioned by the presence, not noticed by the saw operator, of a piece of root attached to the log. During cutting, the log and roll were again thrown upwards, resulting in fracture of the stand. It is possible to cite other parallel cases in which severe damage to frame saws was caused by insufficient tensioning of the saw blade as well as by collision with or jamming by offcuts or short pieces of timber. It has, however, also happened from time to time that ensuing damage to frame saws has arisen when feeding in the material to be cut or by jamming or clogging as a result of omitting to clean out waste materials. It is therefore essential that optimum handling should include the entire course of operations.

Figure 4. During overhaul, an incipient crack on a connecting rod was inexpertly welded. The result was a fatigue failure which led to total loss.

Supervision

The special supervisional duties of operating personnel include paying attention to any changes in running behaviour and unusual noises, all of which must be reported to the management and must be investigated immediately to prevent ensuing severe damage, which arises frequently as a result of loosened screw fasteners of the stand or damaged bearings. If the causes of trouble cannot be established by the user's own personnel, it is absolutely essential to call in the staff of the manufacturer. Also, after sudden overloads of the frame saw or after impaction by a log, the installation should be shut down and the endangered areas of the stand examined for incipient cracks or fractures. In one particular frame saw, an offcut section became jammed in the frame. After clearing the fault, the saw

Figure 5. Failure of the hydraulic clamping equipment caused a log to be pulled up and down by the saw frame. As a result, the stand broke at its upper cross-beam on the feed side.

was started up again and, when doing so, the operator heard an unusual knocking sound. Not much later, a broken screw of the connecting-rod guide led to fracture of the stand and saw frame, as well as to distortion of the connecting rods and damage to the bearings.

Maintenance

A prerequisite for the operational safety and long life of a frame-saw installation is appropriate maintenance, overhaul and attention in accordance with the maintenance schedules of the manufacturers concerned. The fact that 27% of the cases of damage to frame saws are a result of inadequate maintenance is not a good testimonial for the operating personnel. It is therefore desirable, while adhering to the maintenance instructions of the manufacturer, to indicate the following general rules applying to proper maintenance.

Every day before starting operation it is necessary to inspect the lubricating-oil lines and the pressure lines of the hydraulic equipment for leak-tightness and also to check the oil levels of the associated tanks and any oil-ring bearings present. If any leakages or fractures are noted, they must be rectified at once. For topping up, use must always be made of lubricating and hydraulic oils, as well as lubricating greases, selected in accordance with the prescribed quality specifications, and their containers must be clearly marked to prevent mistakes. The lower rolling bearings of the connecting rods must be relubricated after every 4 to 5 hr in operation, after first cleaning the lubricating nipples. In addition, the adjustment slide of the upper roll, as well as the slide of the lower guide for the adjustment of overhang, must be cleaned daily to remove sawdust or dirt and must be lubricated separately, together with any existing mechanical control devices.

At the weekend, the installation should regularly be given a thorough cleaning. In this way it is easier to recognize any faults and incipient damage that may be present and to remedy them promptly. Also, the opportunity should be taken to retighten screwed fasteners and to check (particularly adjusting screws) for deformations and cracks

Figure 6. Breakage of the main shaft of a frame saw at the journal. Examination of the damage revealed that the journal exhibited thermal cracks that could be traced to an earlier known period of running hot. Evidently, insufficient attention was paid at the time to the hot-running condition.

and to renew them if necessary. Furthermore, the saw-frame guides must be checked in regard to the prescribed clearance and re-adjusted on the run-out side. In addition, the upper connecting-rod bearings must be lubricated.

On a monthly basis, a check, after cleaning, should be carried out regularly for excessive clearance of the main and connecting-rod bearings. Adjustment of rolling bearings is effected by tightening locknuts and of plain bearings by removing shims. If, during a bearing check, signs of wear are discovered on the raceways of rolling bearings, they must be replaced. Similarly, if the plain bearings of the main drive show signs of wear, such as pronounced scoring or incipient break-outs, they must also be replaced, while the journals must be smoothed, if necessary. In this connection, one should realize that, with new bearings, repeated checking and correcting of the bearing clearance by removing shims must be performed at weekly intervals.

At least once every three to four months the foundation and the screwed connections of the stand must be checked for loosening or incipient cracks. At the same time, those components of the mechanical roll-lifting and feeding equipment that are subject to wear, such as sprocket wheels, chains, guide elements and racks, must be checked for excessive wear.

It has proved useful to keep a maintenance log in which are entered the various monthly and quarterly overhauls and repairs with replacements of components, as well as details of cases of damage or problems that have occurred in between. In this way, a clear picture of the condition and history of the installation always exists for future planning.

Revision

At intervals of six months up to not more than a year, revision of the installation as a whole should be carried out by special fitters of the manufacturer's staff. For this purpose, many users have concluded maintenance contracts with the manufacturers and have reported favourable results, because these fitters, with their wealth of experience and superior training can more easily recognize existing faults and shortcomings and eliminate them. The revision also involves checking of the central lubricating installation and the hydraulic pumps for the roll-lifting and control equipment, as well as the inspection for cracks of such highly stressed components as connecting rods, saw frames, bearings and, if possible, the main drive shaft.

8.9.4. Guidelines for Repair Work

If cases of unforeseen damage occur suddenly, or if signs of immediately impending damage are detected, the first question that generally arises is the quickest means of effecting repairs. In some cases, the necessary replacement parts and an expert fitter for effecting the repair work are available.

Figure 7. Forced fracture of feed rollers by a broken and jammed piece of timber.

Figure 8. As a result of loosening, a fatigue fracture occurred in the crank-pin of a frame drive. Traces of fretting corrosion are clearly visible.

Figure 9. The location of incipient cracks in this flywheel are marked with chalk. These stress cracks originated at sharp cross-sectional transitions and are an indication of casting stresses.

The situation is more serious if neither replacement parts nor the services of an expert fitter are temporarily available and one is perhaps forced, without a proper knowledge of the interrelated factors and effects of the damage, to decide to carry out provisional repairs at one's own risk. In such circumstances, it is possible that further damage will ensue, as demonstrated by the following two examples.

A plain bearing of the main drive of a multiple-blade frame saw overheated repeatedly and on each occasion, without any further examination, it was quickly babbitted again and fitted in by a neighbouring workshop. One day the main shaft suddenly fractured in the region of this bearing and the now free-running flywheel with connecting rod caused total loss of the frame saw. Subsequent examination showed that the bearing journal had suffered a fatigue failure as a result of thermal cracks sustained during the earlier hot-running conditions.

In another machine, after an incipient crack had been found on a connectingrod shank, the crack was removed by gouging and a repair was made by welding carried out by the local personnel. Some time later, a fatigue fracture occurred at this location as a result of inexpert welding, again leading to total loss of the frame saw.

These two cases of damage indicate that provisional repairs without expert assistance are not worthwhile. With older models of frame saws or with makes that no longer exist, it is frequently necessary and quite possible to obtain specially made replacement parts at a reasonable cost from competent firms and specialized workshops and to have optimum repairs carried out expertly.

Bibliography to 8.9

[1] Schäden an Sägegattern. Der Maschinenschaden 35 (1962), No. 9/10, pp. 160/161.

[2] Schaden an einem Vollgatter durch Zahnbrüche am Ritzel des Vorschubschaltwerkes. Der Maschinenschaden 39 (1966), No. 1/2, p. 29.

[3] Kurbelzapfenbruch an einem Vollgatter. Der Maschinenschaden 40 (1967), No. 3, p. 107.

8.10. Chipboard Presses

Hydraulic platen presses with one or more daylights are used for the manufacture or finishing of chipboard. In the last few years, the demands made on these presses have increased considerably in respect of output and precision. Thus, chipboard presses currently in operation are available with pressing areas of more than 30 m^2 at a pressure of 35 kg/cm^2 and capable of achieving an accuracy of ± 0.2 mm.

From the various types of press constructions produced, three main designs of press frames can be recognized. Apart from these basic types, overlapping of the various designs can often be found.

Welded Design (Profiled Frames)

With press frames of welded design, the uprights consist of rolled-steel profiles, such as U-beams or square billets, and the yokes of heavy plate. Joining of the components is effected by welding seams, mainly fillet welds. The position and quality of the welding seams in relation to the stresses to which they are exposed are of primary importance for the life of the press and its safety in operation.

Pillar Design

Pillar design involves a method of manufacture which largely dispenses with welding seams for the frame.

The press frames comprise pillars connected together by yokes. Frequently, the hydraulic equipment is already installed in these yokes, which are usually cast or fabricated by welding. The pillars are generally made of machined round high-quality steel, and connection between yoke and pillar is effected by screwed fasteners. A prerequisite of this screwed connection is that its high dynamic stressing must be taken into account in its design and manufacture. Furthermore, the correct prestress of the screwed connection must be set up during assembly and maintained during operation.

Sheet-Frame Design

Sheet-frame design has been widely adopted in recent years.

In this design, the frame elements are produced from sheet steel and are connected together in accordance with the particular design of the press. When manufacturing the sheet-steel frame, either the complete component is cut out from steel plate or the yoke and upright plates are butt-welded together and machined and stress-annealed to increase their capacity to withstand dynamic stresses. Sufficient testing is essential for success, apart from properly designed and executed welded connections. In this regard, it is necessary to check not only for welding defects that lie on the surface, but also those that lie deeper down. For this purpose, radiographic and ultrasonic test methods are available.

The press tables are welded or cast. The heating platens are usually made in one piece by machining and the heating channels are drilled out.

Table 1 surveys the distribution of cases of damage in accordance with their causes and locations:

Product Faults

Product faults comprise all damage that can be traced to faults in the presses themselves and are concerned with the direct and indirect conse-

Table 1. Distribution of the causes and locations of damage to chipboard presses.

Cause of damage	Distribution of the number of damage cases [%]	Location of damage	Distribution of the number of damage cases [%]
Product faults	57	Press frames	58
Operational faults	28	Press tables and platens	21
Other faults	15	Hydraulics and control	12
		Others	9

quences of faults in design, manufacture and materials, and also with the consequences of technological shortcomings at the time of manufacture. In addition, damage to presses that has been the result of certain actions, often of the user, after manufacture and commissioning is taken into account.

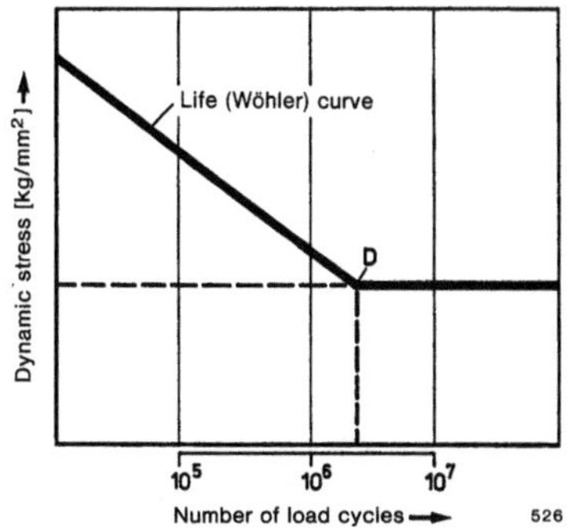

Figure 1. Schematic representation of a life (Wöhler) curve. The line dropping to point D comprises the region of endurance limit where, after a limited number of load cycles, crack formation and finally fracture will occur. The horizontal line indicates the level of the fatigue strength of the material.

Operational Faults

This group mainly covers faults which can be traced to human fallibility (mishandling and maintenance faults). Despite the relatively low number of damage cases in this category, there is more extensive damage in these cases and its elimination is more costly.

Other Faults

Under this heading are taken into account all those faults which cannot be allocated categorically or which have occurred to auxiliary equipment.

Discussion of Damage Locations

The major proportion of damage occurs to press frames and mainly concerns local crack formation. By systematic revisions with suitable test procedures, it is often possible to recognize these faults at an early stage and to adopt the appropriate countermeasures. A substantial proportion of damage to press tables and platens is caused by the effects of foreign bodies and human fallibility. Wear plays a major part in defects of hydraulic and control equipment.

Main Stresses on Platen Presses

In operation, platen presses are stressed thermally and mechanically, chiefly dynamically. Stressing by thermal loading does not present any extraordinary problems, provided that thermal shock is avoided and operating instructions are followed during stopping and starting. In this connection, one exception is the type of press in which the platens are heated and then cooled during one load cycle. With presses of this type, e.g., laminating presses, crack formation has repeatedly been observed at thermally stressed locations on the heating platens after a certain period in operation, which depends on the temperature gradient, the number of load cycles and the design of the heating platens. Here, it is a matter of a region of cracks induced by temperature variations.

Far more of a problem is the part played by mechanical stressing. In operation, the press is mainly subjected to pulsating stresses, i.e., during a press cycle the stress rises to a maximum value which is removed again on unloading. The dynamic characteristics of the materials used are therefore decisive for the security of the design from crack formation. These characteristics are clearly shown by the life (Wöhler) curve (Figure 1). If the structural component is stressed above its fatigue strength, i.e., at the level of its endurance limit, it will develop cracks after a limited number of load cycles, the number of load cycles to rupture decreasing with increasing load. In the case of steel in air (without corrosive or other influences), the limiting number of load cycles at which the load can be sustained for an indefinite number of cycles (fatigue limit) is at least 3,000,000. With platen presses, it is not always possible, mainly for reasons of cost but also for technical reasons, to dimension them in such a way that the fatigue strength of the material is not exceeded at any location in the press, for the majority of presses do not reach the limiting number of load cycles related to fatigue strength in their whole useful lives. It is therefore

possible that certain areas of the press where the fatigue strength is to some extent exceeded will exhibit crack formation only after a longer period in operation. Revisions which enable the consequences of local overstressing, i.e., crack formation, to be detected and remedied are, therefore, clearly necessary, for the safety of presses over their whole service life.
From these considerations and the results of examinations of damage, the following selected requirements for loss prevention with platen presses can be stipulated.

8.10.1. Loss Prevention by the Manufacturer

Under the influence of operational loading, local defects in design and construction can lead to such high stress peaks that the fatigue limit of the material is locally exceeded, resulting in the occurrence of crack formation at these locations—often even after several years in operation. Two types of defects have time and again been found characteristic. One of these concerns fillet welds which run in a direction perpendicular to the direction of force in regions of high stressing, while the other concerns sharp transitions in highly stressed areas. The particular influence of transverse welding seams on reducing fatigue strength has been established by fatigue tests with steels St 37 and St 42, which are frequently used for press frames. Thus, according to these tests, fatigue strength is reduced by transverse seams by more than 50% by comparison with the unwelded base material.
The effects of sharp transitions in combination with a transverse welding seam are made clear by the following example: After a year in operation, the weldments between the yoke plates and the square-section uprights of the frame of a platen press exhibited cracks (Figure 2). The cracks are positioned at the inner transverse welding seam, from which was provided, at a distance of 65 mm, a relief notch in the yoke plate (Figure 3 (top)). To reduce the density of the lines of force, the transverse welding seam was

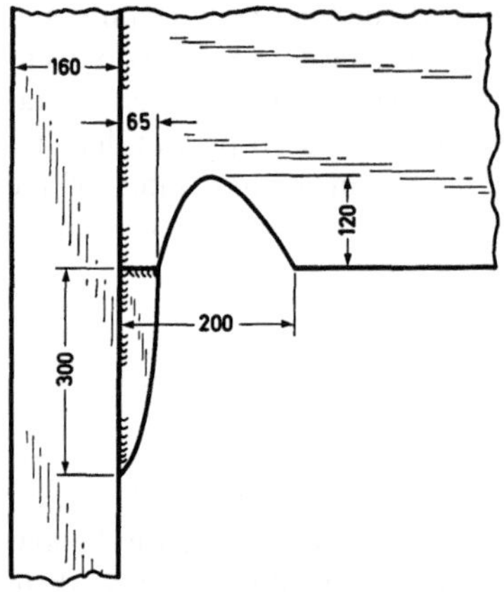

Figure 3. The improved transition area between the uprights and yoke plate. A transverse seam at the lower edge of the wedge was removed.

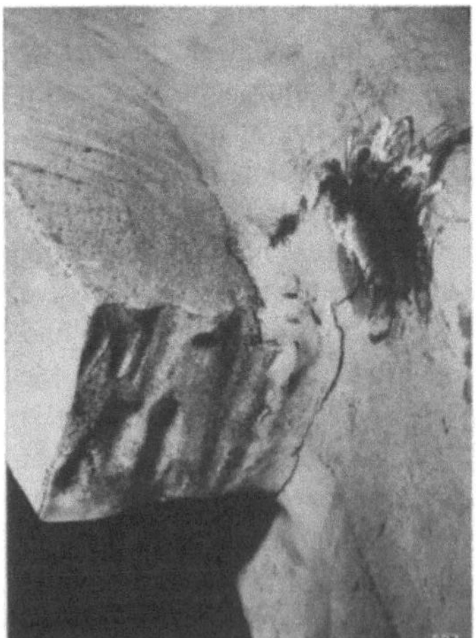

Figure 2. Crack formation at the transverse welding seam between an upright and yoke plate.

removed and a wedge profile was welded into the web between the relief notch and the inner edge of the upright (Figure 3). After this improvement was effected, crack formation no longer occurred.

8.10.2. Guidelines for Loss Prevention during the Erection of Chipboard Presses

During the erection of the press, it often happens that supplementary welding and drilling operations for auxiliary components or pipelines are carried out on the press frame. Time and again, these attachments have been positioned so unfavourably that they have led to exceeding the fatigue strength locally and hence to crack formation. To avoid such occurrences, it is now becoming increasingly common to attach pipelines and similar items to the presses by adhesive bonding.
Here are two typical examples of cases of damage during erection, as observed in practice:
In a preforming press for chipboard, welding seams for auxiliary components were laid on the frame at right-angles to the main direction of stressing (transverse seams). These seams had the effect, as previously mentioned, of reducing fatigue strength and, after a short period in operation, led to cracks up to 115 mm in length in the frame.
In the case of another press, the pillars were used in striking tests on arc-welding electrodes. The burnt-in notches so created (Figure 4) could, however, be detected in good time and removed by grinding.

Figure 4. During erection, the pillar of the press was used for striking tests on arc-welding electrodes.

Figure 5. A warning sign attached to all presses insured by Allianz.

8.10.3. Guidelines for Loss Prevention by the User of Chipboard Presses

Because it repeatedly happens that presses are endangered by the user's personnel through the application, at a later date, of incorrectly positioned and executed welds and holes, Allianz have asked those insured to attach to the presses a yellow warning sign issued by them (Figure 5).
A common cause of damage to presses is the presence of foreign bodies in the materials to be pressed. It has therefore proved useful to scan the surface of the material with suitable equipment which ensures that, in the event of any surface elevations in the "press-cake", the installation will immediately shut down.
However, in addition to foreign bodies, uneven distribution of the material to be pressed can lead to severe damage to platen presses. For example, in the case of a single-daylight chipboard press, a change from automatic to manual control had to be carried out because of a failure in

Figure 6. Crack in a frame, emanating from a transverse welding seam.

Figure 7. A press ram was deformed by a faulty pressing procedure. The damage to the press became obvious only after the stacked-up chipboards exhibited a local increase in height.

the stock-distribution mechanism. During removal under manual control the completed chipboard became jammed after about one-third of it had left the press, with the result that the following press-cake collided with the end of the board, generating a pile of chips. The press operator, unaware of this occurrence, initiated the pressing operation, which deformed the ram of the press so badly that extensive repair work was necessary.

8.10.4. Revision of Chipboard Presses

Constant checking of presses has often proved to be particularly useful for preventing cases of severe damage. It is therefore customary now for many users of presses to have their presses checked at least once a week by an expert specially qualified to carry out this task. In this way, it is often possible, for instance, to detect incipient cracks or leakages and to instigate countermeasures in good time. These checks do not, however, replace regular revisions, which should be performed by suitable test teams equipped with the appropriate instrumentation.

Apart from visual inspections, special methods are employed during revisions for the detection of cracks. The most frequently used method of detecting cracks on accessible surfaces is the magnetic-particle test, in which the location to be examined is magnetized and dusted with a special magnetic powder. A crack then manifests itself

Figure 8. After a few years in operation, it was discovered during a revision of a preforming press for chipboard that the welding seams at the collars of the press cylinder had fractured. The cause in this instance was exceeding the fatigue strength of the weldment.

Figure 9. Ultrasonic testing for cracks during revision of a press.

by accumulations of the powder along a line (bridge formation) above the separation. For the detection of cracks in inaccessible areas, the ultrasonic test is often suitable. With this test it is advantageous if design records are available for the location under examination for correct interpretation of the ultrasonic echo. For special tests, other suitable equipment, such as X-ray apparatus and endoscopes, should be provided.

Regular revision, which has been obligatory for insured platen presses for some years, has clearly proved successful and has resulted in a reduction in the number of defects. However, statistics show that, even today, deficiencies in presses are still found during every second revision. In many cases, a failure to perceive defects would after some time have resulted in extensive damage.

The period of time between revisions of individual presses cannot be specified with general validity, since it depends on the level of loading, the time in operation (e.g., three-shift work), and the age, condition and type of press. However, the interval between revisions should as a rule be not more than one year.

Bibliography to 8.10

[1] Kober, A.: Dauerfestigkeitsuntersuchungen an Blechrahmen von Spanplattenpressen. Der Maschinenschaden 43 (1970), No. 6, pp. 230/237.

[2] Pope, J.A.: Metal Fatigue. London: Chapman & Hall Ltd. 1959.

[3] Niemann, G.: Maschinenelemente. Vol. I. Berlin. Göttingen, Heidelberg: Springer-Verlag 1963.

[4] Münz, D., Schwalbe, K., Mayr, P.: Dauerschwingverhalten metallischer Werkstoffe. Braunschweig: Vieweg & Sohn 1971.

[5] Baade, B.: Wöhlerlinien-Katalog. Published by Institut für Leichtbau, Dresden 1965.

[6] Erker, A.: Sammlung von Dauerfestigkeitswerten. Schweißen und Schneiden 20 (1968), No. 4.

MIX
Papier aus verantwortungsvollen Quellen
Paper from responsible sources
FSC® C105338

If you have any concerns about our products,
you can contact us on
ProductSafety@springernature.com

In case Publisher is established outside the EU,
the EU authorized representative is:
Springer Nature Customer Service Center GmbH
Europaplatz 3, 69115 Heidelberg, Germany

Printed by Libri Plureos GmbH
in Hamburg, Germany